纪念吴文俊先生诞辰100周年

国家出版基金项目
NATIONAL PUBLICATION FOUNDATION

The Complete Works of Wu Wen-Tsun

Mathematics Mechanization I

吴文俊全集·数学机械化卷 I

——Mathematics Mechanization:

Mechanical Geometry Theorem-Proving, Mechanical Geometry Problem-Solving and Polynomial Equations-Solving

吴文俊　著

高小山　编订

科　学　出　版　社
龍　門　書　局

北　京

内 容 简 介

本卷收录了吴文俊的 *Mathematics Mechanization: Mechanical Geometry Theorem-Proving, Mechanical Geometry Problem-Solving and Polynomial Equations-Solving* 一书. 本书是围绕作者命名的"数学机械化"这一中心议题而陆续发表的一系列论文的综述, 试图以构造性与算法化的方式来研究数学, 使数学推理机械化以至于自动化, 由此减轻繁琐的脑力劳动.

全书分成三个部分: 第一部分考虑数学机械化的发展历史, 特别强调在古代中国的发展历史. 第二部分给出求解多项式方程组所依据的基本原理与特征列方法. 作为这一方法的基础, 本书还论述了构造性代数几何中的若干问题. 第三部分给出了特征列方法在几何定理证明与发现、机器人、天体力学、全局优化和计算机辅助设计等领域中的应用.

本书可供数学工作者、数学及计算机专业高年级大学生和研究生以及有关工程人员参阅.

图书在版编目(CIP)数据

吴文俊全集. 数学机械化卷. I /吴文俊著; 高小山编订. —北京: 龙门书局, 2019.5

国家出版基金项目

ISBN 978-7-5088-5550-9

Ⅰ. ①吴… Ⅱ. ①吴… ②高… Ⅲ. ①数学理论-文集 Ⅳ. ①O1-53

中国版本图书馆 CIP 数据核字(2019) 第 074591 号

责任编辑: 赵彦超 李 欣 / 责任校对: 邹慧卿

责任印制: 肖 兴 / 封面设计: 无极书装

科学出版社 出版

北京东黄城根北街 16 号

邮政编码: 100717

http://www.sciencep.com

北京通州皇家印刷厂印刷

科学出版社发行 各地新华书店经销

*

2019 年 5 月第 一 版 开本: 720 × 1000 1/16

2019 年 5 月第一次印刷 印张: 26 3/4

字数: 540 000

定价: 188.00 元

(如有印装质量问题, 我社负责调换)

编 者 序

中国现代数学的崛起，开始于20世纪初，经历了几代人坚苦卓绝的努力. 在这百年奋战中涌现出来的数学家中，吴文俊是最杰出的代表之一. 他早年留学法国，留学期间就已在拓扑学方面做出了杰出贡献，提出了后来以他的名字命名的“吴公式”和“吴示性类”. 回国后提出了“吴示嵌类”等拓扑不变量，发展了统一的嵌入理论. 他关于示性类与示嵌类的研究，已成为20世纪拓扑学的经典，至今还在前沿研究中使用. 20世纪70年代以来，吴文俊院士在汲取中国古代数学精髓的基础上，开创了崭新的现代数学领域——数学机械化. 他发明的被国际上誉为“吴方法”的数学机械化方法，改变了国际自动推理的面貌，形成了自动推理的中国学派，已使中国在数学机械化领域处于国际领先地位. 上述工作无疑属于 20 世纪中国数学赶超国际先进水平的标志性成果，而吴文俊院士博大精深的科学研究，除了拓扑学与数学机械化以外，还跨越了代数几何、博弈论、中国数学史、计算图论、人工智能等众多领域，并在每个领域都留下了这位多能数学家的重要贡献.

吴文俊先生是一位具有强烈爱国精神的数学家. 自 1950 年谢绝法国师友的挽留回到祖国后，半个世纪如一日，为在他深爱的中华故土发展数学事业而鞠躬尽瘁. 除了第一流的科研成果，吴文俊先生长期身处中国数学界领导地位，在团结带领整个中国数学界赶超世界先进水平方面，也做出了不可磨灭的贡献. 特别是，吴文俊先生在担任中国数学会理事长期间，领导中国数学最终成功地加入了国际数学联盟，此举大大提高了我国数学界的国际地位，同时也为我国成功举办 2002 年国际数学家大会铺平了道路.

吴文俊治学严谨，学术思想活跃，无论获得多么高的声誉，他总是勤奋地在科研第一线工作，一生积极进取，锲而不舍，不断取得新的成就. 在开始从事机器证明时，他已近花甲之年，从零开始学习编写计算机程序，每天十多个小时在机房连续工作，终于在几何定理机器证明这一难题上取得成功.

吴文俊先生为中国现代数学的发展建立了丰功伟绩，而他本人却始终淡泊、谦逊. 他处事公正豁达，待人充满善意，受过他帮助的人可以说不计其数. 正因如此，这位有着崇高国际声望而平易近人的学者，受到了每一个认识他的人格外的爱戴与尊敬.

2019 年 5 月 12 日是吴文俊先生百年诞辰. 为了纪念这个特殊的日子，我们编辑出版了《吴文俊全集》，通过系统地收录、整理吴文俊先生的学术著作和论文，纪念吴先生的学术思想及学术成就. 全集共计 13 卷, 包括拓扑学 4 卷、数学机械化 5 卷以及数学史、博弈论与代数几何、数学思想各 1 卷; 同时, 全集还设有附卷, 收录吴文俊先生的同事、学生和其他社会各界人士发表过的与吴先生有关的各类文献资料.

最后, 我们对在全集编辑中给予帮助的各位同事表示衷心感谢; 感谢国家出版基金对于全集出版的资助; 感谢科学出版社编辑人员在出版全集时认真细致的专业精神; 感谢相关出版与新闻机构在版权方面提供的帮助.

李邦河　高小山　李文林

2019 年 3 月

Preface

The present book is actually a collection of essays of the writer since 1977 centered around the subject which has been baptized the name of *mathematics mechanization.* The subject tries to deal with mathematics in a constructive and algorithmic manner so that the reasonings become mechanical, automated, and as much as possible to be intelligence-lacking, with the result of lessening the painstaking heavy brain-labor.

The desire of getting mathematics mechanized or more generally of getting all kinds of reasonings automated, may be traced back to G.W. Leibniz (+1646, +1716). However, the real founding of mathematics mechanization, or more generally automated reasoning, should be, in the writer's opinion, attributed to D.Hilbert (+1862, +1943). The great merit of Hilbert on mathematics mechanization seems to be long time overshadowed by his great success on his existence proof of finite basis theorem and his works on mathematics axiomatization, as well as his controversies with L.Kronecker (+1823, +1891), who may be considered as one of the banner-holders of mathematics mechanization. In spite of these, Hilbert is well-deserved to be the real founder of mathematics mechanization, as reflected by his other kinds of great achievements or influences, for which some examples are listed below:

Example 1. In Hilbert's famous speech on mathematical problems in 1900 the Problem 10 asks for an automated method of diophantine equations-solving. Though the Problem has been solved in the negative in 1970 by Matijasevic, the positive aspects of the negative solution have been shown to be able to bear extremely important consequences to modern pure mathematics. See e.g. the paper of M.Davis, et al in *Proc.Symp. in Pure Math.*, vol.28, (1976).

Example 2. In Hilbert's classic *Grunlagen der Geometrie* in 1899, there is some obscure passage, which, when reinterpreted, is equivalent to an automated method of proving a whole class of geometry theorems of some special type. The well-known theorem of Tarski in 1950 and the latter developments of the writer on mechanical geometry theorem-proving are nothing else but continuation of this method of Hilbert in the narrow realm of elementary geometry. For this reason the writer has baptized the above result of Hilbert, formulated as Theorem 62 in latter editions of

Grundlagen, *Hilbert's Mechanization Theorem*.

Example 3. Hilbert's program about consistency of mathematics implies a program of automated proving of *all* theorems in mathematics by a universal method. The discovery of Goedel has declared the collapse of Hilbert's program in being too ambitious to be realizable. Nonetheless Hilbert's efforts have given rise to a new discipline mathematical logic and show a new way of proving theorems or solving problems class by class under the name of decidability of which Hilbert's Mechanization Theorem furnishes a simple but extremely deep and instructive example. It opens a new way of dealing with mathematics: mechanization of mathematics.

Besides these we may note that there are papers of Hilbert with theorems proved in a constructive and algorithmic manner. As an example, we may cite the proof of the theorem concerning what we call nowadays Hilbert function of an ideal. We may even note that the proof of his finite basis theorem is achieved in forming the finite basis in a somewhat constructive and algorithmic manner.

The present book may be considered as a primary development of mathematics mechanization along the line of thought of Hilbert.

The whole book is divided into three parts. Part I, consisting of Chapters 1 and 2, concerns historical developments of mathematics mechanization, with emphasis on the developments in ancient China. This is quite natural. In fact, in contrast to the modern development of pure mathematics in accordance to the Euclidean axiomatic pattern of ancient Greece, the ancient mathematics of China is highly constructive, algorithmic and computational in character, with most of the results expressed in the form of algorithms. Even in geometry, the ancient Chinese mathematicians paid little attention to geometry theorem-proving. On the contrary, they devoted themselves rather on geometry problem-solving which leads naturally to the solving of polynomial equations. Accordingly, the problem of polynomial equations-solving occupies a central position throughout thousands of years of development of mathematics in ancient China. In present book we follow closely the tradition of our ancient mathematics in focusing on polynomial equations-solving with geometry theorem-proving and geometry problem-solving as its two main applications. This attitude of treatment is well-reflected in the subtitles of the present book.

The Part II, which contains the three chapters 3, 4, and 5, bears the title of Principles and Methods. It aims at the description of the underlying Principle of polynomial equations-solving, with polynomial-coefficients in fields restricted to case of characteristic 0. Based on the general Principle, we show how general Methods of solving such arbitrary polynomial systems may be found. In order to achieve this we shall use the naive notion of *zero-sets* of polynomial sets in forsaking the usual notions of ideals. An analysis of the zero-set of a polynomial set, under the guidelines of the Principle give rise to the fundamental notion of *characteristic set* (abbr. *char-set*) and the further Principles under the names: Well-Ordering Principle, Zero-Decomposition Theorems, and Variety-Decomposition Theorem. These Principles form the basis of actual polynomial equations-solving.

It is to be noted that the underlying Principle of polynomial equations-solving had its origin in ancient Chinese mathematics, notably in the classic *Jade Mirror of Four Elements* in year +1303 of Zhu Szijie. Of course, there are incompleteness and deficiencies in Zhu's work, but uncontestedly it shows the right way of solving arbitrary systems of polynomial equations. To complete the way indicated in Zhu's work we have relied upon the important works of J.F.Ritt as exhibited in his two classics on the algebraic studies of differential equations. Ritt's theory and method are highly constructive, algorithmic, and computational in character which rely in turn heavily on the *early* works of Van der Waerden on algebraic geometry. In contrast to the later developments of modern algebraic geometry, the *early* works of Van der Waerden on algebraic geometry is essentially constructive, algorithmic, and in the main computational. It is based on the very useful notions of *generic point* and specialization, which have unfortunately disappeared in modern algebraic geometry. As a concrete application of Van der Waerden's treatment, we show in Sect. 3 of Chap. 5, how to define the important notions of Chern classes and Chern numbers of algebraic varieties with arbitrary singularities, in basing on the notion of generic points. Though there are diverse alternative definitions for getting such extended notions of Chern classes but not of Chern numbers, our treatment has the peculiarity of being computational, giving for example the remarkable Miyaoka-Yau inequality and its extensions by mere simple computations. We refer this to Sect. 3 of Chap. 5. A further application of this notion of generic pointis is given in Sect. 4 of Chap. 8 in Part III.

Usually polynomial equations are to be solved in complex field. However, to solve in real field is clearly of utmost importance. An example is furnished by optimization problems for which only real solutions are to be considered. The general method given in Chap. 3, which permits to give, theoretically speaking, a complete answer to the problem of polynomial equations-solving in the complex case, furnishes however merely a very incomplete answer with only partial success in the real case. In Sect. 5 of Chap. 5, we point out that, for the important optimization problems, the polynomial case is quite different from the general differential case. In fact, we show that, in the polynomial case of optimization problems, there is a *finite* set of real values, for which the optimal value, supposed exist, is just the optimal value of this finite set. Our method permits also to give partial success for problems involving inequalities. However, the method is far from being satisfactory and much has to be done in the realm of problems in the real field.

Part III, containing chapters 6, 7, and 8, bears the title of Applications and Examples. Chap. 6 concerns applications to polynomial equations-solving. Remark that, though our principles and methods developed in Part II theoretically furnish complete solutions of arbitrary polynomial equations over fields of characteristic 0, the computational complexity is so high that, in actual computations one meets often unsurmountable complications to be able to carry out the computations up to the end. It seems that the only issue is to find some hybrid method of combining our symbolic computations with the usual numerical computations. We give such a hybrid method in Sect. 2 of Chap. 6 with a sound theoretical basis. The writer believes that the method will actually permit us to settle the question for all kinds of problems arising from practical applications but experiments are yet to be done.

Chap. 7 of Part III deals with geometry theorem-proving which is treated as applications of our general principles and methods of polynomial equations-solving. Chap. 8 of Part III deals with diverse problems which, in apparence having nothing to do with polynomial equations, are reduced to some form of polynomial equations-solving and then treated by our general methods.

As a preliminary step toward mathematics mechanization, the present book has to be restricted to the polynomial case for equations-solving, and mainly to elementary geometries for theorem-proving and problem-solving. Only some indications about possible developments in differential geometry and differential equations are slightly touched in the last section, Sect. 5 of Chap. 8. We shall leave the studies of mathematics mechanization in the differential case and in other domains of mathematics in latter occasions.

The writer would like to take this opportunity to express his deep gratitudes to the National Committee of Science and Technology, the National Natural Science

Foundation of China, and the Chinese Academy of Sciences. Without their financial supports and encouragements mathematics mechanization is impossible to be so prosperous in China, today, and also tomorrow.

The writer would like also to express his deep gratitudes to his comrades and colleagues, especially those in Mathematics Mechanization Research Center (MMRC) of Institute of Systems Science, CAS, who have contributed so much to the development of mathematics mechanization, and to the writing of the present book. As they are so numerous and most of their names are scattered throughout the whole book, the writer would like to apologize in mentioning no more their names here in the Preface.

Contents

Chapter 1

Polynomial Equations-Solving in Ancient Times, Mainly in Ancient China

1.1 A Brief Description of History of Ancient China and Mathematics Classics in Ancient China.

Equations-Solving and Theorem-Proving.

There are two main activities of mathematics: theorem-proving and equations-solving. While theorem-proving was originated in ancient Greek mainly in so-called Euclidean geometry, equations-solving was the main concern of mathematics in ancient China. In order to give a brief description of equations-solving of China in ancient times, let us first give a brief chronicle of the history of ancient China.

Brief Chronicle of China.

China turned from tribes dispersed throughout the vast territory into a loosely centralized dynasty in 21st century B.C. and was unified into an empire under the first emperor of Qin in 221B.C. Since that time China was unified and split, split and unified, over and over until the beginning of the 20th century with the collapse of the feudalistic dynasty system. The main periods of ancient China are listed below, the years before and after Christ being indicated by negative and positive numbers respectively.

Xia (about $(-21c, -16c)$).

Yin or Shang (about $(-16c, -11c)$).

Western Zhou (about $-11c, -771$).

Spring and Autumn Period $(-770, -476)$.

Warring-States Period $(-475, -221)$.

Qin Dynasty $(-221, -207)$.

Western Han Dynasty $(-206, +8)$.

Eastern Han Dynasty (+25, +220).

Three Kingdoms Period (+220, +265).

Western Jin (+265, +316).

Eastern Jin and 16 Kingdoms (+317, +420).

Southern and Northern Dynasties (+420, +589).

Sui Dynasty (+581, +618).

Tang Dynasty (+618, +907).

Five Dynasties and Ten Kingdoms Period (+907, +960).

Northern Song Dynasty (+960,+1127) and Liao Dynasty (+938, +1125).

Southern Song Dynasty (+1127,+1274) and Jin Dynasty (+1127, +1234).

Yuan Dynasty (+1271, +1368).

Ming Dynasty (+1368, +1644).

Qing Dynasty (+1644, +1911).

Some Characteristic Features of Chinese Ancient Mathematics.

1. Instead of calculations of pencil-paper type, the ancient Chinese made all computations in manipulating rods on counting boards or anything alike to serve the purpose. Counting rods made of bones or bamboo had been excavated in years 1971 and 1975 which date back to Western Han Dynasty in 1c or 2c B.C. See the photograph in Fig. 1.1.

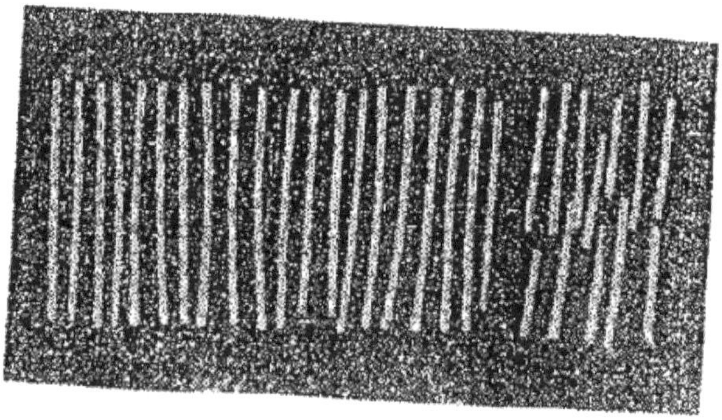

Fig. 1.1.

2. The Chinese already possessed, in very remote times, the most perfect *place-valued* decimal system. This allowed them to represent the integers by properly arranged rods placed in due positions on the board. In particular, the integer 0 in, or as, a decimal integer was just represented by leaving some empty place in the right position. Thus, the integer 22022 will be represented on the board in the form

$$\| \; = \qquad = \; \|$$

Highly efficient arithmetic operations should have been well-developed much earlier than the Qin Dynasty. In fact, the word "arithmetic", the usual terminology for "mathematics", was just a literal translation of Chinese characters "Suan Shu" meaning "counting methods."

3. Results were usually expressed in the form of separate problems, each of which was divided into several items, as follows.

(1) Statement of the problem with numerical data.

(2) Numerical answer to the problem.

(3) "Shu", or the method of arriving at the result. It was most often just what we call today the "algorithm", sometimes also just a formula or a theorem. Note that the numerical values in (1) play no role at all in the method, which was so general that any other numerical value of same type could be substituted equally well. Item (1) thus served just as an illustrative example which shows the motivation of considering such kind of problems to be solved by this "Shu".

(4) Sometimes "Zhu", or demonstrations which explained the reason underlying the method in Item (3). It was actually a "proof" when "Shu" was actually a formula or a theorem.

(5) "Cao", meaning drafts, containing details of the calculations for arriving at the final result, appeared first in some classic of Tang Dynasty. It was then widely added in writings of Song and Yuan Dynasties. This was possible since printing measures were well-developed in these days.

4. The main concern of our ancestors was the solving of problems arising in reality or from necessity which naturally led to the solving of equations. Thus instead of devoting to theorem-proving as in the ancient Greek mathematics, equations-solving was the central theme of Chinese ancient mathematics. In fact, equations-solving may be considered as a golden thread permeating the whole of Chinese ancient mathematics which will be explained in more details in the next section.

5. For geometry instead of developing a deductive system for theorem-proving, measuration and geometry problem-solving were emphasized based on some different way of proof-procedure which will be explained in more details in Sect. 3 of Chap. 2.

6. In a word, the ancient Chinese mathematics was highly computational and *mechanical* in character. It seems that it was developed in its own way, indepen-

dent of Western mathematics which was developed in a way dating back to the ancient Greeks. It may be considered as a *mechanical* system quite different from the *axiomatic* system of Greek origin which governs the present day modern pure mathematics.

Mathematics Classics in Ancient China.

Many important mathematical treatises which had been cited in ancient documents have been unfortunately lost, including the all-the-more important *Zhui Shu (Art of Mending)* by Zu Chongzhi (+429,+500), the celebrated mathematician, astronomer and engineer in Southern and Northern Dynasties. In spite of this the survived treatises were sufficient to give a general view of our ancient mathematics, showing clearly the line of thought and line of development. Below we shall list a number of such treatises which are closely connected with our main themes, viz. polynomial equations-solving and geometry theorem-proving.

[ZB]* Zhou Bi Mathematics Manuel, in probably 1c B.C.

This classic will be cited henceforth as *Zhou Bi*. It was one of the oldest Chinese classics among historical sources in ancient Chinese mathematics available up to the present day. The author of this book is unknown. Researchers hold that the book could not have been written later than 100 B.C. It is not a book on mathematics, but rather one on astronomy. The Chinese character Bi means some instrument *biao* like scaled rod used in astronomical measurements and Zhou is just the dynasty before the unification of China under the First Emperor of Qin. The book was chosen as the first one of ten important textbooks for mathematical education at Imperial College from 6c to 10c. In spite of the astronomical character of the book, it contains important passages on our ancient geometry which brought exceedingly important consequences on our further developments of geometry. We shall discuss this in more details in Sect. 3 of Chap. 2.

[NC]* Nine Chapters of Arithmetic, completed in 2c or 3c B.C.

Henceforth, this classic will be cited as *Nine Chapters.*

It is a comprehensive discourse on traditional Chinese mathematics. Its influence on subsequent developments in traditional Chinese mathematics is no less profound than that of Euclid on Western mathematics. It was the main mathematical textbook studied for more than 1,000 years in ancient China. It was also used as a textbook in some other countries neighboring to China before the introduction of Western learning.

The treatise was not due to a single or a few authors as usually in the present day. It seems rather to be the collection of achievements due to several generations of intelligent scholars, learned people, and officials in charge of astronomical, financial, reconstructive, plannifying, etc. affairs.

Most of the results and methods in the book were already in current use long before the Qin Dynasty. According to Liu Hui, in his Preface of Annotations to this book (see [Liu1]* below) the work appeared originally before Qin Dynasty but damaged in the reign of Qin. Additions and deletions were made by the high officials of Qin and Western Han Dynasties Zhang Cang (−250, −152) and Geng Shouchang (fl. (−57,−54)) to arrive at the completed form as seen by Liu Hui. A tomb in Province Hubei had been discovered and excavated in recent years. In the tomb there was excavated more than one thousand bamboo pieces on which were carved a book on computation and numeration which was actually a predecessor of the *Nine Chapters*. This also gives clear evidence of Liu Hui's Preface. The completion of the work may, therefore, be safely fixed before the death of Zhang Cang.

[Liu1]* Annotations to *Nine Chapters in Arithmetic*, +263 by Liu Hui.

This classic is of utmost importance and will be cited henceforth as *Annotations*.

The *Nine Chapters* and the present *Annotations* may be regarded as the representative masterpieces of ancient China which enjoyed the same popularity as Euclid's *Elements* of ancient Greece. Besides laying down a solid foundation for the theory of traditional mathematics in ancient China, *Annotations* also contains numerous original ideas and Liu's own contributions which had a far-reaching influence for the further development of mathematics in China.

[LCF1]* Annotations to *Nine Chapters in Arithmetic*, by Li Chunfeng, fl. +656 in Tang Dynasty.

Li was an astronomer in Tang Dynasty and was highly esteemed by ancient mathematical circle. His annotations were nevertheless of little value from a mathematical point of view. However, he added in his annotations some passages describing the contribution of Zu Geng, son of the great Zu Chongzhi, in the formulation of a principle equivalent to the later Cavalieri principle, for determining the volume of a sphere. See Sect. 3, Chap. 2 for more details.

Remark that the *Annotations* of Liu and Li are not texts separate from the *Nine Chapters*. All three are binded together in a single book in later times with the annotatations clearly indicated so that one knows readily to whom the annotation is to be attributed.

[Liu2]* Sea Island Mathematical Manuel, 3c A.D., by Liu Hui.

It was compiled as a continuation of *Nine Chapters* but was ultimately separated to become an independent monograph. The classic dealt with problems involving measurements by some Gou-Gu instruments. Henceforth it will be cited as *Sea Island.*

[ZS1]* Sun-Height Theory with Diagrams, 3c A.D., by Zhao Shuang.

[ZS2]* Gou-Gu and Circle-Square Theory with Diagrams, 3c A.D., by Zhao Shuang.

Both [ZS1]* and [ZS2]* were extractions from the annotations to [ZB]* by Zhao Shuang in the period of Three Kingdoms or Jin Dynasty. Only remnants of the writings with damaged diagrams have been preserved. However, from these remnants we can draw important conclusions about our ancient geometry, see more details in Sect. 3, Chap. 2.

[WXT]* Continuation of Ancient Mathematics, +626, by Wang Xiaotong.

It was a classic devoting to cubic equations, see the next section for more details.

By the 13th century, the traditional Chinese mathematics had become comprehensively systematic after more than 1,000 years of development. It was crowned with the highest accomplishments during the Song and Yuan dynasties, the brightest period in China's history of mathematics with fastest development, largest number of works and highest academic level.

Four great mathematicians emerged within a few decades of the later half of the 13th century: Qin Jiushao, Li Ye, Yang Hui and Zhu Shijie. What people call the Song-Yuan mathematics is chiefly represented by the works of these four, to be described below:

[QJS]* Mathematical Treatise in Nine Chapters, +1247, by Qin Jiushao.

This is a comprehensive treatise on Chinese traditional mathematics. It contains the method of solving high degree algebraic equations and the so-called Chinese Remainder Theorem. The geometry part had a lot of flaws which had been pointed out by U.Libbrecht in his *Chinese Mathematics in 13th Century.* Nevertheless it contains the important formula which was named in Western mathematics as Heron's formula. For more details see Sect. 3, Chap. 2.

[LY1]* Sea Mirror of Circular Measurement, +1248, by Li Ye.

[LY2]* New Steps in Computation, +1259, by Li Ye.

These two classics have the great merit of presenting the method in introducing the notion of *unknown* under the name of *heaven's element* for a systematic establishment of algebraic equations from a given problem. It also paves the way to the algebrization of geometry. The discovery of the method should be regarded as one of highest importance in our ancient mathematics.

[YH1]* Detailed Analysis of the *Nine Chapters*, +1261, by Yang Hui.

[YH2]* Method of Computation for Daily Use, +1262, by Yang Hui.

[YH3]* Yang Hui's Method of Computation, +1274-5, by Yang Hui.

These classics contained materials about applied mathematics and simplifications of the counting-rod arithmetical operations which paved the way to the invention of modern abacus. They also contained passages taken from other classics which have been lost and are important to our main theme of polynomial equations-solving.

[Zhu1]* Introduction to Mathematics, +1299, by Zhu Shi-jie.

It was a treatise of tutorial character.

[Zhu2]* Jade Mirror of Four Elements, +1303, by Zhu Shi-jie.

This classic, besides the achievements on higher order arithmetic series and method of higher power finite differences, extended the method of a single unknown, the heaven's element, to a method involving up to four *elements* , or four *unknowns* . It marked the climax of our traditional mathematics. The merit and the influences to even modern mathematics of the method will be more fully discussed in the next section.

Some Recent Literature on Chinese Ancient Mathematics.

History of Chinese ancient mathematics has been vigorously studied in recent years in China with fruitful consequences so that we understand now the real nature of our ancient mathematics much deeper than ever before. Below we list a few of these writings which we have particularly consulted.

[ACTS]* Ancient China's Technology and Science, Foreign Language Press, Beijing, (1983).

[Chen]* Chen Liang-tso, The relation between the proof of the "Kou-Ku" Theorem in Chou-Pei Suan-Ching and the "Chu Ju Hsiang Pu" Principle, (in Chinese), Studies of Sinology, 7 (1989) 255-281.

[Du]* Du Shiran, Mathematical classics, in [ACTS]*, 50-56.

[Guo1]* Guo Shuchun, The numerical solution of higher equations and the *Tianyuan* method, in [ACTS]*, 111-123.

[Guo2]* Guo Shu-chen, Research on the right triangles with integral sides in *Jiu Zhang Suan Shu*, in [KJSWJ]*, 54-66.

[Guo3]* Guo Shu-chen, Liu Hui, giant of mathematics in ancient times, (in Chinese), Science and Technology Press of Shandong Province, (1992).

[He]* He Shaogong, Method for determining segment areas and evaluation of π, in [ACTS]*, 90-110.

[Hoe]* John Hoe, Les systeǹes d'équations polynomes dans le Siyuan Yujian (1303), Inst.Hautes Etudes Chinoise, Paris (1977).

[KJSWJ]* Kejishiwenji (or Collection of Papers on History of Science and Technology), No.8, Special Collection of Papers on History of Mathematics, (in Chinese), Shanghai Science and Technology Press, (1982).

[Kong]* Kong Guangping, Biography of Li Ye, (in Chinese), Hubei Education Press, (1988).

[Lam]* Lam Lay Yong, The geometrical basis of the ancient Chinese square-root method, Isis, 61 Pt.1, No.206, (1970) 92-102.

[L-S]* Lam Lay Yong and Shen Kangshen, Methods of solving linear equations in traditional China, Historia Math., 16 (1989) 107-122.

[LJM1]* Li Jimin, Liu Hui's investigation on the right-triangles with integral sides, in [KJSWJ]*, 51-53.

[LJM2]* Li Jimin, A study on the oriental classic *Nine Chapters of Arithmetic* and their annotations by Liu Hui (in Chinese), Shan Xi People's Education Press, 1990.

[LJM3]* Li Jimin, Textual research on "Shang-Gao" Theorem, Studies in the History of Natural Sciences, 12 (1993) 29-41.

[Libb]* U.Libbrecht, Chinese mathematics in thirteenth century, MIT Press, (1973).

[LKW1]* Ko-Wei Lih, Studies on the chapter "Shang Gao said, method of mathematics is based on circles and spheres" in Zhou Bi Suan-Jing, (in Chinese) Proc. 2nd Symposium on History of Science, (1989) 227-234.

[LKW2]* Ko-Wei Lih, From one gnomen to two gnomens — a methodological study of the method of double differences, 5th Intern. Congress on History of Science in China, 1988, San Diego.

[LKW3]* Ko-Wei Lih, A new interpretation of the proof of Pythagorous theorem in the Zhou Bi Suan-Jing, 18th Intern.Congress of History of Science, 1989, Hamburg and Munich.

[Liu]* Liu Jiemin, Annotations about "Commentary on Yang-ma formula", in [W-B-S-L]* 244-252.

[QBC]* Qian Baocong, History of Chinese mathematics, (in Chinese), Science Press, Beijing, (1962).

[Wag]* D.B.Wagner, An early Chinese derivation of the volume of a pyramid: Liu Hui, third century A.D., Historia Math., 6 (1979) 169-188.

[W-N]* Wang Ling and J.Needham, Horner's method in Chinese mathematics;

its origin in root-extraction procedures of the Han Dynasty, T'oung Pao, 43 (1955) 345-401.

[W-B-S-L]* Wu Wentsun, Bai Shangshu, Shen Kangshen, and Li Di (Eds), Studies on Liu Hui, (in Chinese), Shan Xi People's Education Press and Jiu Zhang Press, (1993).

[WWT1]* Wu Wentsun, On the theory of double differences basic the earth measurements and astronomical observations in ancient Chinese mathematics, criticism about the methodology of historical studies of mathematics, in [KJSWJ]*, 10-30.

[WWT2]* Wu Wentsun, About the source of ancient proofs in *Sea Island Mathematics Manuel*, in [WWT5]*, 162-180, or in [WWT6]*, 54-73.

[WWT3]* Wu Wentsun, The out-in complementary principle, in [ACTS]*, 66-89.

[WWT4]* Wu Wentsun, Recent studies of the history of Chinese mathematics, Proc. Intern. Cong. Math., (1986) 1657-1667.

[WWT5]* Wu Wentsun (Ed.), "*Arithmetic in Nine Chapters*" and its commentarist Liu Hui, (in Chinese), Beijing Normal University Press, Beijing, (1982).

[WWT6]* Wu Wentsun, Selected popular papers of Wu Wentsun (in Chinese), Shandong Education Press, (1986).

[WWT7]* Wu Wentsun, Wu Wentsun on Mathematics-Mechanization (in Chinese), Shandong Education Press, (1996).

[Yan]* Yan Dunjie, A brief history of Chinese numerals, in [KJSWJ]* 31-50.

1.2 Polynomial Equations-Solving in Ancient China.

A. Solving of System of Linear Equations and Notion of Negative Numbers.

An Example from Nine Chapters.

The 8th chapter of *Nine Chapters* bears the title *Fang-Cheng* and is devoted to the solving of systems of linear equations. It contains in all 18 problems. We reproduce the original text for the first problem as follows.

> **Problem 1.** 3 bunches of highest class cereals, 2 bunches of middle class cereals, and 1 bunch of lowest class cereals will produce 39 dou of grains, 2 bunches of highest class cereals, 3 bunches of middle class cereals, and 1 bunch of lowest class cereals will produce 34 dou of grains, 1 bunch of highest class cereals, 2 bunches of middle class cereals, and 3 bunches of lowest class cereals will produce 26 dou of grains.
>
> **Ask** how many dou of grains will produce one bunch each of cereals of highest, middle, and lowest classes.

Answer.

One bunch of highest class cereals, 9 and $\frac{1}{4}$ dou,

One bunch of middle class cereals, 4 and $\frac{1}{4}$ dou,

One bunch of lowest class cereals, 2 and $\frac{3}{4}$ dou.

Fang-Cheng Shu says:

(1) Put 3 bunches of highest class cereals, 2 bunches of middle class cereals, 1 bunch of lowest class cereals, and 39 dou of grains at the right column. Do same with the cereals in the middle and left columns.

(2) Multiply all entities in the middle column by the highest class cereal of right column and

(3) subtract successively that column.

(4) Multiply the next and

(5) subtract successively too.

(6) Then multiply all entities in the left column by the non-zero middle class cereal of middle column and

(7) subtract successively that column.

(8) For the entities in the left column, take the non-zero lowest class cereal as denominator, and the grain as numerator, the quotient will be the grain for lowest class cereals.

(9) To find that for the middle class cereals, multiply the grain of middle column by the denominator and subtract the grain of lowest class.

(10) Divide the remainder by the bunch number of middle class grain and get the grain of middle class cereals.

(11) To find that for the highest class cereals, one does the same as before: Multiply the grain of right column by the denominators and subtract successively the grains of lowest class and middle class cereals.

(12) Divide the remainder by the bunch number of highest class cereal and get the grain of highest class cereals.

(13) The numerators are respectively divided by the denominators to get the respective grains.

Remark that in the above reproduction *dou* is some Chinese unit of dry measure of grains and *Shu* is somewhat equivalent to the present day *algorithm*. The numbers $(1), \cdots, (13)$ are added by the present author to indicate the respective steps of operations.

Detailed Steps for the Example by Fang-Cheng Shu.

Our ancient texts are usually written in very concise form and the sentences above are sometimes difficult to ascertain word by word. However, the various steps of operations (1)-(13) are clear enough to leave any doudt for their exact meanings. Thus, the step (1) consists of arranging the datas in 3 columns on the counting board to have the form

$$\left[\begin{array}{c|c|c} 1 & 2 & 3 \\ 2 & 3 & 2 \\ 3 & 1 & 1 \\ 26 & 34 & 39 \end{array}\right].$$

The next steps consist of successive reductions of the above form as follows:

$$\longrightarrow_{(2)} \left[\begin{array}{c|c|c} 1 & 6 & 3 \\ 2 & 9 & 2 \\ 3 & 3 & 1 \\ 26 & 102 & 39 \end{array}\right] \longrightarrow_{(3)} \left[\begin{array}{c|c|c} 1 & 0 & 3 \\ 2 & 5 & 2 \\ 3 & 1 & 1 \\ 26 & 24 & 39 \end{array}\right]$$

$$\longrightarrow_{(4)} \left[\begin{array}{c|c|c} 3 & 0 & 3 \\ 6 & 5 & 2 \\ 9 & 1 & 1 \\ 78 & 24 & 39 \end{array}\right] \longrightarrow_{(5)} \left[\begin{array}{c|c|c} 0 & 0 & 3 \\ 4 & 5 & 2 \\ 8 & 1 & 1 \\ 39 & 24 & 39 \end{array}\right]$$

$$\longrightarrow_{(6)} \left[\begin{array}{c|c|c} 0 & 0 & 3 \\ 20 & 5 & 2 \\ 40 & 1 & 1 \\ 195 & 24 & 39 \end{array}\right] \longrightarrow_{(7)} \left[\begin{array}{c|c|c} 0 & 0 & 3 \\ 0 & 5 & 2 \\ 36 & 1 & 1 \\ 99 & 24 & 39 \end{array}\right]$$

$$\longrightarrow_{(9)} \left[\begin{array}{c|c|c} 0 & 0 & 3 \\ 0 & 180 & 2 \\ 36 & 0 & 1 \\ 99 & 765 & 39 \end{array}\right] \longrightarrow_{(11)} \left[\begin{array}{c|c|c} 0 & 0 & 540 \\ 0 & 180 & 0 \\ 36 & 0 & 0 \\ 99 & 765 & 4995 \end{array}\right]$$

The steps (8),(10) and (12) (with (13) a summary) give respectively.

For lowest class cereal: $\dfrac{99}{36} = 2$ and $\dfrac{3}{4}$.

For middle class cereal: $\dfrac{765}{180} = 4$ and $\dfrac{1}{4}$.

For highest class cereal: $\dfrac{4995}{540} = 9$ and $\dfrac{1}{4}$.

Introduction of Negative Numbers in Nine Chapters.

In present day Chinese mathematics the two characters *Fang* and *Cheng* together are used to mean an *equation*. In fact the Chinese character *Fang* means *square* in the present day. On the other hand, in ancient time it had some other meaning.

The character *Cheng*, according to the annotations of Liu Hui, means somewhat the arrangement of data in columns. Whatever was the original meaning in ancient time, from the above example we see that *Fang-Cheng* may be interpreted as *square array* and the *Fang-Cheng Shu* is just the manipulation of such square arrays in some definite way to arrive finally at some normal form so that the answer may be readily found. The procedure described in the corresponding *Shu* is of a quite general character. However, it may happen that in the subtraction steps some (positive) number should be subtracted by some larger ones. This happens actually in the Problem 3. It is here that the notions of *positive* and *negative* numbers were introduced. According to the annotations of Liu Hui, these numbers were distinguished by the use of counting rods in different colors, red and black. The rules of addition and subtraction of such numbers were given by the *Positive-Negative Shu* in the *Nine Chapters*. Furthermore, Problem 2 involves the operation of transfering the data and the rules governing such operations were expressed by the *Shu* below:

> Change the positive data into negative ones and the negative ones into positive ones.

The above 3 *Shus* form the substantials of solving the linear systems of equations in the present-day sense.

The achievements in Chap. 8 of *Nine Chapters* were surely outcomes of a long series of successive developments in quite remote times. It seems that the methods in Chaps. 2, 3, and 6 of *Nine Chapters* about various kinds of problems involving proportions and Chap. 7 on surplus and deficiency may be the predecessors of the general method of Chap. 8. The traces of such developments are not clear, however.

B. Quadratic and Cubic Equations-Solving and Numerical Solutions of High Degree Algebraic Equations up to 13th Century.

Square-Root-Extraction Shu in Nine Chapters.

The simplest quadratic equation is in the form $x^2 = A$ with $A > 0$ an integer which is equivalent to the extraction of square root of A. In ancient China such problem of square-root extraction arose from measurements of plots of land and also naturally from Gou-Gu studies, namely determination of one side of a right-angled triangle from the other two sides supposed known. The former and related problems were the subject of Chap. 4 of *Nine Chapters* and the latter and related problems of Chap. 9. In Problems 12-16 of Chap. 4 it was asked the sides of square fields with known areas. The problems were solved by the algorithm called *Root-Extraction Shu.* All the methods of solving for positive roots of high degree algebraic equations in later generations of ancient China were actually based on extensions of this algorithm. In fact, in ancient Chinese classics of mathematics *root-extraction* was synonymous to what we call in present day *equation-solving.* Owing to the significance of this method of square-root-extraction we shall go into details of the corresponding algorithm.

For this purpose let us take Problem 12 of Chap. 4 as an illustrative example. The problem is to determine the side of a square field of known area 55225 units. The algorithm shows how the determination of the side is carried out step by step. However, it is difficult to see the principle underlying the procedure. It is the annotations of Liu Hui showing clearly the underlying principle which had a geometric basis and there was an annexed diagram with regions marked by different colors. The diagram was lost, but based on the annotations, it is easily recovered as shown schematically in Fig. 2.1 below:

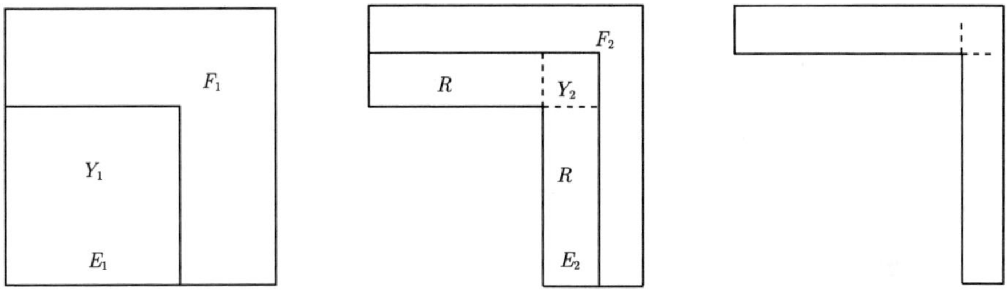

Fig. 2.1.

In Fig. 2.1 $Y, R,$ and B correspond to colors yellow, red, and blue respectively and 1,2, and 3 correspond to Chinese charaters in some calender ordering of ancient China.

The method of determining the square root of 55225 runs as follows.

The given area of the square field F is 55225. Now China was already in possession of a perfect *place-valued* decimal system for integers since very remote times. Owing to this it is first determined that the required side has a length of place-order 10^2. Since $2^2 < 5 < 3^2$, it is estimated that the given area will contain a smaller square Y_1 of side $2 * 10^2$ to be called provisionally *Fa* of the square Y_1. The *estimation* will be denoted by E_1. Remove now the square Y_1 from the given one there will remain a knife-shaped field F_1 whose area is $55225 - (2 * 10^2)^2 = 15225$. The inner side length of the field F_1 is known to be $2 * Fa = 4 * 10^2$ and is called *Ding-Fa* . In order to find the breadth of the field F_1 it is first determined that the breadth is of place-order 10. Since $3 * 4 < 15 < 4 * 4$ it is estimated that the knife-shaped field F_1 will contain a smaller one of the same form ($= R + R + Y_2$ in Fig. 2.1) whose estimated breadth is $E_2 = 3 * 10$. The area of this field is clearly

$Ding\text{-}Fa * E_2 + E_2^2 = 400 * 30 + 30^2 = 12900$. Remove this field we get a still smaller knife-shaped field F_2 for which the area is $15225 - 12900 = 2325$. The inner-side of F_2 is now known to be $Ding\text{-}Fa + 2 * E_2 = 4 * 10^2 + 2 * 3 * 10 = 460$ to be called again by *Ding-Fa* or *Ding-Fa*$_2$. Proceeding as before we find that the further place-order will be 1 and the estimated place-value will be 5. It turns out that, in removing a still smaller knife-shaped field of breadth $5 * 1$ the field F_2 will be exhausted. It follows that the side of the given field F_1 is $2 * 10^2 + 3 * 10 + 5 = 235$.

Notion of Mian or Irrational Numbers.

The algorithm *Root-Extraction Shu* is nothing else but an arithmetization of the above procedure based on geometrical considerations of a quite general character. Besides, in Chap. 4 it was also given the root extraction of a fraction not necessarily an integer. However, what is of utmost significance is some sentence that follows immediately after the above *Shu.* The sentence may be translated literally as follows.

> If one cannot get an exact integer in the extraction procedure, it is *unextractible* and the side will be called *Mian.*

In Liu Hui's annotations, the following passage of interpreting the above sentence is now extremely important. Liu Hui said:

> Some Shus give the root in adding some fractional part by adding 1 to *Ding-Fa* as denominator in case of inexaustible root-extraction. The result may be quite close to the true one but the method is illegitimate to be used. For in extracting an integer to get the root, the square of the root should be returned back to the integer given. Now, without adding 1 to *Ding-Fa* for the denominator of the fractional part, the number to be extracted is usually a little too small. On the other hand if 1 is added to *Ding-Fa* for the denominator of the fractional part, then the number will usually be a little too large. It is impossible to get an exact value. So the only way is to define it straightly as the *Mian* .

Liu Hui compared this to the introduction of *fractions* when division cannot be carried out if one is restricted to integers alone. Liu Hui cited as example the division of 10 by 3 compells one to introduce $\frac{1}{3}$ as a single entity. Thus, in a similar way the unextractible root is defined as the *Mian* which equivalently is an *irrational number* in the modern sense. However, this concept already clearly introduced and explained in the original text of *Nine Chapters* and Liu Hui's *Annotations* was overlooked by all historians of Chinese mathematics for so long a period and was pointed out explicitely only in year 1992 by Li Jimin of Northwest University at Xi-An, see his book [LJM2]*. Li pointed out further that in *Nine Chapters* and *Annotations* there were results about such *Mians* or *irrational numbers* which amount to identities of the form

$$(\sqrt{N})^3 = \sqrt{N^3},\ N = \sqrt{N^2},\ \sqrt{M} * \sqrt{N} = \sqrt{M * N},$$

$$\sqrt{\frac{M}{N}} = \frac{\sqrt{M}}{\sqrt{N}},\ \sqrt{\frac{M}{N}} = \frac{\sqrt{M * N}}{N},\ \sqrt[3]{\frac{M}{N}} = \frac{\sqrt[3]{M}}{\sqrt[3]{N}},\ \sqrt[3]{\frac{M}{N}} = \frac{\sqrt[3]{M * N^2}}{N}.$$

Remark that the last four identities were at the basis of root-extraction of fractions and appeared already in the text of *Nine Chapters*.

Root-Extraction of Fractions and Complete Place-Valued Decimal System.

Besides the introduction of this important concept *Mian* equivalent to *irrational numbers*, Liu Hui also had some suggestions equally important about root-extraction. In fact, owing to our perfect place-valued decimal system about integers since very remote times Liu Hui suggested to proceed further and further for the root-extraction by determining in succession the digits or place-values corresponding to place-orders $\frac{1}{10}, \frac{1}{10^2}$, etc. Liu Hui said:

> In receding to successive places further and further, the fractional rests will be smaller and smaller, then in keeping the red-colored region, the rest thrown away may be neglected to be taken care of.

Thus, what Liu Hui suggested was actually the adoptation of a complete place-valued decimal system to approximate to arbitrary precision any *real number* conceivable. Unfortunately this important suggestion was neglected by Chinese scholars for centuries. However, at latest about 10 centuries later, the system was ultimately adopted and was applied to the approximation of real solutions of algebraic equations. As for Europe, we know it well that it was Stevin (+1548, +1620) who introduced decimal fractions in his *La Disme* in +1585 and was highly praised as an important achievement.

Root-Extraction with Zong.

Returning to Root-Extraction Shu we see that during the procedure we meet the following situation. Let a knife-shaped field F be formed from a square of known area A in removing a smaller square with side-length a at one of the corners. Then F has the known area $A - a^2 = B$, say. The procedure shows that we are able to solve the following problem:

Problem Q. Given a knife-shaped field F of known area B and known inner-side $2 * a$. To determine the breadth x of the field F.

Let us consider the diagram Fig. 2.2 above arisen from Fig. 2.1 with $E_1 = a$ and unknown side of F_1 to be denoted by x. Then we see that the field F of area

$B = A - a^2$ is composed of 2 red subfields R each of area $a*x$ and one yellow subfield Y of area x^2 so that in modern notation we have

$$x^2 + 2*a*x = B, \ (a > 0, B > 0), \text{ or} \tag{2.1}$$

$$(x+a)^2 - a^2 = B, \ (a > 0, B > 0). \tag{2.2}$$

Now move one of the two red subfields to the side of the yellow subfield to get a rectangular field as in Fig. 2.3 above. Then the area of the rectangular field will remain to be the same as the knife-shaped field F. The two sides of the rectangular field will have a difference $2*a$ equal to the inner-side of the knife-shaped field which is called Ding-Fa in the procedure of root-extraction. Hence the procedure in the Root-Extraction Shu permits also to solve the following problem:

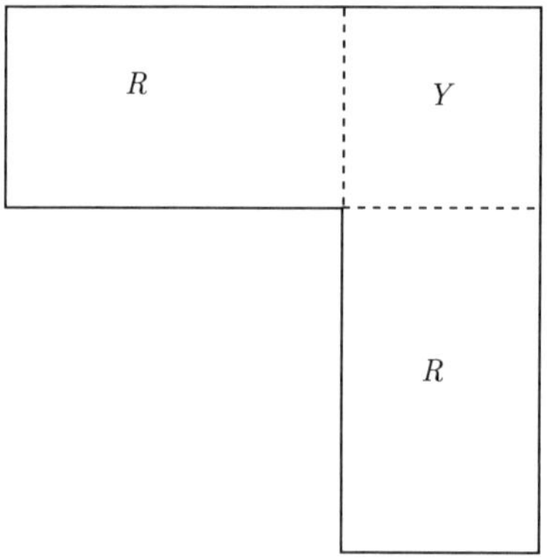

Fig. 2.2.

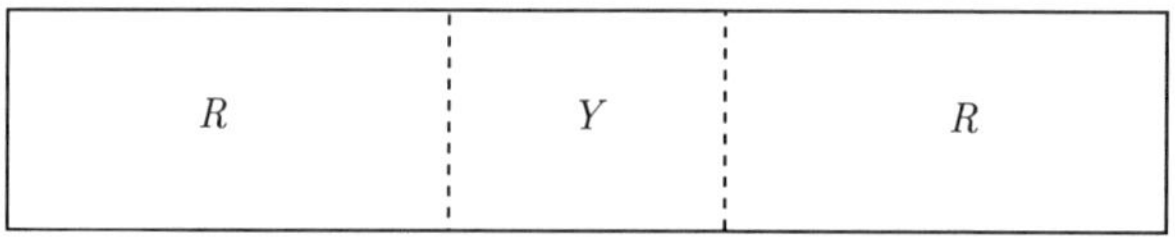

Fig. 2.3.

Problem R. Given a rectangle of known area B and known difference $2*a$ of its two sides. To determine the shorter side of the rectangle.

Our ancestors called the solving of problems **Q, R** or equivalently the solving of the quadratic equations (2.1) or (2.2) *root-extraction with zong*. Here *zong* or *zong-fa* is an extra-data $2*a$ corresponding to the inner side of the knife-shaped field F which is called *ding-fa* in the procedure of the Root-Extraction Shu. The term B in (2.1) is called in our ancient classics *shi* , literally meaning area, which is actually the area of the knife-shaped field F or that of the rectangle.

An Example about Root-Extraction with Zong.

As an illustrative example we reproduce below a City-Width Problem taken from Chap. 9 of *Nine Chapters.*

Problem 20. There is a square-formed city of unknown width. Each side of the city has a gate in the middle. Some one goes northward 20 bus out of the northern gate and reaches a tree. If he goes southward 14 bus out of the southern gate and turns westward 1775 bus, then he just sees the tree. To determine the width of the city.

Answer. City width: 250 bus.

Shu. Take 2 times the product of northern bus out of northern gate and the turning westward bus as shi. Take the sum of northern bus out of northern gate and the southern bus out of the southern gate as zong. By root-extraction with zong, one gets the city width.

Remark that *bu* is some measure of length in ancient time. Let the northern, southern and western bus be denoted by n, s, and w respectively. Then the *Shu* above says that Problem 20 can be solved as a Problem R with shi or area of the rectangle $B = 2 * n * w = 71000$ and zong or side-difference $2 * a = n + s = 34$. In modern notations this amounts to solving of the equation below:

$$\begin{aligned} &x^2 + (n+s) * x = 2 * n * w, \text{ or} \\ &x^2 + 34 * x = 71000. \end{aligned} \tag{2.3}$$

In Sect. 3 of our next Chap. 2 it will be explained how, based on Liu Hui's *Annotations*, the original Problem 20 can be reduced to a Problem R. See the diagrams in Figs. 2.4 and 2.5 below.

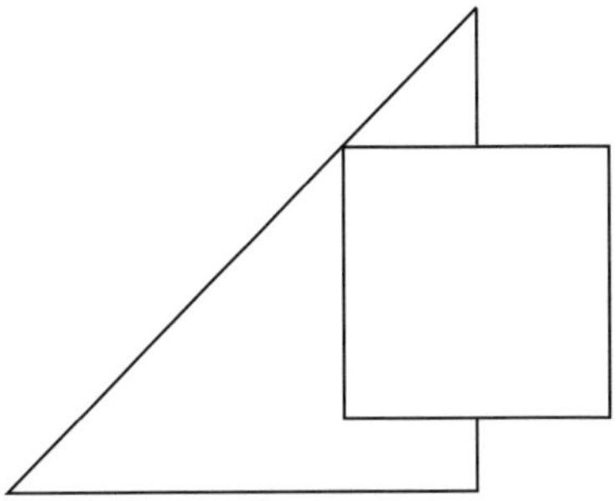

Fig. 2.4.

Fig. 2.5.

Cubic-Root-Extraction Shu in Nine Chapters..

In Chap. 8 of *Nine Chapters* there is also a *Cubic-Root-Extraction Shu* which corresponds to an arithmetization of geometric procedure as exhibited in the following diagram Fig. 2.6 analogous to Fig. 2.1:

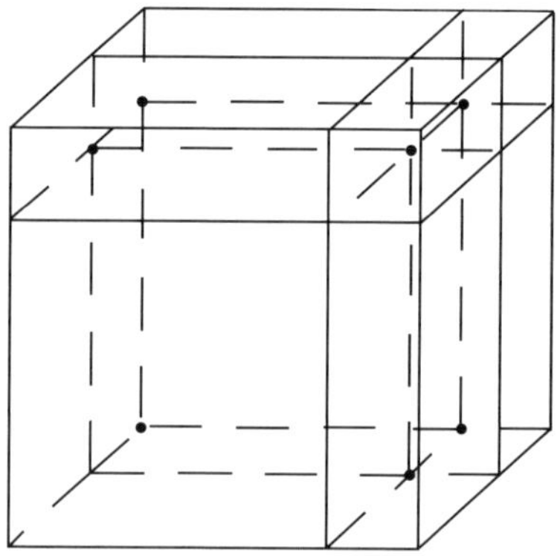

Fig. 2.6.

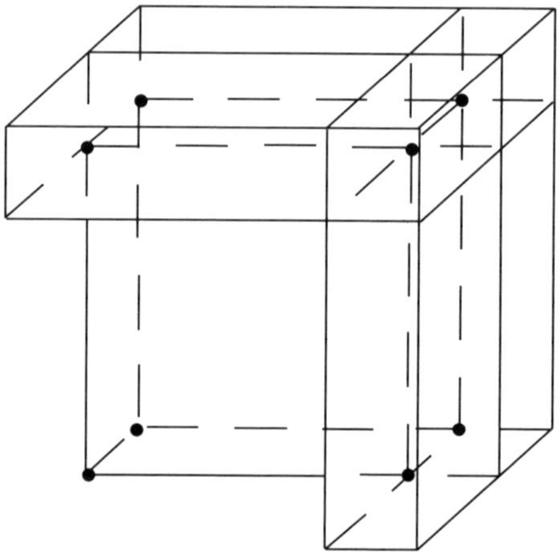

Fig. 2.7.

As before, during the procedure we encounter the following situation. Remove the smaller cube in one of the corners of side-length a from the given cubic solid of volume V in Fig. 2.6, we get a solid S as in the diagram Fig. 2.7 above with known volume $W = V - a^3$. The procedure in the Cubic-Root-Extraction Shu permits to determine the side-length of the yet smaller cube in the opposite corner. Using the modern notations let us denote this side-length by x. Then from Fig. 2.7 we see that the solid S is composed of 7 rectangular parallelipipeds of sides a, a, x three times, a, x, x three times, and x, x, x one time. Hence the procedure in modern notations is equivalent to one of solving a cubic equation of the form below:

$$x^3 + 3 * a * x^2 + 3 * a^2 * x = W,\ (a > 0,\ W > 0),\ \text{or} \tag{2.4}$$

$$(x + a)^3 - a^3 = W,\ (a > 0,\ W > 0). \tag{2.5}$$

Some General Remarks.

In 3c A.D. Zhao Shuang in his work [ZS2]* had given some propositions about Gou-Gu forms (i.e. right-angled triangles). Some historians suggested that one such proposition may be interpreted as giving the solution of a quadratic equation in explicit form. However, our ancestors seem to have little interest in getting any explicit formula for the root of a quadratic equation of the form (2.1) or (2.2) or more general ones. They were rather interested in numerical solutions of quadratic equations to meet real requirements. This should be the same case for cubic equations of form (2.4), (2.5), or any more general ones.

Cubic Equations from Han Dynasty to Tang Dynasty.

After the period of three kingdoms little was known about developments of Chinese ancient mathematics until the time of Zu Chongzhi (+429, +500). In the *Historical Record of Sui Dynasty* it was reported the great achievements of Zu about computation of π, and also his methods of *Kai Cha Mi, Kai Cha Li* , which literally means *extraction difference square, extraction difference cubic* respectively. The historian Qian Baozong explained these as methods of square- and cubic-root-extraction with *zong* equivalent to the solving of equations

$$x * (x + k) = A,\ \textit{or}\ x^2 + k * x = A,\ (k, A > 0),\ \text{and}$$

$$x * (x + k) * (x + l) = V,\ \textit{or}\ x^3 + (k + l) * x^2 + k * l * x = V,\ (k, l, V > 0).$$

Unfortunately the major work of Zu was lost and it is difficult to ascertain the suppositions of Qian. In any way Zu's creations were continued by his son Zu Geng and the latter's work was severely criticized by Wang Xiaotong in his classic [WST]*. The reasons of his criticisms are unclear owing to the missing of relevant writings.

Wang's Classic on Cubic Equations.

The classic [WST]* of Wang Xiaotong was a treatise devoting to cubic equations which appeared in early years of Tang Dynasty in 7c. It contained 20 problems arising mostly from large-scale reconstructions. For solving these problems Wang showed how to establish first cubic equations and then get the required solution by *cubic-root-extraction with zong* . The equations in modern notations will be of the form

$$x^3 + a_1 * x^2 + a_2 * x = a_3,\ (a_1, a_2, a_3 > 0). \tag{2.6}$$

Wang gave in each case the formation of a_1, a_2, a_3 from given data which are very complicate and no reasons were given in Wang's treatise. Once a_i were determined then Wang said simply that one gets the answer by cubic-root-extraction with zong. It seems highly probable that numerical method of solving cubic equations in the form of (2.6) should be well-known to scholars at that time so that no more explanations were necessary.

Numerical Solving of High Degree Algebraic Equations in Song Dynasty.

After the classic of S.T.Wang little developments were known until Northern Song Dynasty (+960, +1038). A breakthrough occured which was due to Jia Xian whose principal writings seems to appear in the period (+1023, +1050). These works were all lost. Fortunately, in the latter works [YH1,2,3] of Yang Hui (living in end of Southern Song Dynasty) the methods and a number of problems in Jia's works were recalled. Owing to works of Yang Hui we know that the methods of Jia were ones of high-order root-extraction or equivalently the solving of equations of the form

$$x^n = A, \ (A > 0) \tag{2.7}$$

in which n is a positive integer which may be greater than 3. The methods of quadratic- and cubic-root-extraction as expressed in the equations of (2.1), (2.2), (2.4), (2.5) were based on an arithmetization of the geometrical considerations as shown in the diagrams of Figs. 2.1, 2.2, 2.6, and 2.7. However, for higher degree equations such geometrical considerations were no more available. On the other hand according to Yang Hui's works, there already occured in Jia's time some diagrams bearing the name of *Root-Extraction Basic Diagrams.* Such diagrams are actually the same as the so-called Pascalian Triangle of 17th century. Thus, it seems that Jia had freed himself from geometrical considerations and, with the aid of the root-extraction basic diagrams, had discovered his methods directly from a generalization of the *arithmetized* Root-Extraction Shu of his ancestors. Furthermore, along with the equation (2.7), the equations of the following type

$$x^4 + 4 * a * x^3 + 6 * a^2 * x^2 + 4 * a^3 * x = D, \ (a > 0, D > 0), \text{ or} \tag{2.8}$$

$$(x + a)^4 - a^4 = D, \ (a > 0, D > 0) \tag{2.9}$$

etc. should also be known how to be numerically solved at the time of Jia.

Further improvements of Jia's methods were made by Liu Yi (in 12c) in his no more existant writing *Discussions on the Old Sources.* Fortunately the methods and a number of problems of Liu's work were again recalled in Yang Hui's works. Among the problems preserved was one of solving an equation

$$-5 * x^4 + 52 * x^3 + 128 * x^2 = 4096. \tag{2.10}$$

Thus, Liu had freed himself from restricting the coefficients of an equation to be positive ones as in (2.8).

In the classic [QJS]* of Qin Jiushao the methods of Jia and Liu had been combined together and developed into a general method called *Positive-Negative Root-Extraction Shu* which permits to solve numerically an algebraic equation of any high degree with numerical coefficients arbitrarilly positive, negative, or zero. The decimal numbers, as suggested by Liu Hui, were used freely in Qin's methods. The Qin's method was actually the same as the so-called Horner's method in +1819. Thus, starting from Root-Extraction Shu in *Nine Chapters*, after about 12 centuries of incessant efforts the ancient Chinese had ultimaletly achieved their goal: Solving numerically an algebraic equation of arbitrarily high degree with arbitrary numerical coefficients.

C. Method of Heaven's Element and Elimination Method of Polynomial Equations-Solving in Song and Yuan Dynasties.

Heaven's Element Shu and Algebrization of Geometry.

During the period of Song and Yuan Dynasties there was a discovery of utmost importance in the introduction of the precise notion of an *unknown* under the name of *Heaven's Element.* A so-called *Heaven's Element Shu* or *Heaven's Element Method* was developed which permitted to turn easily geometrical entities into the form of polynomials and geometrical relations into polynomial equations. Two classics due to Li Ye (+1192, +1279), [LY1]* and [LY2]*, had appeared and survived up to present days. The first one in +1248 was a treatise devoted to some special topic in applying the *Heaven's Element Shu* and the second treatise in +1259 was of an elementary character aimed at popularization of that method. The method may be briefly described as follows. It begins by setting some unknown entity to be determined as *Heaven's Element.* Some other entity is then expressed in polynomials of that heaven's element in two different ways. By equating these two expressions one gets an equation which may then be solved by methods already well-developed at that time. The final equation to be solved may be of quite high degree. Thus, among the 170 problems in [LY1]*, 13 are of degree 3, 13 of degree 4, and one of degree 6. Remark that in spite of the fact that equation-solving had been well-developed since the time of *Nine Chapters*, there was no precise notion of *unknowns* and the establishment of an equation which, usually based on intricate geometrical considerations, was a very difficult task. The introduction of the notion of *heaven's element* rendered such tasks very easily done. In fact, in Qing Dynasty the scholar Zhang Dunren had applied the heaven's element method to establish the cubic equations in the classic [WST]* in a simple manner which were otherwise very complicated. Besides, Li Ye in his treatises introduced the notions equivalent to present day polynomials as well as their manipulations which opened the way toward an algebrization of geometry. Remark that the treatises [QJS]* of Qin Jiushao and [LY1]* were written almost in the same time but they never mentioned each other in any of their writings. It seems that they didn't know each other at all. In fact, Qin lived in the southern part of China while Li in the northern China and these two parts were under reigns of dynasties constantly in warfare. The interest of these scholars were also different. One was interested in the numerical solving of an equation already established, while the other paid his efforts mainly to the establishment of equations. Their works complemented each other and combined together, marked a golden age of mathematics in ancient China.

High Degree Polynomial Equations-Solving and Elimination Method.

Early in *Nine Chapters* there were already methods of solving systems of multivariate linear equations. In Song-Yuan Dynasties there also appeared methods of high degree polynomial equations in a single unknown. It would be natural to unify these two kinds of methods together to consider systems of polynomial equations of high degrees not restricted to a single unknown. This was really the case. Already in the classics of Li Ye there were problems for which equations will be more readily established if one introduces several unknowns instead of a single one. Li Ye himself had not done it in this way and kept use of a single unknown. The advantage of extending the method to more than one unknown should be clear to scholars in latter times. In fact, it was known that there were texts about treatment of pairs of polynomial equations in two unknowns, *Heaven's* and *Earth's Elements*, and even triads of equations in three unknowns which were all lost. Fortunately, there remained a classic [Zhu2]* bearing the title *Jade Mirrors of Four Elements* of Zhu Shijie in +1303 of Yuan Dynasty which treated polynomial equations up to even four unknowns, called the *Heaven's Element, the Earth's Element, the Man's Element*, and the *Matter's Element* respectively. The method is one of elimination of unknowns in succession. It is based on some elimination procedure and runs somewhat as follows. First we establish equations in designating some unknowns to be heaven's element, etc. Suppose there are four such equations in four unknowns. Choose one of such unknowns and eliminate it from the given system to get a triad of equations in three unknowns. Then eliminate a further unknown from this triad to get a pair of equations in two unknowns. Eliminate now one of these two unknowns from the pair to get a single equation in a single unknown. Solve now this equation in that unknown by known methods. Substitute now such a solution in the other equations and solve them in succession to get the other unknowns in succession. The method is clearly a quite general one and depends heavily on manipulations of polynomials and elimination method so that some sort of algebra and algebrization of geometry were developed to meet the purpose.

An Example of Zhu Shijie.

The method of Zhu Shijie may be seen clearly from one of the representative examples in Zhu's classic [Zhu2]* as shown below.

A geometrical problem has led to a set of three equations in three unknowns which in modern notations may be written in the following form:

$$\begin{aligned} x*y*z - x*y^2 - z - x - y &= 0, \\ x*z - x^2 - z - y + x &= 0, \\ z^2 - x^2 - y^2 &= 0. \end{aligned} \tag{2.11}$$

The problem is to determine z. Zhu's method runs now briefly as follows. From geometry x cannot be 0 so x will be divided out freely during the procedure below. In eliminating now first y from the given equations (2.11) and then reducing successively degrees of x in the equations thus obtained we will get finally two equations in x and z alone which are both linear in x. Eliminating then x from these two last equations we will get a single equation in z alone, viz.

$$z^4 - 6*z^3 + 4*z^2 + 6*z - 5 = 0. \tag{2.12}$$

Zhu now chose the positive solution $z = 5$ of the last equation as the solution required. The details of computations may be found in the excellent treatise [Hoe]*.

Comments about Method of Zhu Shijie.

Zhu stops at the solution $z = 5$ for the above problem. However, we may go a step further in taking any one of the two equations in x, z and linear in x and also any one, say the second one of (2.11), to form a system of three equations (2.13) below:

$$\begin{aligned} z^4 - 6*z^3 + 4*z^2 + 6*z - 5 &= 0, \\ x*(7 + 3*z - z**2) - (6 + 7*z + 3*z^2 - z^3) &= 0, \\ y - (x*z - x^2 - z + x) &= 0. \end{aligned} \tag{2.13}$$

We may then solve the equations (2.13) successively in single unknowns z, x, and y in succession to get solutions of the triple (z, x, y). If we start from $z = 5$ we will get one of the solutions to be $(z,x,y)=(5,3,4)$. However, the equation (2.12) has two further solutions $z=-1,+1$ besides $z=5$. If we start from $z=-1$, we will get a solution-triple $(z,x,y)=(-1,1,0)$ which is readily verified to be really a solution of the given system (2.11). On the other hand if we start from $z=1$ then the solution-triple $(z,x,y)=\left(1,\frac{5}{3},-\frac{4}{9}\right)$ will not be one of the given equations (2.11).

For the actual computation at the time of Zhu one had to place the coefficients of various kinds of terms of polynomials in counting rods at definite positions of the counting board. This limited the method to at most four equations in four unknowns and only quite simple ones can be so treated. However, that the method of Zhu enjoys a general character which can be applied to arbitrary systems of equations is quite clear. On the other hand the analysis of the above particular example shows that the method has great defects and requires clarification. It is one of the main purposes of the present book to give a solid mathematical foundation of methods influenced by that of Zhu which will be done in Chap. 3.

Since the Ming Dynasty the brilliant development of Chinese ancient mathematics in algorithmic fashion has declined and has not exerted any more influence to the mathematical world.

1.3 Polynomial Equations-Solving in Ancient Times beyond China and the Program of Descartes.

Publications on History of Mathematics beyond China.

There were innumerous publications on history of mathematics in the western world, touching sometimes the history in Japan, Hindu, or Arabic World, but rarely China. We list below a few of them which we have particularly consulted.

[AA]* A.Aaboe, Episodes from the early history of mathematics, Yale University, (1964).

[Caj]* F.Cajori, A history of mathematics, 2nd Ed., Macmillan, (1919).

[Des1]* R.Descartes, La géomètrie, (1637).

[Des2]* R.Descartes, Rules for the direction of the mind, (English translation), in E.S.Haldane & G.R.T.Ross, The philosophical works of Descartes, (1911).

[Ev]* H.Evens, An introduction to the history of mathematics, 5th Ed., Saunders Colledge Publishing, (1903).

[Kar]* L.C.Karpinski, Robert of Chester's Latin translation of the algebra of al-Khwarizmi, (1915).

[Kl]* M.Kline, Mathematicsal thought from ancient to modern times, Oxford University Press, (1972).

[LWL1]* Li Wenlin, On the mechanical character of Descartes' geometry, Studies in the History of Natural Sciences, 12 (1993) 225-234.

[LWL2]* Li Wenlin, The tendencies of algorithm and deduction for demarcation of periods in the history of mathematics, J. of Dilectics of Nature, v.8, no.2, (1986) 46-50.

[M]* Y.Mikami, The development of mathematics in China and Japan, 2nd Ed., Chelsea, (1974).

[N]* J.Needham, Science and civilization in China, vol. 3, Cambridge University Press, (1959).

[P]* G.Polya, Mathemaical discovery, vol. 1, John Wiley & Sons, (1962).

[Sc]* J.F.Scott, A history of mathematics from antiquity to the beginning of the nineteenth century, Dover Publ., (1958).

[Sm]* D.E.Smith, History of mathematics, vols. 1,2, Dover Publ., (1958).

[S-L]* D.E.Smith & L.M.Latham, The geometry of René Descartes, Dover Publ., (1954).

[St]* D.J.Struik, A concise history of mathematics, Dover Publ., (1948).

[VdW]* B.L.Van der Waerden, A history of algebra, from al-Khwarizmi to Emmy Noether, Springer, (1985).

Systems of Linear Equations.

In ancient times little was known about the solving of systems of linear equations beyond China. As for an elimination method for the solving of systems of linear equations, it is usually attribuled to K.F.Gauss and bears the name of *Gaussian Elimination.* The method occured first in the paper [Gau3] of Gauss.

Quadratic Equations.

In Euclid's *Elements* of 3c B.C., there were propositions about geometrical constructions, notably Prop. 11 of golden ratio, and Props. 28, 29 which, if interpreted in modern form, correspond to solving of some quadratic equations.

The book *Hisab Al-jabr w'al muqabala* in +830 of al-Khwarizmi (c.+813, +850) of Persian origin studied, in detail, the solving of linear and quadratic equations. The quadratic equations were divided into 5 separate types which in modern notations may be written as

$$a * x^2 = b * x,\ a * x^2 = b,\ a * x^2 + c = b * x,\ a * x^2 + b * x = c,\ a * x^2 = b * x + c,$$

in which a, b, c are all positive. The solving methods were geometric in character and resembled very much the ancient Chinese dissection method connected with the *out-in complementary principle*, see the Sect. 3 of next Chap. 2. The title of al-Khwarizmi's book suggests that some of the present-day terminologies may come from the word *al-jabr*. On the other hand D.E.Smith suggested that *al-jabr* has the fundamental idea of the transposition of a negative quantity, and *muqabalah* the transposition of a positive quantity and the simplification of each member, cf. [Sm]*, vol.2, p.389.

Sources of al-Khwarizmi's Work.

In [VdW]*, p.13, Van der Waerden discussed the sources of al-Khwarizmi's work, in particular of his Algebra. He said:

> Three theories have been proposed. He may have used classical Greek sources, or Hindu sources, or popular mathematical writings belonging to the Hellenistic and post-Hellenistic tradition.

To the last of these sources it seems that Van der Waerden means the native Syriac-Persian tradition including that of Hebrew.

In [N]*, p.107 (also p.147), Needham suggested ancient China as a further possible source of Al-Khwarizmi's work. Needham wrote:

> One may indeed legitimately ask whether Arabic mathematics may not have been influenced by the Chinese in this respect. It would be rewarding to study the original works of al-Khwarizmi from this point of view. He was sent as an ambassador to Khazaria by the Caliph between +842 and +847 and Khazaria lay athwart some of the trade routes between China and the West.

About the Khazar kingdom Needham wrote further in an Addenda about Israel and Khazaria ([N]*, p.681-683):

> ··· Chinese was understood in Khazar lands, and that the khanate used Chinese court etiquete. ··· As an example of the role which Khazaria may have played in scientific and technological transmissions, we remember the embassy there of the great algebraist al-Khwarizmi between +842 and +847.

Besides the above suggested possible sources of al-Khwarizmi's work the present writer would like to suggest that al-Khwarizmi's work was al-khwarizmi's own creation. In fact, quite a number of historians in mathematics, e.g. Struik in [St]*, argued in their writings that already in the Babylonian age before 3c B.C. one knew how to solve quadratic equations, even higher degree or simultaneous ones, and knew how to express the roots of a quadratic equation in terms of the coefficiants. If this was really true, then there is no reason not to believe that after more than 1000 years later, almost in the same region of the Babylonians, there would be some mathematical genius to make such a creation as in Al-Khwarizmi's work.

Cubic and Higher Degree Equations.

In the Arabic world Abu'l-Wafa (+940, +998) gave geometrical solutions to some special quartic equations. Later, Omar Khayyam (c.+1044, +1123) gave systematic classification of cubic equations and geometric solutions of cubic equations by considering intersections of circles and conic sections.

In Europe cubic and quartic equations were studied by Scipione del Ferro (+1465, +1526) around +1515. The studies continued until Cardano (+1501, +1576) and Ferrari (+1522, +1565) discovered their solutions of cubic and quartic equations. The solution of cubic equations by Cardano was geometrical in character and marked the rise of European mathematics after hundreds of years of silence under the dark age.

Unlike the ancient Chinese, who were mainly concerned with numerical solutions of algebraic equations to meet the reality requirements, the European mathematicians were rather interested in seeking solutions in closed form in terms of coefficients by means of radicals as in the case of Cardano and Ferrari. Only after about three hundred years of efforts was it ultimately found that such attempts were deemed to failure owing to the impossibility discovered by Abel (+1802, +1829) and Galois (+1811, +1832) which led to the important discovery of Galois Theory and group theory in the 19th century. As for numerical solutions of algebraic equations, this was achieved by W.H.Horner (+1786, +1837) in 1819 and the method bears now his name. As cited by Eves in [Ev]*, p.159, Horner was completely unaware of the fact that he had rediscovered an ancient Chinese computational scheme.

Negative Numbers.

As pointed out by M.Kline in [Kl]*, p.185:

> The Hindus introduced negative numbers to represent debts; in such situations, positive numbers represented assets. The first known use is by Brahmagupta about 628; he also states the rules for the four operations with negative numbers. Bhaskara points out that the square root of a positive number is twofold, positive and negative.

From this passage we see that the Hindus knew very well the nature and the operations about negative numbers at latest in the time of Brahmagupta (c.628). As for the situations in western world, we may cite the following passage of M.Kline ([Kl]*, p.252):

> As for negative numbers, though they had become known in Europe through the Arab texts, most mathematicians of the sixteenth and seventeeth centuries did not accept them as numbers, or if they did, would not

accept them as roots of equations.

Kline said further in [Kl]*, p.593:

> Indeed, even in Descartes' *La Géomètrie* of 1637, negative roots of an algebraic equation were designated as "false roots".

Remark that the ancient Chinese adopted also only positive roots of an algebraic equation, but the situations were different. For, the ancient Chinese largely considered problems arising from reality so that only positive roots will be of interest to them. On the other hand they knew very well the meaning of negative numbers and manipulated these negative numbers together with positive numbers freely as seen already from problems in Chap. 8 of *Nine Chapters*.

Program of Descartes on Problem-Solving.

In Chap. 2 of the classic [P]*, Polya has called our attention to a posthumous work of Descartes, [Des2]*. Polya described a rough outline of the scheme that Descartes expected to be applicable to all types of problems which we reproduce below:

> First, reduce any kind of problem to a mathematical problem.
>
> Second, reduce any kind of a mathematical problem to a problem of algebra.
>
> Third, reduce any problem of algebra to the solution of a single equation.

The third step may be further clarified as shown in the paragraph below which we again reproduce from the above classic of Polya:

> Descartes advises us to set up as many equations as there are unknowns. Let n stand for the number of unknowns, and $x_1, x_2, \cdots, x_n$ for the unknowns themselves; then we can write the desired system of equations in the form
>
> $$r_1(x_1, x_2, \cdots, x_n) = 0,$$
> $$r_2(x_1, x_2, \cdots, x_n) = 0,$$
> $$\cdots\cdots$$
> $$r_n(x_1, x_2, \cdots, x_n) = 0,$$
>
> where $r_1(x_1, x_2, \cdots, x_n)$ indicates a polynomial in $x_1, x_2, \cdots, x_n$, and the left-hand sides of the following equations must be similarly interpreted. Descartes advises us further to reduce this system of equations to one final equation.

Some Comments about Descartes' Program.

It seems that Descartes was absolutely absurd in reducing any kind of problem to one of polynomial equations-solving. In Descartes' time differential calculus had not yet been discovered so it is natural that Descartes had no idea about differential equations and other kinds of equations but the polynomial ones. However, even a problem can be so reduced to polynomial equations, the number of equations may not be the same as the number of unknowns: it may be smaller or greater. Even the numbers of equations and unknowns are the same, the reduction of such a system of polynomial equations to the solving of equations involving a single unknown is not at all evident how to do and is actually entirely non-trivial. Even if it can be so reduced, there may occur redundant solutions or loss of solutions. As pointed by Polya in the above-mentioned book, the reduced equations in single unknowns may furnish eventually the whole set of solutions only "in general", e.g. in the "regular case". However, the clarification of "in general" or "regular case" will be far from being trivial, as may be seen already from the simplest case for which r_i are all linear in the x's. In fact, it is one of our main objects of the present book to achieve such clarifications which will be done in Chap. 3.

However, in spite of the apparent absurdity of Descartes' Program, let us reproduce some paragraph again from the same classic of Polya:

> Descartes' project failed, but it was a great project and even in its failure it influenced science much more than a thousand and one little projects which happened to succeed. Although Descartes' scheme does not work in all cases, it does work in an inexhaustible variety of cases, among which there is an inexhaustible variety of *important* cases.

In fact, now-a-days we see that there are really inexhaustive problems from mathematics, from sciences, from technologies, etc. which will ultimately be reduced to polynomial equations-solving. Among all kinds of systems of equations to be solved, those of polynomial forms are clearly the simplest, and the most fundamental ones, but at the same time a problem entirely non-trivial at all. Among all kinds of problems in mathematics, the problem of polynomial equations-solving should, therefore, be considered as one of primary importance to be settled with highest priority.

Descartes Program and Chinese Ancient Mathematics.

In a recent paper [LWL1]* Li Wenlin has analyzed the classic [Des1]* of Descartes. We cite a few of Li's arguments below:

1. By Descartes' method geometry is arithmetized so that geometrical problems are reduced to the solving of algebraic equations.

2. To each kind of (algebraic) equation, it was given some *standard* geometrical constructions of roots or *standard* method of solving.

3. *Standard* may be considered as some sort of *mechanical* procedure.

4. *La Géomètrie* may be considered as a realization of Descartes' Program in the realm of geometry with complements and improvements.

In fact, from [Des1]* it seems clear that Descartes had the altitude of emphasizing on geometry problem-solving by means of equations-solving rather than geometry theorem-proving, just in the same spirit of our ancestors.

It seems also interesting to compare Descartes' way of achieving this goal with the way of our ancestors. For this sake let us reproduce below a passage from [Des1]*, p.6:

> If, then, we wish to solve any problem, we first suppose the solution already effected, and give names to all the lines that seem needful for its construction, — to those that are unknown as well as to those that are known. Then, making no distinction between known and unknown lines, we must unravel the difficulty in any way that shows most naturally the relation between these lines, until we find it possible to express a single quantity in two ways. This will constitute an equation, since the terms of one of these two expressions are together equal to the terms of the other.

It would be interesting in comparing this paragraph of Descartes in establishing equations for problem-solving with the way of our ancestors in introducing unknowns called Heaven's element, etc. and then establishing equations for problem-solving as explained in last section. In a word, it may be said that Chinese ancient mathematics in the main were developed along the way as indicated in Descartes' Program, and conversely, Descartes' Program may be considered as an overview of the way of developments of Chinese ancient mathematics.

Chapter 2

Historical Development of Geometry Theorem-Proving and Geometry Problem-Solving in Ancient Times

2.1 Geometry Theorem-Proving from Euclid to Hilbert.

Stages of Development of Geometry Theorem-Proving.

Theorem-Proving, as a kind of mental labor, is one of the main activities of mathematics. It was originated in ancient Greek with Euclid's *Elements* as its representative work, which appeared, according to historians, around 300 B.C. However, it seems that the recognition and treatment of theorem-proving had undergone several stages of development since the time of Euclid. This will be explained below.

Stage I. Geometry Theorem-Proving in Euclidean Fashion.

Almost everybody is well-acquanted with the Euclidean fashion of proving geometry theorems from his learning in high schools since boyhood. Such proofs are carried out in the manner as described in the characteristic features (a) and (b) below.

(a) *Proving is carried out in a deductive way by logical reasonings based on an axiomatic system and presented in a Definition-Axiom-Theorem-Proof style.*

(b) *Each individual theorem has an individual proof peculiar to the theorem itself.*

Examples : Pythagoras' Theorem and Heron's Formula.

There is no general guideline of thought for the proof which requires usually high intelligence, and intricate roundabout way of reasonings to find such a proof. Quite often a seemingly trivial theorem is very difficult to find out a proof. In order to compare with the other manner of proving theorems in later developments and in ancient China, we shall give below a few such theorems among the innumerous interesting and beautiful theorems discovered since the time of Euclid.

Example 1.1. Pythagoras' Theorem. In right-angled triangles the square on the side subtending the right angle is equal to the sum of squares on the sides containing the right angle.

The theorem was given as Proposition 47 in Book I of Euclid's *Elements.* According to Proclus (+410 or +412, +485), the theorem was attributed to Pythagoras (about −560, −480). We reproduce here the figure (Fig. 1.1) in *Elements.* It is to be remarked that no notion of *area* appeared in *Elements* and the *equality* in the statement of the theorem is some kind of *dissection-equality* of figures to be discussed later in Sect. 2.

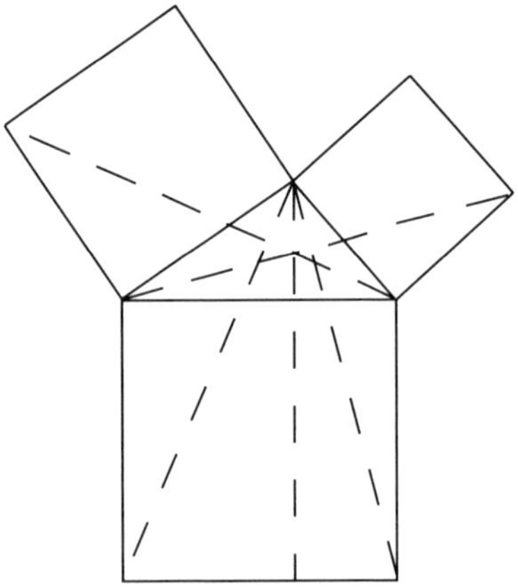

Fig. 1.1.

Example 1.2. Heron's Formula. The area $\triangle$ of a triangle with side-lengths a, b, c is, in setting $s = \dfrac{a+b+c}{2}$, given by

$$\triangle = \sqrt{s*(s-a)*(s-b)*(s-c)}.$$

The formula was usually attributed to Heron of Alexandria (probably around +62). We reproduce here only the figure (Fig. 1.2) associated to the proof, given in the works *Metrica* and *Dioptra* of Heron, according to T.L.Heath, *The thirteen books of Euclid's Elements*, v.2, p.87.

The Classic of Descartes on Geometry.

The appearance of the classic *La Géomètrie* by Descartes in 1637 made possible an utterly new fashion of geometry theorem-proving. In the French edition of the classic [Des2]* Descartes wrote at the very beginning on the margin

Comment le calcul d'Arithmetique se rapport aux operations de Géomètrie.

Descartes explained in the text how geometrical constructions may be carried out corresponding to the main arithmetical operations $a+b, a-b, a*b, a/b, \sqrt{a^2+b^2}$, etc. for segments of lengths a, b. He then explained, with an implicit coordinate system, how the method can be applied to various problems arising from geometry and optics. The creation of this now-a-days so-called analytical geometry has thus paved the way for reducing geometry to algebra or arithmetic in Descartes's own terminology and replacing logical reasonings by mere algebraic computations. This makes the geometry theorem-proving entering an entirely new stage of development.

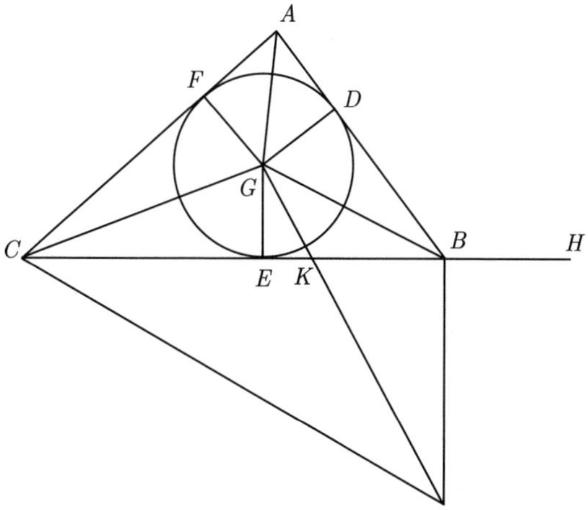

Fig. 1.2.

An Example : A Theorem of Gauss.

As an illustration of Descatesian fashion of geometry theorem-proving let us consider a theorem discovered by C.F.Gauss as follows.

Using modern terminology instead of the original one used by Gauss, let us call by a *complete quadrilateral* a geometrical configuration consisting of 4 lines in the plane meeting each other in 6 different points. Cf. Fig. 1.3 in which the 4 lines forming the complete quadrilateral are AEB, AFD, BCF, CDE. The three pairs of the intersecting points $(A, C), (B, D)$, and (E, F), not lying both on any such lines, are then called diagonal pairs. Gauss, in a short paper [Gau1], proved that the centers of all conics touching the 4 lines of such a quadrilateral are collinear. As the 3 diagonal pairs may be considered each as a degenerate line conic touching these 4 lines, the following theorem will follow as a corollary.

Example 1.3. Gauss-Line Theorem. For a complete quadrilateral the mid-points of the three diagonal pairs are collinear.

Proofs in Euclidean fashion of Gauss Theorem, which require drawing of auxiliary lines with quite involved reasonings, may be found in usual geometry texts and we shall not enter. On the other hand, it seems that Gauss was very proud of his theorem and he considered it to be entirely new. He wrote:

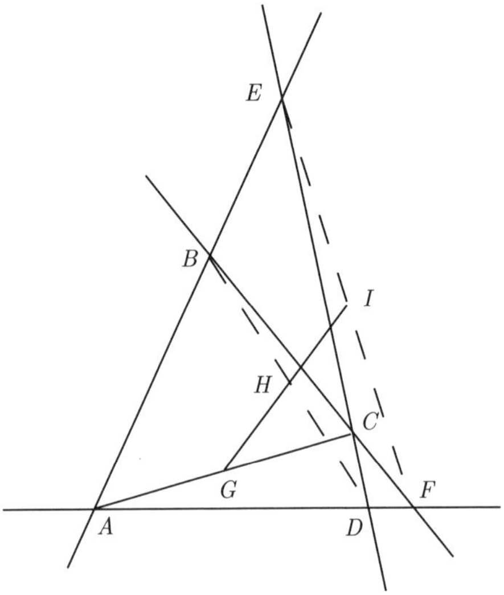

Fig. 1.3.

> Diese allgemeine Eigenschaft eines jeden Vierecks ist meines Wissens nicht bemerkt, ich werde davon unten einen einfachten direkten Beweis geben.

We reproduce now Gauss' own proof below, using Gauss' own notation. Let us take an arbitrary coordinate system and let the various points be $A = (a, a')$, $B = (b, b')$, etc. Similarly let the mid-points of the diagonal pairs $(A, C), (B, D)$ and (E, F) be $G = (g, g'), H = (h, h')$, and $I = (i, i')$ respectively. Then

$$g = \frac{1}{2} * (a + c),\ g' = \frac{1}{2} * (a' + c'),\ \text{etc.}$$

The geometrical condition that (A, E, B) should be collinear will be turned to an algebraic equation, viz.

$$a * (e' - b') + b * (a' - e') + e * (b' - a') = 0.$$

Similarly the conditions that the triple of points $(A, F, D), (B, C, F), (C, E, D)$ are collinear will be turned respectively into algebraic equations

$$a * (f' - d') + d * (a' - f') + f * (d' - a') = 0,$$
$$b * (c' - f') + c * (f' - b') + f * (b' - c') = 0,$$
$$c * (e' - d') + d * (c' - e') + e * (d' - c') = 0.$$

Adding the last 4 equations together, we get

$$(a+c)*(e'+f'-b'-d')+(b+d)*(a'+c'-e'-f')+(e+f)*(b'+d'-a'-c')=0$$

which is equivalent to

$$g*(i'-h')+h*(g'-i')+i*(h'-g')=0,$$

or (G, H, I) are collinear as to be proved.

Descatesian Fashion of Geometry Theorem Proving.

From the above example we see that owing to the method of Descartes geometry theorem-proving had entered a new stage, viz.

Stage II. Geometry Theorem-Proving in Descartesian Fashion.

We see that this Descartesian fashion differs from the Euclidean fashion of first stage in that the characteristic feature (a) has to be replaced by (a)' below:

(a)' *Proving is carried out by mere computations.*

On the other hand the characteristic feature (b) remains unchanged. For each individual theorem one should seek a way of manipulating the algebraic equations involved to arrive at the goal, and the manipulation is peculiar to that theorem itself. Moreover, the discovery of such a way of proof is not at all a trivial one. Ingenuity and cleverness are required in the algebraic manipulation as already shown above in Gauss' proof of his theorem, though usually far less than what is required for a proof in Euclidean fashion. Furthermore, the computational work may be so enormous that it becomes out of control. These are perhaps the reasons why such a powerful method of Descartes of geometry theorem-proving is not taught in courses on analytic geometry in high schools.

Hilbert's Classic on Foundation of Geometry.

A further complete change in the fashion of geometry theorem-proving occured with Hilbert's classic *Grundlagen der Geometrie* appeared in 1899. In this classic Hilbert had shown how to introduce Cartesian coordinate system based on the Euclidean geometry axiomatic system via the introduction of some number system as shown in the following diagram:

$$\textit{Axiomatic System} \longrightarrow \textit{Number System} \longrightarrow \textit{Coordinate System}.$$

Hilbert had thus built a bridge leading from the logical deductive method of Euclid to the computational method of Descartes. Remark that the number system in the sense of Hilbert is in now-a-days terminology just a *field*. Such a field is determined by means of calculus of segments alike to that of Descartes but is based on the axiomatic system of the geometry.

Hilbertian Class of Theorems in Hilbert's Classic.

Now in contrast to the consideration of proving theorems one by one as in either Euclidean or in Descartesian fashion, there appeared, in the above mentioned classic of Hilbert some passages which suggest the proving in considering a whole class of theorems instead of single ones. In fact, in Chap. 6, Sect. 35 of the first edition of this classic in 1899, there was some assertion in italics which was turned into a theorem in latter editions. We reproduce it below:

> Satz 62. Jeder reiner Schnittpunktsatz der in einer ebener Geometrie gilt, in der die Axiome ... und der Pascalschen Satz gueltig sind, stellt sich durch Konstruktion geeigneter Hilfspunkte und Hilfsgeraden als eine Kombination endlich vieler Pascalsche Konfiguration heraus.

The "reiner Schnittpunktsatz" in the above theorem was explained by Hilbert which we again reproduce as follows:

> Unter einer reiner Schnittpunktsatz verstehen wir hier einen Satz, der eine Aussage ueber die vereignete Lage von Punkten und Geraden und ueber die Parallelitaet von Geraden enthaelt, ohne weitere Beziehungen, und etwa Kongruenz oder Senkrechtstehen, zu benutzen.

Hilbert has implicitely imposed a restriction on the class of such pure intersection theorems in requiring that the points and lines involved in the theorem should be entered one by one in succession in a definite order. In fact, Hilbert wrote:

> Man waehle zunaechst ein System von endlich vielen Punkten und Geraden willkuerlich, sodann ziehe man in vorgeschriebener Weise zu gewissen dieser Geraden beliebige Parallellen, waehle auf gewissen der Geraden beliebige Punkte und lege durch gewisse der Punkte beliebige Geraden; wenn man dann in vorgeschriebener Weise Verbindungsgerade, Schnittpunkte sowie Parrallellen durch bereits vorhandene Punkte konstruiert, so gelangt man schliesslich zu einem bestimmten System von endlich vielen Geraden, von denen der Satz aussagt, dass sie durch den naemlichen Punkt hindurchlaufen bzw. parallel sind.

Let us call a pure intersection theorem which may be described in the above manner as a *constructive* one. We see clearly that the whole class of so-called "*konstructive* rein Schnittpunktsatz" or *Hilbertian class* of theorems for short, includes an infinity of theorems, among which are a number of well-known theorems associated with the names of Pappus, Desargues, Pascal, Steiner, von Staudt, etc.

An Example in Hilbertian Class : Desargues Theorem.

As an example let us cite the theorem belonging to the Hilbertian class below:

Example 1.4. Desargues Theorem (Projective Form). If for two triangles in the projective plane the joining lines of three pairs of corresponding vertices are concurrent in a point, then the intersection points of the three pairs of corresponding sides are collinear on a line. The converse is also true.

The theorem was discovered by Desargues (+1593, +1662), a pioneer of projective geometry. It was proved in the Euclidean fashion first in the case of two triangles situated in two different planes in a projective space, and then it was deduced for the case of two triangles in the same plane.

We shall restrict ourselves to the following particular case of the theorem which is of high interest in the study of foundation of euclidean geometry, as shown in the classic of Hilbert.

Example 1.4′. Desargues Theorem (Special Form). If for two triangles in the euclidean plane the joining lines of three pairs of corresponding vertices are either concurrent in a point or parallel to each other, while two pairs of corresponding sides are parallel respectively, then the third pair of corresponding sides are parallel too. Conversely, if the three pairs of corresponding sides are mutually parallel, then the joining lines of the three pairs of corresponding vertices are either concurrent or parallel to each other.

Universal Proof of Theorems in Hilbertian Class.

Now let us come to the striking contribution of Hilbert: In spite of the fact that the whole class of constructive pure intersection theorems or the Hilbertian class contains an infinity of theorems, Hilbert had pointed out that all these theorems may be proved by one and the same universal method. Owing to its far-reaching significance let us reproduce below the whole original passage of Hilbert:

> Die Koordinaten der zunaechst ganz willkuerlich gewaelten Punkte und Geraden betrachten wir als Parameter $p_1, \cdots, p_n$; von den sodann mit beschraenkter Willkuer gewaehlten Punkten und Geraden werden einige Koordinaten als weitere Parameter $p_{n+1}, \cdots, p_r$ festgelegt sein. Die Koordinaten aller Verbindungsgeraden, Schnittpunkte und Parallellen, die nun weiter konstruiert werden, werden dann rational von diesen Parametern abhaengige Ausdruecke $A(p_1, \cdots, p_r)$ sein, und die Aussage des vorgelegten Schnittpunktsatzes wird dargestellt durch die Behauptung, dass gewisse solche Ausdruecke fuer uebereinstimmende Parameterwerte uebereinstimmende Werte ergeben; d.h. der Schnittpunktsatz wird aussagen, dass bestimmte, von gewissen Parametern $p_1, \cdots, p_r$ rational abhaengige Ausdruecke $R(p_1, \cdots, p_r)$ stets verschwinden, sobald wir fuer diese Parameter irgendwelche Elemente der in der vorgelegten Geometrie eingefuerten Streckenrech-

nung einsetzen. Da der Bereich dieser Elemente unendlich ist, so schliessen wir nach einem bekannten Satze der Algebra, dass die Ausdruecke $R(p_1, \cdots, p_r)$ identisch auf Grund der Rechnungsgesetze $\cdots$ verschwinden muessen. Um aber in unserer Streckenrechnung das identische Verschwinden der Ausdruecke $R(p_1, \cdots, p_r)$ nachweisen, genuegt nach dem, was wir oben fuer die Anwendung der Rechnungsgesetze bewiesen, die Anwendung des Pascalschen Satzes, und wir erkennen:

Satz 62. $\cdots$

Thus, while the statement of theorem 62 is itself quite obscure, the passage before this statement makes clear that Hilbert had shown how any of the innumerous theorems in the whole class can be proved by a *universal* method.

Mechanical Procedure of Hilbert's Universal Proof.

Let us describe now Hilbert's universal method in our own way as follows (we write the parameters as u_i instead of p_i):

Hilbert's Universal Proof Procedure:

Step 1. Take coordinate system.

Step 2. Introduce points and lines in succession.

Step 3. Introduce parameters $u_1, u_2, \cdots, u_r$ and bounded variables $x_1, x_2, \cdots, x_s$ in succession so that hypothesis of the geometry theorem in question will be turned successively in the form below:

$$\begin{aligned} x_1 &= \frac{P_1(u)}{Q_1(u)}, \\ x_2 &= \frac{P_2(u, x_1)}{Q_2(u, x_1)}, \\ &\cdots\cdots \\ x_s &= \frac{P_s(u, x_1, \cdots, x_{s-1})}{Q_s(u, x_1, \cdots, x_{s-1})}. \end{aligned} \tag{1.1}$$

In these equations all P_i, Q_i are polynomials in $u = (u_1, u_2, \cdots, u_r)$ and the previous bounded variables $x_1, \cdots, x_{i-1}$. Also convert the conclusion of the theorem in question into the form

$$R(u, x_1, \cdots, x_s) = 0, \tag{1.2}$$

with R a rational function.

Step 4. Eliminate $x_s, x_{s-1}, \cdots, x_2, x_1$ from R in succession by means of (1.1) such that

$$R(u, x_1, \cdots, x_s) \longrightarrow R_1(u, x_1, \cdots, x_{s-1}) \longrightarrow \cdots \longrightarrow R_{s-1}(u, x_1) \longrightarrow R_s(u).$$

Step 5. If $R_s(u) = 0$ then we conclude that the theorem in question is true provided that the conditions

$$Q_i(u, x_1, \cdots, x_{i-1}) \neq 0 \tag{1.3}$$

are observed. Otherwise we conclude that the theorem in question is not true, at least under the conditions (1.3). The theorem is then to be considered as not true at all.

Example : Proof of Desargues Theorem by Hilbert's Universal Method.

We remark that for a theorem in the above-mentioned class to be true some restrictions like (1.3) are usually unavoidable. Accordingly we shall refer to such conditions as *non-degeneracy conditions.* To see its geometrical significance let us consider the proof of Desargues Theorem of Ex. 1.4′ as follows:

Let the two triangles in question be ABC and $A'B'C'$. Consider the case for which the joining lines AA', BB', CC' meet in the same point O and the two pairs of parallel sides be $AB, A'B'$ and $AC, A'C'$ respectively. We have to prove that $BC, B'C'$ are parallel too. See the Fig. 1.4 above.

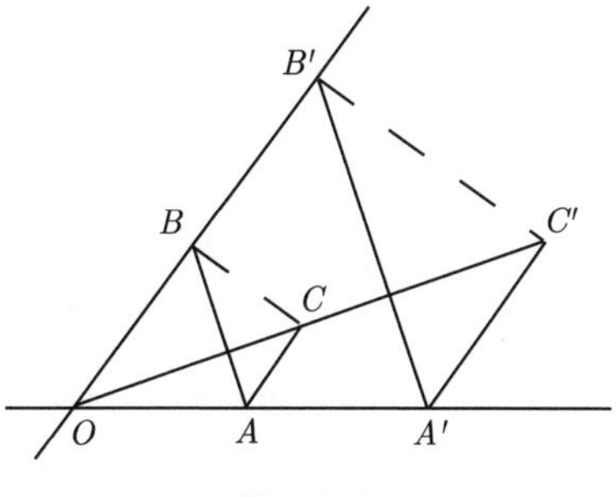

Fig. 1.4.

For the proof let us follow the steps as indicated above. First, choose the point O as origin and two arbitrary lines through it as the oblique X- and Y-axis. Choose now in succession arbitrary points $A = (u_1, 0), A' = (u_2, 0)$ on X-axis, $B = (0, u_3)$ on Y-axes, and $C = (u_4, u_5)$ in the plane. Join AB and through A' draw a line parallel to AB meeting the Y-axis in a point $B' = (0, x_1)$, then

$$x_1 = \frac{u_2 * u_3}{u_1}. \tag{1.4}$$

Join AC, OC and through A' draw a line parallel to AC meeting OC in a point $C' = (x_2, x_3)$. The geometrical conditions, CC' passes through O and AC parallel to $A'C'$, are respectively

$$u_4 * x_3 - u_5 * x_2 = 0, \tag{1.5}$$

$$(u_4 - u_1) * x_3 - u_5 * (x_2 - u_2) = 0. \tag{1.6}$$

Since u_i are considered as independent parameters, we have in particular

$$u_1 \neq 0,\ u_4 \neq 0,\ u_5 \neq 0. \tag{1.7}$$

Under the conditions (1.7) the two equations (1.5),(1.6) are then equivalent to the following ones:

$$\begin{aligned} x_2 &= \frac{u_2 * u_4 * u_5}{u_1 * u_5}, \\ x_3 &= \frac{u_5 * x_2}{u_4}. \end{aligned} \tag{1.8}$$

The equations (1.4),(1.8),(1.9) form the hypothesis of the theorem. The conclusion that BC is parallel to $B'C'$ corresponds to

$$u_4 * (x_3 - x_1) - (u_5 - u_3) * x_2 = 0. \tag{1.9}$$

To prove it let us eliminate x_3, x_2, x_1 in succession from the left-hand side of the last equation. The computation gives

$$\begin{aligned} u_4 * (x_3 - x_1) - (u_5 - u_3) * x_2 &\longrightarrow_{(1.8)_2} u_4 * \left(\frac{u_5 * x_2}{u_4} - x_1\right) - x_2 * (u_5 - u_3) \\ = -u_4 * x_1 + u_3 * x_2 &\longrightarrow_{(1.8)_1} -u_4 * x_1 + \frac{u_2 * u_3 * u_4 * u_5}{u_1 * u_5} \longrightarrow_{(1.4)} 0. \end{aligned}$$

The above Desargues theorem is thus seen to be true provided that (1.7) to be called the *non-degeneracy conditions* are observed.

Some Remarks on Hilbert's Universal Method.

Let us consider more closely the conditions (1.7) for the proof of the above Desargues Theorem. The case $u_1 = 0$ will make A coincident with O and the theorem will become devoid of geometrical meaning. The condition $u_4 = 0$ will make C a point on the Y-axis. The point C' will then be a definite point on the Y-axis. In that case, the theorem is still true in some *degenerate* sense as seen from Fig. 1.5 below. On the other hand, the condition $u_5 = 0$ will make C a point on the X-axis. C' will become then an arbitrary point on the X-axis. In that case we see that the hypothesis remain verified in some *degenerate* sense but the conclusion that BC is parallel to $B'C'$ is no more true. See the Fig. 1.6 below. Hence, the non-degeneracy condition $u_5 \neq 0$ in (1.7) becomes indispensable to make the theorem not untrue.

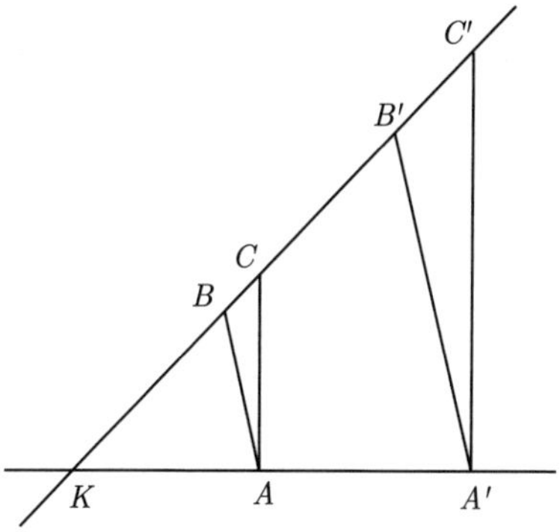

Fig. 1.5.

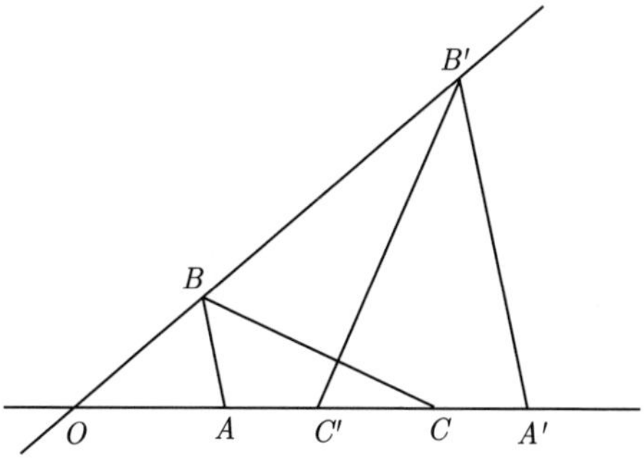

Fig. 1.6.

The non-degeneracy conditions (1.7) are thus indispensable for the theorem in question to be meaningful or to be true. We remark that this is the general situation of a geometry theorem and for discussions and further examples we refer to Chap. 7.

A Further Example : Proof of Gauss Theorem.

To explain this Hilbert's universal method of theorem-proving, let us take again Gauss' theorem in Ex. 1.3 for illustration. We remark that though the theorem is outside the Hilbertian class, the Hilbertian universal method works well still in this case. For further discussions see Sect. 2 of this chapter and also Sect. 2 of Chap. 7.

Consider Fig. 1.3 again. Instead of considering the figure as drawn in one stroke, let us consider it as being drawn step by step by introducing the points and lines one-by-one in succession. Thus, we shall complete the figure in the following way.

First, choose an arbitrary point A in the plane and take it as the origin. Through A choose now two arbitrary lines to be taken as the X-axis and Y-axis of an oblique coordinate system. Next, choose an arbitrary point B on the Y-axis with coordinates

$(0, u_1), u_1$ being a parameter showing its arbitrariness. Choose an arbitrary point C in the plane with coordinates (u_2, u_3) and an arbitrary point D on the X-axis with coordinates $(u_4, 0)$. The u's are again all parameters. Now join C, D by a line which meets the Y-axis at a point E. The coordinates of E are now no-longer arbitrary. Let E be $(0, x_1)$, where x_1 is readily seen to be

$$x_1 = \frac{u_3 * u_4}{u_4 - u_2}. \tag{1.10}$$

Note that here we have implicitely assumed $u_4 - u_2 \neq 0$, implying that DC is not parallel to the Y-axis AB so that E is well-defined and x_1 is well-determined by the above expression. Now join B, C by a line which meets the X-axis AD in a point F. The coordinates of F are not arbitrary. Let F be $(x_2, 0)$ then x_2 will be given by

$$x_2 = \frac{u_1 * u_2}{u_1 - u_3}. \tag{1.11}$$

As before we assume here $u_1 - u_3 \neq 0$ or BC is not parallel to the X-axis. Join now A, C and let their mid-point G be (x_3, x_4) so that

$$x_3 = \frac{1}{2} * u_2, \ x_4 = \frac{1}{2} * u_3. \tag{1.12}$$

Join B, D respectively E, F and let their mid-points be respectively $H = (x_5, x_6)$ and $I = (x_7, x_8)$ so that

$$x_5 = \frac{1}{2} * u_4, \ x_6 = \frac{1}{2} * u_1, \tag{1.13}$$

$$x_7 = \frac{1}{2} * x_2, \ x_8 = \frac{1}{2} * x_1. \tag{1.14}$$

Equations (1.10)-(1.14) form the hypothesis system of Gauss-Line Theorem in question. The conclusion of the theorem corresponds then to the equation below:

$$\begin{vmatrix} x_3 & x_4 & 1 \\ x_5 & x_6 & 1 \\ x_7 & x_8 & 1 \end{vmatrix} = 0. \tag{1.15}$$

For the proof let us start from the left-hand side of the conclusion (1.15) and eliminate $x_8, x_7, \cdots, x_2, x_1$ in succession by means of the hypothesis equations (1.14),..., (1.11),(1.10). The computation runs as follows:

$$\begin{vmatrix} x_3 & x_4 & 1 \\ x_5 & x_6 & 1 \\ x_7 & x_8 & 1 \end{vmatrix} \xrightarrow{(1.14)} \begin{vmatrix} x_3 & x_4 & 1 \\ x_5 & x_6 & 1 \\ \frac{1}{2} * x_2 & \frac{1}{2} * x_1 & 1 \end{vmatrix} \xrightarrow{(1.13)} \begin{vmatrix} x_3 & x_4 & 1 \\ \frac{1}{2} * u_4 & \frac{1}{2} * u_1 & 1 \\ \frac{1}{2} * x_2 & \frac{1}{2} * x_1 & 1 \end{vmatrix}$$

$$\longrightarrow_{(1.12)} \frac{1}{4} * \begin{vmatrix} u_2 & u_3 & 1 \\ u_4 & u_1 & 1 \\ x_2 & x_1 & 1 \end{vmatrix} \longrightarrow_{(1.11)} \frac{1}{4} * \begin{vmatrix} u_2 & u_3 & 1 \\ u_4 & u_1 & 1 \\ \dfrac{u_1 * u_2}{u_1 - u_3} & x_1 & 1 \end{vmatrix}$$

$$\longrightarrow_{(1.10)} \frac{1}{4} * \begin{vmatrix} u_2 & u_3 & 1 \\ u_4 & u_1 & 1 \\ \dfrac{u_1 * u_2}{u_1 - u_3} & \dfrac{u_3 * u_4}{u_4 - u_2} & 1 \end{vmatrix}.$$

The final expression is readily verified to be 0 which shows that the Gauss-Line Theorem is true provided that the denominators in (1.10) and (1.11) are not 0, which hold naturally as already pointed out.

In comparison with previous proofs in Euclidean or Descatesian fashion we see that intelligence and cleverness play almost no role anymore. A layman with only rudiments of knowledge of geometry may well carry out the proof without much effort. What is all the more important is this: the above method of proof works not only for the above particular theorems, but works as well for innumerous theorems of similar kind, in contrast to the Euclidean or Descartesian fashion proofs which work usually for the single theorem in question. We owe both the universal method and its wide applicability as described above to D.Hilbert.

Stage of Hilbertian Fashion of Geometry Theorem-Proving.

From the above we see that Hilbert's method provided us an entirely new way of geometry theorem-proving which marks a 3rd stage of development.

Stage III. Geometry Theorem-Proving in Hilbertian Fashion.

In this Hilbertian fashion the characteristic feature (a)' of proving by computations remains the same as in the 2nd stage of Descartesian fashion. On the other hand the feature (b) has been changed to (b)' below:

(b)' *Proving is carried out not for theorems one by one, but class by class, i.e., a whole class of theorems will be proved by one and the same method.*

2.2 Geometry Theorem-Proving in the Computer Age.

Before entering upon discussions about the further developments of Hilbertian fashion of geometry theorem-proving in the computer age, let us first make some remarks and comments about it.

Enlargement of Hilbertian Class of Geometry Theorem-Proving.

Let us first remark that the proof of Gauss-Line theorem in Hilbertian fashion as given in Sect. 1 shows that the class of constructive pure intersection theorems or the Hilbertian class provable by the Hilbert universal method may be enlarged to include Gauss theorem. In fact, it was D.M.Wang who first pointed out that the class for which the universal method of Hilbert, as formulated in Sect. 1 may be significantly enlarged to include theorems involving perpendicularities or even circles, provided that the hypotheses can be so arranged that the bounded variables may be entered in succession as rational functions in the parameters as well as the previously introduced bounded variables. A list of interesting non-trivial geometry theorems have indeed been proved by Wang in this manner. Cf. e.g. [W-G]. Quite recently Lu, Li, and others have further extended the class to one including almost all known interesting theorems in Euclidean geometry, see the Sect. 2 of Chap. 7.

Constructiveness of Theorems in the Hilbertian Class.

The Hilbert original method, as formulated before, has great limitations not only because it considers class of only theorems involving mainly incidence and parallelism properties alone (though it has been much enlarged by Wang and others) but also because the theorems should be of a *constructive* character in the sense that points and lines, or rather bounded variables, should be introduced one-by-one in succession. That this may or may not easily be carried out, even if we restrict ourselves to theorems involving incidence and parallelism alone, may be seen from the following example due to Bokowski:

Example 3.1. Consider Fig. 2.1. The numbers 1, 2,..., 8 signify 8 points situated in the ordinary 3-space. A square in the figure with corners e.g. 1, 2, 5, 6 will be said to be *coplanar* if the 4 points corresponding to 1, 2, 5, and 6 are coplanar. Then we have the following

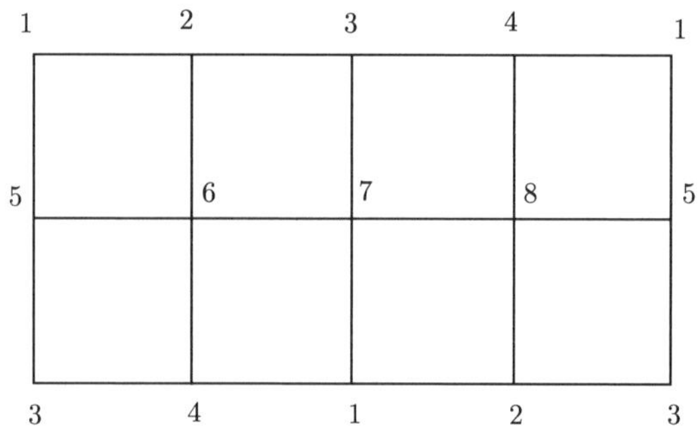

Fig. 2.1.

Bokowski Theorem. If 7 of the 8 squares in Fig. 2.1 are coplanar, then the 8th one will also be coplanar.

It is seen that the above conjecture is hardly constructive at least in appearance so the method of Hilbert can hardly be applied (though it turns out to be so at length). For another example see Ex. 7.3.1 in Sect. 3 of Chap. 7 and also Ex. 7 of [WWT3].

Non-Degeneracy Conditions in Hilbertian Universal Method.

As a further remark let us consider the non-degeneracy conditions (1.3), (1.7) and the non-vanishing of denominators in (1.8) occuring in Hilbert's universal proof procedure. As seen from the proof of Desargues theorem in Sect. 1, we see that such non-degeneracy conditions are usually not clearly stated in statements of theorems to be proved. To see the grave consequence of this fact let us reproduce below some passage in the author's paper [WWT3]:

> ... almost all theorems in ordinary geometry are only generically true, or true under some non-degeneracy conditions usually not explicitly described in the statement of the theorem. This fact is quite fatal to make the usual Euclidean fashion of proving theorems rigorous as one believes to be so, since it is difficult to state clearly the non-degeneracy conditions to be observed and to verify that the previous theorems to be applied in the proof of the theorems in question do not fall into the degeneracy cases under which the previous theorem to be applied might not be true.

Mechanical Character of Hilbertian Fashion of Geometry Theorem-Proving.

Though the class of theorems provable by the Hilbert method is quite limited the importance of the method is almost inestimable. Besides the novelty of proving the theorems not one-by-one but by a single universal method for those in a whole class, the proving method itself is a novel one: it is *mechanical* in character in that one may just follow the instructions step-by-step in a *mechanical* way in order to arrive at the final conclusion of the theorem in question. For the recognition of the merit of the Hilbert method let us reproduce a passage of the author's paper [WWT2] as given below:

> The great merit of Hilbert's classic *Grundlagen der Geometrie* of 1899 is universally recognized as being representative for axiomatization of mathematics, laying in particular a rigorous foundation of the Euclidean geometry. However, another great merit (perhaps greater in the opinion of the present author) of this classic seems hardly to be noticed up to the present. In fact, this classic is also representative for the mechanization of geometry, showing clearly at the same time the way to achieve it.

We remark in passing that the saying "rigorous foundation" may be of some doubt, in view of the occurence of non-degeneracy conditions as mentioned above.

Owing to the mechanical character of the Hilbert method we shall reformulate the Satz 62 of Hilbert's classic as the following

Hilbert Mechanization Theorem. The Hilbertian class of planar constructive pure intersection theorems is *mechanizable*.

Computerwise Stage of Geometry Theorem-Proving.

The geometry theorem-proving in Hilbertian fashion has received an immense impetus since the appearance of computers in the forties. In fact, the very *mechanical* character of Hilbert method of proving permits us to turn the proving procedure directly into a program which can then be run on the computer to see whether the theorem in question is true or not. The running computations of the computer is then itself a *proof* or a *disproof* of the theorem. With computers as powerful tools, geometry theorem-proving entered a new 4th stage of proving by computer.

Stage IV. Geometry Theorem-Proving in Computerwise Fashion.

In this stage characteristic feature (a)' is further strengthened to

(a)" Proving is carried out by computations on a *computer* .

The characteristic feature (b)' of proving class by class will remain unchanged. We shall emphasize in passing that the proving of the four-color theorem on the computer by Haken ([Hak]) cannot be considered as a proof in the present fashion since the proof works for this single theorem alone, contrary to the feature (b)'. In the present author's opinion, it is still a proof in Euclidean or Descartesian fashion, though the great merit of Haken's proof is clearly unestimable.

Mechanical Theorem-Proving.

With the appearance of computer it is natural to ask to turn the algorithmic proving procedure of certain class of theorems, whose existence is already assured in some way as in the case of Hilbertian class, into a program to be run on the computer. This is really the case. In the fifties we saw a period of earnest trials in proving theorems, geometrical ones or not by computer. A breakthrough occured eventually in the very beginning of the sixties. Wang Hao (+1921, +1995), a Chinese mathematical logician, proved on the IBM 704 computer approximately two hundred theorems of propositional calculus in barely 3 minutes, cf. [WH1]. The pioneering achievements were rightly cheered as "seven flies in one blow" and Wang Hao was accordingly awarded the first mechanical or automated theorem-proving prize in 1983.

The development of the stage of theorem-proving in Euclidean fashion to the present stage of proving by computer may be synthesized in the vivid expression of Wang Hao, see [WH2], p.106:

to trade effectively quantitative complexity with qualitative difficulty.

While qualitative difficulty of geometry theorem-proving had been overcome for the most part through the work of Descartes and Hilbert, the quantitative complexity will be seen to be overcome by the use of computers. The success of Wang Hao thus induced great efforts in seeking for more powerful methods of proving more profound theorems by computer. Through the sixties and onwards there appeared thus various kinds of resolution and other methods for this purpose, for a summary of which we refer to [B-L].

However, all the above achievements, though of great theoretical importance, seem to be of little interest for the taste of a working mathematician. From the point of view of a mathematician, a remarkable theorem about geometry theorem-proving discovered by Tarski in 1950 is of highest interest. In comparing with the Hilbertian Mechanization Theorem we shall call this theorem of Tarski also a *mechanization theorem*:

Tarski Mechanization Theorem. Elementary geometry in the ordinary sense is mechanizable.

It would be interesting to see the outcome of a computer in implementing Tarski's method in his mechanization theorem. The results seem to be somewhat discouraging. The method of Tarski is too complicated to be efficient so that actually few theorems of some geometrical interest have ever been proved by means of large present-day computers.

Studies of Mechanical Geometry Theorem-Proving in Recent Years.

In year 1977 the present author discovered some *universal* method of proving geometry theorems for certain classes englobing all theorems of Hilbertian class even enlarged but not all ones in the class considered by Tarski. The method is quite efficient and has been applied to prove and even to discover hundreds of quite non-trivial theorems. We shall describe in later chapters, particularly in Chap. 7, what has been done in this field of mechanical geometry theorem-proving.

2.3 Geometry Problem-Solving and Geometry Theorem-Proving in Ancient China.

Geometry Problem-Solving in Ancient China.

In ancient China geometry was well developed but in some fashion quite different from the Euclidean one. In fact, while geometry theorem-proving was much emphasized in ancient Greece, one considered usually in ancient China geometry problem-solving rather than geometry theorem-proving. Most of geometry problems to be solved arose from practical requirements. We may cite in particular:

(a) Problems arising from measurements, e.g. the measurement of sun height over the ground level or the measurement of hill height over the sea from the shadows of biaos (or scaled rods).

(b) Problems about area and volume determination. In particular a formula appeared in the classic [QJS]* of Qin Jiushao (+1249) which is equivalent to Heron's formula about area of a triangle in terms of the three sides but in a different form.

(c) Problems involving Gou-Gu Form (i.e. a right-angled triangle), e.g. the determination of such forms with 3 sides in *rational* ratios.

(d) Problems arising from actual lifes, e.g. the determination of the extensions of a city or an agricultural field in walking around in some way.

In a word, the development of geometry in ancient China is rather applications-oriented, in contrast to the Euclidean way for which logical basis and logical interrelations are highly emphasized. However, this does not mean that there were no geometry theorem-proving in ancient China. In fact, instead of set of axioms the ancient Chinese formulated rather a few simple and plausible principles and then proved the geometry theorems in the form of formulae or solved the geometry problems by logical reasoning.

Out-In Complementary Principle.

The simplest one of such principles is perhaps the Principle below.

Out-In Complementary Principle. When a planar (or a solid) figure be cut into pieces, which are moved to somewhere else and combined together, the area (or volume) of the figure will remain the same.

Diverse applications to geometry problem-solving and geometry formula-proving have been made by means of this seemingly trivial principle in an elegant and sometimes unbelievably striking manner. Most of the formulae stated above in (a)-(d) were proved in basing on this Principle.

The Gou-Gu Theorem.

In ancient China a right-angled triangle was called a Gou-Gu form with Gou as the shorter and Gu as the longer sides around the right angle, while the hypotenues was called Xian. The Gou-Gu form was a favourite subject to ancient Chinese mathematicians and constitutes the core of traditional Chinese geometry. Thus, the Pythagoras' Theorem was called the Gou-Gu Theorem in our terminology.

In *Nine Chapters* the Chap. 9 was devoted to Gou-Gu Theorem and its applications, containing in all 24 problems. We reproduce Problem 1 below (*chi* below is some unit of length):

> **Problem 1.** We have Gou 3 chis, Gu 4 chis, to ask for Xian.
>
> **Answer.** 5 chis.
>
> **The Gou-Gu Shu** says: Squaring Gou and Gu separately and adding, extracting square root we get the Xian.

In the *Annotations* Liu Hui, after giving the definitions of Gou, Gu and Xian as stated above, indicated a proof of the conclusion in basing on the *Out-In-Complementary Principle.* Liu's annotations shew that there was an unfortunately lost annexed diagram with some square on Gou marked in red color and some square on Gu marked in blue color. Remark that the particular triple of numbers (3,4,5) serves only as an illustrative example while both the statement and proof are of a general character having nothing to do with the particular triple, as was the usual tradition of Chinese writings. Cf. the Sect. 1 of Chap. 1 about this point.

Measurement Instrument Ju in Ancient China.

The Gou-Gu Theorem should, however, be already well known since very remote time. Before we explain this, let us cite some instrument *Ju* which was widely used in measurements in ancient China.

The instrument *ju* was consisting of two arms forming a right angle with one arm Gou or Gu or both scaled. In *Zhou Bi* there was a reply of scholar Shang Gao to Lord Zhou:

> The *ju* laid on the ground with the aid of cords serves to lay out works straight and square . The *ju* recumbent serves to observe heights. The *ju* reversed serves to fathom depths. The *ju* flattened serves to ascertain distances. The *ju* rotated serves to form circles. The *jus* combined serve to form the square.

We remark that Lord Zhou was a famous statesman in our history, reknown for his conquering of Shang Dynasty in establishing the new Western Zhou Dynasty in the year 1122B.C. As for Shang Gao, there are historical data about his life, collected from the local chronicles of Shan Xi Province.

Gou-Gu Theorem in Zhou Bi.

In the classic *Zhou Bi* there was a passage describing a discourse between scholar Shang Gao and Lord Zhou concerning the principle of Gou-Gu Theorem. The particular example of the triple (3,4,5) was cited in the discourse, but as we mentioned several times before, it will serve only for illustrations. In the passage there were some sentences (see below) which were little understood before. However, recent efforts due to Professors Chen Liang-tso and Ko-wei Lih of Taiwan and Li Ji-min of mainland China had clarified the meaning of the passage and made evident that the few sentences may be regarded as a "proof" of the *general* Gou-Gu Theorem, though not so rigorous in the strict sense. In fact, in the period of 3 kingdoms Zhao Shuang (circa +3c) have annotated the *Zhou Bi* which resulted in two separated articles [ZS1]* and [ZS2]* in Sect. 1 of Chap. 1. Along with [ZS1]* there were some diagrams marked with colors which fortunately remained up to the present day. Based on the diagrams which seems to have a remote origin and owing to due interpretations of some Chinese characters in the original passage the above-mentioned mathematicians suggested some proof seemingly the original one of the ancient time. The following Fig. 3.1 and interpretations are taken from the paper [LJM1]*.

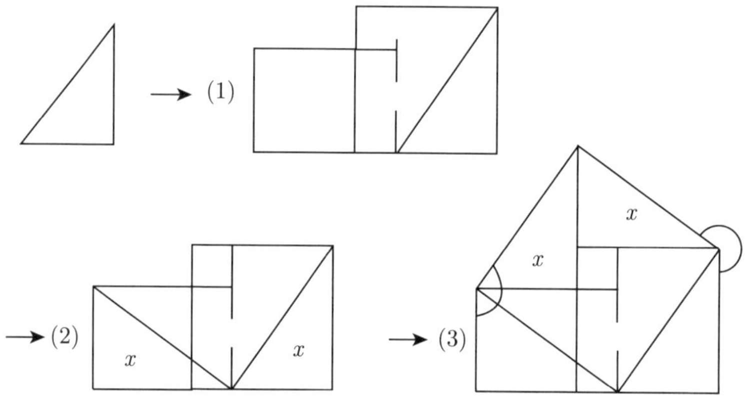

Fig. 3.1.

In Fig. 3.1 (1), (2), (3) correspond to the statements in the passage below, the numeration (1),(2),(3) being added by the present author:

(1) Form squares about the sides.

(2) Take away halves of the ju.

(3) Rotate and combine to form a new square.

Sun-Height Formula in Zhou Bi.

In *Zhou Bi* there was another important geometrical result which has been later developed into more precised form and may be stated as follows.

Let us plant two biaos at a distance apart along a line in north-south direction. Let us measure the shadow lengths of the two biaos at noon-time. Then the height of the sun over the horizontal level-plane will be given by the following formula:

Sun-Height Formula.

$$\textit{Sun-Height} = \frac{\textit{Biao-Distance} * \textit{Biao-Height}}{\textit{Shadow-Difference}} + \textit{Biao-Height}. \tag{3.1}$$

The data given in *Zhou Bi* were too rough to cope with the reality. However, based on the formula some kind of shadow-observatory was built in antiquity named after Lord Zhou. In the present day there was still such an observatory in Province Hunan at the same place believed to be in the time of Lord Zhou but was re-built in Yuan Dynasty by the famous astronomer, mathematician and engineer Guo Shoujing (+1213, +1316).

In *Zhou Bi* the proof of the formula was lacking. In modern times many proofs were given mainly by drawing some auxiliary parallel and then deduced the formula in the usual Euclidean fashion. See the accompanying diagram Fig. 3.2.

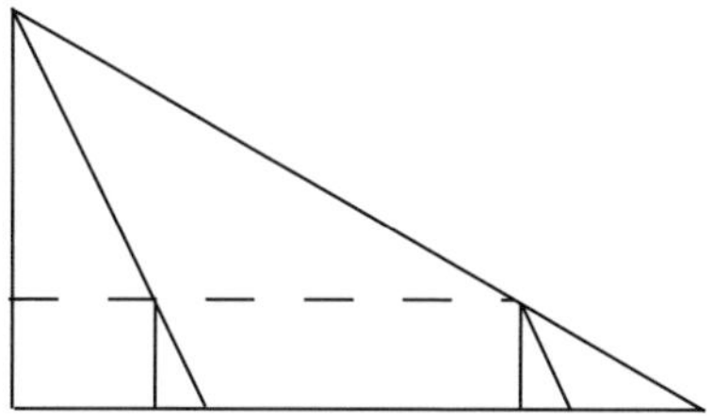

Fig. 3.2.

However, such proofs could not be the original one in our ancient time. In fact, in ancient China the notion of parallels was little noticed and there was no trace of Euclidean geometry until much later in Ming or atmost in Yuan Dynasty.

Recovered Proof of Sun-Height Formula.

Fortunately, in the annotations of *Zhou Bi* by Zhao Shuang there was an article [ZS2]* together with some Sun-Height Diagram marked with different colors. From the diagram somewhat mutilated and the article the present author had given a proof which seems, if not the original one of Zhao Shang, at least it will be nearly so. For this purpose let us first reproduce a literal translation of some passage in the article [ZS2]* with some re-ordering of sentences as follows, the numbers (1)-(9) being added by the author.

(1) Y_1 and Y_2 are equal in areas.

(2) Y_1 is connected with B_3 and Y_2 is connected with B_6.

(3) They are also equal in areas.

(4) B_3 and B_6 are also equal in areas.

(5) Multiply biao-distance by biao-height to be the area of Y_1.

(6) Take shadow-difference as breadth of Y_2

(7) and divide, one gets height of Y_2.

(8) The heigths rise up to same level as sun.

(9) So from diagram biao-height is to be added.

In the above statements Y, B means that the corresponding areas were in colors yellow and blue respectively. From these statements especially the statement (8) we shall complete the mutilated diagram somewhat as in Fig. 3.3 below.

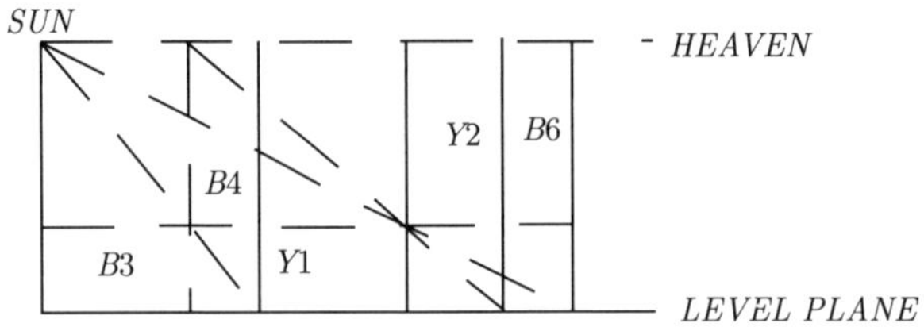

Fig. 3.3.

The proof becomes now quite clear. In fact, as a simple consequence of the Out-In Complementary Principle, we have from statements (1),(2),(3) the following equalities for areas:

$$Y_1 = Y_2,\ B_3 + Y_1 = B_6 + Y_2.$$

Whence we have (4):

$$B_3 = B_6.$$

In the same way we have $B_3 = B_4$ so that we have

$$B_4 = B_6.$$

The statement (5) gives

$$Area\ Y_1 = Biao\text{-}Distance * Biao\text{-}Height.$$

By (1) this is also the area of Y_2. Now

$$\begin{aligned} Breadth\ Y_2 &= Breadth\ (Y_2 + B_6) - Breadth\ B_6 \\ &= Breadth\ (Y_2 + B_6) - Breadth\ B_4. \end{aligned}$$

Whence we get statement (6), viz.

$$Breadth\ Y_2 = Shadow\text{-}Difference.$$

Then we further get as in statements (7) and (8),

$$\begin{aligned} Sun\text{-}height\ over\ Biao &= \frac{Area\ Y_2}{Breadth\ Y_2} \\ &= \frac{Biao\text{-}Height * Biao\text{-}Distance}{Shadow\text{-}Difference}. \end{aligned}$$

As in statement (9), we have then

$$\textit{Sun-Height over Level-Plane} = \textit{Sun-Height over Biao} + \textit{Biao-Height}.$$

This gives the Sun-Height Formula (3.1).

The City-Width Problem Revisited.

Based on the Out-In Complementary Principle the reduction to square-root-extraction with cong of the city-width-problem in the *Nine Chapters* as described in Sect. 2 of Chap. 1 with accompanying diagram Figs. 1.2.4 and 1.2.5 may be proved as follows.

Let us first complete Figs. 1.2.4 and 1.2.5 to the above diagram Fig. 3.4. By Out-In Complementary Principle we have *Area Rect PR* = *Area Rect QS* = $w * n$. The rectangle RT has thus an area $2 * w * n$ while the difference of its two sides is $n + s$. Hence the width of the rectangle, i.e. the width of the city, may be gotten by square-root-extraction with $shi = 2 * w * n$ and $cong = w + s$ as asserted before.

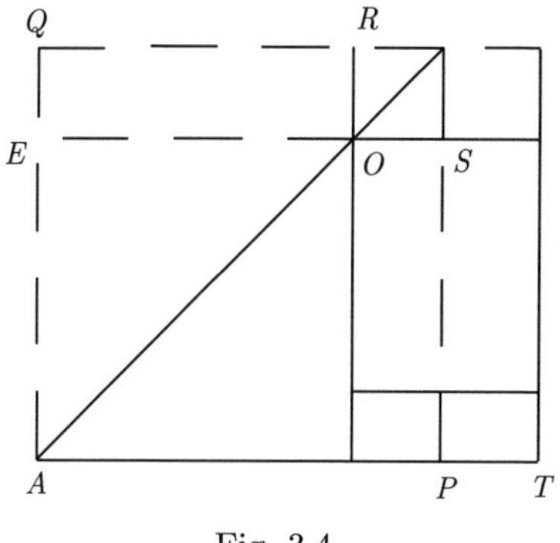

Fig. 3.4.

Applications of Out-In Complementary Principle to Studies of Gou-Gu Forms.

The applications of the *Out-In Complementary Principle* are multifold. Besides those about square-root-extraction, the proof of Gou-Gu Theorem, and the proof of the Sun-Height Formula, we shall cite two further applications (A) and (B) below from the Gou-Gu Chapter of the *Nine Chapters*, cf. [WWT1]*.

(A) Gou-Gu Problems: Among $Gou \pm Gu$, $Xian \pm Gou$, $Xian \pm Gu$, Gou, Gu, $Xian$ suppose two of them are known, to determine the others.

As an example, let us consider the Broken-Bamboo Problem 13 of the Gou-Gu Chapter. Liu Hui in his *Annotations*, first reduced the problem to one of determining $Xian$ and Gu from Gou and $Xian + Gu$, and then gave the solution in the form of the formulae below:

$$Gu = \frac{(Xian + Gu)^2 - Gou^2}{2 * (Xian + Gu)}, \qquad (3.2)$$
$$Xian = (Xian + Gu) - Gu.$$

The proof of Liu Hui was entirely geometrical and was based on the Out-In Complementary Principle. See the right-hand side figure in Fig. 3.5 below.

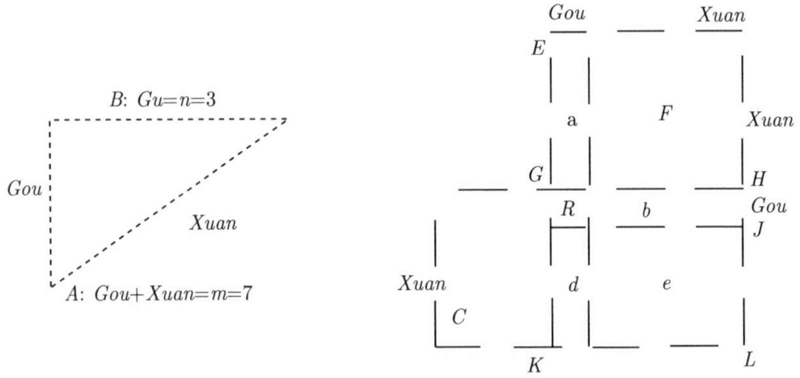

Fig. 3.5.

(B) Gou-Gu Triples.

Problem 14 of the Gou-Gu Chapter in *Nine Chapters* gave rational Gou-Gu Triples in all its generality. To see this let us reproduce the Problem below:

Problem 14.

Two persons start from same position.

Person Jia has a speed-rate 7 while person Yi has rate 3.

Yi goes eastward while Jia goes first southward 10 units and then meets Yi in going north-east slantwise.

To find units traversed by Jia and Yi.

Answer.

Yi goes eastward 10 and half units.

Jia goes slantwise 14 and half units and then meets Yi.

Shu says:

Squaring 7, also 3, taking half of the sum. This will be the slant unit-ratio of Jia.

Subtract this unit-ratio from square of 7, the remainder is the southern unit-ratio.

Multiply 7 by 3 is the eastern unit-ratio of Yi.

As mentioned several times before, the numerical data 3,7 play actually no role in Shu. We may replace 3 and 7 by general values n, m so that we have

$$n = Gu, \quad m = Gou + Xian. \tag{3.3}$$

See Fig. 3.5 above. In (3.2) let us interchange Gou and Gu and multiply the expressions by $2 * (Xian + Gou)$, then we get readily the formula below which gives the general rational Gou-Gu triples::

$$Gou : Gu : Xian \;=\; m^2 - \frac{(m^2 + n^2)}{2} : m * n : \frac{m^2 + n^2}{2}, \tag{3.4}$$

Remark 3.1. In Gou-Gu Chap. 9 of *Nine Chapters* there appeared 8 rational Gou-Gu triples, viz.

3:4:5, 5:12:13, 7:24:25, 8:15:17, 20:21:29, 48:55:73, 60:91:109, 20:99:101.

This shows that the mathematicians at that time should know already how to form such triples in a general way. In fact, these triples correspond to values $m : n =$ 2:1, 3:2, 4:3, 5:3, 7:3, 11:5, 13:7, and 11:9 respectively in the general formula (3.4). The Problem 14 above is the case $m : n = 7 : 3$. Furthermore, in the classic [WXT]* of early Tang Dynasty there appeared 3 further rational Gou-Gu triples 9:40:41, 12:35:37, and 13:84:85, which correspond to $m : n =$ 5:4, 7:5, and 7:6 respectively.

Remark 3.2. That Problem 14 of *Nine Chapters* leads actually to a general formula of Gou-Gu rational triples seems to be first pointed out by Li Jiemin, see e.g. [LJM1,2]*.

Measurements in Liu Hui's Sea-Island.

The *Sea-Island* of Liu Hui in +263 was a treatise devoted to measurements in using mainly the instrument *ju*. The treatise contained in all 9 problems with titles below.

Problem 1. Viewing at Sea-Island.

Problem 2. Viewing at Pine-Tree.

Problem 3. Viewing at City.

Problem 4. Viewing at Valley.

Problem 5. Viewing at Building.

Problem 6. Viewing at River-Mouth.

Problem 7. Viewing at Clean Brook.

Problem 8. Viewing at Port.

Problem 9. [Viewing at] Rectangular-Formed City.

For each of the problems data were gotten by measuring at various positions in using the *ju*, sometimes also using the cords and biaos. Explicit measurement formulae quite intricate in form were then given in terms of these data. In particular the Problem 1 was nothing else but the sun-height measurement as in *Zhou Bi*. However the *sun* had been replaced by a *sea-island* and the measurements were taken on the sea-shore. The problem thus became a realistic one. In recent years some Chinese historians in mathematics, based on indications in existant chronicles and historical records, had even suggested some place in Province Shandong, with data quite near to those given in Liu Hui's treatise, cf. e.g. [Guo3]*.

There were no indications how these measurement formulae were obtained. In Qing Dynasty some mathematicians had given proofs in Euclidean fashion in drawing auxiliary parallels which were very complicate and quite unnatural. Moreover, there were even difficulties in turning the resulting formulae thus gotten to the form given in *Sea-Island* On the other hand the present author had tried to reconstruct such "ancient" proofs in applying the Out-In Complementary Principle. It turns out that the proofs are simple and elegant and lead directly to the intricate formulae given in *Sea-Island*. It is thus highly believable that these are really the "ancient" proofs lost or at least close to the lost ancient proofs. See [WWT2]*.

Qin-Heron Formula of Triangle Area.

In [QJS]* of Qin Jiushao in +1247 there was a problem of finding the area of a scalene triangular plot. Qin named the three unequal sides of the triangle as long-slant, medium-slant, and short-slant, to be denoted as *long, med, short* below. Qin's solution can then be formulated as follows.

$$Area^2 = \frac{1}{4} * \left[short^2 * long^2 - \left(\frac{long^2 + short^2 - med^2}{2}\right)\right]^2. \tag{3.5}$$

Qin said nothing about the source of this formula. The proof of the formula had also been lost. Making use of the results and the methods in *Annotations* of Liu Hui, we may also infer the lost proof to be somewhat as follows (see the diagram Fig. 3.6 above):

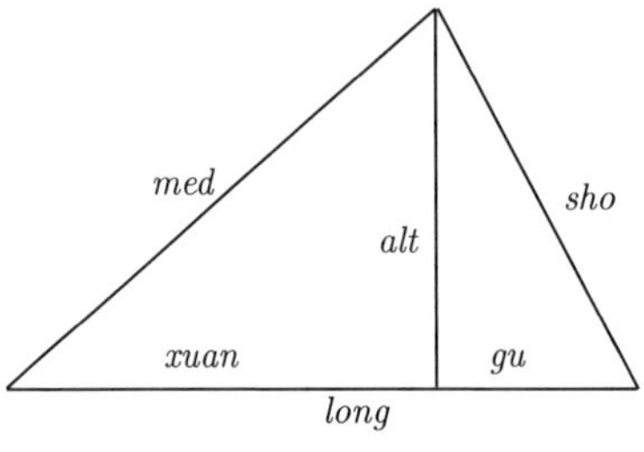

Fig. 3.6.

Draw an altitude of the triangle perpendicular to the long-side, dividing it into two parts. Let the longer and the shorter parts be the *xian* and the *gu* of a right triangle. From *Nine Chapters* we know that the area of a triangle to be $\frac{1}{2} * altitude * long$, therefore our problem becomes one of finding the altitude, then further boils down to finding the *gu* of that right triangle. Since

$$xian + gu = long,$$
$$gou^2 = xian^2 - gu^2 = med^2 - short^2,$$

our problem is the same as that of finding *gu*, given *gou* and the sum of *xian* and *gu*. From Liu Hui's formula (3.2) we have

$$gu = \frac{(xian + gu)^2 - gou^2}{2 * (xian + gu)} = \frac{long^2 - (med^2 - short^2)}{2 * long},$$

$$altitude^2 = short^2 - gu^2 = short^2 - \left(\frac{long^2 + short^2 - med^2}{2 * long}\right)^2.$$

From this we get Qin's formula (3.5).

Remark 3.3. Qin's formula looks rather odd. But the proof traced above is quite natural and perfectly in line with ancient Chinese geometrical tradition. We may even regard it as the original proof, cf. [WWT3]*. Heron's formula in Western geometry, however, is neat in form and good looking, as given in Ex. 1.2 of Sect. 1. It is hardly probable that once a neat formula was known, one would like to turn it into an ugly-looking form. Hence Qin's formula is not likely to have been derived from Heron's, and we may conclude that it had its indigenous origin independent of Heron's influence.

Remark 3.4. The present author owes to Professor Shen Kangshen of Hang-Zhou University and Professor Li Di of Inner-Mongolia Normal University in pointing out that the above proof of Qin-Heron's formula appeared already in some treatise *Essentials of Plane Trigometry* (+1684) of the great scholar Mei Wending (+1633, +1721).

Area and Volume Theory in Ancient China.

For measuration of plots of lands and contents of containers it is of primary importance to know how to determine the area of a planar figure and the volume of a solid. It is thus not surprising that area and volume studies occupy a central position in ancient Chinese geometry.

It is easy to apply the Out-In Complementary principle to obtain the ordinary formula that the area A of a triangle is given by

$$A = \frac{1}{2} * s * h, \tag{3.6}$$

in which s is one of the sides and h is its associated altitude. The area of any polygonal figure, however complicate, will then be easily determined in applying the Out-In Complementary Principle. It follows that the theory of polygonal areas can be established solely on this plausible principle. The case for volumes of polyhedral solids in space is however quite different and is far from being trivial.

Liu Hui Principle for Volumes of Polyhedral Solids.

To establish a satisfactory theoy of volumes of polyhedral solids the critial point is to show that, analogous to (3.6), the volume V of a tetrahedron should be equal to

$$V = \frac{1}{3} * B * h, \tag{3.7}$$

in which B is the area of a base and h is its associated altitude. We have to remark that (3.7) cannot be proved solely by the Out-In Complementary Principle alone and Liu Hui was well awared of it. Liu Hui, in his *Annotations*, had however overcome the difficulties and completely solved the problem in reasoning as follows. Let a rectangular parallelopiped be cut slantwise into two equal parts called Qiandu, and then cut one of the two Qiandus again slantwise into two parts of which one, called Yangma, is a pyramid with square base and the other, called Bienao, is a tetrahedron with 3 successive edges in mutually perpendicular directions. See Fig. 3.7 below.

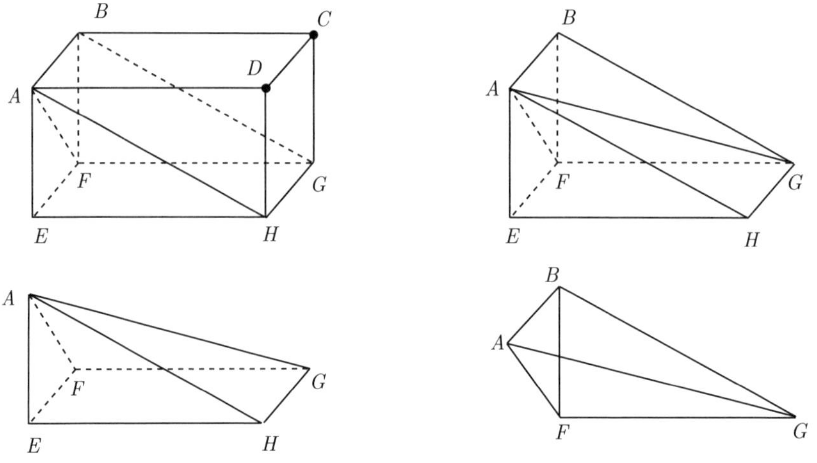

Fig. 3.7.

Cutting now the big YANGMA and BIENAO into pieces of smaller yangma and bienao as in Fig. 3.8 below Liu Hui showed that

$$1 \text{ YANGMA } - 2 \text{ BIENAO} = 2 * (1\ yangma\ - \ 2\ bienao). \tag{3.8}$$

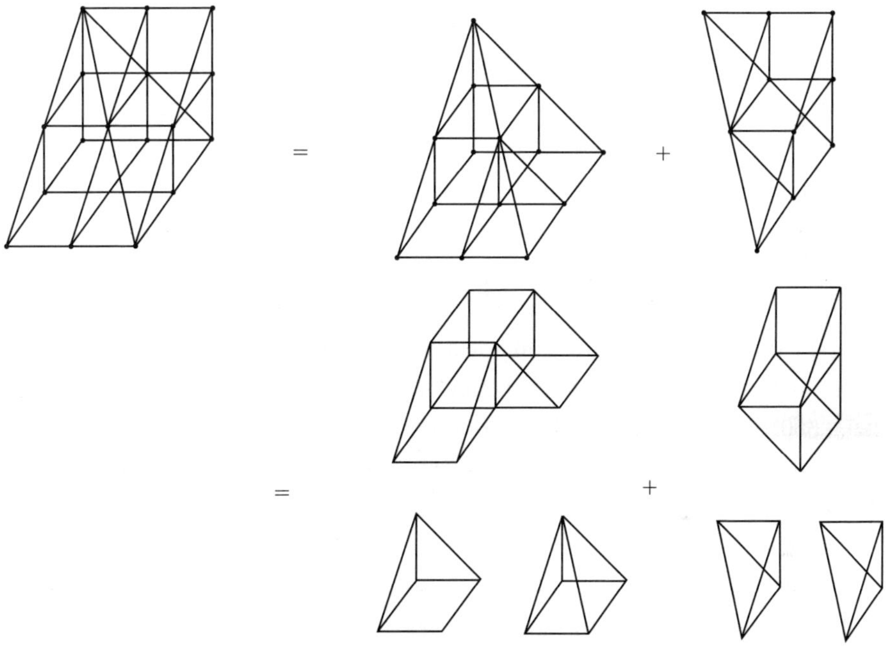

Fig. 3.8.

Continuing further and further, the expression in the parenthesis of right-hand side of (3.8) will become smaller and smaller and can be ultimately neglected, as argued by Liu Hui:

> The more they are cut into smaller halves, the smaller will be the remains. The ultimate smallness is infinitesimal, and infinitesimal is formless. Accordingly it is no need to take into account the remain.

From these arguments it is legitimate to put forth from (3.8) the formula

$$Yangma : Bieonao = 2 : 1. \tag{3.9}$$

Accordingly Liu Hui announced an assertion which we shall baptize as a general Principle in honor of Liu Hui:

Liu Hui's 2:1 Principle.

> Yangma occupies two and Bienao occupies one. That's an invariable ratio.

Remark 3.5. That Liu Hui actually gave a proof of the 2:1 Principle was first pointed out by Wagner, see his important paper [Wag]*. Cf. also [LJM2]* for more detailed discussions.

Based on Liu Hui's 2:1 Principle, combined with the Out-In Complementary Principle, it is easy to establish the critical formula (3.7) for the volume of a tetrahedron. The volumes of various polyhedral solids are then easily deduced from (3.7). In fact, in *Nine Chapters* there is a lot of such beautiful formulae.

Liu-Zu Principle for Volumes of Curved Solids.

For curved solid figures (also for curved planar figures) the above Principles are clearly insufficient to determine the volumes (or curved areas). Thus, the determination of the volume of a sphere in terms of its diameter is a long-standing problem in early days of our history. Liu Hui, in his annotations, suggested that the study of some peculier solid, which is the intersection of two inscribed cylinders in different directions in a cube, may lead to the ultimate solution of the problem. However, it was only much later that Zu Geng, son of the great scholar Zu Chongzhi (+429,+500), in following the suggestion of Liu Hui, had solved this problem. In solving the problem Zu Geng had formulated a Principle as stated below:

Zu Geng Principle.

Since areas in equal height are equal the volumes cannot be unequal.

The Zu Geng Principle furnishes a general method of determining volumes of curved solids in principle. In particular, Zu Geng, in following some suggestions of Liu Hui in his *Annotations*, succeeded in determining the volume of a sphere in terms of its diameter. Remark that this Zu Geng Principle, or more exactly Liu-Zu Principle, was already applied without precise formulation, to the determination of simple curved solids by Liu Hui.

Remark 3.6. As is well-known, this very Principle was later rediscovered in 17c under the name of Cavalieri Principle, which was applied to the determination of various kinds of curved solids, and became one of the cornerstones leading to the discovery of infinitesimal calculus in 17c.

General Principles for Area and Volume Theory in Ancient China.

We have exhibited three general Principles for geometry problem-solving, in particular the establishment of area- and volume-theory of planar and solid geometrical figures, viz.

I. The Out-In Complementary Principle.

II. The Liu Hui's 2:1 Principle.

III. The Zu Geng or Liu-Zu Principle.

In conclusion: The Principle I is sufficient for the establishment of a theory of polygonal areas. The Principle I,II together will be sufficient to establish a satisfactory theory of polyhedral volumes. The Principles I, II, and III together give in principle an effective method of determining the volumes of arbitrary curved volumes.

Volume Theory in Western World.

A comparison with the area and volume theory as developed in Hilbert's classic *Grundlagen der Geometrie* in 1899 may be of some interest. Hilbert, based on the notion of *Zerlegungs-aequivalenz*, (alike to the Out-In Complementary Principle), had shown how the theory of polygonal areas may be founded. However, the method works no more in the case of polyhedral volumes and its solution forms the content of Problem 3 of Hilbert's well-known 23 mathematical problems in 1900. The problem was solved soon by M.Dehn in 1902, see his paper [Dehn1]. Dehn gave even a necessary condition for two polyhedral solids to be *Zerlegungs-aequivalent*, see his paper [Dehn2]. Dehn's work was complemented by J.P.Sydler much later in showing that Dehn's condition is not only necessary, but also sufficient, see his paper [Syd]. Remark that the Dehn's conditions are expressed in a very intricate manner. Besides, a lot of papers were published in Mathematische Annalen shortly after Hilbert's classic in 1899 which showed that the involvement of some sort of limit concept was inavoidable for the establishment of a satisfactory theory of polyhedral volumes. The problem of polyhedral volumes is thus a highly non-trivial one. In comparing the way of Liu Hui and that of Hilbert-Dehn-Sydler for the settlement of the problem, the author has the impression that the work of Liu Hui on volume theory should remain one of the most fundamental, the most important, and the most beautiful *chéf-d'oeuvres* throughout the whole history of mathematics.

Computation of π in Ancient China.

For measuration it is also very important to know the ratio of the circumference of a circle to its diameter, i.e. π. According to known records, the value of π was computed in more and more exact values, as social productivity and science developed further and further. We list a few of such values known in ancient China below:

Perimeter 3, diameter 1, i.e. $\pi = 3$, before Han Dynasty.

$\pi = 3.1547$, according to an inscription on a bronze cylindrical standard measure *Li Jia Liang Hu*, cast on official order in the period of Xin Dynasty (+9, +23).

$\pi = \dfrac{730}{232}$ or 3.1466, in *Ling Xian (The spiritual construction of the universe)* of the astronomer Zhang Heng (+78, +139).

$\pi = \sqrt{10}$ or 3.1622, according to Zhang Heng's formula for the volume of a sphere.

$\pi = \frac{142}{45}$ or 3.1556, due to Wang Fan (+219, +287).

All the above calculations seem to be of empirical character with unknown theoretical basis.

It is again Liu Hui who, in his *Annotations*, pointed out the general principle for getting the value of π as accurate as possible. He pointed out that the very rough estimate of $\pi = 3$ is just the ratio of the perimeter of an inscribed regular hexagon to the diameter. He proved that in starting from the inscribed regular hexagon and doubling the number of sides of the inscribed regular hexagon further and further, one will get the ratio nearer and nearer to the exact value. Moreover, the loss of the value will be bounded by the sum of rectangular areas over the sides of the regular polygons. See the accompanying Fig. 3.9 above in following the annotations of Liu Hui:

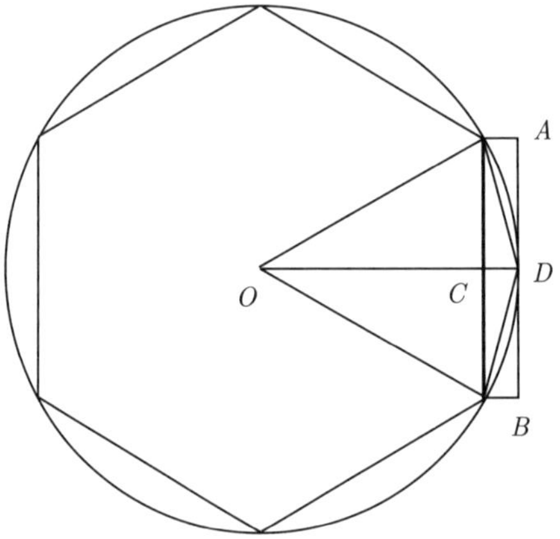

Diagram showing Liu method determing segment areas.

Fig. 3.9.

In taking the number of sides to be $3 * 2^6 = 192$, Liu gave an approximate value of the ratio $> 3.14\frac{64}{625}$ or 3.141024 and $< 3.14\frac{169}{625}$. If the number of sides is $3 * 2^{10} = 3072$, then the approximate value will be slightly less than $\frac{3927}{1250}$ or 3.1416.

What is of highest importance is the limit concept implied in Liu Hui's method. In fact, Liu Hui says:

> The finer we cut the segments, the less will be the loss. Cut further and further until unable to cut further. Then [the polygon] will coincide with

the circle and there will be no more any loss.

In comparing with what Liu Hui had said in establishing the 2:1 Principle, we may see that Liu Hui had a very clear picture of the limit concept and also the limit procedure.

In 5th Century the great scientist Zu Chong-Zi (+429, +500) gave the value of π to be:

$$\frac{22}{7} < \pi < \frac{355}{117}, \text{ or}$$
$$3.1415926 < \pi < 3.1415927,$$

which had been called the *coarse ratio* and the *close ratio* respectively in the Calendrical and Astromical Chapters of *Sui Shu (History of the Sui Dynasty).* The Zu's evaluation of π and other important achievements were recorded in his book *Zhui Shu (The Art of Mending).* Unfortunately the book has long been lost and no one knows how Zu had arrived at this value. Had Zu employed Liu Hui's method, he would have laboured over polygons of $3 * 2^{12} = 12,288$ and $3 * 2^{13} = 24,576$ sides to obtain the *coarse ratio* and the *close ratio* respectively.

Remark 3.7. $\frac{355}{117}$ is the best fractional value for π with both the numerator and denominator less than 1,000. It was obtained in Europe only 1,000 years later by the German Otto and the Dutch Antonisz. It is therefore known as the *Antonisz value* in Western mathematics.

Algebrization of Geometry in Song-Yuan Dynasties.

From the recovered proofs of various formulae in *Sea-Island* and the Qin-Heron Formula we may see the strong *algebraic* character of our ancient geometry. Such algebraic character was turned into a *systematic* method when the *Heaven's Element* notion and *Heaven's-Element Shu* were invented in the period of Song-Yuan Dynasties. In fact, by assigning some undetermined geometric entities as *Heaven's Element, Earth's Element*, etc. which are somewhat like unknown quantities x, y, etc. in now-a-days notation, the other geometric entities may be turned into what we call now *polynomials* and geometrical relations into *polynomial equations* in *Heaven's Element* or x, etc. The two books [LY1,2]* of Li Ye were devoted to applications of the method of Heaven's Element.

It was known that there were books in Song-Yuan Dynasties which extended the method of Heaven's Element to methods involving two, three, or even up to four *Elements*, corresponding to two, three, or four undetermined quantities. All these books had been lost except fortunately the two books [Zhu1,2]* in 1299 and +1303 of Zhu Shijie. The Part B, Sect. 5 of [ZSJ2]* contained 8 problems about Gou-Gu and measurements. Problems 1-3 were variants of the City-Width Problem in *Nine Chapters*. Problems 4-8 were same as the Problems 1-4,7 of *Sea-Island* with slight modifications, e.g. the sea-island was changed to some circular-formed city. The proofs of formulae in *Sea-Island* as described above were now replaced by solving of equations.

As already pointed out in Sect. 2 of Chap. 1, the problem of equations-solving occupies a central position in ancient Chinese mathematics and is in fact the main concern of the present book.

Chapter 3

Algebraic Varieties as Zero-Sets and Characteristic-Set Method

3.1 Affine and Projective Space. Extended Points and Specialization.

Fundamental Notions and Notations.

Notation 1.1.

$\mathbf{K}$:= Basic field of characteristic 0.

$\omega_1, \cdots, \omega_i, \cdots$:= Infinite sequence of independent indeterminates over $\mathbf{K}$.

$\boldsymbol{\Omega}$:= Field of rational functions of ω_i over $\mathbf{K}$.

$\boldsymbol{\Omega}^*$:= The algebraic closure of Ω.

Example 1.1. $\mathbf{K}$ = The rational field $\mathbf{Q}$, the real field $\mathbf{R}$, the complex field $\mathbf{C}$, etc.

Definition 1.1. Ω^* := *Universal field* over $\mathbf{K}$.

Definition 1.2. *Extension field* of $\mathbf{K}$:= Any finite extension field of $\mathbf{K}$ contained in $\boldsymbol{\Omega}^*$.

Remark 1.1. The introduction of the universal field Ω^* is solely for the sake of avoiding the logical difficulties involved in the incessant extensions of fields. Cf. e.g. [H-P2], Book III, Chap. X, Sect. 1. It brings actually no consequence at all to the whole theory developed henceforth.

Notation 1.2. $\mathbf{K}(\xi_1, \cdots, \xi_r)$:= Extension field of $\mathbf{K}$ by adjoining $\xi_1, \cdots, \xi_r \in \boldsymbol{\Omega}^*$ to $\mathbf{K}$:= Minimal extension field of $\mathbf{K}$ containing all of $\xi_1, \cdots, \xi_r$.

Notation 1.3. For an extension field $\mathbf{K}^*$ over $\mathbf{K}$, $TrDeg(\mathbf{K}^*/\mathbf{K})$:= Transcendence degree of $\mathbf{K}^*$ over $\mathbf{K}$ (= maximun number of algebraically independent elements in $\mathbf{K}^*$ over $\mathbf{K}$).

Affine Space and Extended Points.

Notation 1.4. $\mathbf{K}^n$ or $A^n_{\mathbf{K}}$:= Affine space of dimension n or n-dimensional affine space over $\mathbf{K}$.

Definition 1.3. *Points* in $\mathbf{K}^n$:= Elements in $\mathbf{K}^n$, written as $X = (x_1, \cdots, x_n)$, with $x_i \in \mathbf{K}$:= The i-th *coordinate* of X.

Notation 1.5. For $X = (x_1, \cdots, x_n)$, $X_k := (x_1, \cdots, x_k)$ for all $k > 0$ and $\leqslant n$.

Convention. For $k = 0, X_k = X_0$:= The empty set.

Definition 1.4. *Extended point* in $\mathbf{K}^n$:= Point in $\mathbf{K}^{*n}$ with $\mathbf{K}^*$ some extension field of $\mathbf{K}$. *Ordinary point* in $\mathbf{K}^n$:= Point in $\mathbf{K}^n$.

Definition 1.5. *Defining field* of an extended point $\Xi = (\xi_1, \cdots, \xi_n)$ in $\mathbf{K}^n$:= The extension field $\mathbf{K}(\Xi) = \mathbf{K}(\xi_1, \cdots, \xi_n)$.

Definition 1.6. *Dimension* over $\mathbf{K}$ of an extended point Ξ in $\mathbf{K}^n$:= $TrDeg$ $(\mathbf{K}(\Xi)/\mathbf{K})$.

Notation 1.6. *Dimension* over $\mathbf{K}$ of an extended point Ξ in $\mathbf{K}^n$:= $Dim_{\mathbf{K}}\ \Xi$.

Canonical Representation of an Extended Point.

Notation 1.7. For $\Xi = (\xi_1, \cdots, \xi_n)$ an extended point in $\mathbf{K}^n$, $\Xi_i := (\xi_1, \cdots, \xi_i)$, $i = 1, \cdots, n$. For $i = 0$, Ξ_0 := The empty set.

For the extended point $\Xi = (\xi_1, \cdots, \xi_n)$ suppose for $0 < i < c_1$, each ξ_i is transcendental over $\mathbf{K}(\Xi_{i-1})$, while ξ_{c_1} is algebraic over $\mathbf{K}(\Xi_{c_1-1})$, with defining equation

$$E_1 \equiv \xi_{c_1}^{d_1} + f_{11} * \xi_{c_1}^{d_1-1} + \cdots + f_{1d_1} = 0, \ f_{1j} \in \mathbf{K}(\Xi_{c_1-1}).$$

Suppose for $c_1 < i < c_2$, each ξ_i is transcendental over $\mathbf{K}(\Xi_{i-1})$, while ξ_{c_2} is algebraic over $\mathbf{K}(\Xi_{c_2-1})$, with defining equation

$$E_2 \equiv \xi_{c_2}^{d_2} + f_{21} * \xi_{c_2}^{d_2-1} + \cdots + f_{2d_2} = 0, \ f_{2j} \in \mathbf{K}(\Xi_{c_2-1}).$$

This may be continued as far as possible. So the extended point $\Xi = (\xi_1, \cdots, \xi_n)$ in $\mathbf{K}^n$ may be represented by a set of polynomials $E_i, i = 1, \cdots, r$, given below:

$$E_i \equiv x_{c_i}^{d_i} + f_{i1} * x_{c_i}^{d_i-1} + \cdots + f_{id_i}, \ f_{ij} \in \mathbf{K}(X_{c_i-1}), \ j = 1, \cdots, d_i. \tag{1.1}$$

Definition 1.7. For the above extended point $\Xi = (\xi_1, \cdots, \xi_n)$, the polynomials E_i in (1.1) := The *defining polynomials* of Ξ. The set of E_i := The *defining polynomial set* of Ξ. The corresponding set of equations $E_i = 0$:= The *defining equations* of Ξ. The ξ_{c_i} := The *bounded coordinates* of Ξ. The ξ_j with $j \neq$ any c_i := The *free coordinates* of Ξ.

Alternative Representations of an Extended Point.

From the theory of algebraic extension of fields we have the following Lemma:

Lemma 1.1. Any element η in the algebraic extension field $\mathbf{K}(\xi)$ with ξ adjoined algebraically to $\mathbf{K}$ by means of the defining equation

$$E \equiv x^d + f_1 * x^{d-1} + \cdots + f_d = 0, \ f_i \in \mathbf{K}$$

may be written uniquely in the form

$$\eta = \Sigma_i \ k_i * \xi^i,$$

in which $k_i \in \mathbf{K}$, and the summation extends over $i = 0, \cdots, d-1$.

Notation 1.8. For $\Xi = (\xi_1, \cdots, \xi_n)$ with defining polynomials E_i as in (1.1):

$y_i := x_{c_i}, Y_i := (y_1, \cdots, y_i),\ i = 1, \cdots, r.$

$u_1, \cdots, u_{n-r} :=$ The remaining x's arranged in the original order.

$U := (u_1, \cdots, u_{n-r})$. $U_i :=$ The set of u_j occuring in $X_{c_i-1}, i = 1, \cdots, r$.

In applying the Lemma 1.1 successively to the equations $E_i = 0$ in (1.1) we get the following

Proposition 1.1. Any element η in the defining field of $\Xi = (\xi_1, \cdots, \xi_n)$ with defining polynomials E_i of (1.1) is uniquely of the form

$$\eta = \Sigma_e \ k_e * \xi_{c_1}^{e_1} * \cdots * \xi_{c_r}^{e_r}, k_e \in \mathbf{K}(\Xi^{(0)}),$$

in which $\Xi^{(0)}$ is the set of ξ_j with j not equal to any c_i and the summation is to be extended over the r-tuples of integers $(e_1, \cdots, e_r)$ with $0 \leqslant e_i < d_i,\ i = 1, \cdots, r$.

Applying now Prop. 1.1 to elements f_{ij} in E_i and then by clearing of fractions we get the following

Proposition 1.2. The defining polynomials E_i in (1.1) of the extended point $\Xi = (\xi_1, \cdots, \xi_n)$ may be taken to be of the following form:

$$E_i' \equiv f_{i0}' * y_i^{d_i} + f_{i1}' * y_i^{d_i-1} + \cdots + f_{id_i}',\ f_{i0}' \in \mathbf{K}[U_i],\ f_{ij}' \in \mathbf{K}[U_i, Y_{i-1}],\ j = 1, \cdots, d_i, \tag{1.1'}$$

with degree in y_k less than d_k for each f_{ij}' and $k < i$. Moreover, for each $i = 1, \cdots, r$ the polynomial E_i' is irreducible in the field $\mathbf{K}(U_i, Y_{i-1})$ and the coefficients $f_{ij}', j = 0, 1, \cdots, d_i$ have no common factor in $\mathbf{K}[U_i]$.

Remark 1.2. Henceforth the *defining polynomials*, etc. will be either the polynomials E_i in (1.1), or E_i' in (1.1)$'$, according to what representation of the extended point is concerned.

Remark 1.3. If Ξ is an ordinary point in $\mathbf{K}^n$, then the corresponding defininig polynomial set is void and $\mathbf{K}(\Xi) = \mathbf{K}$.

Conjugate Extended Points.

Definition 1.8. Extended points in $\mathbf{K}^n$ having same free coordinates and same defining polynomial set or same defining equations := *Conjugate extended points* in $\mathbf{K}^n$.

Suppose given an extended point $\Xi = (\xi_1, \cdots, \xi_n)$ with the defining polynomial set E_i as in (1.1) or E_i' in (1.1)$'$. Let $\mathbf{K}_0$ be the extension field $\mathbf{K}(\Xi_{c_1-1})$. Let

$\xi_1^{(i_1)}, i_1 = 1, \cdots, d_1$, be the set of zeros of $E_1 = 0$ in some extension field of $\mathbf{K}_0$. Denote now the extension field of $\mathbf{K}_0$ in adjoining all the transcendental elements u_j in X_{c_2-1} and the zero $\xi_1^{(i_1)}$ by $\mathbf{K}_1^{(i_1)}$. For any $i_1 = 1, \cdots, d_1$ let $E_2^{(i_1)}$ be the polynomial in $\mathbf{K}_1^{(i_1)}[x_{c_2}]$ gotten from E_2 or E_2' with coefficients $f_{2j}^{(i_1)}$ the elements in $\mathbf{K}_1^{(i_1)}$ gotten from f_{2j} in replacing $x_i, i = 1, \cdots, c_2 - 1$, by η_i where $\eta_i = \xi_i$ for $i \neq c_1$ and $\eta_{c_1} = \xi_1^{(i_1)}$. The set of zeros of $E_2^{(i_1)} = 0$ will then be denoted by $\xi_2^{(i_1 i_2)}, i_2 = 1, \cdots, d_2$ and the extension field of $\mathbf{K}_1^{(i_1)}$ in adjoining all ξ_i for $c_2 < i < c_3 - 1$ and $\xi_2^{(i_1 i_2)}$ by $\mathbf{K}_2^{(i_1 i_2)}$. This will be continued inductively in an evident manner. We will get thus finally an extension field $\mathbf{K}_r^{(i_1 \cdots i_r)}$, a set of equations $E_r^{(i_1 \cdots i_{r-1})} = 0$, and a set of zeros $\xi_r^{(i_1 \cdots i_{r-1} i_r)}, i_r = 1, \cdots, d_r$ of these equations. The following Proposition is now evident:

Proposition 1.3. The set of extended points $\Xi^{(i_1 \cdots i_r)} = (\eta_1, \cdots, \eta_n)$ where $\eta_j = \xi_j$ for $j \neq$ any c_i and $\eta_{c_1} = \xi_1^{(i_1)}, \eta_{c_2} = \xi_2^{(i_1 i_2)}$, etc. gotten in the above manner form then the totality of conjugate extended points of which one is the given one Ξ. Moreover, the extension fields $\mathbf{K}_r^{(i_1 \cdots i_r)}$ are the defining fields of $\Xi^{(i_1 \cdots i_r)}$ which are isomorphic to each other in a natural way.

Example 1.2. Let t be an indeterminate. Then (t^2, t^3, t^4) and $(t^2, -t^3, t^4)$ are conjugate extended points in $\mathbf{K}^3$ with same first free coordinate t^2 and same defining polynomials

$$E_1 = x_2^2 - x_1^3, \quad E_2 = x_3 - x_1^2.$$

Dimension of an Extended Point.

Proposition 1.4. For an extended point $\Xi = (\xi_1, \cdots, \xi_n)$ with defining polynomial set (1.1) or (1.1)$'$, the dimension over $\mathbf{K}$ of the extended point Ξ is equal to $n - r$.

Proof. Let the defining polynomials of Ξ be as in (1.1) or (1.1)$'$. Then it is clear that the $n-r$ elements ξ_j's with j not equal to any ξ_{c_k} are algebraically independent over $\mathbf{K}$, while each ξ_{c_k} is algebraically dependent on these $n-r$ elements. Hence

$$Dim_{\mathbf{K}}(\Xi) = TrDeg(\mathbf{K}(\Xi)/\mathbf{K}) = n - r \tag{1.2}$$

as to be proved. □

Extreme Cases:

$r = n : Dim_{\mathbf{K}}(\Xi) = 0$.

$r = n$ and all $d_i = 1 :\ \Xi =$ an ordinary point in $\mathbf{K}^n$.

$r = 0 : Dim_{\mathbf{K}}(\Xi) = n$.

Example 1.3. Let $\mathbf{K} = \mathbf{R}$ and $\Xi = (-i, +i),\ \Xi' = (+i, -i)$, where $i = \sqrt{-1}$.

The points Ξ and Ξ' are conjugate extended points in $\mathbf{R}^2$ with same defining polynomials

$$x_1^2 + 1, \;\; x_2 + x_1.$$

Hence $n = r = 2$ and $Dim_{\mathbf{R}}\ \Xi \;=\; Dim_{\mathbf{R}}\ \Xi' \;=\; 0$.

Example 1.4. Let $\Xi = \left(\dfrac{2*t}{1+t^2}, \dfrac{1-t^2}{1+t^2}\right)$, $\Xi' = \left(\dfrac{1-t^2}{1+t^2}, \dfrac{2*t}{1+t^2}\right)$, t being an indeterminate.

Remark first that both $\xi_1 \;=\; \dfrac{2*t}{1+t^2}$ and $\xi_1' \;=\; \dfrac{1-t^2}{1+t^2}$ are transcendental over $\mathbf{K}$. For, suppose that there is a polynomial $P \in \mathbf{K}[x]$ such that $P(\xi_1) = 0$. Let $P = a_0 * x^d + \cdots + a_d$ with $a_i \in \mathbf{K}$. Then clearing of fractions in $P(\xi_1) = 0$, we shall have an equation of the form

$$a_0 * (2*t)^d + \cdots + a_d * (1+t^2)^d \;=\; 0.$$

Setting $t \;=\; 0$ in the above equation we get then $a_d \;=\; 0$. Removing the non-zero factor $2*t$ we get then the equation

$$a_0 * (2*t)^{d-1} + \cdots + a_{d-1} * (1+t^2)^{d-1} \;=\; 0.$$

Setting again $t \;=\; 0$ in the above equation we get then $a_{d-1} \;=\; 0$. Removing the non-zero factor $2*t$ and proceeding as before further and further we shall get $a_i = 0$ for all i or $P \equiv 0$. Hence ξ_1 is transcendental over $\mathbf{K}$. Similarly for ξ_1'.

It is now easy to see that both Ξ and Ξ' have the same single defining polynomial $x_2^2 + x_1^2 - 1$ so that $n = 2, r = 1$. Thus $Dim_{\mathbf{K}}\ \Xi \;=\; Dim_{\mathbf{K}}\ \Xi' \;=\; 1$.

Specialization of an Extended Point.

Definition 1.9. An extended point Ξ' in $\mathbf{K}^n$:= A *specialization* over $\mathbf{K}$ of an extended point Ξ in $\mathbf{K}^n$ if for any polynomial $P \in \mathbf{K}[X]$ with $P(\Xi) = 0$, we have also $P(\Xi') = 0$.

Notation 1.9. For extended points Ξ, Ξ' in $\mathbf{K}^n$, Ξ' is a specialization of Ξ over $\mathbf{K}$:=

$$\Xi \longrightarrow_{\mathbf{K}} \Xi'. \tag{1.3}$$

Notation 1.10. Set of all specializations over $\mathbf{K}$ of an extended point Ξ in $\mathbf{K}^n$:= $Spec_{\mathbf{K}}(\Xi)$.

Proposition 1.5. For any two conjugate extended points in $\mathbf{K}$ each one is the specialization over $\mathbf{K}$ of the other.

Proof. This follows from the very definitions by induction. See also Sect. 3.6 of this chapter. □

Examples of Specializations.

Proposition 1.6. Let $\xi_1, \cdots, \xi_n$ be independent indeterminates. Then any extended point $\Xi' = (\xi'_1, \cdots, \xi'_n)$ in $\mathbf{K}^n$ is a specialization of $\Xi = (\xi_1, \cdots, \xi_n)$.

Proof. Consider any polynomial $P(X) \in \mathbf{K}[X]$. Then

$$P(\Xi) = 0 \ ===> \ P(X) \equiv 0 \ ===> \ P(\Xi') = 0.$$

Hence we have (1.3) as to be proved. □

Proposition 1.7. In $\mathbf{K}^n$ let $\Xi = (\xi_1, \cdots, \xi_n)$ be an extended point for which each ξ_i equals some rational function of parameters $U = (u_1, \cdots, u_d)$ with coefficients in $\mathbf{K}$. Let u'_j be values in any extension field $\mathbf{K}'$ of $\mathbf{K}$ such that each ξ_i has a well-defined value ξ'_i when each u_j is substituted by u'_j. Then $\Xi' = (\xi'_1, \cdots, \xi'_n)$ is a specialization of Ξ over $\mathbf{K}$.

Proof. Let $\xi'_i = \dfrac{p_i}{q_i}$ with $p_i, q_i \in \mathbf{K}[U]$ and prime to each other. By hypothesis $q_i(U') \neq 0$ for each i where $U' = (u'_1, \cdots, u'_d)$. For any polynomial $P(X) \in \mathbf{K}[X]$ we have then

$$\begin{aligned} P(\Xi) = 0 \ &===> P\left(\frac{p_1(U)}{q_1(U)}, \cdots, \frac{p_n(U)}{q_n(U)}\right) = 0 \\ &===> P\left(\frac{p_1(U')}{q_1(U')}, \cdots, \frac{p_n(U')}{q_n(U')}\right) = 0 \\ &===> P(\Xi') = 0. \end{aligned}$$

Hence we have (1.3) as to be proved. □

The following examples show that not every specialization may be achieved by mere substitution as in Propositions 1.6 and 1.7.

Example 1.5. In $\mathbf{K}^3$, the point $(0, 0, 1)$ is a specialization of the extended point $\Xi = \left(\xi_1, \xi_2, \dfrac{\xi_2}{\xi_1}\right)$, where ξ_1, ξ_2 are independent indeterminates over $\mathbf{K}$.

Proof. Consider any polynomial $P(x_1, x_2, x_3) \in \mathbf{K}[x_1, x_2, x_3]$ with $P(\Xi) = 0$ so that

$$P\left(x_1, x_2, \frac{x_2}{x_1}\right) = 0. \tag{1.4}$$

Dividing P by $x_1 * x_3 - x_2$ w.r.t. the variable x_3 we get for some polynomials $Q \in \mathbf{K}[x_1, x_2, x_3]$ and $R \in \mathbf{K}[x_1, x_2]$ such that

$$x_1^d * P(x_1, x_2, x_3) = (x_1 * x_3 - x_2) * Q + R. \tag{1.5}$$

Setting $x_3 = \dfrac{x_2}{x_1}$ in it we get then from (1.5)

$$R \ = \ x_1^d * P\left(x_1, x_2, \frac{x_2}{x_1}\right) \ = \ 0.$$

(1.5) becomes thus

$$x_1^d * P(x_1, x_2, x_3) = (x_1 * x_3 - x_2) * Q.$$

It follows that x_1^d divides Q and $x_1 * x_3 - x_2$ divides P so that for some polynomial $P'(x_1, x_2, x_3) \in \mathbf{K}[x_1, x_2, x_3]$ we have

$$P(x_1, x_2, x_3) \quad = \quad (x_1 * x_3 - x_2) * P'(x_1, x_2, x_3).$$

By direct substitution we have then $P(0, 0, 1) = 0$.

It follows that for any polynomial $P \in \mathbf{K}[x_1, x_2, x_3]$ for which $P(\Xi) = 0$ it would follow also $P(0, 0, 1) = 0$. Hence

$$\left(\xi_1, \xi_2, \frac{\xi_2}{\xi_1}\right) \longrightarrow_{\mathbf{K}} (0, 0, 1)$$

as to be proved. □

Example 1.6. In $\mathbf{K}^2$ we have for t an indeterminate

$$\left(\frac{2 * t}{1 + t^2}, \frac{1 - t^2}{1 + t^2}\right) \longrightarrow \mathbf{K}(0, 1) \text{ and } (0, -1). \tag{1.6}$$

Proof. Consider a polynomial $P(x_1, x_2) \in \mathbf{K}[x_1, x_2]$ with

$$P\left(\frac{2 * t}{1 + t^2}, \frac{1 - t^2}{1 + t^2}\right) = 0.$$

Divide P by $x_1^2 + x_2^2 - 1$, considered as a polynomial in the variable x_1, we have for some polynomials $Q(x_1, x_2) \in \mathbf{K}[x_1, x_2]$ and $R_1(x_2), R_2(x_2) \in \mathbf{K}[x_2]$ a relation of the form

$$P = (x_1^2 + x_2^2 - 1) * Q + x_1 * R_1 + R_2.$$

Substituting $\frac{2 * t}{1 + t^2}, \frac{1 - t^2}{1 + t^2}$ for x_1, x_2 in P, Q, R_1, R_2 we get an identity of the form

$$P\left(\frac{2 * t}{1 + t^2}, \frac{1 - t^2}{1 + t^2}\right) = \frac{2 * t}{1 + t^2} * R_1\left(\frac{1 - t^2}{1 + t^2}\right) + R_2\left(\frac{1 - t^2}{1 + t^2}\right).$$

The left-hand side and hence also the right-hand of the above identity is equal to 0. From degree considerations we have then

$$R_1\left(\frac{1 - t^2}{1 + t^2}\right) = 0, \quad R_2\left(\frac{1 - t^2}{1 + t^2}\right) = 0.$$

Now in Ex. 1.4 we have proved that $\dfrac{1-t^2}{1+t^2}$ is transcendental over $\mathbf{K}$ so that we have $R_1(x_2) \equiv 0, R_2(x_2) \equiv 0$. It follows that

$$P(x_1, x_2) = (x_1^2 + x_2^2 - 1) * Q(x_1, x_2).$$

Hence $P(0, 1) = 0$ and $P(0, -1) = 0$. As P is arbitrary in $\mathbf{K}[x_1, x_2]$, we get (1.6). □

Remark 1.4. That $(0, 1)$ is a specialization of $\left(\dfrac{2*t}{1+t^2}, \dfrac{1-t^2}{1+t^2}\right)$ may be seen directly in substituting 0 for t. However this cannot be done by direct substitution for the case of $(0, -1)$.

Remark 1.5. The method used in the proof of Ex. 1.5 and Ex. 1.6 is a general one and will play an important role in the theory which will be seen in latter sections.

Problem S. Given two extended points in $\mathbf{K}^n$, to decide whether one is a specialization of the other.

The problem will be solved in latter section Sect. 3.6 of this chapter.

Dimension under Specialization.

Proposition 1.8. For extended points $\Xi = (\xi_1, \cdots, \xi_n), \Xi' = (\xi_1', \cdots, \xi_n')$ in $\mathbf{K}^n$ we have

$$\Xi \longrightarrow_{\mathbf{K}} \Xi' \ ===> Dim_{\mathbf{K}}\, \Xi \geqslant Dim_{\mathbf{K}}\, \Xi'. \tag{1.6}$$

Proof. Let $Dim_{\mathbf{K}}\, \Xi' = e$ so that the maximum number of algebraically independent elements in the ξ_i' over $\mathbf{K}$ is e. Suppose for simplicity that these independent ones are the first e ξ_i', viz. $\xi_1', \cdots, \xi_e'$. Then the first e ξ_i, viz. $\xi_1, \cdots, \xi_e$ should also be algebraically independent elements over $\mathbf{K}$. For otherwise there will be some non-zero polynomial $P \in \mathbf{K}[X_e]$ with $P(\Xi_e) = 0$. As Ξ' is a specialization over $\mathbf{K}$ of Ξ, it would follow $P(\Xi_e') = 0$ too and $\xi_1', \cdots, \xi_e'$ would be algebraically dependent, contrary to hypothesis. Hence $Dim_{\mathbf{K}}\, \Xi \geqslant e$ as to be proved. □

Primitive Element of an Extended Point.

Definition 1.10. *Primitive element* of an extended point Ξ in $\mathbf{K}^n$:= An element θ algebraic over $\mathbf{K}$ such that

$$\mathbf{K}(\theta) \equiv \mathbf{K}(\Xi).$$

Proposition 1.9. For Ξ given by (1.1) there are integers $n_1, \cdots, n_r$ such that $\theta = n_1 * x_{c_1} + \cdots + n_r * x_{c_r}$ is a primitive element of Ξ.

Proof. We refer the proof to classical texts like [VdW1] or [H-P1]. Remark that the mere *existential* proofs may be easily turned into *constructive* ones which we omit here. □

Projective Spaces and Projective Points.

It is sometimes convenient to consider *projective spaces* instead of affine spaces. Instead of going into details we shall satisfy ourselves in introducing the notations below:

Notation 1.11.

$A^n_{\mathbf{K}}(X) :=$ Affine n-space $\mathbf{K}^n(X)$.

${}^hX := (x_0 : x_1 : \cdots : x_n)$, not all $x_i = 0$.

$P^n_{\mathbf{K}}({}^hX) :=$ Projective n-space over $\mathbf{K}$ with homogeneous coordinates ${}^hX$.

Definition 1.11. ${}^hX$ with $x_0 = 1 :=$ *Completed homogeneous coordinates* of X. $P^n_{\mathbf{K}}({}^hX) :=$ *Completed projective space* of the affine space $A^n_{\mathbf{K}}(X)$. $x_0 :=$ *Coordinate at infinity* of $P^n_{\mathbf{K}}({}^hX)$. For an extended point $X' = (x'_1, \cdots, x'_n)$ in $A^n_{\mathbf{K}}(X)$, the *homogenized point* of $X' :=$ The homogeneous extended point ${}^hX' = (1 : x'_1 : \cdots, x'_n)$ in $P^n_{\mathbf{K}}({}^hX)$.

Notation 1.12. For a polynomial $P \in \mathbf{K}[X]$ of total degree d in X and a monomial $M = x_1^{i_1} * \cdots * x_n^{i_n}$ in P, ${}^hM :=$ The monomial $x_0^{i_0} * M$ where $i_0 = d - (i_1 + \cdots + i_n)$.

Definition 1.12. For a polynomial $P \in \mathbf{K}[X]$ of total degree d in X, the *homogenized polynomial* of $P :=$ The homogeneous polynomial gotten from P in replacing each monomial M in P by ${}^hM$.

We may naturally define the notions of *extended point, specialization*, etc. for points in projective spaces which we shall not enter. What is important is that in the projective case we have the following Proposition which has no analogue in the affine case. Cf. e.g. [VdW2], p.107.

Proposition 1.10. Specialization-Extension Theorem. Let ξ, ξ' be two extended points in a projective n-space $P^n_{\mathbf{K}}(X)$ and η an extended point in a projective m-space $P^m_{\mathbf{K}}(Y)$ of homogeneous coordinates ${}^hY = (y_0 : y_1 : \cdots : y_m)$. If ξ' is a specialization of ξ over $\mathbf{K}$, then such a specialization can be extended over $\mathbf{K}$ to one of the pair $(\xi, \eta) \longrightarrow_{\mathbf{K}} (\xi', \eta')$ for some point η' in $P^m_{\mathbf{K}}(Y)$.

3.2 Algebraic Varieties and Zero-Sets.

Fundamental Notions and Notations.

Notation 2.1.

$\Omega, \Omega^*, \mathbf{K}, X = (x_1, \cdots, x_n), X_k = (x_1, \cdots, x_k), A^n_{\mathbf{K}}(X)$, etc. as in Sect. 3.1.

Definition 2.1. *Pol* (in X over $\mathbf{K}$) := Polynomial in $\mathbf{K}[X]$.

Definition 2.2. *Polset* (in X over $\mathbf{K}$) := Set of pols in $\mathbf{K}[X]$.

Convention. The polset will be a finite set unless otherwise stated.

Definition 2.3. *Zero* of a pol $P \in \mathbf{K}[X]$:= An extended point Ξ in $\mathbf{K}^n$ such that $P(\Xi) = 0$.

Definition 2.4. *Zero* of a polset $PS \subset \mathbf{K}[X]$:= A common zero of all pols in PS.

Definition 2.5. *Zero-set* of a polset $PS \subset \mathbf{K}[X]$:= Set of all zeros of PS.

Notation 2.2. For $PS \subset \mathbf{K}[X]$, $Zero(PS)$:= Zero-set of PS.

Definition 2.6. Two polsets $PS_1, PS_2 \subset \mathbf{K}[X]$ are *0-equivalent* $:= Zero(PS_1) = Zero(PS_2)$.

Notation 2.3. PS_1, PS_2 0-equivalent $:= PS_1 \sim_0 PS_2$.

Notation 2.4. For polsets $PS, PS_1, PS_2 \subset \mathbf{K}[X]$, a pol $G \in \mathbf{K}[X]$, an extension field $\mathbf{F}$ of $\mathbf{K}$, and an arbitrary domain D in $\mathbf{K}^n$, we write by definition

$$\begin{aligned}
&PS_1 \times PS_2 := \{P_{1i} * P_{2j} \mid P_{1i} \in PS_1, P_{2j} \in PS_2\},\\
&Zero(PS/G) := Zero(PS) \setminus Zero(G),\\
&Zero_{\mathbf{F}}(PS/G) := Zero(PS/G) \cap \mathbf{F}^n.\\
&DZero(PS/G) := Zero(PS) \cap D.
\end{aligned} \tag{2.1}$$

Remark 2.1. The basic field $\mathbf{K}$ is understood to be prescribed and fixed in what follows unless otherwise stated so that it will not appear in the notations. By $\mathbf{F}^n$ in $(2.1)_3$ we mean the set of all points in $\mathbf{K}^n$ with coordinates all in $\mathbf{F}$, Similarly for D in $(2.1)_4$.

Notation 2.5. For pols $P_1, P_2 \in \mathbf{K}[X]$ and a polset $PS = \{A_1, \cdots, A_m\} \subset \mathbf{K}[X]$,

$$P_1 \equiv P_2 \ mod_{\mathbf{K}} \ PS$$

:= There exist pols $Q_1, \cdots, Q_m \subset \mathbf{K}[X]$ such that

$$P_1 - P_2 = \Sigma_i \ Q_i * A_i.$$

Remark 2.2. We write throughout the book + instead of $\cup$ for the union of sets. But we preserve the notation $\setminus$ for the set difference.

Algebraic Varieties as Zero-Sets.

Definition 2.7. An *affine algebraic variety* or simply a *variety* in $\mathbf{K}^n$ or $A^n_{\mathbf{K}}(X)$, (or over bf K in $A^n_{\mathbf{K}}(X)$) := Zero-set of some polset in $\mathbf{K}[X]$.

Definition 2.8. For a variety $V = Zero(PS)$, PS := A *basis* of V.

Definition 2.9. Given a variety $V = Zero(PS)$ with basis PS a polset in $\mathbf{K}[X]$, a *point on* or *of the variety* V := A zero of PS.

Definition 2.10. Given a variety $V = Zero(PS)$ with PS a polset in $\mathbf{K}[X]$, and a pol $G \in \mathbf{K}[X]$, $G = 0$ *on* $V := Zero(PS) \subset Zero(G)$, or any zero of PS is also a zero of G.

Notation 2.6. For V a variety and G a pol, $G = 0|V := G = 0$ on V or $Zero(V) \subset Zero(G)$.

Simple Properties.

Proposition 2.1. Let PS, PS_1, PS_2 be polsets and F_1, F_2, G, G_1, G_2 be pols in $\mathbf{K}[X]$. Then we have:

$$\begin{aligned}
&PS_1 \subset PS_2 ===> Zero(PS_2) \subset Zero(PS_1).\\
&Zero(1) = \varnothing.\\
&Zero(0) = Zero(\varnothing) = \Omega^{*n}.\\
&Zero(PS/1) = Zero(PS).\\
&Zero(PS/0) = \varnothing.\\
&Zero(PS/G_1 * G_2) = Zero(PS/G_1) \cap Zero(PS/G_2).\\
&Zero(PS) = Zero(PS/G) + Zero(PS + \{G\}).\\
&Zero(PS_1 + PS_2 \ / \ G) = Zero(PS_1/G) \cap Zero(PS_2/G).\\
&Zero(PS + \{F_1 * F_2\}) = Zero(PS + \{F_1\}) + Zero(PS + \{F_2\}).\\
&Zero(PS_1) + Zero(PS_2) = Zero(PS_1 \times PS_2).
\end{aligned} \tag{2.2}$$

Proof. Let us prove e.g. the last identity in (2.2). If Ξ is a point of the left-hand side L of the identity, then it is a zero in either $Zero(PS_1)$ or $Zero(PS_2)$ or both. If it is a zero in $Zero(PS_1)$, then it is a zero of all pols $P_i \in PS_1$, so it is a zero of all pols in $PS_1 \times PS_2$ or L is contained in the right-hand side R of the identity. Similarly for the zeros in $Zero(PS_2)$ and hence $L \subset R$. Consider now any point in R which is a zero of all the pols $P_{1i} * P_{2j}$ with $P_{1i} \in PS_1$ and $P_{2j} \in PS_2$. Suppose that this point is not a zero in $Zero(PS_1)$, then it is not a zero of some pol, say P_{1i_1} of PS_1. As it is a zero of all the pols $P_{1i_1} * P_{2j}$ with P_{2j} in PS_2, it should be a zero of all these P_{2j} or it is a zero in $Zero(PS_2)$. Similarly if that point is not a zero in $Zero(PS_2)$, then it should be a zero in $Zero(PS_1)$. In any way that point is a point in L. Hence $R \subset L$ too and we have $L = R$ as to be proved. □

Convention. As we are interested in the study of zeros of polsets but not the polsets themselves, and the appearance of the pol 0 in a polset will not affect its zero-set, we shall suppose that the pol 0 does not appear in any polset, unless otherwise stated.

Some Examples.

Example 2.1. For $P = x_1^2 + x_2^2 - 1 \in \mathbf{K}[x_1, x_2]$, (1,0), (0,1), $\left(\frac{2*t}{1+t^2}, \frac{1-t^2}{1+t^2}\right)$, t an indeterminate, are all in $Zero(P)$. For ξ_1 an indeterminate, and ξ_2 any solution of $x_2^2 - 1 + \xi_1^2 = 0$ in some extension field of $\mathbf{K}(\xi_1)$, (ξ_1, ξ_2) is also a zero of $Zero(P)$.

Example 2.2. Suppose that in taking convenient coordinate systems, the hypothesis of a geometry theorem T is given as $HYP = 0$ where $HYP \subset \mathbf{K}[X]$. Then any zero of the polset HYP is nothing else but a geometrical configuration in the extended space $\mathbf{\Omega}^{*n}$ which verifies the geometrical hypothesis of the theorem T. If the conclusion of the theorem is also given as $CONC = 0$ for some pol $CONC \in \mathbf{K}[X]$, then the theorem will be true if $CONC = 0$ on the variety $Zero(HYP)$ or if $Zero(HYP) \subset Zero(CONC)$. In general, this will be true only when some *non-degeneracy conditions* $ND_i \neq 0$ are observed, where ND_i are pols in $\mathbf{K}[X]$. The truth of the theorem is then lessened to the form

$$Zero(HYP \ / \ \Pi_i \ ND_i) \subset Zero(CONC). \tag{2.3}$$

The task of proving such a theorem is reduced to the finding of non-degeneracy conditions $ND_i \neq 0$ such that the above inclusion relation (2.3) is true. This is the basic underlying principle of our mechanical geometry theorem-proving. For more details see the Chap. 7 in Part III devoting mainly to mechanical geometry theorem-proving.

Example 2.3. Let $n = 3$, $\mathbf{K} = \mathbf{Q}$, p any positive integer, and $P_p = x_1^p + x_2^p - x_3^p$. Then Fermat's conjecture amounts to say that

$$Zero_{\mathbf{Q}}(P_p) = \varnothing \text{ for } p > 2.$$

On the other hand the Gou-Gu rational triple formula (2.3.4) in Sect. 3 of Chap. 2 shows that for $p = 2$,

$$Zero_{\mathbf{Q}}(P_2) = \{(a*(m^2 - n^2), 2*a*m*n, a*(m^2 + n^2)) \mid a, m, n \in \mathbf{Q}\}.$$

Example 2.4. Given pols $P_1, P_2 \in \mathbf{K}[x_1]$ with

$$P_i = a_{i0} * x_1^{d_i} + \cdots + a_{ij} * x_1^{d_i - j} + \cdots a_{id_i}, \ a_{ij} \in \mathbf{K}, \ i = 1, 2.$$

Let R be the resultant of P_1, P_2 w.r.t. x_1. Suppose that a_{10}, a_{20} are not both 0. Then P_1, P_2 will have a common zero if and only if $R = 0$ so that

$$Zero(P_1, P_2) \neq \varnothing <\!\!===\!\!> R = 0, \ (a_{10}, a_{20} \text{ not both } 0). \tag{2.4}$$

In what follows it is to be understood that all pols and polsets are in $\mathbf{K}[X]$ and all varieties are in $\mathbf{K}^n$ and are non-empty unless otherwise stated.

Irreducible Algebraic Variety.

Definition 2.11. An algebraic variety $V_1 = Zero(PS_1)$ is a *subvariety* of an algebraic variety $V_2 = Zero(PS_2) := V_1 \subset V_2$. It is a *proper subvariety* $:= V_1 \subset V_2$ and $V_1 \neq V_2$.

Definition 2.12. An algebraic variety $V = Zero(PS)$ in $\mathbf{K}^n$ is *irreducible* over $\mathbf{K} :=$ V is not the union of two proper subvarieties in $\mathbf{K}^n$. V is *reducible* $:=$ V is not irreducible.

Remark 2.3. The irreducibility of a variety depends on the basic field $\mathbf{K}$. For example, the variety $V = Zero(x_1^2 + 1)$ is irreducible over the field $\mathbf{K} = \mathbf{R}$, but is reducible over $\mathbf{K} = \mathbf{C}$, since $V = Zero(x_1 + i) + Zero(x_1 - i)$, where $i = \sqrt{-1}$.

Some Simple Propositions.

Proposition 2.2. If $V_1 = Zero(PS_1)$ is a proper subvariety of $V = Zero(PS)$, then there is a pol $P_1 \in PS_1$ such that $P_1 \neq 0|V$.

Proof. Suppose the contrary that for all pols $P_1 \in PS_1$ we have $P_1 = 0|V$ or $Zero(PS) \subset Zero(P_1)$. Then we would have $Zero(PS) \subset Zero(PS_1)$ and it would follow $Zero(PS_1) = Zero(PS)$, contrary to hypothesis. □

Proposition 2.3. The intersection and union of two varieties are also varieties.

Proof. This follows directly from $(2.2)_8$ and $(2.2)_{10}$. □

Proposition 2.4. If an irreducible variety V is contained in the union of two varieties V_1 and V_2, then V will be contained wholly in V_1 or wholly in V_2.

Proof. By hypothesis $V \subset V_1 + V_2$. Hence

$$V = V \cap V_1 + V \cap V_2.$$

As V is irreducible $V \cap V_1$, as a subvariety of V, is either same as V or is empty. Similarly $V \cap V_2$ is either same as V or is empty. Hence the non-empty V is same as either $V \cap V_1$ or $V \cap V_2$ or both. In case $V = V \cap V_1$ then $V \subset V_1$ and in case $V = V \cap V_2$ then $V \subset V_2$. Hence the theorem holds. □

Proposition 2.5. Any specialization over $\mathbf{K}$ of a point in a variety is also a point of that variety.

Proof. Let Ξ be a point of $V = Zero(PS)$ so that $P(\Xi) = 0$ for any pol $P \in PS$. If Ξ' is a specialization of Ξ over $\mathbf{K}$, then $P(\Xi') = 0$ too for all $P \in PS$ so that Ξ' is a point of V. □.

Criteria of Irreducibility.

Criterion Theorem 2.1. An algebraic variety $Zero(PS)$ in $\mathbf{K}^n$ is irreducible over $\mathbf{K}$ if and only if whenever there exist a pair of pols $P_1, P_2 \in \mathbf{K}[X]$ such that $Zero(PS) \subset Zero(P_1 * P_2)$, there will be either $Zero(PS) \subset Zero(P_1)$, or $Zero(PS) \subset Zero(P_2)$. In other words from $P_1 * P_2 = 0|Zero(PS)$ it would follow either $P_1 = 0|Zero(PS)$ or $P_2 = 0|Zero(PS)$.

Proof. Suppose first that $Zero(PS)$ is irreducible and $P_1 * P_2 = 0|Zero(PS)$. Now by $(2.2)_9$ we have

$$Zero(PS) \subset Zero(P_1 * P_2) ===> Zero(PS) \subset Zero(P_1) + Zero(P_2).$$

As $Zero(PS)$ is irreducible we have by Proposition 2.4 either $Zero(PS) \subset Zero(P_1)$ or $Zero(PS) \subset Zero(P_2)$. Hence the final condition is necessary for $Zero(PS)$ to be irreducible.

On the other hand suppose that $Zero(PS)$ is non-irreducible so that $Zero(PS)$ is the union of two proper subvarieties $V_1 = Zero(PS_1)$ and $V_2 = Zero(PS_2)$. Then by Proposition 2.1 there is a pol $P_1 \in PS_1$ such that $P_1 \neq 0|Zero(PS)$ and a pol $P_2 \in PS_2$ such that $P_2 \neq 0|Zero(PS)$. Now

$$P_1 \in PS_1 ===> P_1 = 0|V_1 ===> P_1 * P_2 = 0|V_1.$$

Similarly

$$P_1 * P_2 = 0|V_2.$$

Hence

$$P_1 * P_2 = 0 \mid Zero(PS_1) + Zero(PS_2) = V_1 + V_2 = Zero(PS),$$

and the final condition is not observed. The final condition is thus also sufficient for $Zero(PS)$ to be irreducible. □

Criterion Theorem 2.2. An algebraic variety $V = Zero(PS)$ in $\mathbf{K}^n$ is irreducible over $\mathbf{K}$ if and only if there exists on V an extended point Ξ in $\mathbf{K}^n$ such that all points of V are specicializations over $\mathbf{K}$ of that point Ξ, in other words, if and only if

$$Zero(PS) = Spec_{\mathbf{K}}(\Xi). \tag{2.5}$$

Proof. Let us first suppose that such a point Ξ exists which verifies (2.5). Consider any two pols $P_1, P_2 \in \mathbf{K}[X]$ with $P_1 * P_2 = 0|V$. In particular we have then $(P_1 * P_2)(\Xi) = 0$ so that either $P_1(\Xi) = 0$ or $P_2(\Xi) = 0$. By (2.5) any point Ξ' of V is a specialization over $\mathbf{K}$ of Ξ so that either $P_1(\Xi') = 0$ for all $\Xi' \in V$ or $P_2(\Xi') = 0$ for all $\Xi' \in V$. This means that either $P_1 = 0|V$ or $P_2 = 0|V$. By Criterion Theorem 2.1 V is then irreducible.

Suppose now V is irreducible. To see that a point Ξ verifying (2.5) exists let us introduce an equivalence relation $\sim$ among pols in $\mathbf{K}[X]$ such that for $P_1, P_2 \in \mathbf{K}[X], P_1 \sim P_2$ if and only if $P_1 - P_2 = 0|V$. Let us denote the equivalence class of a pol P by $[P]$. Define now $[P_1] + [P_2] = [P_1 + P_2]$ and $[P_1] * [P_2] = [P_1 * P_2]$. This definition is easily seen to be a legitimate one so that the set of all such equivalence classes forms a ring $\mathbf{A}$ with $\mathbf{K}$ as a subring if we identify $[k]$ to k for any $k \in \mathbf{K}$. Moreover, from $[P_1] * [P_2] = 0$ we have $[P_1 * P_2] = [0]$ or $P_1 * P_2 = 0|V$. As V is irreducible it would follow that either $P_1 = 0|V$ or $P_2 = 0|V$, i.e. either $[P_1] = [0] = 0$ or $[P_2] = [0] = 0$. Hence the above ring $\mathbf{A}$ is an integrity domain and we may form its quotient field $\mathbf{A}^*$ which is clearly an extension field of $\mathbf{K}$. Consider now the extended point $\Xi = ([x_1], \cdots, [x_n])$ in $\mathbf{A}^{*n}$. For any pol $P \in PS$ we have now $P(\Xi) = [P(X)] = 0|Zero(PS)$ or $\Xi \in V$. By Proposition 2.5 we have then $Spec_{\mathbf{K}}(\Xi) \subset V$. Consider now any point $\Xi' \in V$. For any $P \in \mathbf{K}[X]$ with $P(\Xi) = 0$ we have then $[P(X)] = 0$ or $P(X) = 0|V$ so that $P(\Xi') = 0$. It follows that Ξ' is a specialization over $\mathbf{K}$ of Ξ whence $V \subset Spec_{\mathbf{K}}(\Xi)$. Hence $V = Spec_{\mathbf{K}}(\Xi)$ or (2.5) is verified for the above point Ξ. □

Notion of Generic Point.

Definition 2.13. Any extended point Ξ in $\mathbf{K}^n$ verifying (2.5) := A *generic point* of the irreducible variety $Zero(PS)$.

With this definition Criterion Theorem 2.2 can then be re-phrased as

Fundamental Theorem 2.3. An algebraic variety in $\mathbf{K}^n$ is irreducible over $\mathbf{K}$ if and only if it has a generic point over $\mathbf{K}$.

The generic point of an irreducible variety is not unique. However, from its very definition we have the following corollary of Theorem 2.3.

Theorem 2.4. For an irreducible algebraic variety in $\mathbf{K}^n$ any two generic points are specifications over $\mathbf{K}$ of each other.

Dimension of an Irreducible Variety.

By Theorem 2.4 and Proposition 1.9 we see that the following definition is legitimate.

Definition 2.14. *Dimension* of an irreducible algebraic variety V in $\mathbf{K}^n$:= Dimension over $\mathbf{K}$ of any generic point of that variety.

Notation 2.7. Dimension of an irreducible variety V over $\mathbf{K}$:= $Dim_{\mathbf{K}}\, V$.

Theorem 2.5. If $V' = Zero(PS')$ is a proper irreducible subvariety of an irreducible variety $V = Zero(PS)$, then

$$Dim_{\mathbf{K}}\, V' < Dim_{\mathbf{K}}\, V. \tag{2.6}$$

Proof. Let Ξ and Ξ' be generic points of V and V' respectively. As $V' \subset V$, Ξ' is a specialization of Ξ. Hence by Proposition 1.9 we have

$$Dim_{\mathbf{K}}\ V' = Dim_{\mathbf{K}}\ \Xi' \leqslant Dim_{\mathbf{K}}\ \Xi = Dim_{\mathbf{K}}\ V.$$

Suppose that V and V' have the same dimension say e and let $\Xi' = (\xi'_1, \cdots, \xi'_n)$ with $\xi'_{i_1}, \cdots, \xi'_{i_e}$ algebraically independent while other ξ'_i are algebraically dependent on them. Then for $\Xi = (\xi_1, \cdots, \xi_n)$, $\xi_{i_1}, \cdots, \xi_{i_e}$ will be algebraically independent too while other ξ_i are algebraicaly dependent on these ξ. By Proposition 2.2 we have by hypothesis a pol $P \in \mathbf{K}[X]$ such that

$$P = 0|V', \text{ while } P \neq 0|V.$$

We have then

$$P(\Xi') = 0, \text{ while } P(\Xi) \neq 0.$$

As $P(\Xi)$ is clearly algebraically dependent on $\xi_{i_1}, \cdots, \xi_{i_e}$ there is a pol

$$F = a_0 * x^s + \cdots + a_s$$

with $a_i \in \mathbf{K}[x_{i_1}, \cdots, x_{i_e}]$ and $a_0 \neq 0, a_s \neq 0$, such that if α_j be the value of a_j when each x_{i_k} is replaced by ξ_{i_k}, then the equation

$$\alpha_0 * x^s + \cdots + \alpha_s = 0$$

is satisfied by $x = P(\Xi)$. It is clear that $\alpha_0 \neq 0$ since $\xi_{i_1}, \cdots, \xi_{i_e}$ are algebraically independent. Let α'_j be now the value of a_j when each x_{i_k} is replaced by ξ'_{i_k}. As Ξ' is a specialization of Ξ we see that $P(\Xi')$ satisfies the equation

$$\alpha'_0 * x^s + \cdots + \alpha'_s = 0.$$

As before we have $\alpha'_0 \neq 0$. On the other hand as $P(\Xi') = 0$ so that 0 satisfies the above equation and we have therefore $\alpha'_s = 0$. Since $a_s \neq 0$ it follows that $\xi'_{i_1}, \cdots, \xi'_{i_e}$ are algebraically dependent which leads to a contradiction. Consequently the supposition of equality of the dimensions cannot be true and we have accordingly (2.6) as to be proved. □

Irreducible Varieties of Special Dimensions.

Theorem 2.6. The variety $Zero(\varnothing)$ or $Zero(0)$ is the only irreducible variety of dimension n in $\mathbf{K}^n$ with $X = (x_1, \cdots, x_n)$ as a generic point in which $x_1, \cdots, x_n$ are independent indeterminates.

Proof. This follows directly from Proposition 1.7 of Sect. 3.1. □ .

Theorem 2.7. The only irreducible varieties of dimension $n-1$ in $\mathbf{K}^n$ are of the form $Zero(P)$ for some non-zero irreducible pol $P \in \mathbf{K}[X]$.

Proof. First let us consider a variety of the form $V = Zero(P)$ in which $P \in \mathbf{K}[X]$ is irreducible. Let x_c appears actually in P but none of x_i with $i > c$ so that $P \in \mathbf{K}[X_c]$. Let $\Xi = (x_1, \cdots, x_{c-1}, \xi_c, x_{c+1}, \cdots, x_n)$ with ξ_c a solution of $P(X) = 0$. Consider any pol $F \in \mathbf{K}[X]$ with $F(\Xi) = 0$ and any point $\Xi' = (\xi'_1, \cdots, \xi'_n) \in Zero(P)$. As P is irreducible F is divisible by P so that $F(\Xi') = 0$. It follows that Ξ' is a specialization over $\mathbf{K}$ of Ξ so that Ξ is a generic point of V and V is thus irreducible over $\mathbf{K}$. Clearly $Dim_{\mathbf{K}}\ V = Dim_{\mathbf{K}}\ \Xi = n-1$.

Conversely let $V = Zero(PS)$ be an irreducible variety of dimension $n-1$ in $\mathbf{K}^n$. Take any pol P from PS and write PS in the form $PS = PS' + \{P\}$. Let $P = P_1 * \cdots * P_r$ in which P_i are all irreducible pols in $\mathbf{K}[X]$. Then

$$\begin{aligned} V &= Zero(PS' + \{P\}) = Zero(PS' + \{P_1 * \cdots P_r\}) \\ &= Zero(PS' + \{P_1\}) + \cdots + Zer(PS' + \{P_r\}). \end{aligned}$$

As V is irreducible we have by Proposition 2.4

$$V \subset Zero(PS' + \{P_i\}) \subset Zero(P_i)$$

for some i. By the previous part $Zero(P_i)$ is irreducible and has $Dim_{\mathbf{K}}\ Zero(P_i) = n-1$. By hypothesis the irreducible varieties V and $Zero(P_i)$ have the same dimension over $\mathbf{K}$. By Theorem 2.5 we have therefore $V = Zero(P_i)$ as to be proved. □

Theorem 2.8. Any irreducible variety V of dimension 0 in $\mathbf{K}^n$ is a finite set of conjugate points over $\mathbf{K}$.

Proof. Let $\Xi = (\xi_1, \cdots, \xi_n)$ be a generic point of the irreducible variety V. Then $Dim_{\mathbf{K}}\ \Xi = 0$ so that Ξ has its defining polset of the form

$$\begin{aligned} P_1 &= x_1^{d_1} + f_{11} * x_1^{d_1 - 1} + \cdots + f_{1d_1}, \\ P_2 &= x_2^{d_2} + f_{21} * x_2^{d_2 - 1} + \cdots + f_{2d_2}, \\ &\cdots\cdots \\ P_n &= x_n^{d_n} + f_{n1} * x_n^{d_n - 1} + \cdots + f_{nd_n}, \end{aligned}$$

in which $f_{ij} \in \mathbf{K}(X_{i-1})$. The point Ξ is a zero of all these pols P_i. It follows that any point $\Xi' = (\xi'_1, \cdots, \xi'_n)$ of V, as a specialization over $\mathbf{K}$ of Ξ is also a zero of all these pols. Conversely, by Proposition 1.6 of Sect. 3.1, any solution Ξ' of these equations is a specialization over $\mathbf{K}$ of Ξ so that Ξ' is a point of V. It follows that the variety $V = Zero(PS)$ with $PS = \{P_1, \cdots, P_n\}$ is consisting of these points Ξ' which are $\Pi_i\ d_i$ in number and are conjugate to each other. This proves the theorem. □

Examples on Generic Points.

Example 2.5. The algebraic variety $V = Zero(x_1 * x_3 - x_2)$ in $\mathbf{K}^3(x_1, x_2, x_3)$ is irreducible over $\mathbf{K}$ with generic point $\left(x_1, x_2, \dfrac{x_2}{x_1}\right)$, in which x_1, x_2 are independent indeterminates. We have $Dim_{\mathbf{K}}\ V = 2$. If V is considered as an algebraic variety in $\mathbf{K}^n(x_1, \cdots, x_n)$ with $n > 3$, then V has a dimension $Dim_{\mathbf{K}}\ V = n - 1$ and has a generic point $\left(x_1, x_2, \dfrac{x_2}{x_1}, x_4, \cdots, x_n\right)$, with $x_1, x_2, x_4, \cdots, x_n$ independent indeterminates. Cf. Ex. 1.5.

Example 2.6. The variety $V = Zero(x_2^2 + x_1^2 - 1)$ in $\mathbf{K}^2(x_1, x_2)$ is irreducible with generic points $\left(\dfrac{2*t}{1+t^2}, \dfrac{1-t^2}{1+t^2}\right), \left(\dfrac{1-t^2}{1+t^2}, \dfrac{2*t}{1+t^2}\right)$, or more generally (x_1, ξ_2) with defining polset $\{x_2^2 + x_1^2 - 1\}$, the t and x_1 being both indeterminates. Cf. Ex. 1.4.

Irreducible Decomposition of a Variety.

Theorem 2.9. A sequence of proper subvarieties

$$V_1 \supset V_2 \supset \cdots$$

should terminate in a finite stage.

Proof. Let $V_i = Zero(PS_i)$ with $PS_i \subset PS_j$ if $i \leqslant j$ and PS be the union of all these PS_i. By Hilbert's finite-basis theorem there is a finite set $FS \subset PS$ such that $Zero(FS) = Zero(PS)$. Let FS be already contained in PS_k. Then the sequence should end at the stage k. □

Theorem 2.10. A variety V can be decomposed into the union of a finite set of proper irreducible subvarieties of V.

Proof. Suppose that the theorem is not true and V is such a variety which does not possess the property (P) below:

(P): The variety can be decomposed into the union of a finite set of proper irreducible subvarieties.

The variety V not possessing the property (P), should be non-irreducible. Let $V = V_1 + V_1'$ with V_1, V_1' both proper subvarieties of V. Then at least one of V_1, V_1', say V_1, should not possess the property (P). For the same reason V_1 should be reducible and will be of the form $V_1 = V_2 + V_2'$ with V_2, V_2' both proper subvarieties of V_1. As V_1 does not possess the property (P) at least one of V_2, V_2', say V_2, will not possess the property (P) too. The process can be continued indefinitely so that we will have an *infinite* sequence of proper subvarieties $V_1 \supset V_2 \supset \cdots$ in contradiction to Theorem 2.9. Hence no varieties not possessing the property (P) can exist and the theorem is thus proved. □

Definition 2.15. A decomposition of a variety V into subvarieties V_i of the form

$$V = V_1 + \cdots + V_m \tag{2.7}$$

is an *irreducible decomposition* of V := Each V_i is an irreducible proper subvariety of V. The decomposition (2.7) is *non-contractible* := Each V_i in the decomposition is not contained in the union of the other V_j's.

The Theorem 2.10 may be sharpened to the following form:

Theorem 2.11. Variety Decomposition Theorem. Any variety V has a non-contractible decomposition into the union of irreducible proper subvarieties.

Proof. By Theorem 2.10 any variety V has an irreducible decomposition of the form (2.7). If some V_i is contained in the union of the other V_j's, we may then delete this V_i from the union. This may be continued until some non-contractible decomposition is arrived. □

Theorem 2.12. The non-contractible irreducible decompositions of a variety into irreducible proper subvarieties are unique up to an order of the subvarieties.

Proof. Let (2.7) and (2.7)′ below be two non-contractible irreducible decompositions of the same variety V.

$$V = V_1' + \cdots + V_{m'}'. \tag{2.7$'$}$$

The irreducible subvariety V_1' is contained in V with a decomposition (2.7). By Proposition 2.4, V_1' should be contained in one of the V_i's, say V_{i_1}. Similarly, V_{i_1}, being contained in V with a decomposition (2.7)′, should be contained in one of the V_i', say V_k'. We have then $V_1' \subset V_k'$. As the decomposition (2.7)′ is non-contractible, it follows that $V_1' = V_k'$. As $V_1' \subset V_{i_1} \subset V_k'$ we have then $V_1' = V_{i_1}$. In the same way we see that $V_2' = V_{i_2}$ for some i_2 which should be different from i_1. This may be continued and we see finally that $m = m'$ and that the set $\{V_1, \cdots, V_m\}$ is just some rearrangement of the set of $\{V_1', \cdots, V_m'\}$. This proves the theorem. □

Definition 2.16. *Irreducible components* of a variety V := The subvarieties V_i occuring in the non-contractible irreducible decomposition of V in the form of (2.7).

Definition 2.17. *Dimension over* **K** of a variety V := The maximum of the dimensions over **K** of the irreducible components of V.

Notation 2.8. *Dimension over* **K** of a variety V := $Dim_{\mathbf{K}}\, V$.

Problems.

Remark 2.4. All the theorems and their proofs in this section are merely *existential* in character. For example, for a variety $Zero(PS)$ it is not clear whether it is irreducible or not or whether a generic point does exist from the knowledge of PS. We set forth therefore some problems which will be settled or studied in latter sections.

Let $PS, PS' \subset \mathbf{K}[X]$.

Problem V1 Decide whether an algebraic variety $Zero(PS)$ is irreducible over $\mathbf{K}$ or not.

Problem V2. If $Zero(PS)$ is irreducible over $\mathbf{K}$, find a generic point of that variety.

Problem V3. If $Zero(PS)$ is non-irreducible, determine a non-contractible irreducible decomposition of $Zero(PS)$.

Problem V4. Decide whether $Zero(PS) \subset Zero(PS')$ or not.

Problem V5. For a given extended point Ξ in $\mathbf{K}^n$ determine a polset $PS \subset \mathbf{K}[X]$ such that the algebraic variety $Zero(PS)$ in $\mathbf{K}^n$ has Ξ as a generic point.

Problem V6. Study the structure of the variety $Zero(PS)$ via the polset PS.

Problem V7. Study the computational aspects of the variety $Zero(PS)$ via the polset PS.

Projective Varieties in a Projective Space.

Let the notations of projective space, homogenized polynomial, etc. be as in the last section Sect. 3.1.

Definition 2.18. A *homogeneous zero* of a homogeneous pol ${}^hP \in \mathbf{K}[{}^hX] :=$ Any extended point ${}^hX'$ with $X' \in P^n_{\mathbf{K}^*}({}^hX)$ for some extension field $\mathbf{K}^*$ of $\mathbf{K}$ such that ${}^hP({}^hX') = 0$. A *homogeneous polset* := A polset consisting of homogeneous pols alone. A *homogeneous zero* of a homogeneous polset := A common homogeneous zero of all homogeneous pols in the polset.

Notation 2.9. For a homogeous polset ${}^hPS \subset \mathbf{K}[{}^hX]$ the set of all homogeneous zeros of ${}^hPS := Zero^h({}^hPS)$.

Definition 2.19. A *projective variety* in $P^n_{\mathbf{K}}({}^hX)$ with *basis* ${}^hPS \subset \mathbf{K}[{}^hX] :=$ The set $Zero^h({}^hPS)$ for some homogeneous polset ${}^hPS$.

Example 2.7. Consider the pols $P_1, P_2 \in \mathbf{K}[x_1]$ in Ex. 2.4 again. Let the resultant of the homogenized pols ${}^hP_1, {}^hP_2$ w.r.t. x_1 be R which is actually the same as R in Ex. 2.4. Then we have

$$Zero^h({}^hP_1, {}^hP_2) \neq \varnothing <===> R = 0.$$

Remark 2.5. In comparing with (2.4) we see that the zero-sets may become simpler in structure if we consider homogeneous zeros of homogenized pols instead of the original ones. We remark that this is also the underlying reason why the Specialization-Extension Theorem (Proposition 1.11) holds in the projective case but not so in the affine case.

Remark 2.6. In projective spaces we may define lines, linear subspaces, Pluecker coordinates, etc. in the usual way. We may also define *generic* lines, *generic* linear subspaces, etc. in projective spaces. We refer these to some classical texts on algebraic geometry, e.g. [VdW2] and [H-P1,2]. Clearly these notions can also be defined in an evident manner for affine spaces.

3.3 Polsets and Ascending Sets. Partial Ordering.

Natural Ordering of Variables.

Notation 3.1. $\mathbf{K}$ and $X = (x_1, x_2, \cdots, x_n)$, $X_k = (x_1, \cdots, x_k)$ as before.

For the variables $x_1, \cdots, x_n$ it will be understood throughout the whole book that the following natural ordering will be taken unless otherwise stated:

$$x_1 \prec x_2 \prec \cdots \prec x_n.$$

Pols and polsets below will be understood to be in $\mathbf{K}[X]$ unless otherwise stated.

Some Indices of a Pol.

Notation 3.2. For a non-zero pol $P \in \mathbf{K}[X], deg_{x_i}(P) :=$ Degree of P in x_i.

Definition 3.1. For a non-zero pol $P \in \mathbf{K}[X]$:

Class of $P :=$ Greatest subscript c for which x_c occurs actually in P, otherwise 0 if P is a non-zero constant of $\mathbf{K}$. In notation: $cls(P)$.

Leading variable of $P := x_c$ if class of P is $c > 0$, otherwise undefined. In notation: $lvar(P)$,

Degree of $P :=$ Degree of P in x_c if class of P is $c > 0$, otherwise $= 0$. In notation: $deg(P)$.

Length of $P :=$ Number of terms of P. In notation: $t(P)$.

Index-Set of $P := [t\ c\ d]$ with $t = t(P), c = cls(P), d = deg(P)$, sometimes also $[t\ u\ d]$, with u standing for the variable x_c.

Normal Form of a Non-Constant Pol P.

Definition 3.2. For a non-constant pol P of class $c > 0$, P is in *normal form* :=

$$P = I * x_c^d + \text{ lower degree terms in } x_c, \tag{3.1}$$

in which

$$c = cls(P) > 0, \quad d = deg(P) > 0,$$

and coefficients of x_c-powers all belong to $\mathbf{K}[X_{c-1}]$.

Definition 3.3. For non-constant pol P in normal form (3.1), $I :=$ *Initial* of P. In notation: $init(P)$.

Definition 3.4. For non-constant pol P of class $c > 0$, formal derivative $\dfrac{\partial P}{\partial x_c}$ $:=$ *Separant* of P. In notation: $sep(P)$.

Convention. For P a non-zero constant in $\mathbf{K}$, we define $init(P) = 1$, while $sep(P)$ will be undefined.

Partial Ordering of Pols.

Let P, Q be non-zero pols.

Definition 3.5. P is of *lower ordering* than Q or Q is of *higher ordering* than $P :=$ Either $cls(P) < cls(Q)$ or $cls(P) = cls(Q)$ and $deg(P) < deg(Q)$. In notation: $P \prec Q$ or $Q \succ P$.

Definition 3.6. P and Q are of *same ordering* $:=$ Neither is $P \prec Q$ nor is $P \succ Q$. In notation: $P \sim Q$.

Notation 3.3. $P \preceq Q :=$ Either $P \sim Q$ or $P \prec Q$.

Proposition 3.1. The binary relation $\prec$ introduces a partial ordering among all non-zero pols in $\mathbf{K}[X]$.

Proof. It is sufficient to prove the transitivity relations that for any non-zero pols P_1, P_2, P_3 we have

$$\begin{aligned}
P_1 \prec P_2,\ P_2 \prec P_3 &===> P_1 \prec P_3,\\
P_1 \sim P_2,\ P_2 \prec P_3 &===> P_1 \prec P_3,\\
P_1 \prec P_2,\ P_2 \sim P_3 &===> P_1 \prec P_3,\\
P_1 \sim P_2,\ P_2 \sim P_3 &===> P_1 \sim P_3.
\end{aligned}$$

Clearly all these follow directly from the definition. □

Definition 3.7. Q is *reduced* w.r.t. $P := cls(P) = c > 0$ and $deg_{x_c}(Q) < deg(P)$.

Convention. For P a non-zero constant in $\mathbf{K}$ every non-zero pol is *non-reduced* w.r.t. P. The constant 0 as a pol is considered as reduced w.r.t. any non-zero pol P, constant or not.

Ascending Set and Triangulated Set.

Definition 3.8. A polset AS is an *ascending set* $:=$ Its pols are all non-zero and can be so arranged in a sequence

$$AS: \quad A_1, A_2, \cdots, A_r \tag{3.2}$$

such that (a_1) and (a_2) below hold true:

(a_1) $cls(A_1) < cls(A_2) < \cdots < cls(A_r)$.

(a_2) $init(A_j)$ is reduced w.r.t. A_i for each pair $i < j$.

The sequence (3.2) is then said to be *in ascending order* .

Abbreviation 3.1. *Asc-set* := Ascending set.

Definition 3.8′. A polset AS is a *triangulated set* or a *weak ascending set* := Its pols are all non-zero and can be so arranged in a sequence (3.2) such that (a_1) alone holds true with the sequence (3.2) in *ascending order.*

Definition 3.9. *Trivial* asc-set or triangulated set := A polset consisting of only one non-zero constant pol.

Remark 3.1. In (3.2) if A_1 is a non-zero constant in $\mathbf{K}$ then there can be no more any of $A_2, \cdots, A_r$ and the asc-set AS is necessarily a trivial one. For a non-trivial asc-set the condition (a_1) can thus be replaced by

$(a_1)'$ $0 < cls(A_1) < cls(A_2) < \cdots < cls(A_r)$.

Definition 3.10. *Initial-product* of a non-trivial asc-set or a triangulated set AS := Product of all initials of pols in AS. In notation: $IP(AS)$.

Definition 3.11. *Initial-separant-product* of a non-trivial asc-set or a triangulated set AS := Product of all initials and separants of pols in AS. In notation: $ISP(AS)$.

<u>*Partial Ordering of Asc-Sets.*</u>

Let

$$\begin{aligned} AS' : &\quad A'_1, A'_2, \cdots, A'_{r'}, \\ AS'' : &\quad A''_1, A''_2, \cdots, A''_{r''}, \end{aligned}$$

be two asc-sets in ascending order.

Definition 3.12. AS' is of *lower ordering* than AS'' or AS'' is of *higher ordering* than AS' := Either (o_1) or (o_2) below holds true:

(o_1) There is some k such that

$$A'_1 \sim A''_1, \ \cdots, \ A'_{k-1} \sim A''_{k-1}, \text{ while } A'_k \prec A''_k.$$

(o_2) $r' > r''$ and

$$A'_1 \sim A''_1, \ \cdots, \ A'_{r''} \sim A''_{r''}.$$

In notation: $AS' \prec AS''$ or $AS'' \succ AS'$

Definition 3.13. The asc-sets AS', AS'' are of *same ordering* := Neither is $AS' \prec AS''$, nor is $AS'' \prec AS'$. In notation: $AS' \sim AS''$

Notation 3.4. $AS' \preceq AS''$:= Either $AS' \prec AS''$ or $AS' \sim AS''$.

Theorem 3.1. The binary relation $\prec$ introduces a partial ordering among all asc-sets.

Proof. It is sufficient to consider the transitivity of the ordering relation which is then proved by considering various different possible cases according to the rules (o_1) or (o_2). □

Remark 3.2. Besides the notion of triangulated set there are other variants for the notion of asc-set in replacing the condition (a_2) by either more stringent or less stringent condition. We shall however adopt the present notion throughout the whole book.

Definition 3.14. A pol Q is *reduced* w.r.t. an asc-set $AS := Q$ is reduced w.r.t. all pols in AS.

Theorem 3.2. The ordering of an asc-set will be lowered by adjoining to it a pol reduced w.r.t. it. More precisely, by adjoining a pol reduced w.r.t. to an asc-set, there may be extracted an asc-set of lower ordering than the original one.

Proof. For each trivial asc-set all non-zero pols are non-reduced w.r.t. it. Hence we may restrict ourselves to non-trivial asc-set as given in (3.2) and consider a non-zero pol R reduced w.r.t. it. Suppose first that $cls(R) < cls(A_1)$. Then R alone forms an asc-set clearly of lower ordering than AS. Suppose next $cls(R) = cls(A_k) = c_k$ for some $k > 0$ and $\leqslant r$. As R is reduced w.r.t. AS we have necessarily $deg_{c_k}(R) = deg(R) < deg(A_k)$ so that $R \prec A_k$. Then the asc-set

$$AS' : A_1, \cdots, A_{k-1}, R$$

is of lower ordering than the given AS. Suppose finally $cls(R) > cls(A_k), k \leqslant n$, but $\leqslant cls(A_{k+1})$ for $k < n$, then the asc-set

$$AS' : A_1, \cdots, A_k, R$$

is again of lower ordering than AS. This proves the theorem. □

Remark 3.3. This theorem gives a means of getting asc-sets of lower ordering than a given asc-set by simply adjoining reduced pols to it. It also shows the role of condition (o_2) in the partial ordering of asc-sets. As any initial of a pol in the asc-set AS is reduced w.r.t. AS, we may take R to be any such initial to get an asc-set of lower ordering. This also shows the role of condition (a_2) in the definition of an asc-set.

Theorem 3.3. A sequence of asc-sets steadily lower in ordering is finite.

Proof. Let us consider a sequence of asc-sets

$$AS_1 \succeq AS_2 \succeq \cdots \succeq AS_q \succeq \cdots \tag{3.3}$$

It is then sufficient for us to prove that for such a sequence there will be some index q such that from q onwards all the succeeding asc-sets are of same ordering. To see this let each AS_q be given in ascending order by

$$AS_q:\ A_{q1},\cdots,A_{qr_q}, \tag{3.4}$$

in which all $r_q \leqslant n$ and $\geqslant 1$. We put $max(r_q) = r$ and shall make induction on r. Consider the sequence consisting of the first pols of these asc-sets, viz.

$$A_{11}, A_{21},\cdots,A_{q1},\cdots. \tag{3.5}$$

Then each pol in the sequence is of higher or same ordering than the following ones. Hence the class of these pols is never increasing so after some stage they will always be the same, say of same class c_1. After that stage the degree in x_{c_1} of the pols will then be never increasing so that after some further stage all the pols will be of same degree in x_{c_1} and thus of same ordering. Let this stage be q_1. If $r = 1$ then all the asc-sets in (3.4) from q_1 onwards will be of same ordering and the theorem is thus proved for $r = 1$. Suppose now the theorem is true for $r = k \geqslant 1$ and let us consider a sequence (3.4) for which $r = k + 1$. Let AS'_q be the asc-set gotten from AS_q by deleting the first pol A_{q1}. Let q_1 be the stage determined from (3.5) as above. Then we have

$$AS'_{q_1} \succeq AS'_{q_1+1} \succeq \cdots \succeq AS'_{q'} \succeq \cdots \tag{3.6}$$

for which the corresponding r_q and r are $r'_q = r_q - 1$ and $r' = max(r'_q) \leqslant k$. If from some stage onwards all $r'_{q'}$ are zero, then the asc-sets in the given sequence (3.3) will have all the same ordering after that stage and the theorem is already proved. Suppose that this is not the case. Then let us delete all such AS'_q in (3.6) for which $r'_q = 0$. We get still an infinite sequence of asc-sets for which by induction hypothesis all the asc-sets after some stage will have all the same ordering. This is then also true for the original sequence and the theorem is thus proved. □

Partial Ordering of Polsets.

Definition 3.15. *Basic set* of a polset PS := Any asc-set of lowest ordering contained in PS.

Abbreviation 3.2. *Bas-set* := Basic-set.

Theorem 3.4. Any two basic sets of a polset are of same ordering.

Proof. This follows directly from the definition. □

From Theorem 3.4 we see that the following definition is legitimate.

Definition 3.16. For any two polsets PS', PS'' let BS', BS'' be any basic sets of PS', PS'' respectively. Then PS' is of *lower, higher ordering* or not of *higher* *ordering* than PS'' or of *same ordering* as $PS'' := BS'$ is of lower, higher ordering or not of higher ordering than BS'' or of same ordering as BS''. In notation: $PS' \prec PS'', PS' \succ PS'', PS' \preceq PS''$ or $PS' \sim PS''$.

The following Theorems 3.5-3.7 follow directly from the corresponding theorems about asc-sets.

Theorem 3.5. The binary relation $\prec$ introduces a partial ordering among all polsets.

Theorem 3.6. The ordering of a polset will be lowered by adjoining to it a pol reduced w.r.t. a basic set of the polset.

Theorem 3.7. A sequence of polsets steadily lower in ordering is finite.

Remainder and Remainder Formula.

Let F, G be non-zero pols and $c = cls(F), I = init(F)$. By simple division we will have for some non-negative integer s and some pols Q, R

$$I^s * G = Q * F + R, \tag{3.7}$$

with R reduced w.r.t. F. The exponent s will be chosen to be as small as possible to render such a formula holds. The pol R is then uniquely determined.

Definition 3.17. R in (3.7) with s as small as possible := *Remainder* of G w.r.t. F. In notation: $R = Remdr(G/F)$.

Remark 3.4. For $R \neq 0$ with some x_e appearing in R for which $e > c$, we have $deg_{x_e}(R) = deg_{x_e}(G)$.

Remark 3.5. In (3.7) if $c = 0$ or F is a non-zero constant in $\mathbf{K}$ then $R = 0$ with $s = 0, Q = \frac{G}{F}$ so that $Remdr(G/F) = 0$. If $c > 0$ and $deg_{x_c}(G) < deg(F)$, then $s = 0, Q = 0$ and $R = G$ so that $Remdr(G/F) = G$ with G reduced w.r.t. F.

Remark 3.6. The remainder concept here, originated from works of J.F.Ritt, is different from the *pseudo-remainder* concept in the current literature, for which s is taken to be $max(deg_{x_c}(G) - deg(F) + 1, 0)$ for $cls(F) = c > 0$. Throughout the present book the remainder is always understood to be in the Ritt sense, unless otherwise stated.

Theorem 3.8. For an asc-set (or triangulated set) AS in ascending order given by (3.2) and any non-zero pol G, there is a uniquely determined pol R reduced w.r.t. AS such that

$$\begin{aligned} &I_G * G \equiv R \; mod_{\mathbf{K}} \; AS, \text{ or} \\ &I_G * G = \Sigma_i \, Q_i * A_i + R, \end{aligned} \tag{3.8}$$

for some pols Q_i and some power-product I_G of initials of pols A_i in AS.

Proof. Let I_i be the initials of the pols A_i in (3.2) and let us set $R_r = G$. By forming successive remainders of G w.r.t. pols A_i of (3.2) in reverse order in applying (3.7), we get successively unique non-negative integers s_i, pols $Q'_i, R_i, i = r, r-1, \cdots, 2, 1$ for which R_i are the corresponding remainders verifying the equations below:

$$\begin{aligned} I_r^{s_r} * R_r &= Q'_r * A_r + R_{r-1}, \\ I_{r-1}^{s_{r-1}} * R_{r-1} &= Q'_{r-1} * A_{r-1} + R_{r-2}, \\ &\cdots\cdots \\ I_1^{s_1} * R_1 &= Q'_1 * A_1 + R_0. \end{aligned} \tag{3.9}$$

Set $R_0 = R$. Then we get an expression of the form

$$I_1^{s_1} * \cdots * I_r^{s_r} * G = \Sigma_i\, Q_i * A_i + R,$$

in which each Q_i is the sum of products of the Q'_j with some power-products of the initials I_k. This is just the formula (3.8). Clearly R is uniquely determined since each R_i is so. By Remark 3.4 we see that R is reduced w.r.t. AS. □

Definition 3.18. The uniquely determined pol R in (3.8) := The *remainder* of G w.r.t. AS. In notation: *Remdr*(G/AS). The formula (3.8) := The *remainder formula* of G w.r.t. AS.

Definition 3.19. For a polset PS the set of remainders of pols in PS w.r.t. the asc-set AS := The *remainder-set* of PS w.r.t. AS. In notation: *Remdr*(PS/AS).

Remark 3.7. For AS an asc-set the remainder R in (3.8) is reduced w.r.t. the asc-set AS. In view of Theorem 3.6 we see that the ordering of a polset will be lowered by adjoining some pol to it which is the non-zero remainder of some pol w.r.t. any basic set of the polset. This forms the underlying principle of the method to be described in the next section which forms the corner-stone of our whole theory.

Squared-Remainder and Squared-Remainder Formula.

For applications on optimization problems and others in latter chapters for which $\mathbf{K} = \mathbf{R}$ we shall modify a little the notions of remainder and remainder formula as follows.

In (3.7) let us restrict s to be a non-negative *even* integer $2 * s'$ chosen to be as small as possible so that (3.7) becomes

$$I^{2*s'} * G = Q' * F + R', \tag{3.7$'$}$$

we may also replace (3.8) by (3.8)$'$ below:

$$\begin{aligned} &I'_G * G \equiv R' \ mod_{\mathbf{K}}\ AS, \text{ or} \\ &I'_G * G = \Sigma_i\, Q'_i * A_i + R'. \end{aligned} \tag{3.8$'$}$$

In $(3.8)'$ I'_G is some product of *even* powers of initials of A_i in AS. Then corresponding to Definition 3.18 we shall lay down the following Definition:

Definition 3.18′. The uniquely determined pol R' in $(3.7)'$:= The *squared-remainder* of G w.r.t. AS. In notation: $Sq\text{-}Remdr(G/AS)$. The formula $(3.8)'$:= The *squared-remainder formula* of G w.r.t. AS.

Some Examples.

Example 3.1. Consider the example of Zhu in Sect. 1.2 of Chap. 1. We have the polset $PS = \{P_1, P_2, P_3\}$ with P_i given below:

$$\begin{aligned}P_1 &= x*y*z - x*y^2 - z - x - y,\\ P_2 &= x*z - x^2 - z - y + x,\\ P_3 &= z^2 - x^2 - y^2.\end{aligned}$$

In accordance to the method of Zhu in solving the corresponding equations, the variables are to be arranged in the order

$$z \prec x \prec y.$$

This amounts to set

$$z = x_1, \ x = x_2, \ y = x_3.$$

We have now

$$\begin{aligned}cls(P_1) &= 3, \ deg(P_1) = 2,\\ cls(P_2) &= 3, \ deg(P_2) = 1,\\ cls(P_3) &= 3, \ deg(P_3) = 2\end{aligned}$$

The only asc-sets contained in PS are thus 3 in number, viz.

$$AS_1 = \{P_1\}, \ AS_2 = \{P_2\}, \ AS_3 = \{P_3\}.$$

The only basic set is clearly the asc-set AS_2.

Example 3.2. Consider the polset $PS = \{P_1, P_2, P_3\}$ with P_i given below:

$$P_i = a_i * x + b_i * y + c_i * z - d_i, \ i = 1, 2, 3$$

in which all a_i, b_i, c_i are non-zero constants in $\mathbf{K}$. Then whatever the ordering of the variables x, y, z may be, there are 3 asc-sets $AS_i, i = 1, 2, 3$ with

$$AS_i = \{P_i\}, \ i = 1, 2, 3.$$

It is clear that any one of these asc-sets is a bas-set of PS. It follows that the basic-set of a polset is in general not unique.

Some Theorems of Hu and Wang.

Let U be a parameter set $(u_1, \cdots, u_d)$ which may eventually be empty. Consider now an asc-set AS over the field $\mathbf{K} = \mathbf{Q}$ given by (3.2) in ascending order with $cls(A_i) = c_i$, the U being treated as parameters.

Notation 3.5. In AS given by (3.2):

$X' :=$ Set of bounded variables $x_{c_i}, i = 1, \cdots, r$.

$X'' :=$ Set of free variables $X \setminus X'$.

Definition 3.20. The asc-set (3.2) over $\mathbf{K}$ is *complete* (respectively *incomplete* w.r.t. X) $:= X'' = \varnothing$ (respectively $\neq \varnothing$).

The following Theorems are actually due to Hu and Wang for the factorization of multivariate pols over arbitrary extension fields to be considered in latter section Sect. 4.5. Cf. [H-W] and [WDM1].

Theorem 3.9. Let $AS = \{A_1, \cdots, A_r\}$ be an asc-set in $\mathbf{Q}[X, U]$ with initial-product IP. If AS is incomplete w.r.t. X then either $Zero_{\mathbf{Q}}(AS/IP)$ is empty or there will be values $U = U^0$ in $\mathbf{Q}$ such that $Zero_{\mathbf{C}}(AS/IP)$ for which $U = U^0$ is infinite.

Proof. Suppose that $Zero_{\mathbf{Q}}(AS/IP) \neq \varnothing$ with such a zero for which $U = U^0$, $X'' = X''^0$. Then from continuity of roots of an algebraic equation on its coefficients $Zero_{\mathbf{C}}(AS/IP)$ will have zeros for which $U = U^0$ while X'' is arbitrarily near to X''^0. As X'' is non-empty the set $Zero_{\mathbf{C}}(AS/IP)$ for which $U = U^0$ will be an infinite set. □

Theorem 3.10. Let AS, IP be as in Theorem 3.9. If AS is complete then either $Zero_{\mathbf{Q(U)}}(AS/IP) = \varnothing$ or is finite and may be determined by Gauss Lemma.

Proof. This follows directly from the definition and the Gauss Lemma. □

3.4 Characteristic Set of a Polset and Well-Ordering Principle.

Problem, Principle, and Method.

In what follows $\mathbf{K}, X = (x_1, \cdots, x_n), X_k = (x_1, \cdots, x_k)$, etc. are as before. Pols and polsets are understood to be in $\mathbf{K}[X]$. By the *zero-structure* of a polset PS is meant the structure of the zero-set of PS, i.e. the structure of the variety $Zero(PS)$.

Problem Z. To study the structure of the variety $Zero(PS)$ via its basis PS.

Principle and Method. The zero-structure of an asc-set may be considered as well-determined. So we shall study the zero-structure of a polset PS via the zero-structures of its asc-sets of lowest ordering, i.e. its basic sets. By Theorems 3.2

and 3.6, the basic sets of the polset will be lowered in ordering by adjoining non-zero remainders of pols in the polset w.r.t. the basic sets. This suggests the method of the formation of basic sets of an ordering as lower as possible in adjoining to the polset non-zero remainders as much as possible until some asc-set of special type, to be called a *characteristic set* below is arrived.

Characteristic Set and Scheme of its Formation.

Let a polset PS be given whose zero-structure is to be studied. In accordance to the Principle and Method as described above let us first set $PS = PS_0$ and choose an arbitrary basic set BS_0 of PS. If there are still pols in PS_0 but not in BS_0, let us form the remainders of these pols w.r.t. BS_0. Suppose that there are non-zero remainders which will form a non-empty set RS_0. Then we form a new polset $PS_1 = PS_0 + BS_0 + RS_0$. Choose now an arbitrary basic set BS_1 of PS_1. By Theorems 3.2 and 3.6 $BS_0 + RS_0$ is of lower ordering than BS_0 and hence PS_1 or BS_1 will be of lower ordering than BS_0. Form now again the set RS_1 of all non-zero remainders, if there are any, of remaining pols of PS_1 w.r.t. BS_1. Suppose RS_1 is non-empty. Then we form the polset $PS_2 = PS_0 + BS_1 + RS_1$, as well as basic set BS_2 and remainder set RS_2 as before and proceed further in the same way so far the successive remainder sets are non-empty. We have then

$$BS_0 \succ BS_1 \succ BS_2 \succ \cdots .$$

By Theorem 3.3 the sequence can only be a finite one so that up to a certain stage m we should have $RS_m = \varnothing$. The final basic set BS_m is then an asc-set specifically related to the given polset PS which will be denoted by CS.

The above procedure can be exhibited in the form of the scheme (S) below:

$$\begin{array}{rllclcl} PS = & PS_0 & PS_1 & \cdots & PS_i & \cdots & PS_m \\ & BS_0 & BS_1 & \cdots & BS_i & \cdots & BS_m = CS \\ & RS_0 & RS_1 & \cdots & RS_i & \cdots & RS_m = \varnothing. \end{array} \tag{S}$$

In the scheme (S) all the remainder sets RS_k are non-empty for $k < m$ and we have put

$$\begin{aligned} BS_i &= \text{ a basic set of } PS_i, \\ RS_i &= Remdr(PS_i \setminus BS_i \ / \ BS_i) \setminus \{0\}, \\ PS_i &= PS_0 + BS_{i-1} + RS_{i-1}. \end{aligned} \tag{4.1}$$

Theorem 4.1. The procedure in the scheme (S) will end in a finite number of steps at some stage m with corresponding remainder set $RS_m = \varnothing$. The corresponding basic-set $BS_m = CS$ verifies then the properties (c_1) and (c_2) below:

(c_1) $Remdr(PS/CS) = \{0\}$.

(c_2) $Zero(PS) \subset Zero(CS)$.

Proof. We have already proved that the procedure should end in a finite number of steps so that at some stage say m we have $RS_m = \varnothing$ as in the scheme (S). To see that $CS = BS_m$ verifies (c_1) and (c_2) let us first remark that $CS \subset PS_m$ and whence

$$Zero(PS_m) \subset Zero(CS). \tag{4.2}$$

According to the procedure we have also

$$Remdr(PS_m/CS) = \{0\}. \tag{4.3}$$

In particular (c_1) is true since $PS = PS_0 \subset PS_m$. To prove (c_2) let us first prove

$$Zero(PS) = Zero(PS_0) = Zero(PS_k),\ k = 1, \cdots, m, \tag{4.4}$$

by induction on k as follows.

Remark that $PS_1 = PS_0 + BS_0 + RS_0$ and consider any pol $R_0 \in RS_0$ which is a non-zero remainder of some pol $P_0 \in PS_0 \setminus BS_0$ w.r.t. BS_0. By remainder formula (3.8) we will have an identity of the form

$$I_{P_0} * P_0 = \Sigma_i\, Q_i * B_i + R_0,$$

in which B_i are the pols in BS_0 and I_{P_0} is some power-product of initials of the pols B_i. From this formula it follows that any zero of PS_0 and BS_0 is also a zero of RS_0 so that

$$Zero(PS_1) = Zero(PS_0 + BS_0 + RS_0) = Zero(PS_0 + BS_0) = Zero(PS_0).$$

This proves (4.4) for $k = 1$. Suppose we have already proved (4.4) for k and let us consider the polset $PS_{k+1} = PS_0 + BS_k + RS_k$. As before from the remainder formula about pols in $PS_k \setminus BS_k$ w.r.t. BS_k we have $Zero(PS_k + BS_k + RS_k) = Zero(PS_k + BS_k) = Zero(PS_k)$. As $Zero(PS_0) = Zero(PS_k)$ we have then

$$\begin{aligned} Zero(PS_{k+1}) &= Zero(PS_0 + BS_k + RS_k) = Zero(PS_k + BS_k + RS_k) \\ &= Zero(PS_k) = Zero(PS_0). \end{aligned}$$

Consequently (4.4) is also true for $k + 1$. It follows from (4.2) that (c_2) is also true for $PS = PS_0$ ($\subset PS_m$). □

Characteristic Set :

Definition 4.1. A *characteristic set* of a polset PS := Any asc-set CS gotten from PS by means of procedures in accordance to some scheme (S). More generally a *characteristic set* of a polset PS := Any asc-set CS verifying the properties (c_1) and (c_2) in Theorem 4.1. Pols in CS := *Chararacteristic pols* of CS.

Abbreviation 4.1. *Char-set* := characteristic set. *Char-pols* := Characteristic pols.

Remark 4.1. If the char-set CS is a trivial one so that it consists of a single non-zero constant, then $Zero(PS) = \varnothing$.

Remark 4.2. For a char-set CS of a polset PS, it is not required that pols in CS should belong to PS. However, from the procedure we see that all pols CS are in the ideal generated by PS as basis.

Remark 4.3. The char-set CS of a polset PS as determined here is not unique since the choice of basic sets in each step of the scheme (S) is not unique. Cf. e.g. Ex. 3.2.

Remark 4.4. The notion and terminology of characteristic set had their origin in the works of J.F.Ritt, cf. [Ritt1,2]. However, as pointed out to the present author by D.Kapur, Ritt defined his characteristic set only for ideals while the characteristic set here is defined for polsets, or for basis of ideals.

Remark 4.5. In the original procedure of scheme (S) we have used instead of $(4.1)_3$ the relation $PS_i = PS_{i-1} + BS_{i-1} + RS_{i-1}$. It is owing to suggestions of Wu Wen-da and Zhi Lihong we adopt the above change which greatly simplifies the computations.

Char-Set Algorithm :

From the scheme (S) a char-set CS of a polset PS will be determined according to the following algorithm.

Algorithm Char-set:

Input: A polset PS.

Output: A char-set CS of PS.

Step 1. Set $PS_0 \longleftarrow PS$

Step 2. Take a basic set BS of PS_0.

Step 3. Form $RS = Remdr(PS_0 \setminus BS/BS) \setminus \{0\}$.

Step 4. If $RS = \varnothing$ then $CS \longleftarrow BS$ and return. Otherwise $PS_0 \longleftarrow PS + BS + RS$ and goto Step 2.

The Step 2 will be achieved by the algorithm below:

Algorithm Bas-Set:

Input: A polset PS.

Output: A basic set BS of PS.

Step 1. Set $PS' = PS$ and $BS = \varnothing$.

Step 2. If $PS' = \varnothing$ then return BS. Otherwise take from PS' a pol B of least class and least degree and set $BS \longleftarrow BS + \{B\}$.

Step 3. Set $PS' \longleftarrow$ the set of pols in PS' which are reduced w.r.t. B.

Step 4. Goto Step 2.

The Well-Ordering Principle :

From Theorem 4.1 may be deduced the Theorem 4.2 below which shows a close connection between the zero-structures of the given polset and its characteristic sets. The theorem will occupy a central position of our whole theory. For this reason we give the theorem the name of a general principle: the Well-Ordering Principle. It will be seen later that this general principle will play important roles in diverse applications.

Theorem 4.2 (Well-Ordering Principle). Let CS be a char-set of a polset PS. Then:

$$\begin{aligned} &Zero(CS/IP) \subset Zero(PS) \subset Zero(CS), \\ &Zero(PS/IP) = Zero(CS/IP), \\ &Zero(PS) = Zero(CS/IP) + \textstyle\bigcup_i Zero(PS + \{I_i\}). \end{aligned} \tag{4.5}$$

In (4.5) I_i are initials of pols C_i in CS, and IP is the initial-product of CS. Moreover, CS is not of higher ordering than PS.

Proof. For any pol $P \in PS$ we have $Remdr(P/CS) = 0$ by property (c_1). Hence by the remainder formula (3.8) for P we have

$$I_P * P = \Sigma_i\, Q_{iP} * C_i, \tag{4.6}$$

in which Q_{iP} are pols and I_P is some power product of I_i. It follows that

$$Zero(CS/IP) \subset Zero(CS/I_P) \subset Zero(P).$$

As this is true for all P in PS, we have

$$Zero(CS/IP) \subset Zero(PS).$$

By property (c_2) we have then $(4.5)_1$ and also $(4.5)_2$. Since

$$Zero(PS) = Zero(PS/IP) + Zero(PS + \{IP\}),$$

we get readily $(4.5)_3$ from $(4.5)_2$. As the final assertion is clear from the very constructions of CS, the theorem is completely proved. □

A Variant of the Char-Set Algorithm.

An alternative procedure for the determination of a char-set according to the diagram (S) is to apply instead of $(4.1)_3$ the following (4.7):

$$PS_i = BS_{i-1} + RS_{i-1}. \tag{4.7}$$

With (4.7) the procedure will still stopped at a certain finite stage say m with $RS_m = \varnothing$. The final basic-set $BS_m = CS$ will not obey in general the properties $(c_1), (c_2)$ and hence the properties (4.5). It will be so only under conditions (4.8) below as in the following theorem.

Theorem 4.3. In the scheme (S) using (4.7) instead of $(4.1)_3$ the final basic-set $BS_m = CS$ will verify the properties $(c_1), (c_2)$ and hence is a char-set of PS if and only if the following is true:

$$Remdr(PS/CS) \ = \ \{0\}. \tag{4.8}$$

Proof. The proof is actually the same as that of Theorem 4.1 and will be omitted. □

Suppose now that (4.8) is not true. Let us form then a new polset

$$PS' = CS + RS, \text{ where}$$

$$RS = Remdr(PS/CS) \setminus \{0\},$$

and apply the char-set algorithm to PS' again, in using always (4.7) instead of $(4.1)_3$. This procedure will be carried further and further as in the diagram below:

$$\begin{aligned} PS \longrightarrow_{(S)} CS \longrightarrow PS' \longrightarrow_{(S)} CS' \longrightarrow \cdots \\ \longrightarrow_{(S)} CS^{(i-1)} \longrightarrow PS^{(i-1)} \longrightarrow_{(S)} CS^{(i)} \longrightarrow \cdots . \end{aligned} \tag{S$'$}$$

In the diagram we have put

$$\begin{aligned} PS^{(i)} &= CS^{(i-1)} + RS^{(i-1)}, \\ RS^{(i)} &= \{Remdr(PS/CS^{(i)}) \setminus \{0\}. \end{aligned} \tag{4.9}$$

Theorem 4.4. The procedure $(\mathrm{S})'$ has to come to an end after a finite number of steps when $RS^{(k)} = \varnothing$ for some k. Then $CS^{(k)}$ verifies the properties $(c_1), (c_2)$ and hence is a char-set of PS.

Proof. This follows from Theorems 3.2 and 3.6 as in the proof of Theorem 4.1. □

Definition 4.2. *Verification* for the procedure in scheme (S)′ using (4.7) instead of $(4.1)_3$:= The verification whether (4.8) is true.

Remark 4.6. Theorem 4.3 shows that in using (4.7) instead of $(4.1)_3$ the final basic set CS will already be a char-set to be sought if the verification is in the affirmative. Only in case that the verification is in the negative then we have to proceed further according to the scheme (S)′.

Remark 4.7. For practical computations it seems that quite often the procedure of scheme (S)′ is to be preferred than that of scheme (S).

It is clear how to duly modify the char-set algorithm when (4.7) is used instead of $(4.1)_3$.

An Extended Form of Well-Ordering Principle for (4.7).

In case of diagram (S) with (4.7) instead of $(4.1)_3$ the formulae (4.5) are no more true, but can be replaced by more complicate ones. In particular $(4.5)_3$ should be replaced by formula (4.10) below as in the following theorem.

Theorem 4.5. With (4.7) for scheme (S) we have

$$Zero(PS) = Zero(CS/K_m) + \bigcup_k Zero(PS_k/K_{k-1}), \tag{4.10}$$

in which the union runs over k from 0 to m and

$$\begin{aligned} J_k &= IP(BS_k), \\ K_k &= J_0 * J_1 * \cdots * J_k, \\ PS_k &= PS + \{J_k\}. \end{aligned} \tag{4.11}$$

Proof. We refer to the original paper [WWT13] for the proof by induction on k. □

An Example of Zhu Shijie :

Example 4.1. Let us consider the example of Zhu Shijie as in Ex. 3.1. Using the notations there with ordering $z \prec x \prec y$ equivalent to $z = x_1, x = x_2, y = x_3$ we have $PS = PS_0 = \{P_1, P_2, P_3\}, BS_0 = \{P_2\}$ for which the pols P_i are reproduced below:

$$\begin{aligned} P_1 &= x * y * z - x * y^2 - z - x - y, \\ P_2 &= x * z - x^2 - z - y + x, \\ P_3 &= z^2 - x^2 - y^2. \end{aligned}$$

It is readily found that

$$\begin{aligned} R_{01} = Remdr(P_1/BS_0) = x * (&-x^4 + 2 * x^3 * z + 2 * x^3 - x^2 * z^2 - 5 * x^2 * z \\ &- x^2 + 3 * x * z^2 + 3 * x * z + x - 2 * z^2 - z - 2), \end{aligned}$$

$$R_{02} = Remdr(P_3/BS_0) = x * (-x^3 + 2 * x^2 * z + 2 * x^2 - x * z^2 - 4 * x * z - 2 * x + 2 * z^2 + 2 * z).$$

Hence $RS_0 = \{R_{01}, R_{02}\} \neq \varnothing$. Let us proceed as in scheme (S) in using (4.7) so that

$$PS_1 = BS_0 + RS_0 = \{P_0, R_{01}, R_{02}\}. \tag{4.12}$$

The procedure will end at stage $m = 5$ and we have the final $RS_5 = \varnothing$ and $PS_5 = BS_5 = CS = \{C_1, C_2\}$ with C_1, C_2 given below:

$$\begin{aligned} C_1 &= x * (z-1)^2 * (z+1) * (z-5), \\ C_2 &= -y - x^2 + x * z + x - z. \end{aligned} \tag{4.13}$$

It is readily verified that all remainders of P_i w.r.t. CS are 0. Hence by Theorem 4.3

$$CS = \{C_1, C_2\} \tag{4.14}$$

is a char-set of PS. By Well-Ordering Principle (4.5) we have then

$$Zero(PS) = Zero(CS/I_1) + Zero(PS + \{I_1\}), \text{ where} \tag{4.15}$$

$$I_1 = (z-1)^2 * (z+1) * (z-5) \tag{4.16}$$

is the initial of C_1.

Collins-Li Phenomenon.

Given two pols $F, G \in \mathbf{D}[x]$ with coefficients in a unique factorization domain $\mathbf{D}$ let us form a sequence

$$F_0, F_1, F_2, \cdots, F_r$$

with $F_0 = F, F_1 = G$, while each of $F_i, (i \geqslant 3)$ is the pseudoremainder of F_{i-1} w.r.t. F_{i-2} and the final F_r is in $\mathbf{D}$. Collins had discovered the important phenomenon that for i sufficiently large, each F_i will have the leading coefficients of its preceding F's as factors. See [Col1] and [B-T]. This remains true when the pseudoremainder is replaced by the remainder in the sense of Def. 3.17.

Li Zimin has extended the discovery of Collins to the case of char-set formation. Thus, it is proved by Li (cf. [LZM1,2,3]) that at least in the *generic* case, the remainders occuring in the procedure of scheme (S) will have necessarily initials of pols in preceding basic sets as factors. For proofs of theorems of Collins and Li we refer to their original papers. We remark only that owing to the inevitable occurence of such factors, the pols in the procedure would become unwillingly too large which render the computations difficult to be carried on further. Moreover, other unforeseen factors may also occur during the procedure, like the factor x and others in R_{01}, R_{02} of Ex. 4.1. Thus it seems worthy to remove all such factors, initials of basic sets or not, during the procedure to render smaller the pols in the procedure. The influence of the removal of factors on the char-set formation is seen below.

Extensions of Char-Set and Well-Ordering Principle.

Definition 4.1$'$. A *characteristic set* of a polset PS modulo a pol F := An asc-set CS' verifying the conditions (c_1') and (c_2') below:

$(c_1)'$ $Remdr(PS/CS') = \{0\}$.

$(c_2)'$ $Zero(PS/F) \subset Zero(CS'/F)$.

Theorem 4.6. Let CS' be a char-set of a polset PS modulo a pol F. Then:

$$\begin{aligned}
&Zero(CS'/IP * F) \subset Zero(PS/F) \subset Zero(CS'/F),\\
&Zero(PS/IP * F) = Zero(CS'/IP * F),\\
&Zero(PS/F) = Zero(CS'/IP * F) + \bigcup_i Zero(PS + \{I_i'\}/F),\\
&Zero(PS) = Zero(PS/F) + Zero(PS + \{F\}).
\end{aligned} \tag{4.17}$$

In (4.17) I_i' are initials of pols C_i' in CS', and IP' is the initial-product of CS'.

The char-set algorithm will be applied as before with factors F_j freely removed during the procedure. The final asc-set $BS_m = CS'$ will then be a char-set of PS w.r.t. $F = \Pi_j F_j$.

Remark 4.7. The Theorem 4.3 and the verification may be analogously formulated for the present case of modulo some factors which we omit the statements.

Another useful extension of Well-Ordering Principle which follows easily from $(4.5)_3$ in applying further (4.17) with F the product of all separants.

Theorem 4.7 (Modified Well-Ordering Principle). Let CS be a char-set of PS with initial-separant product ISP, then

$$Zero(PS) = Zero(CS/ISP) + \bigcup_i Zero(PS + \{I_i\}) + \bigcup_i Zero(PS + \{S_i\}). \tag{4.5$'$}$$

In $(4.5)'$ I_i, S_i are respectively the initials and separants of char-pols in CS.

Some Examples

Example 4.1a.

Let us consider Ex.4.1 again. In PS_1 of (4.12) the pols R_{01}, R_{02} have evident factor x. If we assume $x \neq 0$ then we can remove the factor x from R_{01}, R_{02} to get $R_{01}' = R_{01}/x, R_{02}' = R_{02}/x$. Form now the polset $\{P_2, R_{01}', R_{02}'\}$ and proceed as before in taking (4.7) for the scheme (S). Then the algorithm will stop at stage $m = 4$. The final basic set $BS'_4 = \{B'_1, B'_2, B'_3\}$ are given as follows

$$\begin{aligned}
B'_1 &:= (z-1)^2 * (z+1) * (z-5),\\
B'_2 &:= x * z^2 - 3 * x * z - 7 * x - z^3 + 3 * z^2 + 7 * z + 6,\\
B'_3 &:= -y - x^2 + x * z + x - z.
\end{aligned}$$

By verification the remainders of P_1, P_3 w.r.t. BS'_4 are non-zero. Adjoin now these non-zero remainders to PS and proceed as in diagram $(\mathrm{S})'$ we get the final char-set of PS modulo x to be $CS' = \{C'_1, C'_2, C'_3\}$, where

$$\begin{aligned}
C'_1 &= (z+1) * (z-5),\\
C'_2 &= x * (z-2) - 2 * z + 1,\\
C'_3 &= -y - x^2 + x * z + x - z.
\end{aligned} \tag{4.18}$$

By Well-Ordering Principle modulo factors $(4.17)_3$ we have then

$$\begin{aligned}
Zero(PS) &= Zero(PS/x) + Zero(PS + \{x\}),\\
Zero(PS/x) &= Zero(CS'/x * (z-2)) + Zero(PS + \{z-2\}/x).
\end{aligned} \tag{4.19}$$

Example 4.2. The following example is taken from the paper [CZ-G] of Czapor and Geddes.

Consider a polset $PS = \{P_1, P_2, P_3\}$ with

$$\begin{aligned}
P_1 &= x^2 + a * y * z + d * x + g,\\
P_2 &= y^2 + b * z * x + e * y + h,\\
P_3 &= z^2 + c * x * y + f * z + k,
\end{aligned} \tag{4.20}$$

in which the a, b, etc. are considered as independent parameters. Consider the ordering $x \prec y \prec z$ equivalent to $x = x_1, y = x_2, z = x_3$. For $PS = PS_0$ let us take $BS_0 = \{B_{01}\}$ with $B_{01} = P_2$. Proceed now as in char-set algorithm in using (4.7) and remove the factors

$$F_1 = x, \;\; F_2 = x * b * f - h \tag{4.21}$$

during the procedure, we get then finally a basic set $BS_5 = CS = \{C_1, C_2, C_3\}$ with index-sets of C_i given below:

$$[137 \; x \; 8], \;\; [35 \; y \; 1], \;\; [4 \; z \; 1]. \tag{4.22}$$

It is readily verified that all P_i of PS have remainders 0 w.r.t. CS. Hence CS is a char-set of PS modulo factors F_1, F_2 and we have by Theorem 4.6:

$$\begin{aligned}
Zero(PS) = {}& Zero(CS/I_2 * F_1 * F_2)\\
&+ Zero(PS + \{I_2\}/F_1 * F_2) + Zero(PS + \{F_1\}) + Zero(PS + \{F_2\}).
\end{aligned} \tag{4.23}$$

In (4.23) I_2 is the initial of C_2 and is a pol of index-set [14 x 4]. The other initials of C_1 and C_3 are $I_1 = (a * b * c + 1)^3, I_3 = b * x$ playing no role in (4.23).

If we do not remove any factors during the procedure, then the final basic set will be consisting of pols

$$x^4 * (x * b * f - h)^2 * C_1,\ x^2 * C_2,\ and\ C_3. \tag{4.24}$$

The remainders of the pols P_1, P_3 of PS w.r.t. this basic set will no more be 0 and we have to proceed further as in scheme (S)$'$ to get the char-set. The computations will then be too complicate to be carried out easily. This shows the importance of removing factors, if there are any, in the char-set formation.

In the very beginning we may also take P_1 instead of P_2 as the only pol of the basic set BS_0. Proceed as before and remove the factors

$$F_3 = x * c * e - k,\ F_4 = x^2 + x * d + g \tag{4.25}$$

during the procedure in using still (4.7), then we get a final basic set $CS' = \{C_1', C_2', C_3'\}$ of index-sets

$$[137\ x\ 8],\ [79\ y\ 1],\ [4\ z\ 1]. \tag{4.26}$$

All the remainders of P_i of PS w.r.t. CS' are 0 so that CS' is a char-set of PS modulo $F_3 * F_4$ and we have by Theorem 4.6:

$$\begin{aligned} Zero(PS) = & Zero(CS'/I_2' * I_3' * F_3 * F_4) \\ & + Zero(PS + \{I_2'\}/F_3 * F_4) + Zero(PS + \{I_3'\}/F_3 * F_4) \\ & + Zero(PS + \{F_3\}) + Zero(PS + \{F_4\}). \end{aligned} \tag{4.27}$$

In (4.27) I_2', I_3' are initials of C_2', C_3' respectively for which $I_3' = a * y$ and I_2' is a pol of index-set [36 x 5].

Example 4.3. (Rotation group in 3-space)

Consider a rotation $R = (x_{ij}),\ i, j = 1, 2, 3$, verifying the relation

$$R^{tr} = R^{-1},\ det(R) = +1. \tag{4.28}$$

More explicitly, the rotation group of 3-space is an algebraic variety in $\mathbf{K}^9(x_{ij})$

defined as $Zero(PS)$ where $PS = \{P_1, \cdots, P_9, P_{10}\}$ with

$$\begin{aligned}
P_1 &= x_{11} - (x_{22} * x_{33} - x_{23} * x_{32}),\\
P_2 &= x_{12} + (x_{21} * x_{33} - x_{23} * x_{31}),\\
P_3 &= x_{13} - (x_{21} * x_{32} - x_{22} * x_{31}),\\
P_4 &= x_{21} + (x_{12} * x_{33} - x_{13} * x_{32}),\\
P_5 &= x_{22} - (x_{11} * x_{33} - x_{13} * x_{31}),\\
P_6 &= x_{23} + (x_{11} * x_{32} - x_{12} * x_{31}),\\
P_7 &= x_{31} - (x_{12} * x_{23} - x_{13} * x_{22}),\\
P_8 &= x_{32} + (x_{11} * x_{23} - x_{13} * x_{21}),\\
P_9 &= x_{33} - (x_{11} * x_{22} - x_{12} * x_{21}),\\
P_{10} &= det(R) - 1.
\end{aligned} \tag{4.29}$$

Let us take the ordering

$$x_{11} \prec x_{12} \prec x_{13} \prec x_{21} \prec x_{22} \prec x_{23} \prec x_{31} \prec x_{32} \prec x_{33}. \tag{4.30}$$

Applying now the procedure of scheme (S) with (4.7) and removing factors x_{11}, x_{12} during the procedure we get a char-set modulo factors $CS = \{C_1, \cdots, C_6\}$ with C_i given below:

$$\begin{aligned}
C_1 &= x_{13}^2 + x_{12}^2 + x_{11}^2 - 1,\\
C_2 &= x_{22}^2 * (x_{11}^2 - 1) - 2 * x_{22} * x_{21} * x_{12} * x_{11} + x_{21}^2 * (x_{12}^2 - 1) - x_{12}^2 - x_{11}^2 + 1,\\
C_3 &= x_{23} * x_{13} + x_{22} * x_{12} + x_{21} * x_{11},\\
C_4 &= x_{31} - x_{23} * x_{12} + x_{22} * x_{13},\\
C_5 &= x_{32} + x_{23} * x_{11} - x_{21} * x_{13},\\
C_6 &= x_{33} - x_{22} * x_{11} + x_{21} * x_{12}.
\end{aligned} \tag{4.31}$$

The zero-set of PS is then given by

$$\begin{aligned}
Zero(PS) = {} & Zero(CS \;/\; x_{11} * x_{12} * x_{13} * (x_{11}^2 - 1))\\
& + Zero(PS + \{x_{13}\} \;/\; x_{11} * x_{12}) + Zero(PS + \{x_{11}^2 - 1\} \;/\; x_{11} * x_{12})\\
& + Zero(PS + \{x_{11}\}) + Zero(PS + \{x_{12}\}).
\end{aligned} \tag{4.32}$$

3.5 Zero-Decomposition Theorems.

Problem to be Studied.

The notations and notions **K**, etc. will be as before.

In the preceding section Sect. 3.4 we have introduced for a polset PS the notion of char-set CS whose zero-structure has a close relation with that of PS as exhibited in the two properties $(c_1), (c_2)$ of Theorem 4.1 and the Well-Ordering Principle (Theorem 4.2) which we reproduce below:

$$\begin{aligned}
&Remdr(PS/CS) = \{0\},\\
&Zero(CS/IP) \subset Zero(PS) \subset Zero(CS),\\
&Zero(PS/IP) = Zero(CS/IP),\\
&Zero(PS) = Zero(CS/IP) + \bigcup_i Zero(PS + \{I_i\}).
\end{aligned} \tag{5.1}$$

In (5.1) CS is a char-set of PS, IP is the initial-product of CS, and I_i are the initials of pols in CS. The set $Zero(CS/IP)$ in the right-hand side of $(5.1)_4$ may be considered as a principal part of $Zero(PS)$. The purpose of the present section is to study, through the investigation of the remaining parts of $(5.1)_4$, the problem below.

Problem D. To determine the zero-structure of a polset PS via those of some asc-sets determined from PS.

The solution of Problem D will be given in the form of Zero-Decomposition Theorems to be explained below.

Zero-Decomposition Theorems.

The set $Zero(CS/IP)$ in the right hand side of $(5.1)_4$ gives usually the principal part of $Zero(PS)$. If we treat each member in the union of right-hand side as before and proceed further and further we shall arrive at the following

Theorem 5.1 (Zero-Decomposition Theorem I). There is an algorithm which permits to determine for a given polset PS in a finite number of steps a finite set of asc-sets CS_j such that we have

$$Zero(PS) = \bigcup_j Zero(CS_j/IP_j), \tag{I}$$

$$Remdr(PS/CS_j) = \{0\}. \tag{5.2}$$

In (I) each IP_j is the initial-product of the corresponding asc-set CS_j. Furthermore, each CS_j in (5.2) is not of higher ordering than PS.

Proof. It follows from $(5.1)_2$ that for each member of the union in $(5.1)_4$ we have

$$\begin{aligned}
&Zero(PS + \{I_i\}) = Zero(PS_i), \text{ with}\\
&PS_i = PS + CS + \{I_i\}.
\end{aligned} \tag{5.3}$$

As I_i is reduced w.r.t. CS and CS is not of higher ordering than PS we have by Theorem 3.2 or 3.6

$$PS \succ PS_i. \tag{5.4}$$

Now $(5.1)_4$ may be rewritten as

$$Zero(PS) = Zero(CS/IP) + \bigcup_i Zero(PS_i). \tag{5.5}$$

If we treat each of the polsets PS_i in the same manner as before we shall get alike to (5.3), (5.4), (5.5)

$$\begin{aligned} &Zero(PS_i) = Zero(CS_i/IP_i) + \bigcup_{i_1} Zero(PS_{ii_1}), \\ &PS_{ii_1} = PS_i + CS_i + \{I_{ii_1}\}, \\ &PS_i \succ PS_{ii_1}. \end{aligned} \tag{5.6}$$

In (5.6) CS_i is a char-set of PS_i, IP_i is the initial-product of CS_i, and I_{ii_1} are initials of pols in CS_i. Remark that by Well-Ordering Principle in Sect. 3.4 CS is not of higher ordering than PS and hence this is so for all PS_i. Moreover, as $PS \subset PS_i$, the condition $(5.1)_1$ for CS_i as a char-set of PS_i shows that $(5.1)_1$ is valid for PS w.r.t. CS_i or

$$Remdr(PS/CS_i) = \{0\}. \tag{5.7}$$

Combining now (5.5) and $(5.6)_1$ we get then

$$Zero(PS) = Zero(CS/IP) + \bigcup_i Zero(CS_i/IP_i) + \bigcup_{ii_1} Zero(PS_{ii_1}). \tag{5.8}$$

As before each of CS_i and PS_{ii_1} is not of higher ordering than PS_i and hence also not of higher ordering than PS.

Let us treat now each PS_{ii_1} in the last union as before and proceed further and further, we shall arrive for each stage k at an identity of the form

$$\begin{aligned} Zero(PS) = & Zero(CS/IP) + \bigcup_i Zero(CS_i/IP_i) \\ & + \cdots \\ & + \bigcup_{ii_1\cdots i_k} Zero(PS_{ii_1\cdots i_k}). \end{aligned} \tag{5.9}$$

As in (5.4) and $(5.6)_3$ we have

$$PS \succ PS_i \succ PS_{ii_1} \succ \cdots \succ PS_{ii_1\cdots i_k} \succ \cdots \tag{5.10}$$

By Theorem 3.3 or 3.7 any sequence in (5.10) should be finite so that for k sufficiently large the last union about $PS_{ii_1\cdots i_k}$ in (5.9) should be non-existant and (5.9) may then be rewritten as an identity of the form (I). Moreover, as in $(5.1)_1$ for CS and (5.7) for CS_i, we should have $Remdr(PS/CS_{ii_1}) = \{0\}$ too. By induction we have then (5.2) for each CS_j. Moreover, by induction we see also that each CS_j in (5.2) is not of higher ordering than PS. The theorem is thus completely proved. □

In starting from $(4.5)'$ of Theorem 4.7 in Sect. 3.4 instead of (4.5) or $(5.1)_4$ we will get in the same manner the following theorem:

Theorem 5.2. (Zero-Decomposition Theorem II) There is an algorithm which permits to determine for a given polset PS in a finite number of steps a finite set of asc-sets CS_k such that we have

$$Zero(PS) = \bigcup_k Zero(CS_k/ISP_k), \tag{II}$$

$$Remdr(PS/CS_k) = \{0\}. \tag{5.11}$$

In (II) each ISP_k is the initial-separant-product of the corresponding asc-set CS_k. Furthermore, each CS_k in (5.11) is not of higher ordering than PS.

Definition 5.1. The set of asc-sets CS_j as asserted in the above Theorem 5.1 or 5.2 := A *characteristic series* of PS.

Abbreviation 5.1. *Char-Series* := Characteristic series.

Below we shall give alternative forms of Zero-Decomposition Theorems. For this purpose we shall introduce some other partial ordering of polsets as that in the preceding sections. First we make some preparations.

Subeliminant Sequence of Pol-Pairs.

Let $\mathbf{D}$ be a unique factorization domain and A_m, B_n be two pols in $\mathbf{D}[x]$ of the following form with m, n both > 0:

$$\begin{aligned} A_m &= a_m * x^m + ... + a_i * x^i + ... + a_0,\ a_m \neq 0, \\ B_n &= b_n * x^n + ... + b_i * x^i + ... + b_0,\ b_n \neq 0. \end{aligned} \tag{5.12}$$

Problem E. Find for any integer $g > 0$ and $< min(m,n)$ pols C_{n-g-1}, D_{m-g-1}, E_g of degrees in x less than or equal to $n-g-1, m-g-1, g$ respectively with coefficients in $\mathbf{D}$ such that

$$C_{n-g-1} * A_m + D_{m-g-1} * B_n = E_g. \tag{$5.13)_g$}$$

Solution. Write these pols in the form below with yet unknown coefficients

c_i, d_i, e_i:

$$\begin{aligned}C_{n-g-1} &= c_{n-g-1} * x^{n-g-1} + \cdots + c_i * x^i + \cdots + c_0,\\ D_{m-g-1} &= d_{m-g-1} * x^{m-g-1} + \cdots + d_i * x^i + \cdots + d_0,\\ E_g &= e_g * x^g + \cdots + e_i * x^i + \cdots + e_0.\end{aligned} \tag{5.14}$$

Comparing coefficients of x-powers in $(5.13)_g$ we get then a set of equations

$$\Sigma_i\ (c_i * a_{r-i} + d_i * b_{r-i}) - e_r = 0, \qquad (5.15)_r$$

for $r = 0, 1, \cdots, m + n - g - 1$. In $(5.15)_r$ the summation runs over i from 0 to r, and $a_i = 0$ for $i > m$, $b_i = 0$ for $i > n$, $c_i = 0$ for $i > n - g - 1$, $d_i = 0$ for $i > m - g - 1$, and $e_i = 0$ for $i > g$. The number of equations in (5.15) is $m + n - g$ and the number of unknowns is $n - g$ for c, $m - g$ for d, and $g + 1$ for e, so in all $m + n - g + 1$. The equations (5.15), being linear and homogeneous in these unknowns, have always solutions for (c, d, e). To fix the ideas, we shall pick out a fixed solution in the following way.

Notation 5.1. $ME_g :=$ Matrix of $m+n-g$ rows and $m+n-g+1$ columns for which the r-th row is consisting of the coefficients of $(5.15)_r$ corresponding to the $m + n - g + 1$ unknowns in the order of

$$c_{n-g-1}, ..., c_0, d_{m-g-1}, ..., d_0, e_g, ..., e_0, \tag{5.16}$$

Notation 5.2. $DE_{gi} :=$ The determinant of the matrix formed from ME_g with i-th column deleted, and multiplied by $(-1)^i$.

Definition 5.2. *g-th subeliminant* of $A_m, B_n :=$ The definite solution E_g of $(5.13)_g$ as determined from (5.15):

$$\begin{aligned}&(c_{n-g-1}, ..., c_0, d_{m-g-1}, ..., d_0, e_g, ..., e_0)\\ =&(DE_{g1}, ..., DE_{g,m+n-g+1}).\end{aligned} \tag{5.17}$$

Notation 5.3. $E_g(A_m, B_n)$ or simply E_g when no confusion can arise := *g-th subeliminant* of A_m, B_n.

Remark 5.1. According to Collins and Brown, the pols $E_g = E_g(A_m, B_n)$, with e_i given by (5.17) are all subresultants up to a sign of the pols A_m and B_n. In particular, E_0 is the resultant up to a sign of A_m and B_n. Cf. [Col1,2] and [B-T]. However, for our purposes it is not necessary to know the whole theory of subresultants as exhibited in the works of Collins and Brown. What is important for us is recapitulated in the propositions below for which we give direct proofs independent of their works.

Properties of Subeliminants of Pol-Pairs.

From the very definitions we have the following

Proposition 5.1. The coefficients of the subeliminants E_g are all definite pols in the coefficients of A_m and B_n.

Proposition 5.2. The 0-th subeliminant E_0 is equal up to a sign to the resultant of A_m, B_n.

Proposition 5.3. If $m \leqslant n$, then the last subeliminant E_{m-1} of A_m and B_n is equal up to a sign to the pseudoremainder of B_n w.r.t. A_m.

Proof. It is easy to verify directly from the definition that

$$E_{m-1} = \pm\, a_m^{n-m+1} * B_n + Q * A_m, \tag{5.18}$$

for some pol Q. The Proposition follows then from definition of pseudoremainder. □

As the pol E_0 is the resultant up to a sign of A_m and B_n the vanishing of E_0 is the necessary and sufficient condition for A_m and B_n to have a GCD of positive degree in x.

Proposition 5.4. If the pols A_m and B_n have a GCD of degree $f > 0$, then all the E_g of A_m and B_n with $g < f$ are equal to 0.

Proof. Consider any subeliminant E_g of A_m and B_n satisfying $(5.13)_g$ for $g < f$. As A_m and B_n have both the GCD F_f of degree f as factors, E_g has F_f as a factor too. As E_g is a pol of degree $\leqslant g$, this will only be possible for $E_g = 0$ if g is less than f. □

Proposition 5.5. If $E_0 = 0$ so that A_m, B_n have a GCD of degree $f > 0$, then this GCD can differ from the first non-zero subeliminant E_g of A_m and B_n only by non-zero constant multiple in $\mathbf{D}$.

Proof. Let the GCD be F_f of degree f in x with $f > 0$ and let us set

$$A_m = A'_{m-f} * F_f,\ B_n = B'_{n-f} * F_f, \tag{5.19}$$

in which A'_{m-f} and B'_{n-f} are pols of degree $m - f$ and $n - f$ respectively. It may happen that the GCD F_f is of the same degree as $min(m, n)$ so that one of A_m, B_n is a multiple of the primitive part of the other. Leaving this case aside the pols A'_{m-f} and B'_{n-f} will be both of positive degree in x having no factors of positive degrees in common. Denote the E_0-subeliminant of A'_{m-f} and B'_{n-f} by E'_0 so that for some pols C'_{n-f-1} and D'_{m-f-1} of degree $\leqslant n-f-1$ and $m-f-1$ respectively we have

$$C'_{n-f-1} * A'_{m-f} + D'_{m-f-1} * B'_{n-f} = E'_0. \tag{5.20}$$

As E_0' is the resultant of A_{m-f}' and B_{n-f}' up to a sign which have no more factors of positive degree in common, we have $E_0' \neq 0$. Multiplying both sides of (5.20) by F_f we get then

$$C_{n-f-1}' * A_m + D_{m-f-1}' * B_n = E_0' * F_f. \tag{5.21}$$

On the other hand $(5.13)_g$ for $g = f$ gives

$$C_{n-f-1} * A_m + D_{m-f-1} * B_n = E_f. \tag{5.22}$$

It is known that the degree of E_f should be exactly f. Cf. e.g. [Bo] or [Ak], p.239 for a proof. Eliminate now the highest degree term in the right hand side of (5.21) and (5.22) we will get an equation of the form

$$C_{n-f-1}'' * A_m + D_{m-f-1}'' * B_n = G,$$

with G a pol of degree $< f$. As in Proposition 5.4, G is necessarily 0 and it follows that E_f and $E_0' * F_f$ can differ only by non-zero constant multiples in $\mathbf{D}$. The Proposition is thus proved. □

Exponent and Eliminant of Pol-Pairs.

Consider now the case for which $E_0 \neq 0$.

The subeliminant E_0 of A_m and B_n is given by $(5.13)_g$ for $g = 0$, viz.

$$C_{n-1} * A_m + D_{m-1} * B_n = E_0. \tag{5.23}$$

In the present case A_m and B_n have no factor of positive degree in common. It follows that in $(5.13)_g$ any $E_g = 0$ would imply $C_{n-g-1} = 0$ and $D_{m-g-1} = 0$.

If for some $g > 0$ the pol E_g is of degree 0 and is non-zero in $\mathbf{D}$, then in eliminating E_0, E_g from (5.23) and $(5.13)_g$ we would have an equation of the form

$$(E_g * C_{n-1} - E_0 * C_{n-g-1}) * A_m + (E_g * D_{m-1} - E_0 * D_{m-g-1}) * B_n = 0.$$

As already remarked the coefficients of A_m and B_n should be both 0 or the equations (5.23) and $(5.13)_g$ are in fact proportional to each other. The only subeliminants E_g of interest are therefore those for which $deg(E_g) > 0$, for us particularly that one with smallest g. We lay down thus the following

Definition 5.3. In the case $E_0 \neq 0$ the *exponent* of A_m and B_n := The least integer g, if it exists, for which the pol E_g has a positive degree in x.

Definition 5.4. In the case $E_0 \neq 0$ the *eliminant* of A_m and B_n := The subeliminant E_e when exponent exists and is equal to e.

The above is all what we need from the known theory of subresultants.

Some Integer Sets Associated to a Polset.

Definition 5.5. The *class set* of a polset PS := The set of different positive classes of pols in PS, arranged in a *decreasing* order:

$$(c_1, ..., c_r), \text{ with } c_1 > c_2 > \cdots > c_r > 0. \tag{5.24}$$

Definition 5.6. The *degree set* of a polset PS := The set

$$(d_1, ..., d_r) \tag{5.25}$$

in which each d_i is the least degree of pols in PS having c_i as class in the class set (5.24).

Definition 5.7. The *pol-number set* of a polset PS := The set

$$(n_1, ..., n_r) \tag{5.26}$$

in which each n_i is the number of pols in PS having c_i as class in the class set (5.24).

Notation 5.4.

$cs(PS)$:= The class set of PS.

$ds(PS)$:= The degree set of PS.

$pns(PS)$:= The pol-number set of PS.

Remark 5.2. The above sets may eventually be all empty, which means that PS is consisting of only pols which are non-zero elements of $\mathbf{K}$. In that case $Zero(PS)$ is then also empty. Furthermore, if some non-zero constant occurs in PS, then $Zero(PS)$ will be an empty set.

Definition 5.8. For a polset PS of non-empty class set the *dominant class* of PS := The greatest class c, if it exists, for which the corresponding pol-number is > 1, otherwise 0.

Definition 5.9. For a polset PS of non-empty class set the *dominant degree* of PS := The least degree of the pols of dominant class in PS, if it exists, otherwise undefined.

A Partial Ordering of Polsets.

Definition 5.10. A polset PS is *higher than* a polset PS' or PS' is *lower than* PS := One of the following cases (a) to (d) is true:

(a) The class sets $(c_1, ..., c_r), (c'_1, ..., c'_s)$ of PS, PS' are both non-empty and the least index k for which $c_k \neq c'_k$ is such that $c_k > c'_k$.

(b) The class sets of PS, PS' are both non-empty and identical in the first k classes, while for their degree sets the least index j for which $d_j \neq d'_j$ is $j = k$ with $d_k > d'_k$.

(c) The class sets and the degree sets of PS, PS' are both non-empty and identical in the first k classes and degrees, while for their pol-number sets the least index j for which $n_j \neq n'_j$ is $j = k$ with $n_k > n'_k$.

(d) The class set of PS is non-empty while that of PS' is empty.

Definition 5.11. PS and PS' are *incomparable* to each other := Neither of PS or PS' is higher than the other.

Notation 5.5. $PS > PS'$ or $PS' < PS$:= PS is higher than PS' or PS' is lower than PS.

Notation 5.6. $PS <> PS'$:= PS, PS' are incomparable to each other.

The following lemmas follow now easily from the definitions:

Lemma 5.1. For a triangulated set TS let PS' be the polset obtained from TS by adjoining to it a further pol reduced w.r.t. TS. Then $PS' < TS$ if the class set remains unchanged.

Lemma 5.2. A sequence of polsets steadily decreasing in ordering is finite.

Proof. It is easy and will be omitted. □

Replacement Rules and Order-Lowering of Polsets.

Our alternative char-set algorithm consists now in the replacement of PS by other polsets of lower ordering by the rules given below. We suppose naturally that the class set of PS is non-empty.

Replacement Rule 1. Suppose some pol P in PS of class $c > 0$ is factorizable as $P = P_1 * P_2$ with P_1, P_2 of same class c. Then we make the replacement

$$\begin{aligned} &PS \longrightarrow PS_1 + PS_2, \text{ where} \\ &PS_1 = PS \setminus \{P\} + \{P_1\}, \\ &PS_2 = PS \setminus \{P\} + \{P_2\}. \end{aligned} \tag{5.27}$$

It is clear that

$$Zero(PS) = Zero(PS_1) + Zero(PS_2), \text{ and} \tag{5.28}$$

$$PS_1 < PS, \ PS_2 < PS. \tag{5.29}$$

Replacement Rule 2. Suppose the dominant class c of PS is > 0. Let A be a pol of class c and of a degree equal to the dominant degree and B be any other pol of class c in PS. Let E_0 be the 0-th subeliminant or the resultant up to a sign of A and B w.r.t. x_c. We now separate into several different cases:

Case 1. $E_0 = 0$.

In this case A and B will have a GCD F of class c and degree $f > 0$. Make now the replacement

$$\begin{aligned} &PS \longrightarrow PS_1 + PS_2, \text{ where} \\ &PS_1 = PS \setminus \{A, B\} + \{F\}, \\ &PS_2 = PS \setminus \{A, B\} + \{A/F, B/F\}. \end{aligned} \tag{5.30}$$

Then it is clear that (5.28), (5.29) hold in this case.

Case 2. $E_0 \neq 0$ and exponent of the pair A, B exists and > 0.

Let the exponent be e. Then the eliminant E_e will be of class c and of a positive degree in x_c which is less than the degree of A. We make now the replacement

$$PS \longrightarrow PS_1 = PS \setminus \{A, B\} + \{E_0, E_e\}.$$

It is clear that we have:

$$Zero(PS) \subset Zero(PS_1), \text{ and} \tag{5.31}$$

$$PS_1 < PS, \tag{5.32}$$

Case 3. $E_0 \neq 0$ and exponent of the pair (A, B) is non-existant. Notably this is the case when degree of A in x_c is equal to 1.

In this case we make the replacement

$$PS \longrightarrow PS_1 = PS \setminus \{B\} + \{E_0\}.$$

It is clear that we have still (5.31) and (5.32).

Remark 5.3. In these Replacement Rules the number of classes in the class set cannot diminish. The number of classes will increase only in Case 2 or Case 3 of Rule 2 for which the 0-th eliminant E_0 or E_e has a positive class and does not occur in the original class set.

Replacement Rule 3. Consider any triangulated set TS and a polset PS. Form the remainders of pols in PS w.r.t. TS. Suppose that there are non-zero ones among the remainders. In that case adjoin all such non-zero ones to TS to get a polset PS'. Then by Lemma 5.1 we have in case the class set remains unchanged,

$$PS' < TS, \text{ and}$$

$$Zero(TS + PS) \subset Zero(PS'). \tag{5.33}$$

Remark 5.4. In this Replacement Rule the number of classes in the class set will be increased if it is not unchanged.

Alternative Zero-Decomposition Procedure by Replacement Rules.

Applying the Replacement Rules 1-2 to a given polset PS and any of the successive polsets obtained whenever these Rules are applicable we will get finally a set of triangulated sets TS_i such that

$$Zero(PS) \subset \bigcup_i Zero(TS_i). \tag{T}$$

From this inclusion relation we get clearly

$$Zero(PS) = \bigcup_i Zero(TS_i + PS). \tag{T$'$}$$

Apply now Replacement Rule 3 to each of $TS_i + PS$ occuring in the decomposition (T)$'$ and then repeat the procedure by applying Replacement Rules 1-3 whenever possible. As the number of classes is limited by n, we will finally get a decomposition of the same form as (T) but with the further property that

$$Remdr(PS/TS_i) = \{0\}. \tag{5.34}$$

For each triangulated set TS_i in (T) verifying (5.34) let us replace $Zero(TS_i)$ by

$$Zero(TS_i) = Zero(TS_i/J_i) + \bigcup_j Zero(TS_{ij} + \{I_{ij}\}).$$

In the expression TS_{ij} is the polset obtained from TS_i by removing the leading term of the j-th pol in TS_i, I_{ij} is the initial of that j-th pol, and J_i is the product of all such initials of pols in TS_i. By Lemma 5.1 all of the sets $TS_{ij} + \{I_{ij}\}$ are of an order lower than TS_i so we may repeat the same procedure as far as possible. By Lemma 5.2 we will get finally a relation of the form

$$Zero(PS) \subset \bigcup_k Zero(TS_k/IP_k),$$

with IP_k the initial-product of TS_k, for which (5.34) is true for each TS_k. It follows easily that the above inclusion relation is actually an identity. Hence the above gives an algorithm for arriving at a decomposition of the form

$$Zero(PS) = \bigcup_k Zero(TS_k/IP_k), \tag{III}$$

in which each TS_k is some triangulated set verifying (5.34) while IP_k is the initial-product of TS_k. If the initial of some pol in some TS_k is not reduced w.r.t. the partial asc-set consisting of the pols preceding that pol, then Replacement Rule 3 may be applied to that pol. In this way the decomposition (III) will ultimately be reduced to one of the form (I). This gives thus an alternative algorithm for the decomposition (I).

Some Examples.

Example 5.1. Let us consider Exs. 4.1 and 4.1a.

Consider first (4.15) for which no non-trivial factors have been removed during the procedure. To get a decomposition of $Zero(PS)$ as in Theorem 5.1 we have to determine $Zero(PS + \{I_1\})$. By the char-set algorithm we get for the polset $PS + \{I_1\}$ a char-set

$$CS_1 = \{C_{11}, C_{12}, C_{13}\}, \text{ with} \tag{5.35}$$

$$\begin{aligned} C_{11} &= (z-1)^2 * (z+1) * (z-5), \\ C_{12} &= x * z * (z+1) * (z-5), \\ C_{13} &= -y - x^2 + x * z + x - z. \end{aligned} \tag{5.36}$$

Hence

$$Zero(PS + \{I_1\}) = Zero(CS_1/I_{12}) + Zero(PS + \{I_1, I_{12}\}), \tag{5.37}$$

$$I_{12} = z * (z+1) * (z-5) \tag{5.38}$$

being the initial of C_{12}.

For $PS + \{I_1, I_{12}\})$ we get now a char-set

$$CS_2 = \{C_{21}, C_{22}, C_{23}\} \text{ with} \tag{5.39}$$

$$\begin{aligned} C_{21} &= (z+1) * (z-5), \\ C_{22} &= x * (z-2) - 2 * z + 1, \\ C_{23} &= -y - x^2 + x * z + x - z, \end{aligned} \tag{5.40}$$

which is same as CS' of (4.18). We have now evidently

$$Zero(PS + \{I_1, I_{12}\}) = Zero(CS_2). \tag{5.41}$$

Combining (4.15), (5.37), (5.41) together we get finally the required decomposition

$$Zero(PS) = Zero(CS/I_1) + Zero(CS_1/I_{12}) + Zero(CS_2). \tag{5.42}$$

The zero-sets of the right-hand side of (5.42) are respectively given by

$$\begin{aligned} (z, x, y) &= (z, 0, -z) \text{ with } z \text{ arbitrary}, \\ (z, x, y) &= (1, 0, -1), \\ (z, x, y) &= (-1, 1, 0) \text{ or } (5, 3, 4). \end{aligned} \tag{5.43}$$

Thus, the complete zero-set of PS is composed of a line $x = 0, y + z = 0$ and two isolated points $(-1, 1, 0), (5, 3, 4)$ in the space $\mathbf{K}^3(z, x, y)$.

Let us consider now (4.19) obtained by removing the factor x during the procedure. For $PS + \{x\}$ we get readily a char-set

$$CS_1' = \{C_{11}', C_{12}'\}, \text{ with} \tag{5.44}$$

$$\begin{aligned} C_{11}' &= x, \\ C_{12}' &= y + z. \end{aligned} \tag{5.45}$$

Hence we get

$$Zero(PS + \{x\}) = Zero(CS_1'). \tag{5.46}$$

It is readily seen that

$$\begin{aligned} &Zero(CS' \;/\; x * (z - 2)) = Zero(CS'), \\ &Zero(PS + \{z - 2\}/x) = \varnothing. \end{aligned}$$

In combinig with (4.18),(4.19) we get then the complete decomposition

$$Zero(PS) = Zero(CS') + Zero(CS_1'). \tag{5.47}$$

It is readily seen that the zero-set is composed of a line and two isolated points as before.

Example 5.2. Let us consider Ex. 4.2 of Czapor and Geddes again.

Remark that the a, b, etc. in (4.20) are considered as independent parameters. It follows that in the formula (4.23) we should have

$$Zero(PS + \{I_2\}/F_1 * F_2) = \varnothing, \; Zero(PS + \{F_1\}) = \varnothing, \; Zero(PS + \{F_2\}) = \varnothing. \tag{5.48}$$

In fact, the last two identities may be verified directly in that they will lead to non-zero pols in the independent parameters a, b, etc. which should be considered as non-zero. For the first identity in (5.48) let us remark that by (4.24) we have

$$x^4 * (x * b * f - h) * C_1 = 0 \mid Zero(PS),$$

in which C_1 is a pol of index-set [137 x 8] and is *irreducible*. On the other hand I_2, being a pol of index-set [14 x 4], is of degree 4 in x. Hence the resultant of C_1 and I_2 w.r.t. x is a non-zero pol in the independent parameters a, b, etc. and is thus non-zero. Consequently I_2 and C_1 cannot be both 0 on the variety $Zero(PS)$. It is also evident that I_2 cannot be both 0 with x or $x * b * f - h$ on the variety $Zero(PS)$.

Therefore we should have $Zero(PS + \{I_2\}) = \varnothing$ and *à fortiori* the first identity of (5.48). For the same reason we have

$$Zero(CS/I_2 * F_1 * F_2) = Zero(CS), \tag{5.49}$$

since each of I_2, F_1, F_2 is of degree in x less than that of C_1 and has thus non-zero resultant in a, b, etc. with C_1. It follows from (5.48),(5.49) that (4.23) becomes finally

$$Zero(PS) = Zero(CS). \tag{5.50}$$

This gives a complete determination of $Zero(PS)$ in the *generic* case of independent parameters a, b,etc. We remark that the above reasoning for the determination is a quite general one and may be applied with great advantage in various cases.

Finally let us remark that we may determine $Zero(PS)$ of this example also by means of Zero-Decomposition III and arrive at the same result which we shall not enter.

Example 5.3. Let us consider Ex. 4.3 again with PS given by (4.29) and $Zero(PS)$ given by (4.32). Remark that owing to the relation $C_1 = x_{13}^2 + x_{12}^2 + x_{11}^2 - 1 = 0$ we have for $\mathbf{K} = \mathbf{R}$,

$$Zero_{\mathbf{R}}(PS + \{x_{11}^2 - 1\} \ / \ x_{11} * x_{12}) = \varnothing, \tag{5.51}$$

$$Zero_{\mathbf{R}}(CS \ / \ x_{11} * x_{12} * x_{13} * (x_{11}^2 - 1)) = Zero_{\mathbf{R}}(CS \ / \ x_{11} * x_{12} * x_{13}). \tag{5.52}$$

Taking into account (5.51) and (5.52), we see that in case $\mathbf{K} = \mathbf{R}$ (4.32) becomes

$$Zero_{\mathbf{R}}(PS) = Zero_{\mathbf{R}}(CS \ / \ x_{11} * x_{12} * x_{13}) + \bigcup_{i=1,2,3} Zero_{\mathbf{R}}(PS + \{x_{1i}\}). \tag{5.53}$$

The last three zero-sets are of the same character and may be treated by the general method which we shall not enter.

3.6 Variety-Decomposition Theorems.

Notation 6.1. $\mathbf{K}, X = (x_1, \cdots, x_n)$, etc. as before.

Irreducible Ascending Set.

Consider a non-trivial asc-set in ascending order

$$AS: \quad A_1, A_2, \cdots, A_r, \tag{6.1}$$

in which $(i = 1, \cdots, r)$

$$0 < c_1 < c_2 < \cdots, < c_r, \tag{6.2}$$

$$A_i = I_{i,0} * x_{c_i}^{d_i} + \cdots + I_{i,j} * x_{c_i}^{d_i - j} + \cdots + I_{i,d_i}, \tag{6.3}$$

$$I_{ij} \in \mathbf{K}[X_{c_i - 1}], \ I_{i,0} \neq 0. \tag{6.4}$$

Definition 6.1 The asc-set

$$AS_k: \quad A_1, A_2, \cdots, A_k$$

consisting of the first k pols in (6.1) := The *k-th partial asc-set* of AS.

Definition 6.2. The asc-set AS in (6.1) is *irreducible in stage* k $(1 \leqslant k \leqslant r)$:= The following is true:

For $k = 1, A_1$ is irreducible in $\mathbf{K}[X_{c_1}]$.

For $k > 1, AS$ is irreducible in stage $k-1$. Moreover, let $\mathbf{K}_{c_k - 1}$ be the extension field of $\mathbf{K}$ with x_j, for $1 < j < c_k$ and $j \neq$ any c_i adjoined to $\mathbf{K}$ as independent transcendental elements, and $x_{c_i}, i < k$, adjoined to $\mathbf{K}$ successively by the irreducible equations $A_i = 0$, for $i = 1, \cdots, k-1$. Then A_k, as a pol in x_{c_k}, is primitive in $\mathbf{K}[X_{c_k}]$ and is irreducible over the extension field $\mathbf{K}(X_{c_k - 1})$ in the sense that there exist no two pols $F_{k1}, F_{k2} \in \mathbf{K}[X_{c_k}]$ of class c_k and a pol I in $\mathbf{K}[X_{c_k - 1}]$ such that $I * A_k \equiv F_{k1} * F_{k2} \ mod_{\mathbf{K}} \ AS_{k-1}$.

Definition 6.3. The asc-set AS in (6.1) is *irreducible* := It is irreducible in stage r.

Example 6.1. Consider the asc-set $AS: \ A_1, A_2$ with

$$A_1 = x_2^2 + x_1, \ A_2 = x_3^2 + x_1.$$

Then AS is clearly irreducible in stage 1 but not so in stage 2 since $A_2 \equiv (x_3 + x_2) * (x_3 - x_2) + A_1$ so that $A_2 \equiv (x_3 + x_2) * (x_3 - x_2) \ mod_{\mathbf{K}} \ A_1$.

Theorem 6.1. Let AS be an irreducible asc-set as in (6.1-4). Set $E_i = \dfrac{A_i}{I_{i0}}$, then $ES = \{E_1, \cdots, E_r\}$ is the defining polset of an extended point $\Xi = (\xi_1, \cdots, \xi_n)$ in $\mathbf{K}^n$ satisfying all $A_i = 0$. If P is a non-zero pol reduced w.r.t. AS, then

$$P(\Xi) \neq 0.$$

In particular, for any initial I_{i0} of pols A_i in AS, we have

$$I_{i0}(\Xi) \neq 0.$$

Proof. The first part of the theorem is evident from the very definitions. For the second part of the theorem let P be a pol reduced w.r.t. AS. Then it is sufficient to prove that $P(\Xi) = 0$ would imply $P \equiv 0$. We shall prove this by induction on the class c of P. Let us write now P of class c in the canonical form

$$P = I_0 * x_c^d + \cdots + I_d,$$

with coefficients $I_k \in \mathbf{K}[X_{c-1}]$. As $P(\Xi) = 0$ and AS is irreducible we should have all $I_k = 0$ if c is not equal to any c_i. On the other hand if c equals some c_i then, P being reduced w.r.t. AS so that $d < d_i$, we would have again all $I_k = 0$ for $X = \Xi$. By induction we have then all $I_k \equiv 0$ and hence also $P \equiv 0$. As the above is true for $c \leqslant c_1$ the theorem is proved. □

Theorem 6.2. An irreducible asc-set will determine uniquely the defining polset of conjugate extended points in $\mathbf{K}^n$. Conversely, the defining polset of any extended point in $\mathbf{K}^n$ will determine irreducible asc-sets which will in turn determine the given defining polset.

Proof. For an irreducible asc-set AS given as in (6.1-4) let us set $E_i = A_i/I_{i0}$ for $i = 1, \cdots, r$. Then E_i are irreducible in corresponding extension fields and will determine a set of conjugate extended points. Conversely, let an extended point has a set of defining polynomials E_i as in (1.1). By clearing of fractions of the equations $E_i = 0$ we will get in succession a set of polynomial equations $A_i = 0$ with $AS = \{A_1, \cdots, A_r\}$ an irreducible asc-set in ascending order. □

Example 6.2. Consider the asc-set $AS = \{A_1, A_2\}$ with

$$A_1 = x_2^2 - x_1, \ A_2 = x_1 * x_3 - x_2^3.$$

It is an irreducible asc-set which determines an extended point Ξ with defining pols

$$E_1 = x_2^2 - x_1, \ E_2 = x_3 - x_2.$$

The point Ξ may be equally well determined by the irreducible asc-set $AS' = \{A_1', A_2'\}$ with

$$A_1' = A_1, \ A_2' = x_2 * x_3 - x_1.$$

Remark 6.1. In comparing with Propositions 1.1,1.2 we see from Exs. 6.1,6.2 that the notion of irredicible asc-set, though in essence equivalent to the notion of an extended point, is nevertheless more flexible than the latter one in the representations. We may e.g. refer to a more sophiscate irreducible asc-set (6.23) in Ex. 6.3 below.

Remark 6.2. To see whether an asc-set AS of (6.1) is irreducible or not is not an easy task since it involves the problem of factorization of pols in algebraic extension fields. We shall deal with the latter problem in Chap. 4, Sect. 5.

Definition 6.4. Any irreducible asc-set AS which determines an extended point Ξ in $\mathbf{K}^n$:= A *defining asc-set* of Ξ.

Specialization of Extended Points.

Theorem 6.3. Let $\Xi = (\xi_1, \cdots, \xi_n)$ in $\mathbf{K}^n$ be an extended point with defining asc-set AS. For any pol P we have

$$P(\Xi) = 0 <===> Remdr(P/AS) = 0. \tag{6.5}$$

Proof. Let AS be given as in (6.1) so that $A_i(\Xi) = 0$. By the remainder formula (3.8) we have for any pol P,

$$I_P * P = \Sigma_i\, Q_i * A_i + R, \tag{6.6}$$

in which I_P is some power-product of initials of A_i and R is the remainder $Remdr(P/AS)$. By Theorem 6.1 we have $I_P(\Xi) \neq 0$. Hence from (6.6) we get

$$P(\Xi) = 0 <===> R(\Xi) = 0.$$

As R is reduced w.r.t. AS, we get again by Theorem 6.1

$$R(\Xi) = 0 <===> R = 0.$$

We get therefore (6.5) and the theorem is proved. □

Theorem 6.4. Let Ξ and Ξ' be extended points in $\mathbf{K}^n$ with defining asc-sets AS and AS' respectively. If Ξ' is a specialization of Ξ over $\mathbf{K}$, then

$$Remdr(AS/AS') = 0. \tag{6.7}$$

Conversely, if (6.7) is true and

$$I_{i0}(\Xi') \neq 0 \; or \; Remdr(I_{i0}/AS') \neq 0 \tag{6.8}$$

for any A_i in AS with $init(A_i) = I_{i0}$, then Ξ' is a specialization of Ξ.

Proof. Let AS be as in (6.1).

Suppose first that Ξ' is a specialization over $\mathbf{K}$ of Ξ. As $A_i(\Xi) = 0$ we have $A_i(\Xi') = 0$ too. By Theorem 6.3 we have then $Remdr(A_i/AS') = 0$ and hence (6.7).

Conversely suppose (6.7) is true. By Theorem 6.3 we have then $A_i(\Xi') = 0$. Consider now any pol P with $P(\Xi) = 0$. By Theorem 6.3 again we have then $R \equiv 0$ in (6.6). If (6.8) is true so that $I_P(\Xi') \neq 0$, then from (6.6) it follows that $P(\Xi') = 0$. Hence Ξ' is a specialization over $\mathbf{K}$ of Ξ. □

Remark 6.3. The above theorem partially solves the Problem S in Sect. 3.1.

Algebraic Varieties associated to an Irreducible Asc-Set.

Definition 6.5. For an irreducible asc-set IRR let Ξ be any extended point with IRR as defining asc-set. Then the algebraic variety $Spec_{\mathbf{K}}(\Xi)$ depending only on $IRR :=$ The *algebraic variety associated to* IRR.

Notation 6.2. For IRR an irreducible asc-set, $Var[IRR] :=$ Algebraic variety associated to IRR.

It follows from Def. 6.5 that we have for an irreducible asc-set IRR which defines an extended point Ξ:

$$Var[IRR] = Spec_{\mathbf{K}}(\Xi). \tag{6.9}$$

Theorem 6.5. We have for an irreducible asc-set IRR,

$$Var[IRR] \subset Zero(IRR). \tag{6.10}$$

Proof. Let the pols of IRR be given as in (6.1). For the generic point Ξ of $Var[IRR]$ we have then $A_i(\Xi) = 0$. Now any point Ξ' of $Var[IRR]$ is a specialization of Ξ so that we have $A_i(\Xi') = 0$ and we have thus $\Xi' \in Zero(IRR)$. Hence we have (6.10). □

Definition 6.6. Let IRR be an irreducible asc-set and IP be its initial-product, then a zero of $Zero(IRR/IP) :=$ A *regular zero* of the irreducible asc-set IRR.

Theorem 6.6. Any regular zero of an irreducible asc-set IRR is a point of the variety $Var[IRR]$, i.e.

$$Zero(IRR/IP) \subset Var[IRR]. \tag{6.11}$$

Proof. Let Ξ be a generic zero of $Var[IRR]$. Consider any pol P with $P(\Xi) = 0$. By Theorem 6.3 we have then $Remdr(P/IRR) = 0$. Hence by the remainder formula there will be some power-product of initials I_P of pols in IRR such that $I_P * P$ is equal to a linear combination of pols in IRR with pols as coefficients. Let Ξ' be any regular zero of IRR so that $I_P(\Xi') \neq 0$. Then the above remainder formula will give $P(\Xi') = 0$. It follows that Ξ' is a specialization of Ξ over $\mathbf{K}$ and belongs therefore to $Var[IRR]$. □

Remark 6.4. In general

$$Var[IRR] \neq Zero(IRR). \tag{6.12}$$

In view of (6.10) and (6.12) we use square bracket [ ] for the associated variety but not the parenthesis ( ) to avoid confusion. See Ex. 6.3 below.

Ideal associated to an Irreducible Asc-Set.

Theorem 6.7. Let IRR be an irreducible asc-set. Then the pols having remainder 0 w.r.t. IRR form a prime ideal.

Proof. By Theorem 6.3 the pols P having remainder 0 w.r.t. IRR are characterized by $P(\Xi) = 0$, where Ξ is a generic point of the variety $Var[IRR]$. That all such pols form an ideal and in fact a perfect ideal follows directly from this characterization. □

Notation 6.3. The ideal determined by IRR in Theorem 6.7 := $Ideal[IRR]$.

Definition 6.7. The ideal $Ideal[IRR]$:= The *ideal associated to* the irreducible asc-set IRR.

Remark 6.5.. Let $Ideal(PS)$ be the ideal with the polset PS as basis. Then we have

$$Ideal[IRR] \supset Ideal(IRR), \tag{6.13}$$

but in general

$$Ideal[IRR] \neq Ideal(IRR). \tag*{(6.13)$'$}$$

In view of (6.13) and (6.13)′ we use square bracket [ ] for the associated ideal but not the parenthesis ( ) to avoid confusion. See Ex. 6.3 below.

Finite Basis of $Var[IRR]$ or $Ideal[IRR]$.

The proofs of Theorems 6.8 and 6.9 below will be postponed to latter chapters.

Theorem 6.8. Based on Chow's theory of Chow forms, there is an algorithm which, for any irreducible asc-set IRR, will give in a finite number of steps a finite basis CB of $Var[IRR]$ or $Ideal[IRR]$, which may be determined directly from IRR such that

$$\begin{aligned} Var[IRR] &= Zero(CB), \\ Ideal[IRR] &= Ideal(CB). \end{aligned} \tag{6.14}$$

Definition 6.8. The basis CB in Theorem 6.8 or (6.13) := The *Chow basis* of the ideal $Ideal[IRR]$ or the variety $Var[IRR]$.

Theorem 6.9. Let $V = Var[IRR]$ and $V' = Var[IRR']$ be the algebraic varieties associated to the irreducible asc-sets IRR and IRR' respectively. Let CB be the Chow basis of $Var[IRR]$ or $Ideal[IRR]$. Then V' is a subvariety of V if and only if

$$Remdr(CB/IRR') = 0. \tag{6.15}$$

Proof. By definition of basis we have

$$Var[IRR] = Zero(CB).$$

Let Ξ' be a generic point of V'. Then by Theorem 6.3 (6.15) is equivalent to

$$P(\Xi') = 0, \text{ for all } pols\ P \in CB, \text{ or}$$

$$P = 0 \mid V' \text{ for all } pols\ P \in CB, \text{ or}$$

$$V' \subset Zero(CB).$$

Hence (6.15) is equivalent to $V' \subset V$. □

Remark 6.6. Theorem 6.9 furnishes a complete solution to Problem S in Sect. 3.1.

Variety-Decomposition Theorem.

Theorem 6.10. There is an algorithm which gives, for any polset PS, a finite number of irreducible asc-sets IRR_k and non-zero pols G_k reduced w.r.t. IRR_k in a finite number of steps, such that

$$Zero(PS) = \bigcup_k Zero(IRR_k/IP_k * G_k), \tag{6.16}$$

in which for each k, IP_k is the initial-product of IRR_k. The equation (6.16) is also equivalent to the following ones:

$$Zero(PS) = \bigcup_k Var[IRR_k], \tag{6.17}$$

$$Zero(PS) = \bigcup_k Zero(IRR_k/IP_k). \tag{6.18}$$

Moreover, redundant varieties in (6.17) can be removed by mere computations so that the union will become a non-contractible one and is unique up to an order.

Proof. By Zero-Decomposition Theorem I in Sect. 3.5 we have for some asc-sets CS_j,

$$Zero(PS) = \bigcup_j Zero(CS_j/IP_j), \tag{6.19}$$

or in more general form:

$$Zero(PS) = \bigcup_j Zero(CS_j/IP_j * H_j). \tag{6.19$'$}$$

In (6.19) or (6.19)$'$ H_j are some non-zero pols and IP_j are the initial-products of CS_j. Moreover, each CS_j is not of higher order than PS. If all CS_j are irreducible then we are done. Suppose there is some CS_j which is not irreducible. Let $CS_j = \{C_{j1}, \cdots, C_{js}\}$ in ascending order with C_{ji} of class c_i. Suppose that CS_j is irreducible at stage $k-1$ but no more at stage k. Then $CS'_j = \{C_{j1}, \cdots, C_{j,k-1}\}$ is irreducible and there are pols F_1, F_2 of class c_k, some pol I of class less than c_k, and also pols $Q_i, i = 1, \cdots, k-1$, such that

$$I * C_{jk} = F_1 * F_2 + \Sigma_i\, Q_i * C_{ji},$$

in which the summation is to be extended over $i = 1, \cdots, k-1$. Let $CS_j^+, CS_j^{(1)}, CS_j^{(2)}$ be the polsets gotten from CS_j in adjoining to it the pols I, F_1, F_2 respectively. Then it is clear that

$$Zero(CS_j) = Zero(CS_j^+) + Zero(CS_j^{(1)}/I) + Zero(CS_j^{(2)}/I). \tag{6.20}$$

By Theorems 3.2 or 3.6, all polsets $CS_j^+, CS_j^{(1)}, CS_j^{(2)}$ are of lower ordering than CS_j. Applying now the Zero-Decomposition Theorem I to each member in the right-hand side of (6.20) we may replace them by zero-sets of the form $Zero(CS'/IP' * G')$, in which CS' are asc-sets not of higher ordering than $CS_j^+, CS_j^{(1)}$, or $CS_j^{(2)}$, and hence all of lower ordering than CS_j.

Replacing now each member in the right-hand side of (6.19) or $(6.19)'$ for which CS_j is non-irreducible by these unions as in (6.20), we get a formula still of the form $(6.19)'$ but the members involving non-irreducible CS have been replaced by unions of zero-sets of polsets of lower ordering which can be decomposed as before. If in the new union there are still members with corresponding CS non-irreducible we can replace them by other ones of lower ordering as before. Proceed in this way further and further the procedure will come to an end owing to Theorems 3.3 or 3.7. We get thus finally an expression like $(6.19)'$ in which all the asc-sets CS are irreducible ones. Rewrite these CS by IRR then we get an expression like (6.16). If for some member in the right-hand side of (6.13) G_k is not reduced w.r.t. the corresponding IRR_k, then let $Remdr(G_k/IRR_k) = G'_k$ so that for some power-product I_{G_k} of initials of pols in IRR_k, we have $I_{G_k} * G_k \equiv G'_k \ mod_{\mathbf{K}}\ IRR_k$. If $G'_k = 0$, then $Zero(IRR_k/IP_k * G_k) = \varnothing$ and may be removed from the right-hand side of (6.16). Otherwise we have $Zero(IRR_k/IP_k * G_k) = Zero(IRR_k/IP_k * G'_k)$ and we may replace the member $Zero(IRR_k/IP_k * G_k)$ by $Zero(IRR_k/IP_k * G'_k)$. Hence we may take the expression (6.16) to be one for which all G_k are non-zero and reduced w.r.t. the corresponding IRR_k as to be asserted.

For such an expression (6.16) we see by Theorem 6.6 that

$$Zero(PS) \subset \bigcup_k Var[IRR_k]. \tag{6.21}$$

Let Ξ_k be now a generic point of $Var[IRR_k]$. By Theorem 6.1 and 6.6 we have then $\Xi_k \in Zero(IRR_k/IP_k * G_k)$. Hence $\Xi_k \in Zero(PS)$ by (6.16) and therefore $Var[IRR_k] \subset Zero(PS)$. It follows that

$$\bigcup_k Var[IRR_k] \subset Zero(PS). \tag{6.22}$$

Combining (6.21) and (6.22) we get then (6.17). We have also (6.18) as in the above proof in taking all $G_k = 1$.

Finally by Theorems 6.8 and 6.9 we can decide by computations whether one variety in the right-hand side of (6.14) is contained in some other one which may then be removed if it is so. In this way we will get a non-contractible expression in the form of (6.17). □

Definition 6.9. The theorem corresponding to a non-contractible union as in (6.17) := The *Variety-Decomposition Theorem* in what follows. Then the decomposition in the right-hand side of (6.17) unique up to an order := The *irreducible decomposition* of the variety $Zero(PS)$ with each member $Var[IRR_k]$:= An *irreducible component.*

Remark 6.7. In comparing with the Variety-Decomposition Theorem as described in Theorem 2.10 or Theorem 2.11, we see that the description here is more general and the proof is now a *constructive* one, in contrast to the previous proofs which are of a purely *existential* character.

Some Examples.

Example 6.3. Desargues Theorem. Let $ABC, A'B'C'$ be two triangles such that the corresponding sides are parallel to each other. Suppose that AA', BB' intersect in a point O. Then CC' will also pass through O.

This is one part of the theorem considered in Ex. 2.1.4′. To study the geometrical configurations verifying the hypothesis of the theorem let us take now coordinates so that $O = (0, 0)$ and

$$\begin{aligned} A &= (x_1, 0), \quad B = (0, x_3), \quad C = (x_4, x_5), \\ A' &= (x_2, 0), \quad B' = (0, x_6), \quad C' = (x_7, x_8). \end{aligned}$$

Then the hypothesis is equivalent to $HYP = 0$ where HYP is the polset $HYP = \{H_1, H_2, H_3\}$ with

$$\begin{aligned} H_1 &= x_1 * x_6 - x_2 * x_3, \\ H_2 &= x_4 * (x_8 - x_6) - x_7 * (x_5 - x_3), \\ H_3 &= (x_4 - x_1) * x_8 - x_5 * (x_7 - x_2). \end{aligned} \tag{6.23}$$

For the sake of avoiding too much uninteresting computations let us consider x_1, x_2, x_3 as non-zero independent constant parameters so that we are actually working in the affine space $A^5_{\mathbf{K}}(x_4, \cdots, x_8)$. The polset HYP is not in the form of an asc-set. By Zero-Decomposition Theorems we find easily

$$Zero(HYP) = \bigcup_i Zero(AS_i \ / \ IP_i), \tag{6.24}$$

in which i runs from 1 to 5 and IP are the initial-products of the asc-sets AS_i given below:

$$\begin{aligned}
AS_1 : \quad & x_1 * x_6 - x_2 * x_3, \\
& (x_1 * x_3 - x_1 * x_5 - x_3 * x_4) * x_7 + x_2 * x_4 * x_5 + (x_4 - x_1) * x_4 * x_6, \\
& (x_4 - x_1) * x_8 - (x_7 - x_2) * x_5; \\
AS_2 : \quad & x_1 * x_5 + x_3 * x_4 - x_1 * x_3, \ x_1 * x_6 - x_2 * x_3, \\
& (x_4 - x_1) * x_8 - x_5 * (x_7 - x_2); \\
AS_3 : \quad & x_4, \ x_5 - x_3, \ x_1 * x_6 - x_2 * x_3, \ x_1 * x_8 + x_3 * (x_7 - x_2); \\
AS_4 : \quad & x_4 - x_1, \ x_5, \ x_1 * x_6 - x_2 * x_3, \ x_1 * x_8 + x_3 * (x_7 - x_2); \\
AS_5 : \quad & x_4 - x_1, \ x_1 * x_6 - x_2 * x_3, \ x_7 - x_2, \ x_1 * x_8 - x_2 * x_5.
\end{aligned} \tag{6.25}$$

It turns out that all the above asc-sets AS_i are irreducible ones so that (6.24) will be sharpened to (6.26) below:

$$Zero(HYP) = \bigcup_i Var[AS_i], \tag{6.26}$$

in which i runs from 1 to 5. Each variety $Var[AS_i]$ has a generic zero $\Xi^{(i)}$ easily found to be given as follows:

$$\begin{aligned}
\Xi^{(1)} &= \left(x_4, \ x_5, \ \frac{x_2 * x_3}{x_1}, \ \frac{x_2 * x_4}{x_1}, \ \frac{x_2 * x_5}{x_1}\right); \\
\Xi^{(2)} &= \left(x_4, \ \frac{(x_1 - x_4) * x_3}{x_1}, \ \frac{x_2 * x_3}{x_1}, \ x_7, -\frac{(x_7 - x_2) * x_3}{x_1}\right); \\
\Xi^{(3)} &= \left(0, \ x_3, \frac{x_2 * x_3}{x_1}, \ x_7, -\frac{(x_7 - x_2) * x_3}{x_1}\right); \\
\Xi^{(4)} &= \left(x_1, \ 0, \ \frac{x_2 * x_3}{x_1}, x_7, \ -\frac{x_3 * (x_7 - x_2)}{x_1}\right); \\
\Xi^{(5)} &= \left(x_1, \ x_5, \ \frac{x_2 * x_3}{x_1}, x_2, \ \frac{x_2 * x_5}{x_1}\right).
\end{aligned} \tag{4.27}$$

It is easy to see that $\Xi^{(5)}$ is a specialization over $\mathbf{K}$ of $\Xi^{(1)}$, and $\Xi^{(3)}, \Xi^{(4)}$ are both specializations over $\mathbf{K}$ of $\Xi^{(2)}$ so that

$$Var[AS_5] \subset Var[AS_1],\ Var[AS_3] \subset Var[AS_2],\ Var[AS_4] \subset Var[AS_2].$$

On the other hand for Ξ_1, Ξ_2 neither one is the specialization over $\mathbf{K}$ of the other. From (6.26) we get therefore the non-contractible irreducible decomposition below:

$$Zero(HYP) = Var[AS_1] + Var[AS_2]. \tag{6.28}$$

The following diagrams Figs. 6.1,..., 6.5 give also a schematic images of the geometric configurations in the various zero-sets of (6.24):

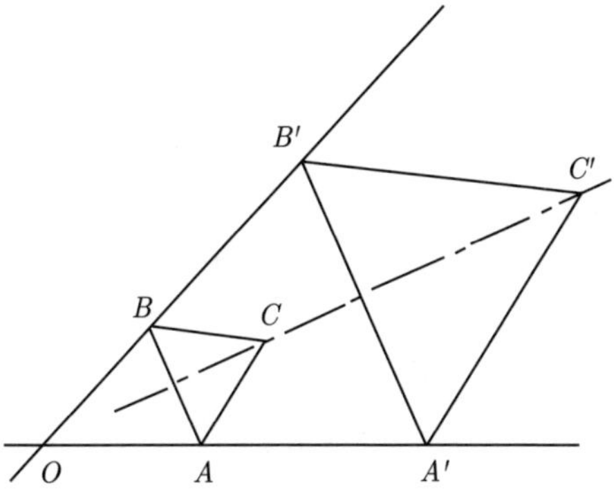

Fig. 6.1.

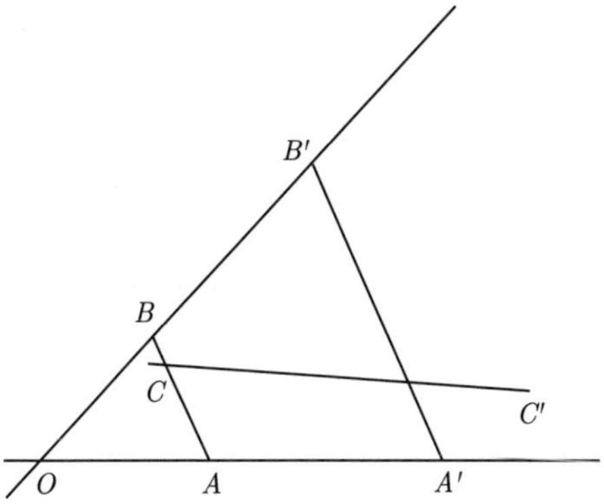

Fig. 6.2.

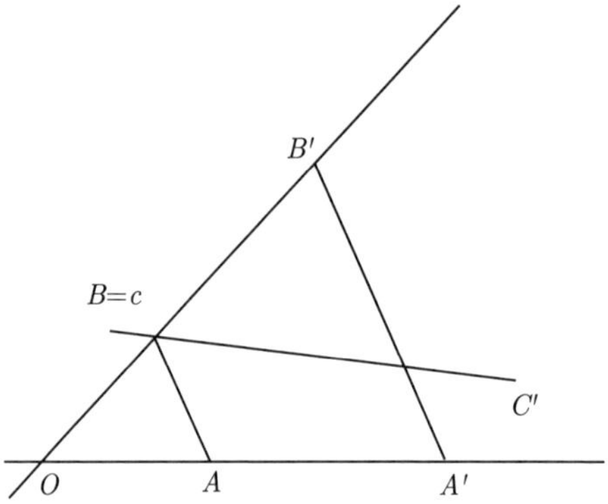

Fig. 6.3.

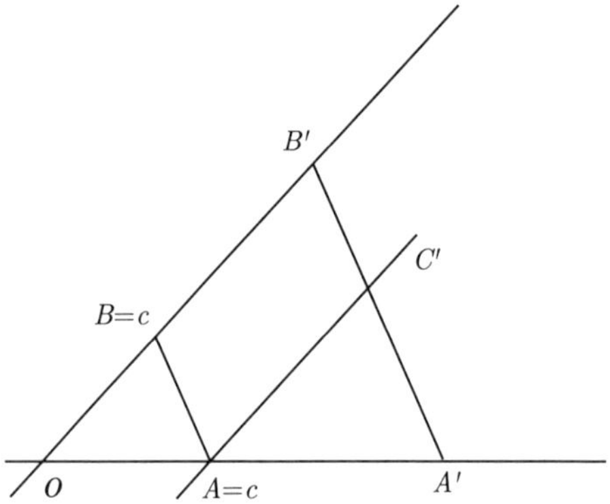

Fig. 6.4.

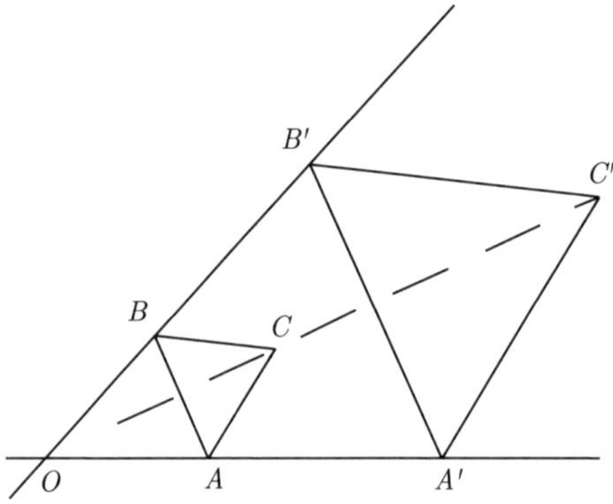

Fig. 6.5.

Example 6.4.

Let us consider the irreducible variety $Var[AS_1]$ in the preceding example Ex. 6.3. It is clear that

$$Var[AS_1] \subset Zero(AS_1).$$

Let IRR be the asc-set gotten from AS_1 in taking e.g. $x_1 = 2, x_2 = 4, x_3 = 2$. Then the asc-set IRR is still irreducible and $Var[IRR]$ has a generic point

$$\Xi^{(0)} = (x_4, x_5, 4, 2 * x_4, 2 * x_5) \in \mathbf{K}^5(x_4, \cdots, x_8),$$

and we have still

$$Var[IRR] \subset Zero(IRR).$$

Consider now the point

$$X^{(0)} = (1, 1, 4, 3, 1) \in \mathbf{K}^5(x_4, \cdots, x_8).$$

It is readily verified that $X^{(0)} \in Zero(IRR)$ but is not a specialization over $\mathbf{K}$ of $\Xi^{(0)}$ so that it is not in $Var[IRR]$. It follows that we have here

$$Var[IRR] \subset Zero(IRR) \textit{ but } Var[IRR] \neq Zero(IRR).$$

In a similar way we have

$$Ideal[IRR] \supset Ideal(IRR) \textit{ but } Ideal[IRR] \neq Ideal(IRR).$$

Remark 6.8. The irreducible asc-set IRR or more generally AS_1 in (6.23) is the char-set of the hypothesis set HYP of the Desargues Theorem of Ex. 6.3. See also Ex. 7.3.2 in Sect. 3 of Chap. 7. That $X^{(0)}$ is not in $Var[IRR]$ or $Var[AS_1]$ has an evident geometrical meaning. See Ex. 7.3.2 and the diagram Fig. 6.6 below.

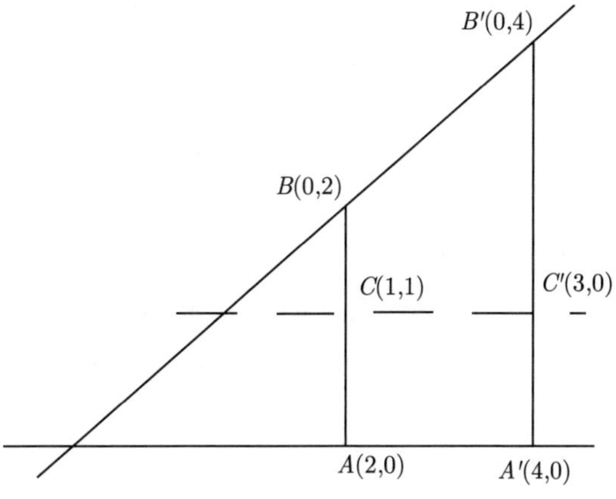

Fig. 6.6.

Chapter 4

Some Topics in Computer Algebra

4.1 Tuples of Integers.

Integer Tuples.

Let n be a positive integer fixed throughout this chapter unless otherwise stated.

Notation 1.1.

$N_n :=$ Set of integers $\{1, 2, \cdots, n\}$.

For a non-negative integer i, $i^+ :=$ The totality of integers greater than or equal to i.

Definition 1.1. An *n-tuple* or simply a *tuple* := An ordered sequence of n non-negative integers

$$t = (I_1, \cdots, I_n).$$

i-th coordinate of $t := I_i$.

Coor-Sum of $t :=$ Sum of i-th coordinates of t for $i \in N_n$.

Notation 1.2.

$Coor_i(t) := i$-th coordinate of tuple t.

$|t| :=$ Coor-sum of tuple t.

Definition 1.2. *Zero-tuple* or *0-tuple* := The tuple with all coordinates 0.

Notation 1.3.

0_n or simply $0 :=$ The zero-tuple $(0, \cdots, 0)$.

$1_i :=$ The tuple $(\delta_{i1}, \cdots, \delta_{ij}, \cdots, \delta_{in})$ with δ_{ij} its j-th coordinate.

Definition 1.3. For any two tuples u and v, u is a *multiple* of v or v is a *divisor* of u :=

$$Coor_i(u) \geqslant Coor_i(v), i = 1, \cdots, n.$$

Notation 1.4.

$u \gg v := u$ is a multiple of v, or

$v \ll u := v$ is a divisor of u.

Definition 1.4. *The Product* $uv = vu$ of two tuples u and v:= The tuple with

$$Coor_i(uv) = Coor_i(u) + Coor_i(v), \ i = 1, \cdots, n.$$

Definition 1.5. *The Least Common Multiple* of two tuples u and v := The tuple l with

$$Coor_i(l) = Max(Coor_i(u), Coor_i(v)), \ i = 1, \cdots, n.$$

Definition 1.6. *The Greatest Common Divisor* of two tuples u and v := The tuple g with

$$Coor_i(g) = Min(Coor_i(u), Coor_i(v)), \ i = 1, \cdots, n.$$

Notation 1.5.

$LCM(u, v)$:= Least common multiple of tuples u, v.

$GCM(u, v)$:= Greatest common divisor of tuples u, v.

Definition 1.7. *Quotient* of a tuple u by a divisor v of u := The tuple q with

$$Coor_i(q) = Coor_i(u) - Coor_i(v), i = 1, \cdots, n.$$

Notation 1.6.

$\frac{u}{v}$:= quotient of tuple u by its divisor v.

Ordering of Tuples.

Definition 1.8. For two tuples u and v, u is *higher than* v or v is *lower than* u := There is some $k > 0$ and $\leqslant n$ such that

$$Coor_i(u) = Coor_i(v), i > k,$$
$$Coor_k(u) > Coor_k(v).$$

Notation 1.7.

$u > v$:= u is higher than v, or

$v < u$:= v is lower than u.

Theorem 1.1. Any sequence of tuples steadily decreasing in order is finite.

Proof. For $n = 1$ the theorem is evident. Suppose now the theorem is true for $(n-1)$-tuples and consider the case of n-tuples. For this let us consider the sub-sequence of n-tuples of the given sequence for which the last coordinate reaches the minimum value. Delete the last coordinate to get a sequence of $(n-1)$-tuples steadily decreasing in order. By induction hypothesis this sequence of $(n-1)$-tuples or the subsequence of n-tuples of minimum last coordinate is finite. As the n-tuples of last coordinate greater than the minimum are clearly finite in number, the given sequence is finite too. □

Autoreduced Tuple-Set.

Definition 1.9. A set of tuples T is *autoreduced* := No t in T is a multiple of another t' in T.

Theorem 1.2. Any autoreduced set of tuples is finite.

Proof. For $n = 1$ the autoreduced set can have only one 1-tuple so the theorem is trivially true. Suppose now the theorem is true for $(n-1)$-tuples and let us consider the case of n-tuples. For this let t be any n-tuple of the given autoreduced set taken to be fixed in what follows. Let T_{ij} where $0 \leqslant j \leqslant Coor_i(t)$ be the subset consisting of all such tuples in the given set for which the i-th coordinate is equal to j. Then the given set, being autoreduced, is the union of the not necessarily disjoint sets T_{ij} which are clearly autoreduced and finite in number. Let T'_{ij} be the set of $(n-1)$-tuples obtained from T_{ij} in deleting the i-th coordinate from each of the n-tuple. By induction hypothesis each T'_{ij} is finite. Hence each set T_{ij} is finite and so is the given autoreduced set. □

Ordering of Autoreduced Sets.

Let us consider any two autoreduced sets AS and AS' with tuples arranged in increasing order:

$$\begin{aligned} (AS): \ & t_1 < t_2 < \dots < t_r, \\ (AS)': \ & t'_1 < t'_2 < \dots < t'_s. \end{aligned} \tag{1.1}$$

Definition 1.10. The autoreduced set (AS) is *higher than* the autoreduced set $(AS)'$, or $(AS)'$ *lower than* (AS) := Either of the two following cases holds:

(a) There is some $k \leqslant r$ and $\leqslant s$ such that

$t_i = t'_i$ for $i < k$, while $t_k > t'_k$.

(b) $r < s$ and $t_i = t'_i$ for $i \leqslant r$.

Notation 1.8.

$(AS) > (AS)' := (AS)$ is higher than $(AS)'$, or

$(AS)' < (AS) := (AS)'$ is lower than (AS).

Theorem 1.3. Any sequence of autoreduced sets steadily decreasing in order is finite.

Proof. Let the sequence be

$$(S): \ AS_1 > AS_2 > \cdots$$

and suppose the contrary that it is infinite. For each autoreduced set AS_i let its tuples be arranged in increasing order. By Theorem 1.1 the sequence of the first tuples in AS_i should be consisting of the same tuple, say t_1, from a certain stage onwards. Denote the corresponding infinite sequence of autoreduced sets from that stage onwards by

$$(S_1): \; AS_{11} > AS_{12} > \cdots$$

Again by Theorem 1.1 the sequence of second tuples in AS_{1i} should be consisting of the same tuple, say t_2, from a certain stage onwards. Denote the corresponding infinite sequence of autoreduced sets from that stage onwards by

$$(S_2): \; AS_{21} > AS_{22} > \cdots$$

The above reasoning can be repeated indefinitely so that we get an infinite sequence of tuples

$$(T): \; t_1 < t_2 < \cdots .$$

which is clearly an autoreduced set. This contradicts however the Theorem 1.2 and hence (S) is finite. □

From the very definition of the ordering we have also

Theorem 1.4. Let AS be an autoreduced set and u be a tuple which is not a multiple of any tuple in AS. Let AS' be the autoreduced set obtained by adjoining u to AS and then removing all tuples in AS which are multiples of u. Then AS' is an autoreduced set of lower ordering than AS.

Proof. Let the tuples of AS be arranged in increasing order as in $(1.1)_1$. Then either $t_k < u < t_{k+1}$ for some k with $0 < k < r$ or otherwise $u < t_1$ or $t_r < u$. In any case the tuple set gotten from AS in adjoining u to AS and removing all multiples of u is an autoreduced set which is of lower ordering than AS by the very definition.

□

Completion and Complete Tuple-Set.

Definition 1.11. For any finite set of tuples T, the *Maxim* of T := The tuple $Max(T) = (Max_1(T), ..., Max_n(T))$, with

$Max_i(T) \; = \; Max\{Coor_i(t) \mid t \in T\}$,

Definition 1.12. For any finite set of tuples T, the *completion* of T := The set of tuples

$\{u \mid u \ll \; Max(T) \text{ and } u \gg t \text{ for some } t \text{ in } T\}$.

Notation 1.9.

$Comp(T)$:= the completion of tuple-set T.

Definition 1.13. A finite set of tuples CT is *complete* := CT is the competion of itself, or $CT = Comp(CT)$.

Theorem 1.5. For any complete tuple-set CT there is a unique autoreduced tuple-set $T \subset CT$ such that $CT = Comp(T)$. Moreover, there is an algorithm to determine T from CT in a finite number of steps.

Proof. Let us arrange the tuples in CT in a decreasing definite order $t_1 > t_2 > \cdots > t_N$. Starting from $i = 1$ up to N, let us delete t_i whenever it is a multiple of some tuple in CT of lower ordering than t_i. Then the final remaining tuple-set T is clearly an autoreduced set required. That such an autoreduced tuple-set is unique may be seen as follows.

Suppose that T', T'' are two autoreduced tuple-sets contained in CT such that $CT = Comp(T') = Comp(T'')$. Let us arrange the tuples of T' in decreasing order: $t'_1 > \cdots > t'_{n'}$ and the tuples of T'' in decreasing order: $t''_1 > \cdots > t''_{n''}$. Suppose that $t''_1 \neq t'_1$. Now t'_1, being in CT, should be the multiple of some tuple say t''_i of T''. Similarly t''_i should be a multiple of some tuple say t'_j of T'. Then t'_1 will be a multiple of t'_j . As T' is autoreduced we have necessarily $t'_1 = t'_j$ and hence also $t'_1 = t''_i$. As $t''_1 \neq t'_1$, it follows that $t''_1 \neq t''_i$ so that $t''_1 > t''_i$ or $t''_1 > t'_1$. In the same manner we will have $t'_1 > t''_1$ if $t'_1 \neq t''_1$. This leads to a contradiction and therefore $t'_1 = t''_1$. The same reasoning shows that $t'_2 = t''_2$, etc. That $n' = n''$ is then clear so that $T' = T''$ as to be proved. □

Definition 1.14. *Contraction* of a complete tuple-set CT := The unique autoreduced tuple-set $T \subset CT$ such that $Comp(T) = CT$.

Notation 1.10.

$Cont(CT)$:= The contraction of the complete tuple-set CT.

The following two theorems follow directly from the very definitions.

Theorem 1.6. For an autoreduced tuple-set T we have

$$Cont\ Comp(T) = T.$$

Theorem 1.7. For a complete tuple-set CT we have

$$Comp\ Cont(CT) = CT.$$

Multipliers and Non-Multipliers.

Definition 1.15. For any finite set of tuples T and any tuple $t \ll Max(T)$, the integer i ($\geqslant 1, \leqslant n$) is a *multiplier* of t w.r.t T :=

$$Coor_i(t) \;=\; Max_i(T).$$

Otherwise i is a *non-multiplier* of t w.r.t. T :=

$$Coor_i(t) \;<\; Max_i(T).$$

Notation 1.11. For any finite set of tuples T and any tuple $t \ll Max(T)$:

$Mult(t/T)$:= Set of all multipliers of t w.r.t. T,

$Nult(t/T)$:= Set of all non-multipliers of t w.r.t. T,

Tuple-Decomposition Theorem.

Definition 1.16. The *total multiple set* of a finite set of tuples T := The set of all multiples of some tuple in T.

Definition 1.17. For $t \ll Max(T)$, the *total multiple set* of t w.r.t. T := The set of all multiples tv of t with

$COOR_i(v) = 0$ for i in $Nult(t/T)$.

Notation 1.12.

$TMS(T)$:= Total multiple set of tuple-set T.

$TMS(t/T)$:= Total multiple set of t w.r.t. tuple-set T.

Theorem 1.8. Let T be a finite set of tuples. For any tuple $u \in TMS(T)$ there is a unique tuple $t \in Comp(T)$ such that $u = tv$ for some tuple v with

u in $TMS(t/T)$ or

$COOR_i(v) = 0$ for i in $Nult(t/T)$.

Moreover, if u is itself in T, then $t = u$.

Proof. t is determined as $COOR_i(t) = Min(COOR_i(u), MAX_i(T))$. □

Definition 1.18. For a finite set of tuples T the *kernel* of a tuple $u \in TMS(T)$:= The unique tuple $t \in Comp(T)$ determined as in Theorem 1.8. On the other hand u := A *cover* of $t \in Comp(T)$.

Notation 1.13.

$Ker(u/T)$:= Kernel of u w.r.t. T.

Theorem 1.9. Tuple-Decomposition Theorem. For any finite tuple set T the total multiple set of T is the disjoint union of sets $TMS(t/T)$ with t running over $Comp(T)$.

Proof. For any tuple $u \in TMS(T)$ we have $u \in TMS(t/T)$ where $t = Ker(u/T) \in Comp(T)$. Hence

$$TMS(T) = \bigcup_t TMS(t/T),$$

in which the union is extended over $t \in Comp(T)$. Clearly these $TMS(t/T)$ are disjoint from each other. The theorem is thus proved. □

The total multiple set of t w.r.t. T may be made more precise as in the following theorem easily proved.

Theorem 1.10. For a tuple $t = (i_1, \cdots, i_j, \cdots, i_n) \in Comp(T)$, we have

$$TMS(t/T) = \{(i'_1, \cdots, i'_j, \cdots, i'_n) \mid (*)\}, \tag{1.2}$$

in which $(*)$ means

$$\begin{aligned} &i'_j = i_j, \ if\ i_j \in Nult(t/T), \\ &i'_j \in i_j^+, \ if\ i_j \in Mult(t/T). \end{aligned} \qquad (1.2)'$$

An Example.

Example 1.1. Consider a tuple set $T = \{t_1, t_2, t_3\}$ with

$$t_1 = (0, 0, 2),\ t_2 = (0, 1, 1),\ t_3 = (1, 0, 1).$$

Then we have $t_1 > t_2 > t_3, Max(T) = (1, 1, 2)$ and

$$\begin{aligned} Comp(T) &= \{1_1 1_2 t_1,\ 1_2 t_1,\ 1_1 t_1,\ t_1,\ 1_1 t_2,\ t_2,\ t_3\}, \\ &= \{(1,1,2),\ (0,1,2),\ (1,0,2),\ (0,0,2),\ (1,1,1),\ (0,1,1),\ (1,0,1)\}. \end{aligned}$$

The multiplier set of the above tuples in above order are respectively

$$\{1,2,3\},\ \{2,3\},\ \{1,3\},\ \{3\},\ \{1,2\},\ \{2\},\ \{1\},$$

The corresponding non-multiplier sets are then respectively

$$\varnothing,\ \{1\},\ \{2\},\ \{1,2\},\ \{3\},\ \{1,3\},\ \{2,3\}.$$

The total multiple sets are by (1.2) and (1.2)′ in some evident abridged notation:

$$\begin{aligned} TMS(1,1,2) &= \{(1^+, 1^+, 2^+)\}, \\ TMS(0,1,2) &= \{(0, 1^+, 2^+)\}, \\ TMS(1,0,2) &= \{(1^+, 0, 2^+)\}, \\ TMS(0,0,2) &= \{(0, 0, 2^+)\}, \\ TMS(1,1,1) &= \{(1^+, 1^+, 1)\}, \\ TMS(0,1,1) &= \{(0, 1^+, 1)\}, \\ TMS(1,0,1) &= \{(1^+, 0, 1)\}. \end{aligned}$$

Other Orderings of Tuples.

Remark 1.1. The ordering of tuples given above are the so-called *inverse lexicographic ordering*. Sometimes other orderings are more convenient than the above one. One of such orderings is the *coor-sum ordering* for which a tuple u is higher than a tuple v if either $|u| > |v|$ or $|u| = |v|$ and then $u > v$ in the above inverse lexicographic ordering. The further concepts and results with such coor-sum ordering can be developed in a parallel way as before and offer no essential difference. We shall stick to the inverse lexicographic ordering in all what follows unless otherwise stated.

4.2 Well-Arranged Basis of a Polynomial Ideal.

Polynomials in the Ring $\mathbf{K}[x_1, \cdots, x_n]$.

Throughout this chapter there will be fixed besides the integer n, also a set of variables $X = (x_1, ..., x_n)$, and a field $\mathbf{K}$ of characteristic 0 unless otherwise stated. As in the previous chapter, a *pol* will be meant a polynomial in $\mathbf{K}[X]$, and a *polset* is meant a finite set of *non-zero* pols unless otherwise stated.

Definition 2.1. A *monom* := A power-product in x_i of the form

$$X^t = x_1^{i_1} * \cdots * x_j^{i_j} \cdots * x_n^{i_n}, \ t = (i_1, \cdots, i_n).$$

It is clear that for any two tuples u, v we have

$$X^u * X^v = X^{uv}.$$

Definition 2.2. *Degree-Tuple* of the monom $X^t := t$.

Definition 2.3. A non-zero pol P is in *Normal Form* :=

$$P = a_1 * X^{t_1} + a_2 * X^{t_2} + \cdots + a_r * X^{t_r}, \tag{2.1}$$

with a_i non-zero in $\mathbf{K}$, and the degree-tuples t_i in decreasing order, viz.

$$t_1 > t_2 > \cdots > t_r. \tag{2.1$'$}$$

Such a normal form is clearly unique for P.

Definition 2.4. For a pol P in the above normal form:

$a_1 * X^{t_1}$:= *Leading term* of P,

X^{t_1} := *Leading monom* of P,

$a_1 :=$ *Leading coefficient* of P,

$t_1 =$ *Leading degree-tuple* of P.

Notation 2.1.

$Lterm(P) :=$ Leading term of P,

$Lmonom(P) :=$ Leading monom of P,

$Lcoef(P) :=$ Leading coefficient of P,

$Ldeg(P) :=$ Leading degree-tuple of P.

A Partial Ordering of Pols.

Definition 2.5. For two non-zero pols P_1 and P_2: P_1 is *higher than, lower than*, or *incomparable to* $P_2 :=$ The leading degree-tuple of P_1 is higher than, lower than, or identical to that of P_2 respectively.

Notation 2.2. For two non-zero pols P_1, P_2:

$P_1 > P_2 := P_1$ is higher than P_2.

$P_1 < P_2 := P_1$ is lower than P_2.

$P_1 <> P_2 := P_1, P_2$ are incomparable in order.

Autoreduced Polsets.

By means of the degree-tuples of pols in a polset we shall transfer various concepts concerning tuple-sets to polsets as follows.

Definition 2.6. For a polset PS the *degree-tuple-set* of $PS :=$ The set of leading degree-tuples of pols in PS.

Notation 2.3.

$DTS(PS) :=$ Degree-tuple-set of polset PS.

Definition 2.7. A polset PS is *autoreduced* $:=$ The degree-tuple-set of PS is autoreduced.

Definition 2.8. Let AS be an autoreduced polset and $T = DTS(AS)$ be its degree-tuple-set. A non-zero pol P is *reduced w.r.t.* $AS :=$ For each term in P, the corresponding degree-tuple is not a multiple of any tuple in T.

Definition 2.9. An autoreduced polset AS is *reduced* $:=$ Each pol of AS is reduced w.r.t. the autoreduced polset formed from AS by removing that pol.

A Partial Ordering of Autoreduced Polsets.

Definition 2.10. For two autoreduced polsets AS_1 and AS_2: AS_1 is *higher than, lower than*, or *incomparable to* $AS_2 :=$ The degree-tuple-set of AS_1 is higher than, lower than, or identical to that of AS_2 respectively.

Notation 2.4. For two autoreduced polsets AS_1, AS_2:

$AS_1 > AS_2 := AS_1$ is higher than AS_2.

$AS_1 < AS_2 := AS_1$ is lower than AS_2.

$AS_1 <> AS_2 := AS_1, AS_2$ are incomparable in order.

Theorem 2.1. Any two autoreduced polsets of lowest order contained in a given polset are incomparable in order.

Proof. It follows directly from the very definitions. □

A Partial Ordering of Polsets.

The Theorem 2.1 permits us to introduce a partial ordering of arbitrary polsets as in the following definition.

Definition 2.11. For two polsets PS_1 and PS_2: PS_1 is *higher than, lower than,* or *incomparable to* PS_2 := Any autoreduced subset of lowest order contained in PS_1 is higher than, lower than, or identical to any autoreduced subset of lowest order contained in PS_2.

Notation 2.5. For two polsets PS_1, PS_2:

$PS_1 > PS_2 := PS_1$ is higher than PS_2.

$PS_1 < PS_2 := PS_1$ is lower than PS_2.

$PS_1 <> PS_2 := PS_1, PS_2$ are incomparable in order.

Theorem 2.2. Given a polset PS let R be a pol reduced w.r.t. any autoreduced subset of lowest order contained in PS. Then the polset gotten from PS in adjoining to it the pol R is of lower order than the original polset PS.

Proof. This follows from the very definitions and Theorem 1.4. □

Remark 2.1. The above theorem gives a means of lowering the order of a polset.

Rests w.r.t. an Autoreduced Polset.

For an autoreduced polset AS consisting of non-zero pols A_i there may be different ways of putting a pol P into a form

$$\begin{aligned} P &= \Sigma_i\ E_i * A_i\ +\ R, \text{ or} \\ P &\equiv R\ mod_{\mathbf{K}}\ AS, \end{aligned} \tag{2.2}$$

in which E_i, R are pols and R, if not zero, is reduced w.r.t. AS. See the example below:

Example 2.1. Let $AS = \{A_1, A_2\}$ be an autoreduced set with

$$A_1 = x_1 * x_2 - 1,\ A_2 = x_2^2 - 1.$$

Then the pol

$$P = x_1 * x_2^2 + x_2^2 + x_1^2 * x_2$$

can be written in two ways as

$$P = x_1 * A_1 + (x_1 + 1) * A_2 \ + (2 * x_1 + 1), \text{ or}$$
$$P = (x_1 + x_2) * A_1 + A_2 + (x_1 + x_2 + 1).$$

Hence R may be either $2 * x_1 + 1$ or $x_1 + x_2 + 1$ and is not a unique one.

In order to get a unique R from P let us proceed in the following way. Let T be the degree tuple-set of AS. Write P in the normal form. Let $c * X^t, c$ in $\mathbf{K}$, be the non-zero term in P, if it exists, such that t is of *highest* order with t a multiple of some tuple in T. Once t is fixed, then u will be chosen to be the *highest* tuple in T for which t is a multiple. Write now $t = uv$ and let the pol in AS having u as its leading-degree-tuple be

$$A_i = a * X^u + A_i',$$

with $a * X^u$ as its leading term, $a \neq 0$ being in $\mathbf{K}$. Set

$$P_1 = P - \frac{c}{a} * X^v * A_i.$$

Then P_1 is such a pol that the term of highest order in P_1 having its degree-tuple t_1 a multiple of some tuple in T, if it exists, is of lower order than t above. We can then apply the same procedure to P_1 as above to get a pol P_2. By Theorem 1.1 the procedure should stop in a final number of steps and one will arrive at a pol P_s which is reduced w.r.t. AS. This pol P_s will then be the R required.

Definition 2.12. The *rest* of a pol P w.r.t. an autoreduced polset AS := The unique pol R reduced w.r.t. AS gotten from P in the above manner. *Subrest* of P w.r.t. AS := Any non-zero pol R reduced w.r.t. AS which verifies (2.2).

Notation 2.6.

$Rest(P/AS)$:= Rest of P w.r.t. autoreduced polset AS.

Example 2.1′. For AS and P in Ex. 2.1 we have $Rest(P/AS) = 2 * x_1 + 1$ but $\neq x_1 + x_2 + 1$. Both $2 * x_1 + 1$ and $x_1 + x_2 + 1$ are subrests of P w.r.t. AS.

We summarize the above in the form of the theorem below:

Theorem 2.3. Given an autoreduced polset AS there is uniquely determined for each pol P a pol R reduced w.r.t. AS such that

$$\begin{aligned} P &\equiv R \ mod_{\mathbf{K}} \ AS, \text{ in which} \\ R &= Rest(P/AS). \end{aligned} \tag{2.3}$$

Definition 2.13. The formula (2.2) with R given by (2.3) := The *rest formula* of P w.r.t. the autoreduced polset AS.

Remark 2.2. In view of Theorem 2.2 and Remark 2.1 we see that in order to lower the order of a polset it is sufficient to enlarge the polset by adjoining non-zero rests of some autoreduced subset of lowest order contained in that polset. This leads to the notion of *well-arranged basis* of a polset and its construction as given below.

Reduction of an Autoreduced Polset.

Theorem 2.4. There is an algorithm which permits to turn any given autoreduced polset AS into a reduced autoreduced polset AS' with identical degree-tuple-set.

Proof. Let us arrange the pols of AS in decreasing order $A_1 > A_2 > \cdots > A_N$. Set $A'_N = A_N$ and $AS'_N = \{A'_N\}$. From $i = N-1, N-2, \cdots, 1$ let us form successively the rest of A_i w.r.t. the autoreduced polset $AS'_{i+1} = \{A'_{i+1}, \cdots, A'_N\}$ with resulting pol A'_i. Then the final polset AS'_1 is the reduced autoreduced polset AS' required. □

Definition 2.13. *Reduction* of an autoreduced polset AS := The procedure described in the proof of Theorem 2.4.

Well-Arranged Basis of a Polynomial Ideal.

Notation 2.7. The polynomial ideal with a polset PS as basis := $Ideal(PS)$.

Definition 2.14. A *well-arranged basis* of a polynomial ideal $Ideal(PS)$ with basis PS := An autoreduced polset WAS which is a basis of the ideal such that $Ideal(WAS) = Ideal(PS)$.

Given an arbitrary finite polset PS of non-zero pols let us form now a scheme (A) below:

$$\begin{array}{rcccl} PS = PS_0 & PS_1 & \cdots & PS_r & \\ AS_0 & AS_1 & \cdots & AS_r & \\ RS_0 & RS_1 & \cdots & RS_r & = \varnothing. \end{array} \tag{A}$$

The scheme is formed in the following manner:

For each i AS_i is an autoreduced polset of lowest order with pols chosen from PS_i such that the degree-tuple of any remaining pol in PS_i is a multiple of the degree-tuple of some pol in AS_i. Each RS_i is then the polset of all non-zero rests, if there exists, of the pols in $PS_i \setminus AS_i$ w.r.t. AS_i. The polset PS_{i+1} is just the union of the previous AS_i and RS_i:

$$PS_{i+1} = AS_i + RS_i. \tag{2.4}$$

From the construction we see by Theorem 2.2 that the autoreduced sets AS_i are steadily decresing in order:

$$AS_0 > AS_1 > \cdots$$

The degree tuple-sets of AS_i are thus steadily decreasing in order. By Theorem 1.3 the sequence is finite so that the procedure has to stop at a certain stage with its corresponding rest-set $RS_r = \varnothing$ as shown in the diagram (A).

Theorem 2.5. The final autoreduced set AS_r in the scheme (A) forms a well-arranged basis for the ideal $Ideal(PS)$ with PS as a basis so that $Ideal(PS) = Ideal(AS_r)$.

Proof. Let AS_0 be consisting of pols P_i and the other pols in PS_0 be Q_j so that $Ideal(PS)$ has a basis consisting of pols P_i and Q_j. Let $R_j = Rest(Q_j/AS_0)$. Then by the rest formulae of Q_j w.r.t. AS it is clear that $Ideal(PS_0) = Ideal(PS_1)$. Continuing in the same way we have

$$Ideal(PS) = Ideal(PS_0) = Ideal(PS_1) = \cdots = Ideal(PS_r).$$

As $Rest(P/AS_r) = 0$ for any pol $P \in PS_r$ we have $Ideal(PS_r) = Ideal(AS_r)$ too so that $Ideal(PS) = Ideal(AS_r)$. AS_r is thus a well-arranged basis as to be proved.

□

Example 2.2. Let us consider Exs. 4.2 and 5.2 in chapter 3 of Czapor and Geddes again. For the polset $PS = \{P_1, P_2, P_3\}$ with P_i given by (4.20) in chapter 3 let us take again the ordering $x \prec y \prec z$. Then the autoreduced polset $WAS_0 = \{W_1, W_2, W_3\}$ with

$$W_1 = P_3,\ W_2 = P_1,\ W_3 = P_2 \tag{2.5}$$

is an autoreduced polset arranged in decreasing order which is already a well-arranged basis of the ideal $Ideal(PS)$. The leading degree-tuple-set of WAS_0 is $T = \{t_1, t_2, t_3\}$ with

$$t_1 = (0, 0, 2) > t_2 = (0, 1, 1) > t_3 = (1, 0, 1)$$

as given in Ex. 1.1.

Groebner Basis of a Polynomial Ideal.

In current literature the notion of *Groebner basis* of a polynomial ideal is widely applied for which we shall give only the definition and the property which is of particular interest to the main purposes of the present book.

Definition 2.15. For a polynomial ideal $ID \subset \mathbf{K}[X]$ an autoreduced polset GB is a *Groebner basis* of $ID :=$ GB is a basis of ID and all subrests of any pol P in ID w.r.t. GB are equal to 0.

Definition 2.16. For a polynomial ideal $ID \subset \mathbf{K}[X]$ an autoreduced polset GB is a *reduced Groebner basis* of $ID :=$ GB is a Groebner basis of ID and is reduced as an autoreduced polset.

Theorem 2.6. (Buchberger) For a polset $PS \subset \mathbf{K}[X]$ let GB be a Groebner basis of $Ideal(PS)$. Then $Zero(PS) = \varnothing$ if and only if for some non-zero constant $k \in \mathbf{K}$, we have $k \in GB$. Moreover, $Zero(PS)$ is non-empty and finite if and only if it is non-empty and for each $i, 1 \leqslant i \leqslant n$, there is a pol in GB with leading monom some power of x_i.

For the proof we refer to the original paper [Bu1] of Buchberger.

Remark 2.3. The (reduced) Groebner basis of an ideal possesses many agreeable properties which we shall not enter. On the other hand it will be introduced in latter sections some basis of a polynomial ideal which will be proved to be essentially identical to the reduced Groebner basis and enjoy the same agreeable properties but is introduced in an entirely different way from the current way of Groebner basis studies.

4.3 Well-Behaved Basis of a Polynomial Ideal.

A Principle of Basis Construction based on Riquier-Janet Theory.

For a polynomial ideal (or an algebraic variety) with a given basis PS, the Groebner basis GB requires that all pols in the given ideal should have their rests w.r.t. GB to be zero. As the pols in an ideal are infinite in number, the latter requirements are at first sight hardly possible to be fulfilled. The criterion of Buchberger shows however that one may restrict the requirement to certain finite set of *Spols* and Buchberger's algorithm is based on this criterion. On the other hand the present author has discovered an alternative method of forming Groebner basis while studying the theory of Riquier and Janet on partial differential equations. Cf. [Riq], [Jan]. When the partial differential equations are linear with constant coefficients in a single dependent variable, then the partial differential equations may be made to correspond to ordinary polynomial ones and we are actually concerned with an ordinary polynomial ideal. The procedure of Riquier-Janet will furnish us then a certain basis of the ideal which enjoys some well-behaved property. For this reason we have called the basis thus obtained a *well-behaved basis* of the given ideal. It turns out that a subbasis of this well-behaved basis is just a Groebner basis of the given ideal. The essence of this alternative method consists in, instead of forming rests of pols as in scheme (A) for well-arranged basis, and instead of forming rests of Spols as in scheme (B) for Groebner basis, rather forming the rests of products with single variables of pols in a definite manner. We shall exhibit the method in details in what follows and refer to the original paper [WWT18].

Completion of Autoreduced Polset.

Below AS will be an autoreduced polset with degree-tuple set T unless otherwise stated.

Notation 3.1. For any $t \in T$, $A_t :=$ The pol in AS with t as its degree-tuple.

Definition 3.1. For any $u \in Comp(T)$ the *completed pol of AS relative to u* := The pol $X^v * A_t$ where $u = tv$ with t the highest tuple in T which is a divisor of u.

Notation 3.2. For $u \in Comp(T)$, $H_u :=$ The completed pol of AS relative to u.

Definition 3.2. *Completion* of AS := The polset consisting of all completed pols H_u of AS relative to all $u \in Comp(T)$.

Definition 3.3. *Completed polset* := Completion of some autoreduced polset.

Definition 3.4. *Contraction* of a completed polset := The autoreduced polset for which the completed polset is its completion.

Notation 3.3. $Comp(AS) :=$ Completion of AS, AS being an autoreduced polset.

Notation 3.4. $Cont(CPS) :=$ Contraction of CPS, CPS being a completed polset.

The two following theorems are evident from the very definitions.

Theorem 3.1. For any autoreduced polset AS:

$$Cont\ Comp(AS) = AS. \tag{3.1}$$

Theorem 3.2. For any completed polset CPS:

$$Comp\ Cont(CPS) = CPS. \tag{3.2}$$

M-Products and M-Pols w.r.t. to a Completed Polset.

Below CPS with degree-tuple-set CT will be a completed polset with completed pols $H_u, u \in CT$ and contraction AS unless otherwise stated.

Definition 3.5. An *M-Product* of CPS := A product of the form $X^v * H_u$ in which H_u is the pol of CPS with degree-tuple $u \in CT$, and each i with $Coor_i(v) \neq 0$ is a multiplier of u, or uv is a cover of u in the sense of Def. 1.18.

Definition 3.6. An *M-pol* of CPS := A finite linear combination of M-products of CPS with coefficients in $\mathbf{K}$.

Canonical Representation of Pols w.r.t. a Completed Polset.

Theorem 3.3. Any pol P can be uniquely represented in the form

$$\begin{aligned} P &= M + N, \\ M &= \Sigma_{u,v}\, c_{uv} * X^v * H_u. \end{aligned} \tag{3.3}$$

In (3.3) M is an M-pol, $c_{uv} \in \mathbf{K}$ is nonzero, and $X^v * H_u$ are mutually different M-products. On the other hand N is reduced w.r.t. $AS = Cont(CPS)$.

Proof. Suppose that P is not reduced w.r.t. AS. Then in P there will be a term $a * X^w$ of highest order not reduced w.r.t. AS with $a \neq 0$ in $\mathbf{K}$. By the Tuple-Decomposition Theorem of Sect. 4.1 there will be a unique $u \in CT$ such that $w = uv$ is a cover of u. Let H_u be the completed pol of CPS with leading degree-tuple u and leading term $Lterm(H_u) = b * X^u$. Set

$$P_1 = P/a - X^v * H_u/b, \text{ or}$$

$$P = c_1 * M_1 + b_1 * P_1, \ (c_1 = a/b, b_1 = a),$$

with $M_1 = X^v * H_u$ an M-product. If P_1 is not reduced w.r.t. AS then there will be a term $a_1 * X^{w_1}$ not reduced w.r.t. AS which is of highest order in P_1 with $a_1 \neq 0$ in $\mathbf{K}$. Apply now the preceding procedure to P_1 we get then a pol P_2 so that $P_1 = c_2 * M_2 + b_2 * P_2$, with b_2, c_2 in $\mathbf{K}$ and M_2 an M-product of lower ordering than M_1. The procedure can be continued to get pols P_3, etc. By Theorem 1.1 on the leading degree-tuples we will arrive in a finite number of steps at some pol P_r which is reduced w.r.t. AS. We may then write P in the form (3.3) with M an M-pol and $N = b_r * P_r$ reduced w.r.t. AS as required. That the decomposition of form (3.3) is unique follows also easily from the Tuple-Decomposition Theorem 1.9 of Sect. 4.1. □

Definition 3.7. (3.3) := *Canonical Representation* of P w.r.t. the completed polset CPS.

Definition 3.8. The pols M and N in (3.3) := The *M-part* and the *N-part* respectively of the pol P w.r.t. the completed polset CPS.

N-pols and N-polset of a Completed Polset.

Consider any u in CT with corresponding completed pol H_u and any non-multiplier i of u. Then $1_i u = w$ is also in CT and the decomposition of $x_i * H_u$ into the M- and N-parts can be put in the form

$$x_i * H_u = a * H_w + M_{iu} + N_{iu}, \tag{3.4}$$

in which $a \neq 0$ is in $\mathbf{K}$, M_{iu} is an M-pol with each M-product in it of lower order than H_w or $x_i * H_u$, and N_{iu} is the N-part of $x_i * H_u$. Note that N_{iu} is reduced w.r.t. $AS = Cont(CPS)$. Owing to its importance we shall lay down the following

Definition 3.9. The N-part N_{iu} of pol $x_i * H_u$ in (3.4) := The *N-pol* of the completed polset CPS relative to the tuple $u \in CT$ and the non-multiplier $i \in Nult(u/CT)$.

Scheme (W) for the Formation of a Basis of a Polynomial Ideal.

Consider now a finite polset PS and let us form the scheme (W) below:

$$\begin{array}{llll} PS = PS_0 & PS_1 & \cdots & PS_s \\ \quad\; WAS_0 & WAS_1 & \cdots & WAS_s \\ \quad\; NS_0 & NS_1 & \cdots & NS_s = \varnothing. \end{array} \tag{W}$$

The scheme is formed in the following way:

For each i WAS_i is a well-arranged basis of the ideal $Ideal(PS_i)$, determined from PS_i as in Sect. 4.2 with scheme (A) applied to PS_i, and NS_i is the set of all non-zero N-pols of $Comp(WAS_i)$, if there are any. Finally, the polset PS_{i+1} is the union given below:

$$PS_{i+1} = WAS_i + NS_i. \tag{3.5}$$

As in the case of scheme (A), the sequence of autoreduced sets WAS_i is steadily decreasing in order so that the above procedure will end in a certain stage with corresponding $NS_s = \varnothing$ as shown in the diagram (W).

Theorem 3.4. The final well-arranged set WAS_s in the scheme (W) is a basis of the ideal $Ideal(PS)$, or

$$Ideal(PS) = Ideal(WAS_s).$$

Proof. By Theorem 2.1 we have $Ideal(PS_i) = Ideal(WAS_i)$. Now each pol N in NS_0 is the N-part of some pol $x_k * H_u$ with H_u the completed pol of $CPS_i =$

$Comp(WAS_i)$ relative to the tuple u in $CT_i = Comp(T_i)$ where T_i is the degree-tuple-set of WAS_i and $k \in Nult(u/CT_i)$ so that $x_k * H_u = M_{ku} + N_{ku}$ with M_{ku}, N_{ku} the M-part and N-part respectively. From the expression it follows that $N_{ku} \in Ideal(PS_i)$ for each pair (k, u) so that we have

$$Ideal(PS_i) = Ideal(PS_i + NS_i) = Ideal(WAS_i + NS_i) = Ideal(PS_{i+1}).$$

Consequently

$$Ideal(PS) = Ideal(PS_0) = Ideal(PS_1) = \cdots = Ideal(PS_s) = Ideal(WAS_s)$$

as to be proved. □

Well-Behaved Basis of a Polynomial Ideal.

Definition 3.10. A *well-behaved basis* of a polynomial ideal $Ideal(PS)$:= A completed polset WB with its contraction a basis of the ideal $Ideal(PS)$ observing the following property:

For any pol H_u in the completed polset WB and any non-multiplier i of u in the degree-tuple-set CT of WB, the N-part of $x_i * H_u$ w.r.t. WB is 0.

Definition 3.11. A well-behaved basis WB of a polynomial ideal is *reduced* := The contraction of WB is a reduced autoreduced polset.

From previous theorems we have as direct consequences the following two theorems:

Theorem 3.5. There is an algorithm which permits to determine in a finite number of steps a well-behaved basis of a given polynomial ideal $Ideal(PS)$ with arbitrary basis PS.

Proof. In fact, by Theorem 3.4 and the construction the completion of the final well-arranged polset WAS_s in the scheme (W) for PS is a well-behaved basis of the ideal $Ideal(PS)$. □

Theorem 3.6. A well-behaved basis WB of a polynomial ideal $Ideal(PS)$ possesses the following *well-behaved property*:

(WBP). Any pol $P \in \mathbf{K}[X]$ has a *unique* canonical representation w.r.t. the well-behaved basis WB in the form of (3.3).

Definition 3.12. *N-part* of a pol $P \in \mathbf{K}[X]$ w.r.t. the well-behaved basis WB of a polynomial ideal := The pol N when P is uniquely expressed in the canonical representation given in the form of (3.3).

Remark 3.1. The main reason that we lay emphasis on the completed well-behaved basis rather than its contraction, viz. the final autoreduced polset in the formation by scheme (W), is owing to the above well-behaved property for which the *uniqueness* of the representation is not possessed by its contraction. There is also some other advantage which will be seen in the next section.

An Example.

Example 3.1. Let us consider Ex. 2.2 of Czapor and Geddes again. In order to simplify the computations we shall consider only the special case for which $a = b = c = g = h = k = 1$ so that for the given polset $PS = \{P_1, P_2, P_3\}$ the pols P_i are simplified as follows:

$$\begin{aligned} P_1 &= x^2 + y * z + d * x + 1, \\ P_2 &= y^2 + z * x + e * y + 1, \\ P_3 &= z^2 + x * y + f * z + 1. \end{aligned} \tag{3.6}$$

We proceed to form well-arranged basis WAS_i successively according to scheme (W) in starting from $PS_0 = PS$. The well-arranged basis $WAS_0 = \{W_1, W_2, W_3\}$ of PS_0 with

$$W_1 = P_3,\ W_2 = P_1,\ W_3 = P_2 \tag{3.7}$$

is given as in Ex. 2.2. The completed H-pols of WAS_0 arranged in descending order are seen to be 7 in number, viz.

$$H_1 = y * x * W_1,\ H_2 = y * W_1,\ H_3 = x * W_1,\ H_4 = W_1,$$

$$H_5 = x * W_2,\ H_6 = W_2,\ H_7 = W_3.$$

Let the N-part of an H-pol H w.r.t. a non-multiplier i be denoted by $N(H/x_i)$. Most of the N-parts may be directly seen to be 0 by the very definition and the only non-zero N-parts are readily found to be the following ones, viz.

$$\begin{aligned} N_1 &= -N(H_5/z) \\ &= -z * H_5 + H_1 - f * H_5 + x^2 * H_7 + d * x * H_7 + H_7 \\ &= 2 * y^2 * x^2 + y^2 * x * d + y^2 + y * x^2 * e + y * x * d * e + y * x + y * e \\ &\quad -x^3 * f - x^2 * d * f + x^2 + x * d - x * f + 1. \\ N_2 &= N(H_6/z) \\ &= z * H_6 - H_2 + f * H_6 - x * H_7 - d * H_7 \\ &= z - 2 * y^2 * x - y^2 * d - y * x * e - y * d * e - y \\ &\quad +x^2 * f + x * d * f - x - d + f. \end{aligned}$$

$$
\begin{aligned}
N_3 =& \ N(H_7/y) \\
&= y * H_7 - H_5 \\
&= y^3 + y^2 * e + y - x^3 - x^2 * d - x. \\
N_4 =& \ N(H_7/z) \\
&= z * H_7 - H_3 - y * H_6 - e * H_6 + f * H_7 \\
&= z + y^2 * f - 2 * y * x^2 - y * x * d + y * e * f - y \\
&\quad -x^2 * e - x * d * e - x - e + f.
\end{aligned}
$$

The polset $PS_1 = WS_0 + NS_0$ is thus consisting of 7 pols, W_i and N_j. Proceeding further we find that for $r = 6$ the polset PS_6 is consisting of 4 pols Q_i below:

$$
\begin{aligned}
Q_1 =& \ z + y^2 * f - 2 * y * x^2 - y * x * d + y * e * f - y \\
&\quad -x^2 * e - x * d * e - x - e + f, \\
Q_2 =& \ y^2 * d * f + y^2 * f^2 + 2 * y^2 + y * G_1 + G_2, \\
Q_3 =& \ y * x^2 * G_3 + G_5, \\
Q_4 =& \ y * x * G_4 + G_6,
\end{aligned}
$$

in which $G_1, \cdots, G_6$ are pols in d, e, and f alone. The number of terms of Q_3 and Q_4 are respectively 90 and 314. To make computations not too complicate we shall further specify to the case $f = e$ which will not influence the computations already done. It turns out that in this case of $f = e$ the pols Q_i will be simplified to the following ones:

$$
\begin{aligned}
Q_1 =& \ z + y^2 * e - 2 * y * x^2 - y * x * d + y * e^2 - y \\
&\quad -x^2 * e - x * d * e - x, \\
Q_2 =& \ y^2 * d * e + y^2 * e^2 + 2 * y^2 \\
&\quad +4 * y * x^3 + 2 * y * x^2 * d - 2 * y * x^2 * e \\
&\quad -y * x * d * e - y * x * e^2 + 2 * y * x \\
&\quad +y * d * e^2 + y * e^3 + 2 * y * e \\
&\quad +2 * x^3 * e + 2 * x^2 * d * e - 2 * x^2 * e^2 + 2 * x^2 - 2 * x * d * e^2 \\
&\quad +d * e - e^2 + 2, \\
Q_3 =& \ F_1 * F_2, \\
Q_4 =& \ F_1 * F_3.
\end{aligned}
$$

The pols Q_3 and Q_4 split into factors F_i with

$$F_1 = 2*x^2 + x*d - x*e + 2,$$
$$F_2 = y*d*e - y*e^2 - 2*x^3 - 3*x^2*d - x*d^2 - x - e,$$
$$F_3 = 2*x^4 + 3*x^3*d + 2*x^3*e$$
$$+x^2*d^2 + 2*x^2*d*e + x^2*e^2 + x^2 + x*d*e^2 + 2*x*e + e^2.$$

The polset $PS_6 = \{Q_1, \cdots, Q_4\}$ is now already an autoreduced one and may be taken as the corresponding well-arranged set WAS_6. Let us denote its completed H-pols by

$$H_{1ij} = x^i * y^j * Q_1, i \leqslant 6, j \leqslant 2,$$
$$H_{2i} = x^i * Q_2, i \leqslant 6,$$
$$H_{3i} = x^i * Q_3, i \leqslant 4,$$
$$H_4 = Q_4.$$

The variables corresponding to the non-multipliers are then respectively at most

x, y	for H_{1ij},
x, z	for H_{2i},
x, y, z	for H_{3i},
y, z	for H_4.

To determine the N-pols let us consider first $N(H_{30}/y)$ where $H_{30} = Q_3$. By direct computation we find:

$$(d*e+e^2+2)*e*(d-e)*y*H_{30} = M_2*Q_4 + M_3*Q_3 + e^2*(d-e)^2*F_1*Q_2,$$

in which M_2, M_3 are pols in x of degree 2 and 3 respectively. As all terms $x^i * Q_j$ occuring in the right-hand side of the above equation are M-products it follows that

$$N(H_{30}/y) = 0.$$

Consider now $N(H_4/y)$ where $H_4 = Q_4$. Write F_2 in the form $F_2 = y*e*(d-e) - M$ with M a pol in x of degree 3. Then by simply rewriting we get

$$e*(d-e)*y*H_4 = F_3*Q_3 + M*Q_4,$$

in which all terms $x^i * Q_4$ in the right-hand side are M-products. We have thus again

$$N(H_4/y) = 0.$$

That other N-pols are all 0 may be deduced from the above ones or directly by rewriting in the similar way almost without computation. In conclusion it follows that $WAS_6 = \{Q_1, \cdots, Q_4\}$ is already a well-arranged basis of the the ideal $Ideal(PS)$ with its completion a well-behaved basis of the ideal.

Some Remarks on Zero-Set Formation.

We have seen in Ex. 3.5.2 that even in the general case of the example of Czapor and Geddes for which $a, \cdots, k$ are independent parameters, the zero-set of PS can be completely determined by its char-set, as shown in (3.5.50) with relative ease. However, this is not so immediate if well-behaved basis or its equivalent Groebner basis is used.

To see this let us try to determine the zero-set $Zero(PS) = Zero(WAS_6)$ via $WAS_6 = \{Q_1, \cdots, Q_4\}$ in the much simpler case of $a = b = c = g = h = k = 1, e = f$. The pols Q_3, Q_4 have been factored into the facors F_i. As F_1, F_3 are easily seen to be prime to each other (d, e are independent indeterminates) so $Q_4 = 0$ has 6 zeros of x, 2 from $F_1 = 0$ and 4 from $F_3 = 0$. For each zero of $F_1 = 0, Q_3$ will be 0 too and $Q_2 = 0, Q_1 = 0$ will give 2 zeros of WAS_6. On the other hand for each zero of $F_3 = 0$ we have $F_1 \neq 0$ and the resultant of Q_2 and F_2 w.r.t. y is found to be 0 in taking account of $F_3 = 0$. It follows that such a zero of F_3 will be extended to only one zero of WAS_6 determined by $F_2 = 0$ and $Q_1 = 0$. In all we have then 8 zeros of WAS_6 or PS. We remark that in the present case each zero of $Q_4 = 0$ can be extended to at least one zero of WAS_6. We see thus there are some complications in the determination of $Zero(PS)$ via the polset WAS_6 even in the present simple case which is incomparably simpler than the case in general. Cf. e.g. [WWT18].

4.4 Properties of Well-Behaved Basis and its Relationship with Groebner Basis.

Recall from Preceding Section.

Notation 4.1. Besides $\mathbf{K}, X = (x_1, \cdots, x_n)$, etc. as before we set:

$ID = Ideal(PS) :=$ A given polynomial ideal with a basis PS.

$WB :=$ A well-behaved basis of ID,

$CT :=$ The complete degree-tuple-set of WB,

$W_1, \cdots, W_s =$ The pols in WB arranged in decreasing order,

$u_i (\in CT) :=$ The leading degree-tuple of W_i,

$GB :=$ The contraction of WB,

$T = Cont(CT) :=$ Degree-tuple-set of GB,

$G_1, \cdots, G_r :=$ The pols in GB arranged in decreasing order,

For the basis WB and GB we have the relations

$$\begin{aligned} WB &= Comp(GB), \\ GB &= Cont(WB). \end{aligned} \tag{4.1}$$

We shall assume that GB is reduced, by further reductions if necessary and WB is then defined by $(4.1)_1$ so that WB is still a well-behaved basis of ID verifying (4.1). The purpose of this section is to study properties of WB or GB and in particular we shall prove, as shown in Theorems 4.4, 4.5 below, that GB, when reduced, is nothing else but the reduced Groebner basis up to non-zero constant multiples in $\mathbf{K}$.

For this purpose let us first remark that by Theorems 3.3, 3.6 any pol P will have a unique canonical representation

$$\begin{aligned} P &= C_1 * W_1 + \cdots + C_s * W_s + N_P, \\ C_i &= \Sigma_{k_i}\ a_{ik_i} * M_{ik_i}, \end{aligned} \tag{4.2}$$

in which C_i are M-pols, M_{ik_i} are M-products of WB relative to $u_i \in CT$, $a_{ik} \in \mathbf{K}$ are all non-zero, k_i runs over some set correspoding to various products of multipliers of u_i, and N_P is the N-part of P w.r.t. WB.

For each W_i let t_i be the divisor of u_i of highest order in $CT, v_i = \frac{u_i}{t_i}$, and G_j be the pol in GB with leading degree-tuple t_i. Then

$$W_i = X^{v_i} * G_j. \tag{4.3}$$

In grouping terms the formula (4.2) may now be rewritten in the form

$$\begin{aligned} P &= D_1 * G_1 + \cdots + D_r * G_r + N_P, \\ D_j &= \Sigma_{(j)}\ C_i * X^{v_i}, \end{aligned} \tag{4.4}$$

in which $\Sigma_{(j)}$ means that summation is to be extended over those i with same t_i as leading degree-tuple of G_j.

Membership Problem of a Polynomial Ideal.

The following problem is fundamental for ideal theory.

Problem M. To decide in an algorithmic way or to give a criterion whether a polynomial P is a member of the given ideal ID or not, i.e. whether $P \in ID$ or not.

The Problem M was first time solved by Buchberger by means of the Groebner basis, see e.g. [Bu3]. We shall instead give an alternative solution of the membership problem by means of the well-behaved basis as in the following theorem.

Theorem 4.1. A pol is in the ideal ID if and only if it is an M-pol of its well-behaved basis WB, or its N-part is 0.

Proof. It is clear that any M-pol of WB is in the ideal ID. Conversely, let CT be the degree-tuple set of WB. For any u in CT let H_u be the corresponding completed pol. It is enough to prove that any product of the form $M * H_u$ with M a monom and u in CT is an M-pol. We shall prove this by induction on the order of $M * H_u$ as well as on the number of x's in the monom M as follows.

If each i for which x_i appears in the monom M is a multiplier of u then $M * H_u$ is already an M-pol and nothing is to be proved. Suppose therefore $M = M' * x_i$ with i a non-multiplier of u. As WB is the well-behaved basis of the ideal, the N-pol relative to u and i is 0 so that (4.2) for $x_i * H_u$ may be written as

$$x_i * H_u = a * H_w + MP,$$

in which $w = ui, a\ (\in \mathbf{K}) \neq 0$, and MP is an M-pol of lower order than H_w or $x_i * H_u$. It follows that $M * H_u = a * M' * H_w + M' * MP$, of which $M' * MP$ is of lower order than $M * H_u$ and M' has a smaller number of x's than M. By induction $M' * H_w$ and each term in $M' * MP$ are M-pols and so is $M * H_u$. The theorem is thus proved. □

Theorem 4.2. The rest of any pol P w.r.t. the contraction of the well-behaved basis WB coincides with the N-part of P w.r.t. WB.

Proof. The rest is determined as a pol R with $P = \Sigma_k C_k * G_k + R$, in which G_k are the pols in $GB = Cont(WB), C_k$ are pols too, and R is reduced w.r.t. GB. By Theorem 4.1 $\Sigma_k C_k * G_k$ is an M-pol so that R is the N-part of P as to be proved.

□

By Theorems 4.2 the Theorem 4.1 can be restated in the following alternative form:

Theorem 4.3. A pol P belongs to an ideal ID if and only if its rest w.r.t. the contraction of a well-behaved basis WB of ID is 0.

Relation between Well-Behaved Basis and Groebner Basis.

The two following theorems show that the well-behaved basis and the Groebner basis of a polynomial ideal may be determined one from the other.

Theorem 4.4. The contraction of a well-behaved basis of a polynomial ideal is a Groebner basis of that ideal.

Proof. By Theorem 4.1 a pol will belong to a polynomial ideal with a well-behaved basis WB if and only if the N-part of that pol w.r.t. WB is zero. By Theorem 4.3 this will be so if and only if the rest of that pol w.r.t. the contraction GB of WB is zero. The same is true for subrests of that pol since they are reduced w.r.t. GB. By very Definition 2.15 the contraction GB is thus a Groebner basis of the ideal, as to be proved. □

Theorem 4.5. The completion of a Groebner basis of a polynomial ideal is a well-behaved basis of that ideal.

Proof. Similar to the proof of the preceding theorem. □

Remark 4.1. The theorems 4.4 and 4.5 show that for a polynomial ideal well-behaved basis and Groebner basis are actually concepts equivalent to each other. We have however lay emphasis on the former concept rather than the latter one owing to the well-behaved property (WBP) enjoyed by the former concept as stated in Theorem 3.6. A further advantage of the former concept may be seen from the formation of syzygies below.

Degree-Tuple-Set of a Polynomial Ideal.

The following definition and theorem are originated from Macaulay's work. Cf. [Mac]:

Definition 4.1. The *degree-tuple-set* of a polynomial ideal := The set of leading degree-tuples of all pols belonging to that ideal.

Theorem 4.6. The degree-tuple-set of a polynomial ideal is the total multiple set of the degree-tuple-set of a well-behaved basis.

Proof. In the unique expression (4.2) for any pol P w.r.t. a well-behaved basis WB of a polynomial ideal we shall have by Theorem 4.1, $N_P = 0$ if and only if P belongs to that ideal. From the Tuple-Decomposition Theorem of Sect. 4.1 we see that the leading degree-tuples of the various products $P_{ik} = M_{ik} * W_i$ are different from each other. So the leading degree-tuple of the pol P is the highest one among them. As each product P_{ik} is in the ideal, the degree-tuple set of the ideal is the set of degree-tuples of all such products P_{ik}. The latter one is however just the total multiple set of the degree-tuple set CT of the basis WB. This proves the theorem.

□

Syzygies of Basis of a Polynomial Ideal.

Definition 4.2. A *syzygy* w.r.t. a basis $PS = \{P_1, \cdots, P_h\}$ of a polynomial ideal $ID = Ideal(PS)$:= The coefficient vector $(g_1, \cdots, g_h)$ of a relation of the form

$$g_1 * P_1 + \cdots + g_h * P_h = 0, \tag{4.5}$$

with $g_i \in \mathbf{K}[X]$.

Theorem 4.7. The set of all syzygies w.r.t. a basis of a polynomial ideal forms a vector space of finite dimension over the polynomial ring $\mathbf{K}[X]$ which is independent of the basis of the ideal. Moreover, a finite basis of the vector space may be determined directly from a well-behaved basis of the ideal.

Proof. That the syzygies w.r.t. a basis PS of a polynomial ideal $ID = Ideal(PS)$ form a vector space over the ring $\mathbf{K}[X]$ is quite clear. Consider now any other basis $QS = \{Q_1, \cdots, Q_k\}$ of the same ideal $ID = Ideal(QS)$. Then each P_i is a linear combination of Q_j with coefficients in the ring $\mathbf{K}[X]$. Hence any syzygy w.r.t. the basis PS will be turned into a syzygy w.r.t. the basis QS. The reverse is also true so the vector spaces w.r.t. the two basis are isomorphic under the above correspondence.

To see that the vector space is of finite dimension over $\mathbf{K}[X]$ let us consider the well-behaved basis WB of the ideal ID. Consider now any pol W_j in the basis WB with leading degree-tuple $u_j \in CT$ and any non-multiplier k of u_j. As $x_k * W_j$ is in the ideal ID there is by Theorem 4.1 a relation of the form

$$x_k * W_j = \Sigma_i \ M_{kji} * W_i \tag{4.6}$$

in which M_{kji} are all M-pols of WB w.r.t. u_i. We can rewrite the above relation in the form

$$\Sigma_i \ S_{kji} * W_i = 0. \tag{4.7}$$

Then we shall see that the vectors $S_{kj} = (S_{kj1}, \cdots, S_{kjs})$ form a basis of the vector space of syzygies w.r.t. the basis WB so the vector space has a finite dimension.

To see this let us consider any syzygy equation w.r.t. WB of the form

$$\Sigma_i \ S_i * W_i = 0. \tag{4.8}$$

Denote the left-side of (4.8) by S. From the proof of Theorem 4.1 we see that S can be shown to be 0 by successive reductions in the form of (4.6) or (4.7). Hence $S = 0$ is a consequence of equations (4.7) or S is a linear combination of S_{kj} with pols as coefficients, as to be proved. □

Uniqueness of Reduced Well-Behaved Basis or Reduced Groebner Basis.

Theorem 4.8. The reduced well-behaved basis WB of an ideal ID is uniquely determined up to constant multiples.

Proof. We have shown how to determine from a given basis of the ideal ID by schemes (A) and (W), by a further reduction if necessary, a reduced autoreduced basis $GB = \{G_1, \cdots, G_r\}$ with its completion a reduced well-behaved basis WB of ID.

Consider now any other reduced autoreduced basis $GB' = \{G'_1, \cdots, G'_s\}$ with completion also a well-behaved basis. There is no loss of generality in assuming that all the pols in GB and GB' have been normalized to have their leading coefficients $= 1$. We are to prove that GB' coincides with GB.

To see this let us arrange the pols in GB and GB' both in decreasing order, viz.

$$GB: \ G_1 > G_2 > \cdots > G_r,$$

$$GB': \ G'_1 > G'_2 > \cdots > G'_s.$$

By Theorems 4.1 and 4.2 we have $Rest(G'_1/GB) = 0$ and by the corresponding rest formula we see that the leading degree-tuple of G'_1 should be a multiple of the leading-degree-tuple of some pol in GB, say G_i. In the same way, the leading-degree-tuple of G_i should be a multiple of the leading-degree-tuple of some G'_j of GB'. As GB' is autoreduced this will only be possible that G'_j coincides with G'_1. Then G'_1 will have same leading-monom as G_i. Applying the same reasoning to G_1 we see that G_1 should have same leading monom as some G'_k of GB'. This is only possible when $G_i = G_1, G'_k = G'_1$ and G_1, G'_1 have same leading-monoms.

Applying now the same reasoning to G_2 and G'_2 we see that they should have same leading monoms. Continuing we see then GB and GB' should have same number of pols or $r = s$ and each pair G_i and G'_i should have same leading monoms.

Consider now the last two pols G_r and G'_r in GB and GB'. As G'_r has same leading monom as G_r and G'_r has rest 0 w.r.t. GB we see that G'_r should be identical to G_r. Let us consider the pair G_{r-1} and G'_{r-1}. As the rest of G'_{r-1} w.r.t. GB is 0 we should have an identity of the form

$$G'_{r-1} = G_{r-1} + M_r,$$

in which M_r is an M-pol constructed from G_r. Now G'_{r-1} and G_{r-1} have the same leading monoms and no other monoms in G'_{r-1} and G_{r-1} can be multiples of the leading monom of G_r. It follows from the Tuple-Decomposition Theorem that this will be only possible when M_r=0 or G'_{r-1} is identical to G_{r-1}. Applying now the same reasoning to the other pairs of pols in GB and GB' successively in the reverse order we see that all the pairs should be identical to each other. The theorem is thus proved. □

Projection of a Polynomial Ideal.

Definition 4.3. *Projection* of an ideal $ID \subset \mathbf{K}[X]$ on the affine space $A^m_{\mathbf{K}}(X_m)$ with $m < n$:= The ideal of all pols in ID which are contained in the polynomial ring $\mathbf{K}[X_m]$ where $X_m = (x_1, \cdots, x_m)$.

Notation 4.1. $Proj_m\ ID$:= The projection of the ideal ID on the affine space $A^m_{\mathbf{K}}(X_m)$.

Theorem 4.9. Let WB be a reduced well-behaved basis of an ideal $ID \subset \mathbf{K}[X]$. Then the polset

$$WB' = WB \cap \mathbf{K}[X_m] \tag{4.9}$$

is a reduced well-behaved basis of the ideal

$$ID' = Proj_m\ ID \subset \mathbf{K}[X_m].$$

Proof. Let CT be the degree-tuple-set of WB and CT' that of WB'. Consider now any pol P in ID'. Let us consider P as a pol in ID and write it in the form (4.2). By Theorem 4.1 we have $N_P = 0$. By the Tuple-Decomposition Theorem we see that in (4.2) for each term in H_u we should have $COOR_k(u) = 0$ for $k > m$. Let $Max'(CT)$ be the m-tuple gotten from $Max(CT)$ by deleting the last $n - m$ coordinates. It is clear that $Max(CT') \ll Max'(CT)$. It follows that for each H_u in (4.2) for which $COOR_k(u) = 0$ for $k > m$, each i with x_i occuring in M_u which is a multiplier of u w.r.t. WB should also be a multiplier of u w.r.t. WB'. Hence the N-part of P, considered as a pol in ID', is 0 too w.r.t. WB'. By Definition 3.10 WB' is then a reduced well-behaved basis of the ideal ID'. This completes the proof of the theorem. □

Projection of a Prime Ideal.

In the case of a prime ideal its projection can be determined in a way as given in the following theorem:

Theorem 4.10. For a prime ideal $ID \subset \mathbf{K}[X]$ let IRR be the irreducible asc-set determining a generic point of the irreducible variety $Zero(ID)$ so that

$$Ideal(ID) = Ideal[IRR].$$

Let $m < n$ and IRR' be the subset of IRR consisting of those pols which involve X_m alone but none of the other x_i for $i > m$. Then

$$Proj_m\ Ideal(ID) = Ideal[IRR']. \tag{4.10}$$

Proof. W.r.t. the natural ordering

$$x_1 \prec \cdots \prec x_m \prec \cdots \prec x_n$$

let IRR be given in ascending order as

$$A_1, \cdots, A_r.$$

Let IRR' be the partial asc-set of IRR consisting of $A_1, \cdots, A_s$ with $s < r$. For any pol $P \in \mathbf{K}[X]$ to be in $Ideal(ID)$ it is by Theorem 3.6.7 necessary and sufficient that

$$Remdr(P/IRR) = 0.$$

A pol $P \in \mathbf{K}[X_m]$ considered as a pol in $\mathbf{K}[X]$ the above condition becomes

$$Remdr(P/IRR') = 0$$

or by Theorem 3.6.7 again $P \in Ideal[IRR']$. Hence we get (4.10) and the theorem is proved. □

Syzygy-Ideal of a Polset.

The following is an important application of Theorem 4.10. First we shall introduce some general notions.

Defintion 4.4. A *syzygy-pol* of a polset $PS = \{P_1, \cdots, P_m\} \subset \mathbf{K}[X] :=$ A pol $S \equiv S(y_1, \cdots, y_m) \in \mathbf{K}[Y], Y = (y_1, \cdots, y_m)$, such that

$$S(P_1, \cdots, P_m) = 0.$$

It is clear that the set of all such syzygy-pols will form an ideal in $\mathbf{K}[Y]$.

Defintion 4.5. The *syzygy-ideal* of a polset $PS = \{P_1, \cdots, P_m\} \subset \mathbf{K}[X] :=$ The ideal in $\mathbf{K}[Y]$ formed by all syzygy-pols of PS in $\mathbf{K}[Y]$, where $Y = (y_1, \cdots, y_m)$.

Notation 4.2. Syzygy-ideal of a polset $PS \subset \mathbf{K}[X] := Syz\ PS$.

The following theorem settles now the problem of determining the syzygy-ideal of a given polset.

Theorem 4.11. For a given polset $PS = \{P_1, \cdots, P_m\} \subset \mathbf{K}[X]$ let us consider the polset $QS = \{Q_1, \cdots, Q_m\} \subset \mathbf{K}[X, Y]$ where $Q_i = y_i - P_i(X),\ i = 1, \cdots, m$. W.r.t. the ordering of variables

$$y_1 \prec \cdots \prec y_m \prec x_1 \prec \cdots \prec x_n \tag{4.11}$$

let the char-set CS of the polset QS be given in ascending order as

$$C_1, \cdots, C_t.$$

Let $CS' : C_1, \cdots, C_s,\ (s < t)$ be the partial asc-set of CS consisting of those pols which involve $y_1, \cdots, y_m$ alone but none of $x_1, \cdots, x_n$. Then both asc-sets CS and CS' are irreducible and we have

$$Syz\ PS = Ideal[CS']. \tag{4.12}$$

Proof. The variety $Zero(QS)$ is irreducible since it has an evident generic point

$$(x_1, \cdots, x_n, y_1, \cdots, y_m) = (\xi_1, \cdots, \xi_n, \eta_1, \cdots, \eta_m)$$

in which $\xi_1, \cdots, \xi_n$ are independent transcendental elements while $\eta_i = P_i(\xi_1, \cdots, \xi_n),\ i = 1, \cdots, m$. The asc-set CS is thus an irreducible one representing some generic point of the irreducible variety $Zero(QS)$ w.r.t. the ordering (4.11). The partial asc-set CS' is then also irreducible and the present theorem follows from Theorem 4.10. □

A Theorem of Chou-Schelter-Yang.

Let IRR be an irreducible asc-set which determines an irreducible variety $Var[IRR]$ with corresponding ideal $Ideal[IRR]$. In the latter Chap. 5 we shall show how a basis CB called the *Chow basis* of the variety $Var[IRR]$, or what is the same a basis of the ideal $Ideal[IRR]$, can be determined directly from the irreducible asc-set IRR such that

$$Ideal[IRR] = Ideal(CB).$$

On the other hand Chou, Schelter, and Yang have shown how to determine a basis of $Ideal[IRR]$ (or what is the same a basis of $Var[IRR]$)) in the form of Groebner basis directly from IRR. See their joint paper [C-S-Y]. In terms of well-behaved basis we shall restate their theorem in the following form:

Theorem 4.12. (Chou-Schelter-Yang). Let IRR be an irreducible asc-set in $\mathbf{K}[X]$ with initial-product IP. Let z be a new variable to be arranged in an order behind all x_i in X. Let IRR^+ be the asc-set consisting of IRR as well as the pol $IP * z - 1$. Then the projection of the well-behaved basis WB^+ of the ideal $Ideal(IRR^+)$ on $A^n_{\mathbf{K}}(X)$ is a well-behaved basis WB of the ideal $Ideal[IRR]$:

$$WB = WB^+ \cap \mathbf{K}[X].$$

Proof. Let us consider any pol P in $Ideal[IRR]$. By Theorem 3.6.7 we have then $Remdr(P/IRR) = 0$. Hence for some non-negative integers s_i we have

$$I_1^{s_1} * \cdots * I_r^{s_r} * P \equiv 0 \ mod_{\mathbf{K}} \ IRR.$$

in which $I_1, \cdots, I_r$ are the initials of pols in IRR. Let $s = max_i \ s_i$, then in multiplying the above congruence by $I_1^{s-s_1} * \cdots * I_r^{s-s_r} * z^s$, we shall have

$$(IP * z)^s * P \equiv 0 \ mod_{\mathbf{K}} \ IRR.$$

From this we deduce readily

$$P \equiv 0 \ mod_{\mathbf{K}} \ IRR^+.$$

It follows that P is in the ideal $Ideal(IRR^+)$. As P is an arbitrary pol in $Ideal[IRR]$, we have

$$Ideal[IRR] = Proj_n \ Ideal(IRR^+),$$

in which $Proj_n$ means projection on the affine space $A^n_{\mathbf{K}}(X)$. The present theorem follows therefore from Theorem 4.9. □

Some Remarks.

Remark 4.2. Most of above theorems are actually discoveries of Buchberger and various researchers working on the study of Groebner basis. The (reduced) well-behaved basis of a polynomial ideal ID is in fact nothing else but the completion of usual (reduced) Groebner basis of ID. What we have done is merely to restate and reprove these theorems in the form of well-behaved basis as given above. The treatements and proofs are however done along the line of thoughts of previous sections, giving thus alternative proofs of these known theorems about Groebner basis different from the known ones. Moreover, the proofs are so given in order that they may be readily transferred to the differential case as described in the beginning of Sect. 4.3 for which the corresponding notion of Groebner basis is non-existant. Furthermore, the above treatment shows that any pol in $\mathbf{K}[X]$ will have a *unique* expression w.r.t. a well-behaved basis, of a polynomial ideal in the form of equation (4.2), which is a property more precise than the corresponding known one for a reduced Groebner basis under the usual known treatment. Cf. e.g. [Wall]. A further advantage of the well-behaved basis may be seen from Theorem 4.7 for the determination of a basis of the syzygy vector space of a polynomial ideal.

Remark 4.3. A comparison of the development of well-behaved basis of a polynomial ideal in this chapter and that of the char-set of a polset in the previous chapter may be of interest. W.r.t. a fixed ordering of variables in question we first give a patial ordering of pols; then a partial ordering of some special kinds of polsets, autoreduced sets in the one case and asc-sets in the other case; then a partial ordering of general polsets according to such special kinds of polsets contained therein. We lower the ordering of polsets in successively adjoining some rests in the one case and some remainders in the other case to arrive at the final goal. The difference in the final result is due mainly to the first step: The partial ordering of pols is in accordance to the *leading monoms* in the first case while it is in accordance to the *leading variables* in the other case. The further steps are then rather natural ones and are essentially parallel to each other.

4.5 Factorization and GCD of Multivariate Polynomials over Arbitrary Extension Fields.

Extension Field of Rational Field.

Notation 5.1.

Consider a variable x and sets of variables

$U := (u_1, \cdots, u_d)$.

$Y := (y_1, \cdots, y_r)$.

$X := (x_1, \cdots, x_N)$.

Ordering: $u_1 \prec \cdots \prec u_d \prec y_1 \prec \cdots \prec y_r \prec x_1 \prec \cdots \prec x_N \prec x$.

Irreducible asc-set $AS := \{A_1, \cdots, A_r\} \subset \mathbf{Q}[U, Y]$ with

$$A_i = I_{id_i} * y_i^{d_i} + \cdots + I_{i0}, \text{ with}$$

$$i \in I = (1, \cdots, r), \ I_{ij} \in \mathbf{Q}[U, y_1, \cdots, y_{i-1}],$$

I_{id_i} being reduced w.r.t. A_j for $j < i$.

$\mathbf{K}$:= The rational function field $\mathbf{Q}(U)$.

$\mathbf{K}^* :=$ The extension field $\mathbf{Q}(U, Y) = \mathbf{K}(Y)$ with Y defined by AS.

$T :=$ The set of all r-tuples of integers $t = (t_1, \cdots, t_r)$ with $0 \leqslant t_i < d_i, i \in I$.

For $t \in T$, $Y^t := y_1^{t_1} * \cdots * y_r^{t_r}$.

$\mathbf{B} :=$ The basic field $\mathbf{K}$ or $\mathbf{Q}$.

For $PS \subset \mathbf{B}[X], G \in \mathbf{B}[X], Zero_{\mathbf{B}}(PS/G) :=$ Set in $\mathbf{B}^N$ of zeros $Zero(PS/G) \cap \mathbf{B}^N$.

For $P_1, P_2 \in \mathbf{B}[x, Y]$, $P_1 = P_2\ mod_{\mathbf{B}}\ AS := P_1 - P_2$ is a linear combination of A_i in AS with pols in $\mathbf{B}[x]$ as coefficients.

Factorization Problem and GCD Problem.

Factorization Problem F. Given a pol $F \in \mathbf{Q}[x, U, Y]$ to be considered as a pol in $\mathbf{K}^*[x]$. To factorize F in $\mathbf{K}^*[x]$.

GCD Problem G. Given two pols $F, G \in \mathbf{Q}[x, U, Y]$ to be considered as pols in $\mathbf{K}^*[x]$. To determine the GCD of F, G over $\mathbf{K}^*$.

It seems that the above problems, fundamental as they are, were little touched in the literature in the general case, with the exception of perhaps some papers of Trager, and Hu and Wang, cf. [Tr], [H-W], [WDM3]. To see the delicate points of the general problem let us first consider some simple examples below:

Example 5.1. Factorize

$$F = u * x^2 - 2 * y * x - u + 1 \in \mathbf{Q}[x, u, y]$$

as a pol in x with y satisfying the equation

$$A \equiv y^2 - u = 0.$$

The solution is given by

$$F = u * \left(x + \frac{u - y}{u}\right) * \left(x - \frac{u + y}{u}\right)\ mod_{\mathbf{K}}\ A$$

which gives a factorization of F over $\mathbf{K}^*$. If we write the above equation as one in $\mathbf{Q}[x, u, y]$, then we would have

$$u * F = (u * x + u - y) * (u * x - u - y)\ mod_{\mathbf{Q}}\ A.$$

The factorization of F becomes thus obscure owing to the factor u in the left-hand side of the equation.

Example 5.2. Factorize

$$F = x^2 - u \in \mathbf{Q}[x, u, y]$$

as a pol in $\mathbf{K}^*[x]$ with y satisfying the equation

$$A \equiv u^3 * y^2 - 1 = 0.$$

The solution is given by

$$F = (x + u^2 * y) * (x - u^2 * y) \; mod_{\mathbf{Q}} \; A.$$

The factors have powers of u higher than that in original F.

The above simple examples show the complexity of factorization (and also GCD determination) of multivariate polynomials over arbitrary extension fields. They suggest that to deal with the Problems F and G, it is more natural to treat them over the field $\mathbf{K} = \mathbf{Q}(U)$ or the field $\mathbf{K}(Y) = \mathbf{K}^*$. The aim of this section is to give an alternative method for the solving of Problems F and G in modifying slightly the method of Hu and Wang. For this purpose we make first some preparations below.

Some Preliminaries.

The following fact is well known:

Lemma 5.1. Any element $q \in \mathbf{K}^*$ can be uniquely written in the form

$$\Sigma_{t \in T} \; q_t * Y^t \in \mathbf{K}[Y], \; q_t \in \mathbf{K}. \tag{5.1}$$

From Lemma 5.1 we get immediately the following lemma as corollary:

Lemma 5.2. Any pol $P \in \mathbf{K}^*[x]$ can be uniquely written in the form

$$P = \Sigma_{t \in T} \; p_t * Y^t \in \mathbf{K}[x, Y], \; p_t \in \mathbf{K}[x], \text{ or} \tag{5.2}$$

$$P = \Sigma_k \; p_k * x^k, \; p_k \in \mathbf{K}^*, \text{ with} \tag{5.3}$$

$$p_k = \Sigma_{t \in T} \; p_{kt} * Y^t, \; p_{kt} \in \mathbf{K}. \tag{5.4}$$

To avoid confusions we introduce the following definitions and notations.

Definition 5.1. The form in (5.1) of $q \in \mathbf{K}^*$:= The *canonical form* of q.

Definition 5.2. The form in (5.3) and (5.4) of $P \in \mathbf{K}^*[x]$:= The *canonical form* of P.

Notation 5.2.

$[q]$:= Canonical form of $q \in \mathbf{K}^*$.

$[P]$:= Canonical form of $P \in \mathbf{K}^*[x]$.

For example, for the coefficients of y_i-powers in A_i we have

$$[I_{ij}] = \Sigma_{t\in T}\ I_{ijt} * Y^t,\ I_{ijt} \in \mathbf{K}. \tag{5.5}$$

With these definitions and notations computations in the realm of $\mathbf{K}^*[x]$ can be reduced to computations in $\mathbf{K}[x, Y]$ as in the following lemma.

Lemma 5.3. For pols $P_1, P_2 \in \mathbf{K}^*[x]$, $P_1 = P_2$ as pols in $\mathbf{K}^*[x]$ is equivalent to

$$[P_1] = [P_2]\ \ mod_{\mathbf{K}}\ AS \tag{5.6}$$

as pols in $\mathbf{K}[x, Y]$

Factorization of Multivariate Polynomials over Rational Coefficients.

There are well-known methods for the factorization of multivariate pols over $\mathbf{Q}$ or $\mathbf{Z}$ as coefficients which are quite efficient and are available on computers in various package-forms. Below we shall describe a method of factorization due to Hu and Wang in their joint paper [H-W] which may also be applied to the solving of Problems F and G in most general case. The method is by induction on the number s of variables involved. In fact, if $s = 1$, then the pol over $\mathbf{Q}$ or $\mathbf{Z}$ in a single variable x may well be factorized by Gauss Lemma. Suppose we know already how to factorize in case $s = d$ and let us consider a pol in $d + 1$ variables. We may take any one of the variables involved as x and the other d's as $u_j \in U$. The problem becomes thus the factorization of a pol $f \in \mathbf{Q}[x, U]$. Remark that by induction hypothesis we know already how to factorize pols in $\mathbf{Q}[U]$.

This being understood let the pol to be factorized be $f \in \mathbf{Q}[x, U]$ given below:

$$f = a_n * x^n + \cdots + a_i * x^i + \cdots + a_0,\ a_i \in \mathbf{Q}[U]. \tag{5.7}$$

We shall see whether f is factorizable as $g * h$ with $g, h \in \mathbf{Q}[x, U]$ given by

$$\begin{aligned} g &= x_1 * x^p + \cdots x_{p+1},\ p > 0, \\ h &= x_{p+2} * x^q + \cdots + x_{p+q+2},\ q > 0,, \end{aligned} \tag{5.8}$$

in which $p + q = n$ and $x_i,\ i = 1, \cdots, N$ where $N = n + 2$ are all pols in $\mathbf{Q}[U]$ yet to be determined. Comparing coefficients of various powers of x in both sides of $f = g * h$ we get a system of pol-equations $PS = 0$ with $PS \subset \mathbf{Q}[U, X]$. Consider X

as unknown variables while U as known parameters we get by Zero-Decomposition Theorem a char-series $CS_k \subset \mathbf{Q}[U, X]$ with initial-products IP_k such that

$$Zero(PS) = \bigcup_k Zero(CS_k/IP_k). \tag{5.9}$$

Suppose that for some $k = k_0$, the asc-set CS_{k_0} is incomplete in the sense of Def. 3.3.19. Then by Theorem 3.3.9 there will be rational values $u_j^0 \in \mathbf{Q}$, such that, denoting by $CS_{k_0}^0, IP_{k_0}^0$ the pols gotten from CS_{k_0}, IP_{k_0} in substituting u_j^0 for u_j, we shall have $IP_{k_0}^0 \neq 0$ and $Zero_{\mathbf{C}}(CS_{k_0}^0/IP_{k_0}^0)$ will be an infinite set. Clearly we may choose u_j^0 to be such that, in denoting by a_i^0 the values in substituting u_j^0 for u_j in a_i, we have also $a_n^0 \neq 0$. Take any solution-set in complex values of $Zero_{\mathbf{C}}(CS_{k_0}^0/IP_{k_0}^0)$, we see that it will give a decomposition of $f^0 = a_n^0 * x^n + \cdots + a_0^0 \in \mathbf{Q}[x]$ into the product of two pols $g^0, h^0 \in \mathbf{C}[x]$. Such decompositions will be infinite in number, corresponding to the infinite number of solution-sets of $Zero_{\mathbf{C}}(CS_{k_0}^0/IP_{k_0}^0)$, which is clearly absurd.

It follows that all the asc-sets CS_k in the decomposition (5.9) should be complete ones in the sense of Def. 3.3.19. By induction hypothesis and Theorem 3.3.10, for $\mathbf{K} = \mathbf{Q}(U)$, each of $Zero_{\mathbf{K}}(CS_k/IP_k)$ is either empty or is finite which can be determined by Gauss Lemma. If all of them are empty so that $Zero_{\mathbf{K}}(PS)$ is empty then f is non-factorizable into products of two pols in $\mathbf{Q}[U, x]$ of fixed degrees $p > 0, q > 0$ with $p + q = n$ in x. Otherwise each of the finite number of possible solutions in $Zero_{\mathbf{K}}(PS)$ will give rise to such a factorization.

Trying all possible pairs (p, q) with $p > 0, q > 0, p + q = n$ we are able to decide whether f is factorizable or not and in the case of being factorizable, determine all such possible factorizations.

Alternatively, we may consider whether f will have a factor g in (5.8) of degree p in x with $0 < p < n$. In fact, g will be a factor of f if and only if $R \equiv Remdr(f/g) = 0$ as a pol in x. Set all coefficients of powers of x in R to be 0 we get a system of pol-equations in x and U which may be treated in the same manner as above.

We summarize now the above in the form of theorem below:

Theorem 5.1. There is an algorithm to factorize any multivariate pol with coefficients in $\mathbf{Q}$ or $\mathbf{Z}$ into irreducible factors in reducing it inductively to problems of decomposing a univariate pol with coefficients in $\mathbf{Q}$ or $\mathbf{Z}$ into irreducible factors.

Remark 5.1. An alternative method of factorizing a multivariate pol over $\mathbf{Q}$ or $\mathbf{Z}$ avoiding induction will be given in Sect. 6.2 of Chap. 6.

<u>*Solving of Problem G.*</u>

For the determination of the GCD of two known pols $F, G \in \mathbf{K}^*[x]$ we shall give a method requiring only the solving of *linear* pol-equations which relies on the following lemma.

Lemma 5.4. Let $F, G \in \mathbf{K}[x, Y]$ be of degree $m, n > 0$ in x respectively. For $e \leqslant min(m, n)$ let C_{m-e}, D_{n-e} be a pair of pols in $\mathbf{K}[x, Y]$ of degrees in x equal to $m-e, n-e$ respectively whose coefficients are yet to be determined. Consider the equation

$$D_{n-e} * F = C_{m-e} * G \text{ in } \mathbf{K}^*[x]. \tag{5.10}$$

Suppose there exist no pair (C_{m-e}, D_{n-e}) satisfying (5.10) for any $e > d$ while pair (C_{m-d}, D_{n-d}) satisfying the corresponding equation in (5.10) does exist so that we have

$$D_{n-d} * F = C_{m-d} * G \text{ in } \mathbf{K}^*[x]. \tag{5.11}$$

Then F, G have a GCD $H \in \mathbf{K}[x, Y]$ of degree d in x given by

$$H = \left[\frac{F}{C_{m-d}}\right] = \left[\frac{G}{D_{n-d}}\right] \ mod_{\mathbf{K}}\ AS. \tag{5.12}$$

Proof. The pair (C_{m-d}, D_{n-d}) cannot have a common factor over $\mathbf{K}^*$ in positive degree of x. For, suppose such a factor $E_c \in \mathbf{K}^*[x]$ of positive degree c in x exists. We may then set

$$C_{m-d-c} = \frac{C_{m-d}}{E_c}, \ D_{n-d-c} = \frac{D_{n-d}}{E_c} \text{ in } \mathbf{K}^*[x].$$

Dividing both sides of (5.11) by E_c we will get (5.10) for $e = d+c > d$ in contradiction to the assumption. Hence C_{m-d}, D_{n-d} have no common factor of positive degree in x so that by (5.11) C_{m-d} divides F and D_{n-d} divides G as pols in $\mathbf{K}^*[x]$ and we have (5.12) as to be proved. □

Theorem 5.2. There is an algorithm which permits to determine the GCD over $\mathbf{K}^*$ of any pols $F, G \in \mathbf{K}[x, Y]$ by means of solving of linear pol-equations.

Proof. Let

$$\begin{aligned} F &= a_m * x^m + \cdots + a_0, \ a_m \neq 0, \ a_i = \Sigma_{t \in T}\ a_{it} * Y^t, \ a_{it} \in \mathbf{K} = \mathbf{Q}(U), \\ G &= b_n * x^n + \cdots + b_0, \ b_n \neq 0, \ b_j = \Sigma_{t \in T}\ b_{jt} * Y^t, \ b_{jt} \in \mathbf{K} = \mathbf{Q}(U), \end{aligned} \tag{5.13}$$

with $a_i, b_j \in \mathbf{K}[Y]$ already known. Let

$$\begin{aligned} C_{m-e} &= c_{m-e} * x^{m-e} + \cdots + c_0, \ c_i = \Sigma_{t \in T}\ c_{it} * Y^t, \ c_{it} \in \mathbf{K} = \mathbf{Q}(U), \\ D_{n-e} &= d_{n-e} * x^{n-e} + \cdots + d_0, \ d_j = \Sigma_{t \in T}\ d_{jt} * Y^t, \ d_{jt} \in \mathbf{K} = \mathbf{Q}(U), \end{aligned} \tag{5.14}$$

be two pols in $\mathbf{K}[x, Y]$ verifying (5.10) with $c_{m-e} \neq 0, d_{n-e} \neq 0$. Now (5.11) is equivalent to

$$\begin{aligned} &R \equiv Remdr(C_{m-e} * G - D_{n-e} * F/AS) = 0, \text{ where} \\ &R = r_{m+n-e} * x^{m+n-e} + \cdots + r_0, \ r_i = \Sigma_{t \in T}\ r_{it} * Y^t, \ r_{it} \in \mathbf{K}. \end{aligned} \tag{5.15}$$

Whence we get a set of equations

$$r_{it} = 0, \ i = 0, 1, \cdots, m + n - e, \ t \in T, \tag{5.16}$$

in which r_{it} are *linear* functions of c_{it}, d_{jt} with coefficients pols in coefficients a_{it}, b_{jt} and I_{ijt} of F, G, I_{ij}. Owing to the *linearity* in c_{it}, d_{jt} it is easy to see whether the system (5.16) has solutions of (c_i, d_j) in $\mathbf{Q}(U) = \mathbf{K}$ or not for any e given. We start now from $e = min(m, n)$. If there exist no solutions of c_i, d_j in $\mathbf{K} = \mathbf{Q}(U)$ we then proceed to the case of $e - 1$. In proceeding in the same way further and further we will ultimately get by Lemma 5.4 the GCD of F, G as required. □

Solving of Problem F.

We shall solve now the Problem F in all its generality in reducing it to problems in pol-equations solving.

Lemma 5.5. For $F \in \mathbf{K}^*[x]$ to have a factor $H \in \mathbf{K}^*[x]$ with degree in x of H positive, it is necessary and sufficient that, as pols in $\mathbf{K}[x, Y]$

$$Remdr([F]/AS^+) = 0, \tag{5.17}$$

in which AS^+ is the asc-set formed of AS and the pol $[H]$.

Proof. By the remainder formula we have for some pols $Q_0, \ Q_i, \ R \in \mathbf{K}[x, Y]$, and some power-product IP of $I_{id_i}, i \in I$,

$$\begin{aligned} &IP * I_H^s * [F] = \Sigma_i\ Q_i * A_i \ + \ Q_0 * [H] \ + \ R, \\ &I_H = Init(H) \in \mathbf{K}[Y], \ R = Remdr([F]/AS^+). \end{aligned} \tag{5.18}$$

Suppose first that $R = 0$. Then (5.18) can be written as an equation in $\mathbf{K}^*[x]$ below:

$$F \ = \ \frac{Q_0}{I_H^s * IP} * H \text{ in } \mathbf{K}^*[x]. \tag{5.19}$$

Now $I_H \in \mathbf{K}[Y]$ so that $\dfrac{Q_0}{I_H^s * IP}$ is a pol in $\mathbf{K}^*[x]$. Hence (5.19) shows that H is a factor of F in $\mathbf{K}^*[x]$.

Conversely, suppose that F has H as a factor in $\mathbf{K}^*[x]$. Then for some pol $F' \in \mathbf{K}^*[x]$, we should have by Lemma 5.3

$$[F] = [F'] * [H]\ mod_{\mathbf{K}}\ AS.$$

This shows that $Remdr([F]/AS^+) = 0$ or the converse is also true. □

Theorem 5.3. There is an algorithm which permits to decide for any pol $F \in \mathbf{K}^*[x]$ with

$$[F] = a_m * x^m + \cdots + a_0,\ a_i = \Sigma_{t\in T}\ a_{it} * Y^t, a_{it} \in \mathbf{K}, \tag{5.20}$$

whether F is irreducible over $\mathbf{K}^*$ or not, and if it is reducible, to decompose F into irreducible factors over $\mathbf{K}^*$.

Proof. Let H be an arbitrary pol in $\mathbf{K}^*[x]$ of degree d in x where $0 < d \leqslant \left[\frac{m}{2}\right]$. Let the canonical form of H be

$$[H] = h_e * x^e + \cdots + h_0,\ h_i = \Sigma_{t\in T}\ h_{it} * Y^t,\ h_{it} \in \mathbf{K}. \tag{5.21}$$

the unknowns h_{it} with coefficients in $\mathbf{K}$ are in number

$$N = (e+1) * d_1 * \cdots * d_r. \tag{5.22}$$

These unknowns will be arranged in some definite order to be denoted by $X = (x_1, \cdots, x_N)$ below.

Start now from a polset $HS = \varnothing$ and $e = 1$ and proceed from $e = 1$ to $e = \left[\frac{m}{2}\right]$ as follows.

Step 1. By Theorem 5.2, the necessary and sufficient condition for F to have the factor H in $\mathbf{K}^*[x]$ is

$$R \equiv Remdr([F]/AS^+) = 0. \tag{5.23}$$

Here AS^+ is as before the asc-set gotten from AS by adjoining to it the pol $[H]$, the basic field being $\mathbf{K} = \mathbf{Q}(U)$. Now R is a pol in $\mathbf{K}[x]$ in the form

$$R = r_{e-1} * x^{e-1} + \cdots + r_0,\ r_j = \Sigma_{t\in T}\ r_{jt} * Y^t, \tag{5.24}$$

with R already in its canonical form. The computation of the remainder shows that $r_{jt} \in \mathbf{K}[X]$, the coefficients of h_{it}-powers in r_{jt} being well-determined rational functions of known a_{it} and I_{ijt} in $\mathbf{K}$. For (5.23) to be true it is clearly necessary and sufficient that

$$r_{jt} = 0,\ j = 0, 1, \cdots, e-1,\ \text{and}\ t \in T. \tag{5.25}$$

Step 2. Set $RS = \{r_{jt}\} \subset \mathbf{K}[X]$. By Zero-decomposition Theorem (over the basic field $\mathbf{K}$ in the unknowns X) there will be a finite set of asc-sets $CS_j \subset \mathbf{K}[X]$ with initial-products IP_j such that

$$Zero(RS) = \bigcup_j Zero(CS_j/IP_j).$$

Step 3. As in the proof of Theorem 5.1 each CS_j should be complete in the variables X and either $ZS_j = Zero_{\mathbf{K}}(CS_j/IP_j)$ is empty or is finite which can be determined by Gauss Lemma. If all ZS_j are empty then no H of the form (5.31) can be a factor of F. In this case replace $e \longleftarrow e+1$ and goto Step 1 if $e < \left[\frac{m}{2}\right]$, and return *irreducible* if $e = \left[\frac{m}{2}\right]$.

Step 4. Otherwise some ZS_j is non-empty. Then any zero in such a ZS_j will give a factor H of F. Put this factor in HS, replace $F \longleftarrow \frac{F}{H}$ and goto Step 1.

In the above way it will be shown either F is irreducible or is reducible as a product of pols in the polset HS. □

Some Examples.

Example 5.3. This example is taken from [W-R], Ex. D17: To factorize f in 3 variables x, y, z given below.

$$\begin{aligned} f &= x^4 + (-z+3)*x^3 + (z*3 + (y-3)*z - y^2 - 13)*x^2 \\ &\quad +(-z^4 + (y^2 + 3*y + 15)*z + 6)*x \\ &\quad +y*z^4 + 2*z^3 + (-y^3 - 15*y)*z - 2*y^2 - 30. \end{aligned} \tag{5.26}$$

Solution. Let us consider f as a pol in x with y, z treated as u_1, u_2 and suppose we know how to factorize pols in two variables y, z. We shall see whether f can be factorized as $f = g*h$ with

$$\begin{aligned} g &= x^2 + x_1*x + x_2, \\ h &= x^2 + x_3*x + x_4, \end{aligned} \tag{5.27}$$

in which $x_i \in \mathbf{Q}[y,z], i = 1,2,3,4$, are to be determined. In comparing the coefficients of powers of x we get from $f = g*h$ a system of equations $PS = 0$ with $PS = \{P_0, P_1, P_2, P_3\}$, where

$$\begin{aligned} P_0 &= -x_4*x_2 + z^4*y + 2*z^3 - z*y^3 - 15*z*y - 2*y^2 - 30, \\ P_1 &= -x_4*x_1 - x_3*x_2 - z^4 + z*y^2 + 3*z*y + 15*z + 6, \\ P_2 &= -x_4 - x_3*x_1 - x_2 + z^3 + z*y - 3*z - y^2 - 13, \\ P_3 &= -x_3 - x_1 - z + 3. \end{aligned} \tag{5.28}$$

The char-set of PS is $CS = \{C_1, C_2, C_3, C_4\}$ complete in $X = (x_1, \cdots, x_4)$ with $C_3 = P_3, C_4 = P_2$, and C_2 of an index-set $[16\ x_2\ 1]$. On the other hand C_1 is of an index-set $[72\ x_1\ 6]$ and is of the form

$$C_1 = x_1^4 + e_3 * x_1^3 + e_2 * x_1^2 + e_1 * x_1 + e_0,\ e_i \in \mathbf{Q}[y, z].$$

Now e_0 is seen to have factors $3, z$ and e_0' so by Gauss Lemma the only possible solutions of $C_1 = 0$ in $\mathbf{K} = \mathbf{Q}(y, z)$ are $\pm 3, \pm z, \pm e_0'$. Let IP be the initial-product of CS. It is readily verified that $Zero(PS + \{IP\}) = \varnothing$ while $Zero_{\mathbf{K}}(CS/IP)$ is consisting of two solutions

$$\begin{aligned}(x_1, x_2, x_3, x_4) &= (-z, z * y + 2, +3, z^3 - y^2 - 15),\\ &= (3, z^3 - y^2 - 15, -z, z * y + 2).\end{aligned}$$

It follows that

$$f = (x^2 - x * z + z * y + 2) * (x^2 + 3 * x + z^3 - y^2 - 15). \tag{5.29}$$

Alternatively, we may try to see whether f has a factor g as given in $(5.8)_1$. Now $Remdr(f/g) = 0$ will give rise to two equations in x_1, x_2 with pols in y, z as coefficients which will be reduced to the pair $C_1 = 0, C_2 = 0$ and we get $g = x^2 - x * z + z * y + 2$, or $g = x^2 + 3 * x + z^3 - y^2 - 15$ as above. □

Example 5.4. Consider Example 5.2 again. Try the factor

$$H = x + (h_1 * y + h_0),\ h_1, h_0 \in \mathbf{K}.$$

We have

$$R \equiv Remdr(F/AS^+) = u^3 * h_1 * h_0 * y + h_1^2 + u^3 * (h_0^2 - u).$$

Set

$$RS = \{r_1, r_0\},\ \text{where}$$
$$r_1 = u^3 * h_1 * h_0,\ r_0 = h_1^2 + u^3 * (h_0^2 - u).$$

Then we find for the ordering $u \prec h_0 \prec h_1$,

$$Zero(RS) = Zero(CS_1) + Zero(CS_2) + Zero(CS_3),\ \text{where}$$
$$\begin{aligned}CS_1 &= \{h_0, h_1 - u^2\},\\ CS_2 &= \{h_0, h_1 + u^2\},\\ CS_3 &= \{h_0^2 - u, h_1\}.\end{aligned}$$

The asc-set CS_3 gives no solution of h_0, h_1 in $\mathbf{K}$ while CS_1, CS_2 give possible solutions in $\mathbf{K}$ and hence factors $x + u^2 * y$ and $x - u^2 * y$ as shown in Ex. 5.2.

Example 5.5. Find GCD in $\mathbf{K}^*[x]$ of

$$F = x^2 - u,\ G = x^2 + 2 * u^2 * y + u$$

with y satisfying $A = 0$ as in Example 5.2.

Solution. Let us try to determine

$$C_1 = x + (c_{11} * y + c_{10}),\ c_{1i} \in \mathbf{K},$$
$$D_1 = x + (d_{11} * y + d_{10}),\ d_{1i} \in \mathbf{K}$$

such that

$$D_1 * F = C_1 * G \ in\ \mathbf{K}^*[x], \text{ or}$$

$$R \equiv Remdr(D_1 * F - C_1 * G/AS^+) = 0 \text{ where}$$

$$R = (r_{21} * y + r_{20}) * x^2 + (r_{11} * y + r_{10}) * x + (r_{01} * y + r_{00})\ in\ \mathbf{K}^*[x], \text{ with}$$
$$r_{21} = d_{11} - c_{11} - 2 * u^2,\ r_{20} = d_{10} - c_{10},$$
$$r_{11} = -2 * u^3 * c_{10},\ r_{10} = -2 * u^2 - 2 * c_{11},$$
$$r_{01} = -d_{11} - c_{11},\ r_{00} = -d_{10} - c_{10}.$$

The equations $r_{ij} = 0$ give

$$d_{10} = c_{10} = 0,\ c_{11} = -u^2,\ d_{11} = u^2.$$

Hence

$$C_1 = x - u^2 * y,\ D_1 = x + u^2 * y,$$

and the GCD is given by

$$H = \frac{F}{C_1} = \frac{G}{D_1} = x + u^2 * y \in \mathbf{K}^*[x].$$

□

Chapter 5

Some Topics in Computational Algebraic Geometry

5.1 Some Important Characters of Algebraic Varieties. Complex and Real Varieties.

Throughout this chapter let $\mathbf{K}$, and $A^n_{\mathbf{K}}(X)$ over $\mathbf{K}$ with coordinates $X = (x_1, \cdots, x_n)$, $X_k = (x_1, \cdots, x_k)$, be as in Chaps. 3 and 4. All algebraic varieties considered below will be in the affine space $A^n_{\mathbf{K}}(X)$ unless otherwise stated.

General Linear Transformation of an Affine Space.

Notation 1.1.

$X' := (x'_1, \cdots, x'_n)'$; $X'_k := (x'_1, \cdots, x'_k)'$.

$A'^n_{\mathbf{K}}(X') :=$ An affine space of dimension n over $\mathbf{K}$ with coordinates X'.

$(\ )' :=$ Points in $A'^n_{\mathbf{K}}(X')$ w.r.t. coordinates X'.

$\mathcal{V} := \{v_i \mid i = 1, \cdots, r\} :=$ Set of independent parameters.

$N := \{0, 1, \cdots, n\}$; $N_0 := \{1, \ldots, n\}$.

$u_{ij},\ i \in N_0,\ j \in N :=$ Set of independent parameters.

$\mathcal{U} := \{u_{ij} \mid i \in N_0,\ j \in N\}$.

$\mathcal{U}_0 := \{u_{ij} \mid i, j \in N_0\}$.

$\triangle := \{\delta_{ij} \mid i \in N_0, j \in N\}$.

$\triangle_0 := \{\delta_{ij} \mid i, j \in N_0\}$.

$T : A^n_{\mathbf{K}}(X) \longrightarrow A'^n_{\mathbf{K}}(X') :=$ An invertible linear transformation defined by

$$x'_i = \Sigma_{j \in N_0}\ u_{ij} * x_j + u_{i0},\ i \in N_0, \tag{1.1}$$

$T' : A'^n_{\mathbf{K}}(X') \longrightarrow A^n_{\mathbf{K}}(X) :=$ The inverse linear transformation of T.

$T_0 : A^n_{\mathbf{K}}(X) \longrightarrow A'^n_{\mathbf{K}}(X') :=$ An invertible linear transformation defined by

$$x'_i = \Sigma_{j \in N_0}\ u_{ij} * x_j,\ i \in N_0, \qquad (1.1)_0$$

$T'_0 : A'^n_{\mathbf{K}}(X') \longrightarrow A^n_{\mathbf{K}}(X) :=$ The inverse linear transformation of T_0.

Definition 1.1. T *(resp.* T_0*)* $: A^n_{\mathbf{K}}(X) \longrightarrow A'^n_{\mathbf{K}}(X')$ defined by (1.1) (resp. $(1.1)_0$) := A *general linear transformation* (resp. *general 0-linear transformation)* of $A^n_{\mathbf{K}}(X)$ in $A'^n_{\mathbf{K}}(X')$. T' *(resp.* T'_0*)* $: A'^n_{\mathbf{K}}(X') \longrightarrow A^n_{\mathbf{K}}(X) :=$ The *inverse* of T (resp. T_0).

Dimension and Specialization under Field Extension.

Proposition 1.1. Let $\Xi = (\xi_1, \cdots, \xi_n)$, $X^0 = (x^0_1, \cdots, x^0_n)$ be extended points in $A^n_{\mathbf{K}}(X)$. If X^0 is a specialization of Ξ over $\mathbf{K}$, then X^0 is also a specialization of Ξ over $\mathbf{K}(\mathcal{V})$.

Proof. Consider any pol $F \in \mathbf{K}(\mathcal{V})[X]$ for which $F(\Xi) = 0$. We may write F in the form of $F = \frac{P}{Q}$ with $P \in \mathbf{K}[\mathcal{V}, X]$, $Q \in \mathbf{K}[\mathcal{V}]$, with P, Q having no common factors in $\mathcal{V}$ alone. Then we have

$$P(\mathcal{V}, X) \neq 0, \ \textit{while} \ P(\mathcal{V}, \Xi) = 0.$$

Let us write P in the form

$$P(\mathcal{V}, X) = \Sigma_I \ \mathcal{V}^I * P_I(X),$$

in which I runs over a finite set of non-negative integer tuples $I = (i_1, \cdots, i_r)$, $\mathcal{V}^I = v_1^{i_1} * \cdots v_r^{i_r}$, and $P_I \in \mathbf{K}[X]$. Then $P(\mathcal{V}, \Xi) = 0$ implies $P_I(\Xi) = 0$ for all tuples I. As X^0 is a specialization of Ξ over $\mathbf{K}$, we have $P_I(X^0) = 0$ for all tuples I. It follows that $P(\mathcal{V}, X^0) = 0$ and hence $F(X^0) = 0$ too for any $F \in \mathbf{K}(\mathcal{V})[X]$ with $F(\Xi) = 0$. This shows that X^0 is a specialization of Ξ over $\mathbf{K}(\mathcal{V})$ as to be proved. □

Proposition 1.2. The dimension of an extended point $\Xi = (\xi_1, \cdots, \xi_n)$ in $A^n_{\mathbf{K}}(X)$ is invariant under field extension of $\mathbf{K}$ to $\mathbf{K}(\mathcal{V})$, or

$$Dim_{\mathbf{K}(\mathcal{V})} \ \Xi = Dim_{\mathbf{K}} \ \Xi. \tag{1.2}$$

Proof. Let $Dim_{\mathbf{K}} \ \Xi = d$. Then any $d+1$ elements $\xi_{i_1}, \cdots, \xi_{i_{d+1}}$ in Ξ are algebraically dependent over $\mathbf{K}$. À fortiori $\xi_1, \cdots, \xi_{d+1}$ are also algebraically dependent over $\mathbf{K}(\mathcal{V})$ or

$$Dim_{\mathbf{K}(\mathcal{V})} \ \Xi \leqslant d. \tag{1.3}$$

Now $Dim_{\mathbf{K}} \ \Xi = d$ implies that there are d elements in Ξ which are algebraically independent over $\mathbf{K}$. Suppose for simplicity that these elements are $\xi_1, \cdots, \xi_d$. If $\xi_1, \cdots, \xi_d$ are algebraically dependent in $\mathbf{K}(\mathcal{V})$, then there will be some non-zero pol

$F \in \mathbf{K}(\mathcal{V})[y_1, \cdots, y_d]$ and hence also some non-zero pol $P \in \mathbf{K}[\mathcal{V}, y_1, \cdots, y_d]$ such that

$$P(\mathcal{V}, y_1, \cdots, y_d) \neq 0, \text{ while } P(\mathcal{V}, \xi_1, \cdots, \xi_d) = 0.$$

Write P in the form

$$P(\mathcal{V}, y_1, \cdots, y_d) = \Sigma_I \mathcal{V}^I * P_I(y_1, \cdots, y_d),$$

with I running over some finite set of non-negative integer tuples $I = (i_1, \cdots, i_r)$, $\mathcal{V}^I = v_1^{i_1} * \cdots * v_r^{i_r}$, and $P_I \in \mathbf{K}[y_1, \cdots, y_d]$. Then we have $P_I(\xi_1, \cdots, \xi_d) = 0$ for all I. Now $P(\mathcal{V}, y_1, \cdots, y_d) \neq 0$ implies at least for one such $I = I_0$ we have

$$P_{I_0}(y_1, \cdots, y_d) \neq 0, \text{ while } P_{I_0}(\xi_1, \cdots, \xi_d) = 0.$$

This shows that $\xi_1, \cdots, \xi_d$ are algebraically dependent over $\mathbf{K}$, contrary to what we have supposed before. It follows that $\xi_1, \cdots, \xi_d$ are algebraically independent over $\mathbf{K}(\mathcal{V})$ or

$$Dim_{\mathbf{K}(\mathcal{V})}\, \Xi \geqslant d. \tag{1.4}$$

Combining with (1.3) we have then

$$Dim_{\mathbf{K}(\mathcal{V})}\, \Xi = d = Dim_{\mathbf{K}}\, \Xi,$$

as to be proved. □

Pols and Polsets under General Linear Transformations.

Definition 1.2. The *transform* of a set S of extended points in $A^n_{\mathbf{K}}(X)$ under the general linear (or 0-linear) transformation T (or T_0) := The set of transforms under T (or T_0) of the extended points in S.

Notation 1.2. For S a set of extended points in $A^n_{\mathbf{K}}(X)$, $T(S)$:= The transform of S under T, and $T_0(S)$:= The transform of S under T_0.

Definition 1.3. The *transform* of a pol $P(X) = P(x_1, \cdots, x_n) \in \mathbf{K}[X]$ under the general linear transformation T or the general 0-linear transformation T_0 := The pol $P' \in \mathbf{K}(\mathcal{U})[X']$ or $\mathbf{K}(\mathcal{U}_0)[X']$ defined as $P'(X') = P(T'(X'))$ or $P'(X') = P(T_0'(X'))$.

Notation 1.3. For $P \in \mathbf{K}[X]$, $T(P)$ (or $T_0(P)$) := The transform of pol P under T (or T_0).

Definition 1.4. The *transform* of a polset $PS \subset \mathbf{K}[X]$ under the general linear transformation T or the general 0-linear transformation T_0 := The polset $PS' \subset$

$\mathbf{K}(\mathcal{U})[X']$ defined as

$$PS' = \{T(P) \mid P \in PS\} \subset \mathbf{K}(\mathcal{U})[X'] \text{ or}$$
$$PS' = \{T_0(P) \mid P \in PS\} \subset \mathbf{K}(\mathcal{U}_0)[X'].$$

Notation 1.4. For $PS \subset \mathbf{K}[X]$, $T(PS)$ (or $T_0(PS)$) := The transform of polset PS under T (or T_0).

Dimension and Specialization under General Linear Transformations.

Lemma 1.1. Let $Y^0 = (y_1^0, \cdots, y_d^0)$ be a set of elements in some extension field of $\mathbf{K}$, and $P_i \in \mathbf{K}[\mathcal{V}, Y]$, $i = 1, \cdots, m$, a set of pols in the independent parameters $\mathcal{V} = (v_1, \cdots, v_r)$ and $Y = (y_1, \cdots, y_d)$. Suppose that $\mathcal{P}_i(\mathcal{V}, Y^0)$, $i = 1, \cdots, m$, are also algebraically dependent over $\mathbf{K}(\mathcal{V})$. Then for any elements $\mathcal{V}^0 = (v_i^0, \cdots, v_r^0)$ in $\mathbf{K}$, $\mathcal{P}_i(\mathcal{V}^0, Y^0), i = 1, \cdots, m$, are also algebraically dependent over $\mathbf{K}$.

Proof. As $P_i(\mathcal{V}, Y^0)$, $i = 1, \cdots, m$, are algebraically dependent over $\mathbf{K}(\mathcal{V})$, so there will be some non-zero pol in $\mathbf{K}(\mathcal{V})[Z]$ with $Z = (z_1, \cdots, z_m)$ independent parameters which vanishes for $z_i = P_i(\mathcal{V}, Y^0)$. By clearing of fractions we may suppose that this is a pol $F(\mathcal{V}, Z) \in \mathbf{K}[\mathcal{V}, Z]$. We may further suppose without loss of generality that F has not any factor of the form $v_i - v_i^0$, since otherwise we may remove such a factor beforehand. With this understood we have then

$$F(\mathcal{V}, Z) \neq 0, \text{ while } F(\mathcal{V}, \mathcal{P}) = 0,$$

where $\mathcal{P} = \{P_i(\mathcal{V}, Y^0) \mid i = 1, \cdots, m\}$. As F has no factors of the form $v_i - v_i^0$, so in setting $\mathcal{V} = \mathcal{V}^0$ in the last expression we have

$$F(\mathcal{V}, Z) \neq 0, \text{ while } F(\mathcal{V}, \mathcal{P}^0) = 0,$$

in which $\mathcal{P}^0 = \{P_i(\mathcal{V}^0, Y^0) \mid i = 1, \cdots, m\}$. Write $F(\mathcal{V}^0, Z)$ as a pol $F^0(Z) \in \mathbf{K}[Z]$ we have then

$$F^0(Z) \neq 0, \text{ while } F^0(\mathcal{P}^0) = 0.$$

This shows that $P_i(\mathcal{V}^0, Y^0)$, $i = 1, \cdots, m$, are algebraically dependent over $\mathbf{K}$ as to be proved. □

Proposition 1.3. Let Ξ, X^0 be extended points in $A^n_{\mathbf{K}}(X)$ with X^0 a specialization of Ξ over $\mathbf{K}$. Then $T(\Xi) = \Xi'$, $T(X^0) = X'^0$ are extended points in $A^n_{\mathbf{K}(\mathcal{U})}(X')$ with X'^0 a specialization of Ξ' over $\mathbf{K}(\mathcal{U})$. The same is true for general 0-linear transformations.

Proof. Consider any pol $F' \in \mathbf{K}(\mathcal{U})[X']$ for which $F'(\Xi') = 0$. Let $F = T'(F') \in \mathbf{K}(\mathcal{U})[X]$. Then $F(\Xi) = F'(\Xi') = 0$. As X^0 is a specialization of Ξ over $\mathbf{K}$, so by Prop. 1.1 X^0 is also a specialization of Ξ over $\mathbf{K}(\mathcal{U})$. From $F(\Xi) = 0$ we have then $F(X^0) = 0$. It follows that $F'(X'^0) = F(X^0) = 0$ for any non-zero pol $F' \in \mathbf{K}(\mathcal{U})[X']$ satisfying $F'(\Xi') = 0$. This shows that X'^0 is a specialization of Ξ' over $\mathbf{K}(\mathcal{U})$, as to be proved. The proof in the case of T_0 is actually the same. □

Proposition 1.4. The dimension of an extended point in an affine space is invariant under a general linear (or 0-linear) transformation. More precisely, for an extended point X^0 in $A^n_{\mathbf{K}}(X)$,

$$Dim_{\mathbf{K}(\mathcal{U})}\ T(X^0) = Dim_{\mathbf{K}(\mathcal{U}_0)}\ T_0(X^0) = Dim_{\mathbf{K}}\ X^0. \tag{1.5}$$

Proof. As in the proof of the preceeing Proposition we shall consider the case of T alone. Let $X^0 = (x^0_1, \cdots, x^0_n)$ in the affine space $A^n_{\mathbf{K}}(X)$ with $T(X^0) = X'^0 = (x'^0_1, \cdots, x'^0_n)'$ in the affine space $A'^n_{\mathbf{K}(\mathcal{U})}(X')$. Let $Dim_{\mathbf{K}}\ X^0 = d$. By Prop. 1.2 we have

$$Dim_{\mathbf{K}(\mathcal{U})}\ X^0 = Dim_{\mathbf{K}}\ X^0 = d$$

too. There is no loss of generality in supposing that $x^0_1, \cdots, x^0_d$ are algebraically independent over $\mathbf{K}(\mathcal{U})$. Consider now d pols in $\mathbf{K}[\mathcal{U}, X]$ defined by

$$P_i(\mathcal{U}, X) = \Sigma_{j\in N_0}\ u_{ij} * x_j + u_{i0},\ i = 1, \cdots, d.$$

For $X = X^0$ and $\mathcal{U} = \triangle$ we get

$$P_i(\triangle, X^0) = x^0_i,\ i = 1, \cdots, d.$$

Now x^0_i, $i = 1, \cdots, d$, are algebraically independent over $\mathbf{K}(\mathcal{U})$. By Lemma 1.1 we see that

$$P_i(\mathcal{U}, X^0) = x'^0_i,\ i = 1, \cdots, d,$$

are also algebraically independent over $\mathbf{K}(\mathcal{U})$. Hence we have

$$Dim_{\mathbf{K}(\mathcal{U})}\ T(X^0) \geqslant d. \tag{1.6}$$

On the other hand we have by Prop. 1.1

$$Dim_{\mathbf{K}(\mathcal{U})}\ X^0 = d$$

so that any $d+1$ elements in X^0 are algebraically dependent over $\mathbf{K}(\mathcal{U})$. As any element in $T(X^0) = X'^0$ is algebraically dependent on X^0 over $\mathbf{K}(\mathcal{U})$, any $d+1$ elements in X'^0 are algebraically dependent over $\mathbf{K}(\mathcal{U})$ too. It follows that

$$Dim_{\mathbf{K}(\mathcal{U})}\ T(X^0) \leqslant d. \tag{1.7}$$

Combining (1.6), (1.7) we get then

$$Dim_{\mathbf{K}(\mathcal{U})}\ T(X^0) = d = Dim_{\mathbf{K}}\ X^0.$$

This proves the proposition. □

Algebraic Varieties under General Linear Transformations.

Theorem 1.1. The transform of an algebraic variety in the affine space $A^n_{\mathbf{K}}(X)$ under the general linear (or 0-linear) transformation T (or T_0) is an algebraic variety in the affine space $A'^n_{\mathbf{K}(\mathcal{U})}(X')$ (or $A'^n_{\mathbf{K}(\mathcal{U}_0)}(X')$).

Proof. Let the variety in question be $V = Zero(PS)$ where $PS \subset \mathbf{K}[X]$. For the general linear transformation T and its inverse T' let $T(PS) = PS' \subset \mathbf{K}(\mathcal{U})[X']$. By direct verification we have

$$T(Zero(PS)) = Zero(PS'). \tag{1.8}$$

Hence $T(V)$ is an algebraic variety in $A'^n_{\mathbf{K}(\mathcal{U})}(X')$ as to be proved. The proof in the case of T_0 is actually the same. □

Theorem 1.2. The transform of an irreducible algebraic variety $A^n_{\mathbf{K}}(X)$ with a generic point Ξ in the affine space $A^n_{\mathbf{K}}(X)$ under the general linear (or 0-linear) transformation T (or T_0) is an irreducible algebraic variety in $A'^n_{\mathbf{K}(\mathcal{U})}(X')$ (or $A'^n_{\mathbf{K}(\mathcal{U}_0)}(X')$) with a generic point $\Xi' = T(\Xi)$ (or $\Xi'_0 = T_0(\Xi)$). .

Proof. Let the irreducible variety be $V = Zero(PS)$ with a generic point Ξ as an extended point in $A^n_{\mathbf{K}}(X)$. Let the transform of V under T be, by the preceding theorem, the variety $V' = T(V) = Zero(PS')$ where $PS' = T(PS)$. Let $\Xi' = T(\Xi)$ and consider any point $X'^0 = T(X^0)$ in V' where X^0 is in V. As X^0 is a specialization of Ξ over $\mathbf{K}$, so by Prop. 1.3 X'^0 is a specialization of Ξ' over $\mathbf{K}(\mathcal{U})$. It follows that V' is an irreducible variety with $\Xi' = T(\Xi)$ as a generic point as to be proved. The proof is actually the same in the case of T_0. □

Dimension of Algebraic Variety under General Linear Transformations.

Theorem 1.3. The dimension of an algebraic variety is invariant under general linear (or 0-linear) tranformations.

Proof. Consider first the case of an irreducible variety V. Let Ξ be a generic point of V. Then by Theorem 1.2 $V' = T(V)$ is an irreducible variety in $A'^n_{\mathbf{K}(\mathcal{U})}(X')$ and $\Xi' = T(\Xi)$ is a generic point of V'. By Prop. 1.4 we have $Dim_{\mathbf{K}(\mathcal{U})}\ \Xi' = Dim_{\mathbf{K}}\ \Xi$. By Def. 2.14 in Chapter 3 we have then $Dim_{\mathbf{K}(\mathcal{U})}\ V' = Dim_{\mathbf{K}}\ V$ and the theorem is proved in this irreducible case. Consider now an arbitrary variety V in $A^n_{\mathbf{K}}(X)$. By Theorems 2.11 or (6.17) in Theorem 6.10 in Chapter 3 V will have a non-contractible decomposition into irreducible components $V = \bigcup_i V_i$ in which V_i are all irreducible proper subvarieties of V. By Theorem 1.3 the variety $V' = T(V)$ has then an irreducible decomposition $V' = \bigcup_i V'_i$ where each $V'_i = T(V_i)$ is an irreducible variety and the decomposition is clearly non-contractible too. By Def. 2.17 in Chapter 3 the dimension of V and V' are respectively the maximum of dimensions over $\mathbf{K}$ and $\mathbf{K}(\mathcal{U})$ of V_i and V'_i. Hence the general case follows from the above particular irreducible case and the theorem is completely proved in the case of T. The proof in the case of T_0 is actually the same. □

Intersection of an Irreducible Variety with a Generic Hyperplane.

In what follows when we speak about the dimension of an algebraic variety, it will mean the dimension over $\mathbf{K}$ or over the extension field $\mathbf{K}(\mathcal{U})$ or $\mathbf{K}(\mathcal{U}_0)$ which will not be explicitely stated as it may be readily seen from the context.

Theorem 1.4. The intersection of an irreducible algebraic variety of dimension d with a generic hyperplane is an irreducible variety of dimension $d-1$ for $d>0$ and is empty for $d=0$.

Proof. Let the irreducible variety be V of dimension d. If $d=0$, then V is consisting of a finite number of points so the theorem is trivially true. Let therefore $d>0$ and let $\Xi = (\xi_1, \cdots, \xi_n)$ be a generic point of V with $Dim_{\mathbf{K}}\ \Xi = d$. Let the generic hyperplane H_1 be given by

$$u_{11} * x_1 + \cdots + u_{1n} * x_n = u_{10} \tag{1.9}$$

and let us consider a general 0-linear transformation T_0 defined by $(1.1)_0$. Under T_0 the hyperplane H_1 will become H'_1 of an equation

$$x'_1 = u_{10}. \tag{1.10}$$

By Theorem 1.2 the point $\Xi' = T_0(\Xi) = (\xi'_1 \cdots, \xi'_n)'$ is a generic point of $V' = T_0(V)$ and is of dimension d over $\mathbf{K}(\mathcal{U})$. Suppose that $\xi'_{i_1}, \cdots, \xi'_{i_d}$ are the d elements in Ξ' which are algebraically independent over $\mathbf{K}(\mathcal{U}_0)$. We may then make a permutation of variables

$$(x'_1, \cdots, x'_n) \longrightarrow (v'_1, \cdots, v'_d, y'_1, \cdots, y'_r), \ with\ r = n - d\ and$$

$$v'_j = x'_{i_j}, \ j = 1, \cdots, d.$$

It is clear that ξ'_1 is transcendental over $\mathbf{K}(\mathcal{U}_0)$ so we may take v'_1 to be x'_1. W.r.t. the ordering

$$x'_1 \prec \cdots \prec x'_n$$

let the defining polset of the extended point Ξ' in $A'^n_{\mathbf{K}(\mathcal{U}_0)}(X')$ be given by the irreducible asc-set

$$AS': \ A'_1, \ \cdots, \ A'_r, \ r = n - d, \text{ with}$$

$$A'_i = I'_i * y'^{d'_i}_i + \text{ lower degree terms in } y'_i, \ i = 1, \cdots, r,$$

in which all the coefficients in A'_i of y'_i-powers are in $\mathbf{K}(\mathcal{U}_0)[v'_1, \cdots, v'_d, y'_1 \cdots, y'_{i-1}]$ while I'_i is in $\mathbf{K}(\mathcal{U}_0)[v'_1, \cdots, v'_d]$. Let A''_i be now the pol gotten from A'_i in replacing v'_1 by u_{10} and A''_0 be the pol

$$A''_0 = v'_1 - u_{10},$$

the u_{10} being considered as a fixed value in some extension field of $\mathbf{K}(\mathcal{U}_0)$. Then

$$AS'': \ A''_0, \ A''_1, \ \cdots, \ A''_r$$

is clearly an irreducible asc-set representing an extended point $\Xi'' = (\xi''_1, \cdots, \xi''_n)'$ in $A'^n_{\mathbf{K}(\mathcal{U}_0)}(X')$. It is clear that Ξ'' is of dimension $d-1$ over $\mathbf{K}(\mathcal{U}_0)$ and Ξ'' is in the hyperplane H'_1 since $\xi''_1 = u_{10}$. Denote the initial of of A''_i by $I''_i, \ i = 1, \cdots, r$, gotten from I'_i in replacing v_1 by u_{10}. It is also clear that $I''_i(\Xi'') \neq 0, \ i = 1, \cdots, r$, so that Ξ'' is in the variety V'. It follows that Ξ'' is in the intersection of V' and H_1. The irreducible variety

$$Var[AS''] = Spec_{\mathbf{K}(\mathcal{U}_0)} \ \Xi''$$

with Ξ'' as a generic point is thus of dimension $d-1$ and is wholly contained in the intersection of V' and H'_1.

Conversely, let $X''^0 = (x''^0_1, \cdots, x''^0_n)$ be an arbitrary point in the intersection of V' and H'_1. Then $H'_1(X''^0) = 0$ implies that $x''^0_1 = u_{10}$. Consider now an arbitrary non-zero pol $F \in \mathbf{K}(\mathcal{U}_0)[X']$ with $F(\Xi'') = 0$. As u_{10} is transcendental over $\mathbf{K}(\mathcal{U}_0)$ the equation $F(\Xi'') = 0$ is equivalent to $F(\Xi') = 0$. As X''^0 is in V' of which Ξ' is a generic point it follows that $F(X''^0) = 0$. This shows that X''^0 is a specialization of Ξ'' over $\mathbf{K}(\mathcal{U}_0)$ so that X''^0 is in the variety $Var[AS'']$. Consequently

$$V' \cap H_1' = Var[AS''] \tag{1.11}$$

and is an irreducible variety of dimension $d-1$ in $A^n_{\mathbf{K}(\mathcal{U}_0)}(X')$ as to be proved. □

<u>*Degree of an Irreducible Algebraic Variety.*</u>

Theorem 1.5. The intersection of an irreducible algebraic variety of dimension d with a generic linear space of dimension r is empty if $r < n-d$ and is an irreducible algebraic variety of dimension 0 consisting of a finite number of conjugate points if $r = n-d$.

Proof. As a generic linear space of dimension r is the intersection of $n-r$ generic hyperplanes the first part of the theorem follows inductively from the previous theorem. The last part of the theorem follows then from Theorem 3.2.8. □

Definition 1.5. *Degree* of an irreducible algebraic variety of dimension $d :=$ The number of conjugate points in the intersection of the variety with a generic linear space of dimension $n-d$.

Theorem 1.6. The degree of an irreducible algebraic variety is invariant under general linear (or 0-linear) transformations.

Proof. In view of Theorems 1.2-1.5 this follows directly from the geometric significance of the notion of degree. □

Theorem 1.7. The degree of an irreducible variety of dimension 0 is seen to be the number of conjugate points in that variety. The degree of an irreducible hypersurface $H = 0$ with $H \in \mathbf{K}[X]$ is seen to be the total degree in X of the pol H.

Proof. This follows directly from the very definitions. □

<u>*Tangent Spaces of an Irreducible Algebraic Variety.*</u>

Let $Zero(H)$ be any hypersurface in $A^n_{\mathbf{K}}(X)$ with $H \in \mathbf{K}[X]$ and $X^0 = (x_1^0, \cdots, x_n^0)$ be any point of $Zero(H)$.

Definition 1.6. *Quasi-tangent space* of $Zero(H)$ at $X^0 :=$ The linear space through X^0 defined by

$$\Sigma_{j \in N_0} \frac{\partial H}{\partial x_j}(X^0) * (x_j - x_j^0) \ = \ 0, \tag{1.12}$$

where $\dfrac{\partial H}{\partial x_j}(X^0)$ means the value of $\dfrac{\partial H}{\partial x_j}$ at the point X^0.

Remark 1.2. The quasi-tangent space will be the whole affine space $A^n_{\mathbf{K}}(X)$ in case all the partial derivatives are zero at the point X^0.

Definition 1.7. *Quasi-tangent space* of an algebraic variety V at a point X^0 of V := The intersection of all quasi-tangent spaces at X^0 of hypersurfaces containing V.

Definition 1.8. *Tangent space* of an irreducible variety V at a point X^0 of V := The quasi-tangent space of V at X^0 if it is of dimension $d = Dim_{\mathbf{K}}\ V$.

Remark 1.3. The definitions 1.7 and 1.8 do not hold for the variety of dimension n, viz. the affine space $A^n_{\mathbf{K}}(X)$ itself. In this case the *quasi-tangent space* and the *tangent space* at a point will be by definition the affine space $A^n_{\mathbf{K}}(X)$ itself.

Theorem 1.8. The quasi-tangent space of an irreducible algebraic variety V of dimension d at a generic point $\Xi = (\xi_1, \cdots, \xi_n)$ is of dimension d and is the tangent space of V at that point Ξ.

Proof. If $d = n$ so that $V = A^n_{\mathbf{K}}(X)$ then the theorem is trivially true by Remark 1.3. Consider therefore the case of $d < n$ and let the generic point Ξ have the defining asc-set in ascending order

$$AS: \quad A_1, \ \cdots, \ A_r, \text{ with} \tag{1.13}$$

$$\begin{aligned} A_i = &I_{i0} * x_{c_i}^{d_i} + \cdots + I_{id_i}, \ i = 1, \cdots, r, \ r = n - d, \\ &0 < c_1 < \cdots < c_r \leqslant n, \ \ I_{ij} \in \mathbf{K}[X_{c_i - 1}]. \end{aligned} \tag{1.14}$$

Let us form the $(r \times n)$-matrix

$$M = \left[\frac{\partial A_i}{\partial x_j}\right], \ i = 1, \cdots, r; \ j = 1, \cdots, n, \tag{1.15}$$

with an $(r \times r)$-submatrix

$$M_0 = \left[\frac{\partial A_i}{\partial x_{c_j}}\right], \ i, j = 1, \cdots, r. \tag{1.16}$$

As $\dfrac{\partial A_i}{\partial x_{c_j}} = 0$ for $j > i$ and $\dfrac{\partial A_i}{\partial x_{c_i}} = S_i$ is the separant of A_i, we have

$$det\ M_0 = S_1 * \cdots * S_r. \tag{1.17}$$

As each S_i is reduced w.r.t. the irreducible asc-set AS, we have by Theorem 3.6.1 $S_i \neq 0$ at Ξ so that

$$det\ M_0 \neq 0 \text{ at } \Xi. \tag{1.18}$$

Whence

$$Rank\ M = r \text{ at } \Xi. \tag{1.19}$$

Now each $Zero(A_i)$ is a hypersurface containing the generic point Ξ of V so containing also the whole of the variety V. The quasi-tangent space of $Zero(A_i)$ at Ξ is

$$\Sigma_{j\in N_0} \frac{\partial A_i}{\partial x_j}(\Xi) * (x_j - \xi_j) = 0, \tag{1.20$_i$}$$

As the partial derivative for $j = i$, being the separant of A_i, is non-zero at Ξ, the quasi-tangent space (1.20) is actually the tangent space of $Zero(A_i)$ at Ξ. The matrix of the coefficients of x_j of these tangent spaces is just the matrix M at the point Ξ, which, by (1.19), is of rank $r = n - d$. The intersection of these tangent spaces at Ξ is thus a linear space of dimension $n - r = d$.

Consider now an arbitrary hypersurface $Zero(H)$ containing V. Then $Zero(H)$ contains Ξ so that $H(\Xi) = 0$. By Theorem 3.6.3 we have then

$$Remdr(H/AS) = 0.$$

By Remainder Formula (3.3.8) of Sect.3.3 there will exist some pols Q_i and some power-product IP_H of initials I_{i0} of A_i such that

$$IP_H * H = \Sigma_i\ Q_i * A_i,$$

the summation being extended over $i = 1, \cdots, r$. Differentiating we get

$$IP_H * \frac{\partial H}{\partial x_j} = -\frac{\partial IP_H}{\partial x_j} * H + \Sigma_i \frac{\partial Q_i}{\partial x_j} * A_i + \Sigma_i\ Q_i * \frac{\partial A_i}{\partial x_j},\ j \in N_0.$$

At the point Ξ these equations reduce to

$$IP_H(\Xi) * \frac{\partial H}{\partial x_j}(\Xi) = \Sigma_i\ Q_i(\Xi) * \frac{\partial A_i}{\partial x_j}(\Xi),\ j \in N_0.$$

As each initial I_{i0} is reduced w.r.t. AS we have by Theorem 3.6.1 $I_{i0}(\Xi) \neq 0$ and therefore $IP_H(\Xi) \neq 0$ too. The equation of the tangent space of $Zero(H)$ at Ξ is

$$\Sigma_{j\in N_0} \frac{\partial H}{\partial x_j}(\Xi) * (x_j - \xi_j) = 0.$$

It is thus seen to be linearly dependent on the equations of the tangent spaces of $Zero(A_i)$ at Ξ. The quasi-tangent space, i.e. the intersection of all quasi-tangent spaces of hypersurfaces containing V is thus just the linear space defined by $(1.20)_i, i = 1, \cdots, r$ and is of dimension d. The quasi-tangent space of V at Ξ is thus the tangent space of V at Ξ as to be proved. □

Theorem 1.9. The quasi-tangent space of an irreducible algebraic variety V of dimension d at an arbitrary point is of dimension $\geqslant d$.

Proof. Let Ξ be a generic point of V. Then a hypersurface $Zero(H)$ containing V is characterized by $H(\Xi) = 0$. Hence the pols H for such hypersurfaces form clearly an ideal ID. By Hilbert's finite basis theorem the ideal ID will have a finite basis say $(H_1, \cdots, H_m)$. The quasi-tangent space of $Zero(H_k)$ at an arbitrary point $X^0 = (x_1^0, \cdots, x_n^0)$ of V is given by

$$\Sigma_{j \in N_0} \frac{\partial H_k}{\partial x_j}(X^0) * (x_j - x_j^0) = 0, \ k = 1, \cdots, m. \qquad (1.21)_k$$

Let us form the $(m \times n)$-matrix of the x_j-coefficients

$$M' = \left[\frac{\partial H_k}{\partial x_j}(X^0)\right], \ k = 1, \cdots, m; \ j = 1, \cdots, n. \qquad (1.22)$$

For any other hypersurface $Zero(H)$ containing V we have for some pols Q_k

$$H = \Sigma_k \, Q_k * H_k.$$

Differentiating we have

$$\frac{\partial H}{\partial x_j} = \Sigma_k \frac{\partial Q_k}{\partial x_j} * H_k + \Sigma_k \, Q_k * \frac{\partial H_k}{\partial x_j}.$$

At any point X^0 of V the above equations become

$$\frac{\partial H}{\partial x_j}(X^0) = \Sigma_k \, Q_k * \frac{\partial H_k}{\partial x_j}(X^0).$$

The equation of the quasi-tangent space of $Zero(H)$ at X^0 is thus seen to be linearly dependent on the equations of those of $Zero(H_k)$ at X^0. The quasi-tangent space of V at X^0 is thus just the intersection of all the hyperplanes in $(1.21)_k$, $k = 1, \cdots, m$. The dimension of this quasi-tangent space at X^0 is then equal to $n - r'(X^0)$ where $r'(X^0)$ is the rank of the matrix M' in (1.22) at the point X^0. Consider the point X^0 to be the generic point Ξ of V. By Theorem 1.8 we have $r'(\Xi) = n - d$. It follows that for all $((n-d+1) \times (n-d+1))$-submatrices of M' will its determinant equal to zero at the generic point Ξ. As any point X^0 of V is a specialization of Ξ, such submatrices will have their determinants equal to zero at X^0 too. The rank of the matrix at X^0 is therefore $\leqslant n-d$ and so the dimension of the quasi-tangent space at an arbitrary point X^0 of V is $\geqslant d$ as to be proved. □

Simple Points and Singular Points of an Irreducible Algebraic Variety.

Definition 1.9. A point X^0 of an irrducible algebraic variety V is *simple* := The tangent space of V at X^0 is well-defined, or, := The quasi-tangent space of V at X^0 is of dimension $d = Dim_{\mathbf{K}}\ V$.

Definition 1.10. A point X^0 of an irreducible algebraic variety V is *singular* := The tangent space of V at X^0 is undefined, or, := The quasi-tangent space of V at X^0 is of dimension $> d = Dim_{\mathbf{K}}\ V$.

Definition 1.11. The *singular locus* of an irreducible algebraic variety V := The set of all singular points of V.

Theorem 1.10. The singular locus of an irreducible algebraic variety V is a proper subvariety of V.

Proof. As in the proof of Theorem 1.9, let $(H_1, \cdots, H_m)$ be a basis of the ideal of pols H for which the hypersurface $Zero(H)$ contains the variety V. Form the matrix M' as in (1.22). Then a singular point X^0 will be characterized by $Rank\ M'(X^0) < n-d$, where d is the dimension of V over $\mathbf{K}$, or

$$det\ M'_0(X^0) = 0,$$

where M'_0 runs over the set of all $((n-d)\times(n-d))$-submatrices of M'. The singular locus is thus the subvariety defined as the zero-set of all such $det\ M'_0$ as well as the pols in PS where $Zero(PS) = V$. The subvariety is a proper one since the generic point of V is clearly not in it. □

Theorem 1.11. The singular locus of an irreducible algebraic variety $V = Zero(PS)$ is contained in the union of proper subvarieties $Zero(PS + \{S_i\})$ where S_i are the separants of pols in the irreducible asc-set defining a generic point of the variety V.

Proof. Let the defining asc-set of a generic point of V be as in (1.13),(1.14). If the separants S_i of the pols A_i are all none-zero at a point X^0 of V then the determinant of the submatrix M_0 of (1.16) will be non-zero at X^0 so the rank of the matrix M in (1.14) will be $r = n-d$ where d is the dimension of V over $\mathbf{K}$. The dimension of the quasi-tangent space at X^0 will then be less than or equal to d. By Theorem 1.9 the dimension can only be equal to d so that X^0 will be a simple point instead of a singular one. Hence at a singular point of V some of the separants should be zero as to be proved. □

Topology of Complex Algebraic Varieties.

Let the basic field **K** be now the complex field **C**.

Definition 1.12. For $PS \subset \mathbf{C}[X]$, the complex part $Zero_{\mathbf{C}}(PS)$ of the variety $Zero(PS)$ of dimension $d :=$ A *complex variety* in $A^n_{\mathbf{C}}(X)$ with *complex basis PS* and *complex dimension d*.

Theorem 1.12. Let $Zero(PS)$ be an algebraic variety of dimension d in $A^n_{\mathbf{C}}(X)$ with $PS \subset \mathbf{C}[X]$. Then the complex variety $Zero_{\mathbf{C}}(PS)$ is a subspace of real topological dimension $2 * d$ of the complex affine space $\mathbf{C}^n(X)$. Moreover, the whole space $\mathbf{C}^n(X)$ can be triangulated into a simplicial complex with the given variety as a subcomplex of dimension $2 * d$ in the topological sense.

Proof. See [Lef], Chap.VIII, Sect.3, or [VdW], Kap.IV, Anhang. □

Theorem 1.13. Given an irreducible complex algebraic variety U of complex dimension $d > 0$ in $\mathbf{C}^n(X)$ the subset U' of its singular points is a proper complex subvariety of complex dimension at most $d-1$. The part $U \setminus U'$ is then a topological manifold of real dimension $2 * d$ in topological sense and the closure of $U \setminus U'$ is the whole of the variety U.

Proof. For the proof see e.g. [Lef], Chap.VIII, Sects.1,2.

Topology of Real Algebraic Varieties.

The case of $\mathbf{K} = \mathbf{R}$ is more delicate See the simple example below:

Example 1.1. In the real plane $A^2_{\mathbf{R}}(x_1, x_2)$ the real point-set $V = Zero_{\mathbf{R}}(x_1^2 + x_2^2) = Zero_{\mathbf{R}}(x_1, x_2)$ is consisting of a single point $(0,0)$. On the other hand $Zero_{\mathbf{C}}(x_1^2 + x_2^2)$ is a complex variety consisting of two complex lines $x_1 \pm \mathbf{i} * x_2 = 0$ ($\mathbf{i} = \sqrt{-1}$), while the complex variety $Zero_{\mathbf{C}}(x_1, x_2)$ is consisting of a single point $(0,0)$.

In view of this we lay down the following definitions.

Definition 1.13. For $PS \subset \mathbf{R}[X]$, the real part $Zero_{\mathbf{R}}(PS)$ of the variety $Zero(PS) :=$ A *real variety* in $\mathbf{R}^n(X)$ with *real basis PS*.

Definition 1.14. For a point set S in the real space $\mathbf{R}^n(X)$ the *ideal* of $S :=$ The ideal consisting of pols in $\mathbf{R}[X]$ which vanish at all points of S.

Notation 1.5. Ideal of a point set $S \subset \mathbf{R}^n(X) := Ideal(S)$.

By Hilbert basis theorem $Ideal(S)$ has always some finite basis.

Definition 1.15. For a real variety $V = Zero_{\mathbf{R}}(PS)$ in $\mathbf{R}^n(X)$ with basis $PS \subset \mathbf{R}[X]$, PS is *real-complete* of $V := V = Zero_{\mathbf{R}}(Ideal(V))$.

Example 1.2. For the real variety V in Ex. 1.1 $\{x_1, x_2\}$ is a real-complete basis of V while the basis $\{x_1^2 + x_2^2\}$ is not a real complete one.

Definition 1.16. For a real variety $V = Zero_{\mathbf{R}}(PS)$ with real basis PS in $\mathbf{R}^n(X)$ the *complexification* of V := The complex variety $Zero_{\mathbf{C}}(PS)$ in $\mathbf{C}^n(X)$ with PS considered as a polset in $\mathbf{C}[X]$.

The following theorems about the topology of real varieties are all due to Whitney. We refer the proofs of these theorems to his paper [Whit].

Theorem 1.14. For a real variety $V = Zero_{\mathbf{R}}(PS)$ in $\mathbf{R}^n(X)$ with a real-complete basis $PS \subset \mathbf{R}[X]$ let the complexification $V^* = Zero_{\mathbf{C}}(PS)$ of V be decomposed into a non-contractible union of irreducible complex varieties V_i^*. Then each V_i^* is the complexification of a real variety V_i with a real-complete basis $PS_i \subset \mathbf{R}[X]$ such that $V_i^* = Zero_{\mathbf{C}}(PS_i)$. Moreover, the union of V_i form a non-contractible decomposition of V.

Theorem 1.15. A real variety $V = Zero_{\mathbf{R}}(PS)$ in $\mathbf{R}^n(X)$ with a basis $PS \subset \mathbf{R}[X]$ is the union of a finite collection of mutually disjoint real topological manifolds each having a real dimension at most equal to the dimension of the algebraic variety V in the usual sense.

Definition 1.17. Any finite collection of real topological manifolds as in Theorem 1.15 := A *Whitney decomposition* of the real variety V.

Theorem 1.16. If the real basis PS of a real variety $V = Zero(PS)$ in $\mathbf{R}^n(X)$ is real-complete, then there is at least one real topological manifold in the Whitney decomposition of V having its real topological dimension equal to the dimension of the algebraic variety $Zero(PS)$ or the complex dimension of the complexification of the real variety V.

5.2 Algebraic Correspondence and Chow Form.

A Simple Example of Algebraic Correspondence.

Example 2.1. Consider in the plane $A^n_{\mathbf{K}}(x, y)$ the X-axis L_x and Y-axis L_y and a correspondence between points $(x, 0)$ (or, x on L_x) and $(0, y)$ (or, y on L_y) defined by the geometrical requiement that x, y lie on some line passing through a fixed point (a, b) in the plane which is not on L_x or L_y. The correspondence may then be defined by the equation

$$(x - a) * (y - b) - a * b = 0,$$

which is *algebraic* in x, y and so the correspondence may be said to be *algebraic.*

It is readily seen that the point a on L_x has no correspondence on L_y and the point b on L_y has no correspondence on L_x. Let us complete L_x (resp. L_y) to a projective line P_x (resp. P_y) in homogeneous coordinates $(x_0 : x_1)$ (resp. $(y_0 : y_1)$) so that the equation defining the correspondence will become

$$(x_1 - a * x_0) * (y_1 - b * y_0) - a * b * x_0 * y_0 = 0.$$

Then to each point on P_x (resp. P_y) will correspond a point on P_y (resp. P_x) without exception. It is for this reason that one usually deals with algebraic correspondence between *projective* spaces instead of affine ones. In spite of this we shall mainly treat the algebraic correspondence in the *affine* spaces and deal with the *projective* case as evident generalizations only if necessary.

Some Notations.

Notation 2.1.

$X := (x_1, \cdots, x_n)$; ${}^hX := (x_0 : x_1 : \cdots : x_n)$, not all $x_i = 0$.

$Y := (y_1, \cdots, y_n)$; ${}^hY := (y_0 : y_1 : \cdots : y_n)$, not all $y_i = 0$.

For $I = (i_1, \cdots, i_n)$, $X^I := x_1^{i_1} * \cdots * x_n^{i_n}$.

For $K = (k_1, \cdots, k_m)$, $Y^K := y_1^{k_1} * \cdots * y_m^{k_m}$.

$A_{\mathbf{K}}^n(X)$:= Affine n-space over $\mathbf{K}$ with coordinates $X = (x_1, \cdots, x_n)$.

$P_{\mathbf{K}}^n({}^hX)$:= Completed projective n-space of $A_{\mathbf{K}}^n$ with homogeneous coordinates ${}^hX = (x_0 : x_1 : \cdots : x_n)$.

$A_{\mathbf{K}}^m(Y)$:= Affine m-space over $\mathbf{K}$ with coordinates $Y = (y_1, \cdots, y_n)$.

$P_{\mathbf{K}}^m({}^hY)$:= Completed projective m-space of $A_{\mathbf{K}}^m$ with homogeneous coordinates ${}^hY = (y_0 : y_1 : \cdots : y_n)$.

$\mathcal{A}$:= The product space $A_{\mathbf{K}}^n \times A_{\mathbf{K}}^m$.

$\mathcal{P}$:= The product space $P_{\mathbf{K}}^n \times P_{\mathbf{K}}^m$.

For $P \in \mathbf{K}[X]$, ${}^hP$:= The homogenized pol of P by means of the variable at infinity x_0.

For $P \in \mathbf{K}[Y]$, ${}^hP$:= The homogenized pol of P by means of the variable at infinity y_0.

For $P \in \mathbf{K}[X, Y]$, ${}^hP$:= The homogenized pol of P by means of the variables at infinity x_0, y_0 separately for X and Y.

For $PS \subset \mathbf{K}[X]$, ${}^hPS$:= The polset consisting of homogenized pols of pols in PS by means of the variable at infinity x_0.

For $PS \subset \mathbf{K}[Y]$, ${}^hPS$:= The polset consisting of homogenized pols of pols in PS by means of the variable at infinity y_0.

For $PS \subset \mathbf{K}[X, Y]$, ${}^hPS$:= The polset consisting of homogenized pols of pols in PS by means of the variables at infinity x_0, y_0 separately for X and Y.

Algebraic Correspondence.

Definition 2.1. $\mathcal{C}$ is an *algebraic correspondence* between $A^n_{\mathbf{K}}(X)$ and $A^m_{\mathbf{K}}(Y)$:= $\mathcal{C}$ is an algebraic variety in the product space $\mathcal{A}$.

Definition 2.2. Algebraic correspondence $\mathcal{C}$ is *irreducible* := $\mathcal{C}$ as a variety in $\mathcal{A}$ is irreducible.

Definition 2.3. $PS \subset \mathbf{K}[X, Y]$ is the *defining polset* of algebraic correspondence $\mathcal{C}$:= $\mathcal{C} = Zero(PS)$.

Definition 2.4. *Projective completion* of algebraic correspondence $\mathcal{C}$:= The algebraic variety in $\mathcal{P}$ defined as $Zero^h({}^hPS)$ where PS is the defining polset of $\mathcal{C}$.

Below $\mathcal{C}$ will be an *irreducible* algebraic correspondence between $A^n_{\mathbf{K}}(X)$ and $A^m_{\mathbf{K}}(Y)$ with defining polset PS unless otherwise stated.

Notation 2.2. (Ξ, Υ) := A generic point of $\mathcal{C}$.

Definition 2.5. *Object variety* of $\mathcal{C}$:= The irreducible variety $Spec_{\mathbf{K}}(\Xi)$ in $A^n_{\mathbf{K}}(X)$.

Definition 2.6. *Image variety* of $\mathcal{C}$:= The irreducible variety $Spec_{\mathbf{K}}(\Upsilon)$ in $A^m_{\mathbf{K}}(Y)$.

Notation 2.3. $Obj(\mathcal{C})$:= Object variety of $\mathcal{C}$; $Im(\mathcal{C})$:= Image variety of $\mathcal{C}$.

Remark 2.1. It is clear that $Obj(\mathcal{C}), Im(\mathcal{C})$ are independent of the generic point (Ξ, Υ) chosen of $\mathcal{C}$.

Theorem 2.1. If $(X^0, Y^0) \in \mathcal{C}$, then $X^0 \in Obj(\mathcal{C})$ and $Y^0 \in Im(\mathcal{C})$.

Proof. Consider any pol $F \in \mathbf{K}[X]$ such that $F(\Xi) = 0$. Then $F(\Xi, \Upsilon) = 0$ if F is considered as a pol in $\mathbf{K}[X, Y]$. As (Ξ, Υ) is a generic point of $\mathcal{C}$, we have $F(X^0, Y^0) = 0$ or $F(X^0) = 0$. As this is true for arbitrary $F \in \mathbf{K}[X]$, X^0 is a specialization over $\mathbf{K}$ of Ξ. Hence $X^0 \in Obj(\mathcal{C})$. Similarly we have $Y^0 \in Im(\mathcal{C})$. □

Some Characteristic Numbers of Irreducible Algebraic Correspondence.

Notation 2.4.

$Im_{\Xi}(\mathcal{C}) := Spec_{\mathbf{K}(\Xi)}\,(\Xi, \Upsilon)$, an irreducible variety in $A^{n+m}_{\mathbf{K}(\Xi)}(X, Y)$.

$Obj_{\Upsilon}(\mathcal{C}) := Spec_{\mathbf{K}(\Upsilon)}\,(\Xi, \Upsilon)$, an irreducible variety in $A^{n+m}_{\mathbf{K}(\Upsilon)}(X, Y)$.

$a(\mathcal{C}) := Dim_{\mathbf{K}}\ Obj(\mathcal{C})$.

$b(\mathcal{C}) := Dim_{\mathbf{K}(\Xi)}\ Im_{\Xi}(\mathcal{C})$.

$c(\mathcal{C}) := Dim_{\mathbf{K}}\ Im(\mathcal{C})$.

$d(\mathcal{C}) := Dim_{\mathbf{K}(\Upsilon)}\ Obj_{\Upsilon}(\mathcal{C})$.

Counting Principle of Irreducible Algebraic Correspondence.

Theorem 2.2. (Counting Principle). Between the characteristic numbers of an irreducible algebraic correspondence $\mathcal{C}$ we have

$$a(\mathcal{C}) + b(\mathcal{C}) = c(\mathcal{C}) + d(\mathcal{C}). \tag{2.1}$$

Proof. The transcendence degree of $\mathbf{K}(\Xi, \Upsilon)$ over $\mathbf{K}$ may be counted in two different ways as

$$\begin{aligned} TrDeg(\mathbf{K}(\Xi, \Upsilon)/\mathbf{K}) &= TrDeg(\mathbf{K}(\Xi, \Upsilon)/\mathbf{K}(\Xi)) + TrDeg(\mathbf{K}(\Xi)/\mathbf{K}) \\ &= TrDeg(\mathbf{K}(\Xi, \Upsilon)/\mathbf{K}(\Upsilon)) + TrDeg(\mathbf{K}(\Upsilon)/\mathbf{K}). \end{aligned}$$

According to definition of dimension in terms of the transcendence degree we get therefore the theorem. □

Remark 2.2. The concept of algebraic correspondence may be naturally generalized to ones between projective spaces or Cartesian products of affine and projective spaces. The *Counting Principle* will remain valid with due modifications. Instead of going into details we shall give below some applications of the Counting Principle for such general algebraic correspondences.

Intersection of an Irreducible Algebraic Variety with a Hypersurface.

Theorem 2.3. Let H be a generic hypersurface in $A^n_{\mathbf{K}(a_I)}(X)$ defined by $H(X) \equiv \Sigma_I\ a_I * X^I = 0$ in which a_I are the generic coefficients of the pol H. Then the intersection of an irreducible algebraic variety V of dimension d in $A^n_{\mathbf{K}}(X)$ with H is an irreducible algebraic variety of dimension $d-1$ in $A^n_{\mathbf{K}(a_I)}(X)$.

Proof. The hypersurface $H(X) = 0$, supposed of degree e, will have $max\ |I| = e$. The number of coefficients a_I is then $E = \begin{pmatrix} n+e \\ e \end{pmatrix}$. Let $P^{E-1}_{\mathbf{K}}(\alpha)$ be the projective space over $\mathbf{K}$ of dimension $E-1$ with α equal to the ratios of a_I taken as homogeneous coordinates. Define now an algebraic correspondence $\mathcal{C}$ between $A^n_{\mathbf{K}}(X)$ and $P^{E-1}_{\mathbf{K}}(\alpha)$ to be consisting of points (X^0, a^0) such that X^0 is in V while $a^0 = (ratios\ of\ a^0_I)$ with $H^0(X^0) = 0$, where $H^0 = \Sigma_I\ a^0_I * X^I$, the a^0_I's being not all 0. Let a_0 be the constant term in $H(X)$ and a^0_0 that in $H(X)$ so that $a_0 = -\Sigma_{I'}\ a_{I'} * X^{I'}$ and $a^0_0 = -\Sigma_{I'}\ a^0_{I'} * X^{I'}$, in which the summation is over only those I' in I differ from the 0-tuple $(0, \cdots, 0)$. The correspondence $\mathcal{C}$ is then seen to be irreducible since it has an evident generic point (Ξ, α^0) for which Ξ is a generic point of V and $\alpha^0 = (ratios\ of\ \alpha_I^0)$ is a point in $P_{\mathbf{K}}^{E-1}(\alpha)$ with $\Sigma_I\ \alpha_I^0 * \Xi^I = 0$ so that $\alpha_0^0 = -\Sigma_{I'}\ \alpha_{I'}^0 * X^{I'}$, It is clear that

$$a(\mathcal{C}) = d,\ b(\mathcal{C}) = E - 2.$$

It is also clear that $d(\mathcal{C}) \leqslant d - 1$ since a generic hypersurface cannot contain V as a whole. By the Counting Principle we have then $c(\mathcal{C}) \geqslant E - 1$. It follows that $c(\mathcal{C}) = E - 1$ which means that the image variety is the whole of projective space $P_{\mathbf{K}}^{E-1}(\alpha)$. By Counting Principle again we should have $d(\mathcal{C}) = d - 1$. This means that the intersection of V with the generic hypersurface $\Sigma_I\ a_I * X^I = 0$ is of dimension $d - 1$ over the field $\mathbf{K}(a_I)$ as to be proved. □

Consider the special case of a hyperplane for which $e = 1$. As a generic linear space L of dimension r in $A_{\mathbf{K}}^n(X)$ is the intersection of $n-r$ generic linear hyperplanes in $A_{\mathbf{K}}^n(X)$, we may apply the above theorem in the case $e = 1$ successively to get as a corollary the following theorem:

Theorem 2.4. The intersection of an irreducible algebraic variety V of dimension d over $\mathbf{K}$ in $A_{\mathbf{K}}^n(X)$ with a generic linear space L of dimension r in $A_{\mathbf{K}}^n(X)$ is void if $r < n - d$. On the other hand if $r \geqslant n - d$ then the intersection is an algebraic variety of dimension $d + r - n$ over the extension field $\mathbf{K}(\mathcal{L})$ where $\mathcal{L}$ is the set of generic parameters determining the linear space L.

Remark 2.3. In comparing Theorems 2.3 and 2.4 with Theorems 1.4, 1.5 and the proofs there, it is clearly seen the powerfulness of the present method of algebraic correspondence.

Dimension of Grassmannian Variety.

Notation 2.5.

$\Omega(n, d) :=$ *Grassmannian variety* of linear spaces of dimension d in $P_{\mathbf{K}}^n({}^hX)$.

$\Omega(n; 0, d) :=$ *Grassmannian variety of composite elements* (S_0, S_d) consisting of pairs of a point S_0 and a linear space of dimension d through S_0 in $P_{\mathbf{K}}^n({}^hX)$.

Theorem 2.5. The point sets $\Omega(n, d)$ and $\Omega(n; 0, d)$ are all irreducible non-singular algebraic varieties in respective projective spaces of sufficient high dimension and have dimensions over $\mathbf{K}$ given below

$$Dim_{\mathbf{K}}\ \Omega(n, d) = (n - d) * (d + 1); \tag{2.2}$$

$$Dim_{\mathbf{K}}\ \Omega(n; 0, d) = (n - d) * d + n. \tag{2.3}$$

Proof. That the point sets are usual projective algebraic varieties are readily seen by means of Grassmann coordinates for linear subspaces. See e.g. [H-P1], Chap. VII. They are irreducible since they possess evident generic points consisting of generic linear subspace of dimension d in the first case and the pair of a generic point and a generic linear subspace of dimension d through that generic point in the second case. That they have no singularities follow readily from the evident homogeneity of these varieties.

We now prove the dimension formulae (2.2) and (2.3) by the Counting Principle as follows.

Let the Grassmannian variety $\Omega(n,d)$ be contained in the projective space $P^G_{\mathbf{K}}(\pi)$ of sufficiently high dimension G in homogeneous coordinates π as above. Consider an irreducible algebraic correspondence $\mathcal{C}$ between a $(d+1)$-fold projective space $P^{n,d+1} = (P^n_{\mathbf{K}}({}^hX))^{d+1}$ and the projective space $P^G_{\mathbf{K}}(\pi)$ defined by its generic point Ξ as follows. For each $j = 0, 1, \cdots, d$, $\Xi^{(j)}$ is a generic point in the j-th component of the product space $P^{n,d+1}$. Let S_d be the linear space of dimension d in $P^n_{\mathbf{K}}(X)$ determined by these $d+1$ points $\Xi^{(j)}$, to be considered as an extended point in $P^G_{\mathbf{K}}(\pi)$. The generic point defining the correspondence $\mathcal{C}$ is then given by $\Xi = ((\Xi^{(0)}, \Xi^{(1)}, \cdots, \Xi^{(d)}), S_d)$.

It is now clear that the image variety of $\mathcal{C}$ is $\Omega(n,d)$ in $P^G_{\mathbf{K}}(\pi)$ and

$$a(\mathcal{C}) = n*(d+1),\ b(\mathcal{C}) = 0,\ d(\mathcal{C}) = d*(d+1).$$

By Counting Principle we have then $c(\mathcal{C}) = (n-d)*(d+1)$ or (2.2).

Similarly let $P^{G'}_{\mathbf{K}}(\pi')$ be a projective space of sufficiently high dimension G' which contains the composite Grassmannian variety $\Omega(n;0,d)$. Consider now an irreducible algebraic correspondence $\mathcal{C}_0$ between the $(d+1)$-fold projective space $P^{n,d+1}$ as before and the product space $P^n_{\mathbf{K}}({}^hX) \times P^{G'}_{\mathbf{K}}(\pi')$ determined by a generic point $\Xi_0 = ((\Xi'(0), \Xi'^{(1)}, \cdots, \Xi'^{(d)}), (\Xi'^{(0)}, S'_d))$, for which S'_d is a genenic linear space of dimension d passing through $\Xi'(0)$ and the composite element $(\Xi'(0), S'_d)$ is to be considered as an extended point in $P^{G'}_{\mathbf{K}}(\pi')$. It is readily seen that the image variety is $\Omega(n;0,d)$ and

$$a(\mathcal{C}_0) = n*(d+1),\ b(\mathcal{C}_0) = 0,\ d(\mathcal{C}_0) = d^2.$$

It follows by Counting Principle that $Dim_{\mathbf{K}}\ \Omega(n;0,d) = c(\mathcal{C}_0) = (n-d)*d+n$ or (2.3). The theorem is thus completely proved. $\square$

Chow Form of Irreducible Algebraic Variety V.

In what follows we adopt the following notations:

Notation 2.6.

$V = Zero(PS) :=$ An *irreducible* algebraic variety in $A^n_{\mathbf{K}}(X)$, for $PS \subset \mathbf{K}[X]$.

$\Xi = (\xi_1, \cdots, \xi_n) :=$ A generic point of V.

$d := Dim_{\mathbf{K}}\ V,\ r := n - d,\ g := Deg\ V.$

$N := (0, 1, \cdots, n),\ N_0 := (1, \cdots, n).$

$L := (1, \cdots, g).$

$J := (0, 1, \cdots, d),\ J_0 = (1, \cdots, d).$

$U_j := (u_{j0}, u_{j1}, \cdots, u_{jn}), j \in J,\ u_{jk}\ (j \in J, k \in N)$ being independent indeterminates.

$(U_j, Z) := u_{j1} * z_1 + \cdots + u_{jn} * z_n + u_{j0}$ for any n-tuple of elements $Z = (z_1, \cdots, z_n)$ in any extension field of $\mathbf{K}$.

$\mathcal{U} := \{u_{jk} \mid j \in J, k \in N\}$; $\mathcal{U}_0 := \{u_{jk} \mid j \in J, k \in N_0\}$.

Consider now $d + 1$ elements $\zeta_j\ (j \in J)$ in $\mathbf{K}(\Xi, \mathcal{U}_0)$ defined by

$$\zeta_j = -\Sigma_{k \in N_0}\ u_{jk} * \xi_k,\ j \in J. \tag{2.4}$$

As Ξ is of dimension d over $\mathbf{K}$, the $d + 1$ elements ζ_j are algebraically dependent over $\mathbf{K}(\mathcal{U}_0)$. Hence for indeterminates $u_{j0}, j \in J$, there is a pol

$$F(u_{00}, u_{10}, \cdots, u_{d0}) \in \mathbf{K}(\mathcal{U}_0)[u_{00}, u_{10}, \cdots, u_{d0}],$$

such that F becomes 0 for $u_{j0} = \zeta_j, j \in J$. With $u_{j0} = \zeta_j, j \in J$, so (2.4) becomes then

$$(U_j, \Xi) = 0,\ j \in J. \tag{2.4$'$}$$

By clearing of fractions if necessary, we may also take F to be a pol

$$CF(\mathcal{U}) \in \mathbf{K}[\mathcal{U}], \tag{2.5}$$

unique up to a non-zero factor in $\mathbf{K}$. Moreover, from general theory about algebraic extension of fields we have the following

Lemma 2.1. If ϕ is a pol in $\mathbf{K}[\mathcal{U}]$ such that ϕ becomes 0 for $u_{j0} = \zeta_j, j \in J$, then ϕ has CF as a factor.

Definition 2.7. *Chow Form* of $V :=$ The pol CF of (2.5) which is unique up to non-zero factors in $\mathbf{K}$.

An Alternative Definition of Chow Form.

Consider d hyperplanes L_j ($j \in J_0$) in $A^n_{\mathbf{K}}(X)$ defined by

$$(U_j, X) = 0, \ j \in J_0. \tag{2.6}$$

By Theorem 1.5 of Sect. 5.1, or Theorem 2.4 of this section, the $(n-d)$-dimensional generic linear space S_{n-d} defined as the intersection of these $L_j, j \in J_0$, will intersect V in g conjugate points $\Xi^{(l)} = (\xi_1^{(l)}, \cdots, \xi_n^{(l)})$, $l \in L$. Then

$$P = \Pi_{l\in L}\ (U_0, \Xi^{(l)})\ \in \mathbf{K}(\mathcal{U}_0)[U_0] \tag{2.7}$$

is homogeneous of degree g in $u_{00}, u_{01}, \cdots, u_{0n} \in U_0$. By clearing of fractions in multiplying P by some pol $\rho \in \mathbf{K}[\mathcal{U}_0]$ we get from P a pol

$$\rho * P \in \mathbf{K}[\mathcal{U}]. \tag{2.8}$$

Comparing (2.7), (2.8) with (2.4) in taking Ξ to be any one of $\Xi^{(l)}$, we see by Lemma 2.1 that $\rho * P$ has CF as a factor. As P is irreducible as a pol in $\mathbf{K}(\mathcal{U}_0)[U_0]$ it follows that $\rho * P$ can differ from CF by almost a non-zero factor in $\mathbf{K}$. Hence we have the following

Theorem 2.6. If $\Xi^{(l)} = (\xi_1^{(l)}, \cdots, \xi_n^{(l)})$, $l \in L$, are the g conjugate points of intersection of a generic $(n-d)$-dimesional linear space S_{n-d} with the irreducible algebraic variety V, then the Chow form CF of V is given by (2.7) and (2.8) up to a non-zero constant factor in $\mathbf{K}$:

$$CF = \rho * \Pi_{l\in L}\ (U_0, \Xi^{(l)}), \tag{2.9}$$

in which $\rho \in \mathbf{K}[\mathcal{U}_0]$.

By Definition 2.7 of the Chow form CF the sets of variables $U_j = (u_{j0}, u_{j1}, \cdots, u_{jn})$ play the same role for each $j \in J$. As CF in (2.9) is homogeneous of degree g in the variables of U_0, so this should be also true for each set of variables in $U_j, j \in J_0$. Hence we have the following

Theorem 2.7. The Chow form $CF(\mathcal{U})$ of an irreducible algebraic variety of degree g is homogeneous of degree g in each set of variables $U_j = (u_{j0}, u_{j1}, \cdots, u_{jn})$ for $j \in J$.

A Theorem of Hodge.

Consider the $(n-d-1)$-dimensional linear space S_{n-d-1} defined by the equations

$$(U_j, X) = 0, \ j \in J. \tag{2.10}$$

Definition 2.8. *Dual Grassmann coordinates* of S_{n-d-1} := The set of determinants $(p^{(i_0 i_1 \cdots i_d)})$ where

$$p^{(i_0 i_1 \cdots i_d)} = \begin{vmatrix} u_{0i_0} & u_{0i_1} & \cdots & u_{0i_d} \\ u_{1i_0} & u_{1i_1} & \cdots & u_{1i_d} \\ \cdots & \cdots & \cdots & \cdots \\ u_{di_0} & u_{di_1} & \cdots & u_{di_d} \end{vmatrix}, \quad i_j \in N.$$

We have now the following theorem due to W.V.D.Hodge, see [H-P2], Chap. X.

Theorem 2.8. The Chow form of V is a pol in the dual Grassmann coordinates $p^{(i_0 i_1 \cdots i_d)}$:

$$CF(\mathcal{U}) = G(p^{(i_0 i_1 \cdots i_d)}). \tag{2.11}$$

Proof. We may suppose $p^{(01\cdots d)} \neq 0$. Then S_{n-d-1} of (2.10) may equally well be defined by $d+1$ equations in ${}^hX = (x_0 : x_1 : \cdots : x_n)$ below:

$$p^{(01\cdots d)} * x_j + \Sigma_k \, p^{(0\cdots,j-1,k,j+1,\cdots,d)} * x_k = 0, \ j \in J, \tag{2.12}$$

in which the summation is to be taken over k with $k \in N$ and $k > d$. Treating (2.12) as in (2.4)′ or (2.6) in getting $CF(\mathcal{U})$, we get a pol $G(p^{(i_0 i_1 \cdots i_d)})$ as to be asserted.

□

Remark 2.3. The dual Grassmann coordinates $p^{(i_0 i_1 \cdots i_d)}$ of a linear space are not independent so that the form of G, unlike that of F, are not unique up to non-zero factors of $\mathbf{K}$.

Chow Form and Algebraic Correspondence.

Notation 2.7.

For $j \in J$, ${}^hU_j := (u_{j0} : u_{j1} : \cdots : u_{jn})$.

$\mathcal{M}$:= The mulptiple product space $A^n_{\mathbf{K}}(X) \times (P^n_{\mathbf{K}}({}^hU_0) \times \cdots \times P^n_{\mathbf{K}}({}^hU_d))$.

$\mathcal{C}$:= The algebraic correspondence in $\mathcal{M}$ consisting of sets $(X^0, (L^0_0, L^0_1, \cdots, L^0_d))$ in which X^0 is a point in V and for $j \in J$, L^0_j is a hyperplane in $A^n_{\mathbf{K}}(X)$ passing through that point, defined by $(U_j, X) = 0$ with $(U_j, X^0) = 0$, to be considered as a point in the projective space $P^n_{\mathbf{K}}({}^hU_j)$.

Theorem 2.9. The correspondence $\mathcal{C}$ is irreducible and the image variety of $\mathcal{C}$ is given by

$$CF(\mathcal{U}) = 0, \tag{2.13}$$

where CF is the Chow form of V.

Proof. $\mathcal{C}$ is irreducible since it has clearly a generic point $(\Xi, ({}^hU_0, {}^hU_1, \cdots, {}^hU_d))$ where Ξ is a generic point of V and for each $j \in J$, ${}^hU_j$ corresponds to an equation

$(U_j, X) = 0$ with $(U_j, \Xi) = 0$ so that $u_{j0} = -\Sigma_{k\in N_0} u_{jk} * \xi_k$, defining thus a generic hyperplane through Ξ. Clearly

$$a(\mathcal{C}) = d,\ b(\mathcal{C}) = (d+1) * (n-1),\ d(\mathcal{C}) = 0.$$

By Counting Principle we have therefore

$$c(\mathcal{C}) = (d+1) * n - 1.$$

The image variety of $\mathcal{C}$ is thus a hypersurface in the product space $P^n_{\mathbf{K}}({}^hU_0) \times \cdots P^n_{\mathbf{K}}({}^hU_d)$ defined by some equation

$$F_0(U_0, U_1, \cdots, U_d) = 0,$$

F_0 being an irreducible pol of $\mathbf{K}[\mathcal{U}]$ which is homogeneous in each set of variables $U_j = (u_{j0}, \cdots, u_{jn}), j \in J$.

Let the linear space of dimension $n-d$ defined by $(U_j, X) = 0,\ j \in J_0$ intersect V in the g generic points $\Xi^{(l)}, l \in L$. Then $F_0(U_0, U_1, \cdots, U_d) = 0$ if and only if for one $l \in L$, $(U_j, \Xi^{(l)}) = 0,\ j \in J$. It follows by Lemma 2.1 that F_0 is divisible by the Chow form CF. As F_0 is irreducible it will differ from CF by almost a non-zero factor of $\mathbf{K}$. This proves the theorem. □

From the final part of the proof of the above theorem we have as corollary the following theorem:

Theorem 2.10. A set of $d+1$ hyperplanes in $A^n_{\mathbf{K}}(X)$ defined by $(U_j, X) = 0,\ j \in J_0$ will have a point in common with the irreducible variety V of dimension d if and only if $F_0(U_0, U_1, \cdots, U_d) = 0$, or $CF(\mathcal{U}) = 0$, where CF is the Chow form of V.

Chow Basis of an Irreducible Algebraic Variety.

Theorem 2.11. A point Ξ^0 is on the irreducible algebraic variety V of dimension d if and only if any generic linear space of dimension $n-d-1$ passing through Ξ^0 will have points in common with V.

Proof. Suppose first that Ξ^0 is not on the variety V. Let us take an arbitrary point on V and let S_{n-1} be a generic hyperplane through Ξ^0 but not through that point. Then S_{n-1} will intersect V in an algebraic variety each of its irreducible components will have a dimension less than or equal to $d-1$. Take now on each of the components an arbitrary point and let S_{n-2} be a generic linear space of dimension $n-2$ through Ξ^0 which is contained in S_{n-1} and contains none of the points chosen.

Again S_{n-2} will intersect V in an algebraic variety each of its irreducible components will have a dimension less than or equal to $d-2$. Proceeding in the same manner we will get a generic linear space S_{n-d} through Ξ^0 of dimension $n-d$ which will intersect V in at most a finite set of points. Take now a generic hyperplane through Ξ^0 but none of these points then this hyperplane will intersect S_{n-d} in a linear space S_{n-d-1} of dimension $n-d-1$ which passes through Ξ^0 but has no points in common with V.

Suppose on the contrary Ξ^0 is on the variety V, then any line space of dimension $n-d-1$ passing through E^0 will have points in common with V, e.g. Ξ^0. This proves the theorem. □

Lemma 2.2. A generic hyperplane through a point

$$Z = (z_1, \cdots, z_n)$$

is of the form $(U, X) = 0$ in which U is the matrix-product

$$U =^h Z^{\cdot} S, \text{ with } {}^h Z = (1, z_1, \cdots, z_n),$$

and S a skew-symmetric $(n+1)$-matrix of independent indeterminates s_{ik}, for $i < k$, while $s_{ki} = -s_{ik},\ i, k \in N$.

Proof. This follows in applying Grassmann coordinates. □

Notation 2.8.

For $j \in J, S^{(j)} = (s_{ik}^{(j)}), i, k \in N :=$ Skew symmetric $(n+1)$-matrices of independent indeterminates $s_{ik}^{(j)}$, for $i < k$, while $s_{ki}^{(j)} = -s_{ik}^{(j)}$, for $i, k \in N$.

For $j \in J, U_j^0 = (u_{j0}^0, u_{j1}^0, \cdots, u_{jn}^0) :=$ The matrix-product ${}^h Z^{\cdot} S^{(j)}$.

$\mathcal{U}^0 :=$ Set of $\{u_{ji}^0\},\ i \in N,\ j \in J$.

$T :=$ Tuples of non-negative integers $(t_{ik}^{(j)} \mid j \in J,\ i, k \in N,\ i < k)$.

For $t = (t_{ik}^{(j)}) \in T,\ S^t :=$ Product of $(s_{ik}^{(j)})^{t_{ik}^{(j)}}$ for $j \in J,\ i, k \in N,\ i < k$.

Theorem 2.12. A point $Z = (z_1, \cdots, z_n)$ belongs to V if and only if

$$CF(\mathcal{U}^0) = 0 \tag{2.14}$$

identically in $s_{ik}^{(j)},\ i < k$, where CF is the Chow form of V.

Proof. This follows from Lemma 2.2 and Theorems 2.10, 2.11 if the $d+1$ hyperplanes through a point $Z = (z_1, \cdots, z_n)$ are taken as $(U_j^0, X) = 0,\ j \in J$. □

As a corollary we have the following theorem which recovers the defining equations of V from its Chow form.

Theorem 2.13. Let

$$CF(\mathcal{U}^0) = \Sigma_{t \in T}\ C_t * S^t. \tag{2.15}$$

Then

$$V = Zero(CB), \text{ where} \tag{2.16}$$

$$CB = \{C_t \mid t \in T\} \subset \mathbf{K}[X] \tag{2.17}$$

so that V may be defined by the equations

$$CB = 0, \text{ or}$$

$$V = Zero(CB). \tag{2.18}$$

Definition 2.9. $CB :=$ *Chow Basis* of the irreducible variety V.

Definition 2.10. The ratios of coefficients of the Chow form CF of an irreducible variety V arranged in a definite order := The *Chow coordinates* of the variety V.

Remark 2.5. For fixed d, g let us consider the Chow coordinates of an irreducible variety of dimension d and degree g as a point in some projective space. Such points will define an irreducible algebraic variety in this projective space called *Chow variety* corresponding to the fixed values of d and g. These notions originated from Chow permit to study not only the properties of algebraic varieties but also algebraic systems of algebraic varieties which we shall not enter. Cf. e.g. [Ch-VdW], [VdW2], [H-P2].

Remark 2.6. The above introduction of Chow form, Chow basis, and Chow coordinates follow closely those given in [Ch-VdW] and [VdW2] with due modifications in taking account of [H-P2], Chap. X. However, the treatment is quite involved and examples show that the method is quite ineffective even in simple cases. Moreover, redundant pols may appear in the basis. It seems thus some methods simplifing the matter are requied.

5.3 Chern Classes and Chern Numbers of an Irreducible Algebraic Variety with Arbitrary Singularities.

Grassmannian of Composite Elements.

Throughout this section $\mathbf{K}$ will be a fixed field of characteristic 0 unless otherwise stated.

Definition 3.1. *Grassmannian of composite elements* $GR(n;0,d)$ in $P_{\mathbf{K}}^n :=$ the set of composite elements (L_0, L_d) for which L_0 is a point in $P_{\mathbf{K}}^n$, and L_d is a linear subspace of dimension d in $P_{\mathbf{K}}^n$ which passes through the point L_0.

Theorem 3.1. $GR(n;0,d)$ is a *non-singular* algebraic variety over $\mathbf{K}$ of dimension

$$D = (n-d) * (d+1) + d \tag{3.1}$$

which can be realized in a projective space of sufficiently high dimension N.

For the proof we refer to the preceding section.

Henceforth the non-singular variety $GR(n;0,d)$ will be denoted by G with fixed n, d unless otherwise stated. It will be supposed to be a subvariety in $P_{\mathbf{K}}^N$ for N sufficiently large.

Schubert Cycles of Grassmannian $GR(n;0,d) = G$.

Consider in the projective space $P_{\mathbf{K}}^n$ a sequence $\mathcal{S}$ of linear subspaces S_i of dimension i such that

$$\mathcal{S} : S_0 \subset S_1 \subset \cdots \subset S_n = P_{\mathbf{K}}^n. \tag{3.2}$$

Definition 3.2. A *Schubert cycle* of *Ehresmann symbol* $\begin{bmatrix} a_0 & & & \\ b_0 & b_1 & \cdots & b_d \end{bmatrix}$ w.r.t. $\mathcal{S}$ in which

$$0 \leqslant a_0 \leqslant n,\ 0 \leqslant b_0 < b_1 < \cdots < b_d \leqslant n \tag{3.3}$$

:= The subvariety of G consisting of composite elements (L_0, L_d) for which

$$Dim_{\mathbf{K}}\ (L_d \cap S_{b_j}) \geqslant j,\ j = 0, 1, \cdots, d. \tag{3.4}$$

In particular, for $j = 0$, $(3.4)_0$ means $L_0 \in S_{a_0}$.

In what follows all Schubert cycles will be refered to the fixed sequence $\mathcal{S}$ in (3.2) so that the phraseology "w.r.t. $\mathcal{S}$" will be omitted unless otherwise stated.

Notation 3.1. (i, j) with $i < j$ in an Ehresmann symbol := A sequence of consecutive integers from i to j.

Definition 3.3. A *fundamental Schubert cycle* := A Schubert cycle of Ehresmann symbol $\begin{bmatrix} a_0 & & & \\ b_0 & b_1 & \cdots & b_d \end{bmatrix}$ for which a_0 equals some b_j. In particular, we have

$$GR(n;0,d) = \begin{bmatrix} n \\ (n-d, n) \end{bmatrix}. \tag{3.5}$$

Definition 3.4. *Dual* of a fundamental Schubert cycle $\begin{bmatrix} b_i \\ b_0 & \cdots & b_d \end{bmatrix}$:= The fundamental Schubert cycle $\begin{bmatrix} n-b_i \\ n-b_d & \cdots & n-b_0 \end{bmatrix}$.

Notation 3.2. The dual of

$$\begin{bmatrix} b_i \\ b_0 & \cdots & b_d \end{bmatrix} := \delta \begin{bmatrix} b_i \\ b_0 & \cdots & b_d \end{bmatrix} = \begin{bmatrix} n-b_i \\ n-b_d & \cdots & n-b_0 \end{bmatrix}. \tag{3.6}$$

Remark 3.1. The dual of a dual is clearly the identity.

Defintion 3.5. *Codimension* of a Schubert cycle := The dimension of its dual.

Notation 3.3. *Dimension* or *codimension* of a Schubert cycle of Ehresmann symbol $E := Dim_{\mathbf{K}}\ E$ or $Codim_{\mathbf{K}}\ E$.

Theorem 3.2. The dimension of a fundamental Schubert cycle

$$E = \begin{bmatrix} b_i \\ b_0 & b_1 & \cdots & b_d \end{bmatrix} \tag{3.7}$$

is given by

$$Dim_{\mathbf{K}}\ E = Codim_{\mathbf{K}}\ Dual\ E = \Sigma_j\ (b_j - j) + i, \tag{3.8}$$

the summation being from $j = 0$ to $j = d$.

The proof is elementary and will be omitted.

<u>*Algebraic-Equivalence Group of Algebraic Varieties and Chow Ring.*</u>

Let an irreducible variety V of dimension d_V in some projective space $P_{\mathbf{K}}^N$ be given. Based on the theory of Chow form the formal algebraic sum of irreducible subvarieties contained in V (or *cycles*) of fixed dimension r can be given an equivalence relation. The totality of *algebraic-equivalence classes* thus endowed a natural additive operation will be turned into an abelian group. This group will be called the *algebraic-equivalence group* of V in the dimension r or *codimension* $d_V - r$ and will be denoted by $Alg_r(V)$. In particular we have

$$Alg_0(V) \approx \mathbf{Z} \tag{3.9}$$

so that we may identify $Alg_0(V)$ with $\mathbf{Z}$. The direct sum of all these groups from $r = 0$ to $r = d_V$ will be denoted by $A_*(V)$. If V is *non-singular* then the *intersection* operation will induce a multiplicative operation $*$ in $A_*(V)$ to turn it into a ring with unity such that

$$\begin{aligned} &Alg_r(V) * Alg_s(V) \subset Alg_{r+s-d_V}(V), \text{ for } r + s \geqslant d_V, \text{ or} \\ &Codim_{\mathbf{K}}\ (A * B) = Codim_{\mathbf{K}}\ A + Codim_{\mathbf{K}}\ B, \text{ for} A, B \in Alg_*(V). \end{aligned} \tag{3.10}$$

We omit the details of proofs of the above facts which may be found in treatises or relevant papers on algebraic geometry, e.g. [H-P2].

Definition 3.6. The ring determined as in (3.10) of the non-singular variety V of dimension d over $\mathbf{K}$ as above := The *intersection ring* or the *Chow ring* of the *non-singular* variety V.

Algebraic-Equivalence Group of $GR(n;0,d)$.

Theorem 3.3. (Ehresmann) For $\mathbf{K}=\mathbf{C}$ the interiors of fundamental Schubert cycles from a cell decomposition of $GR(n;0,d)$ as a topological manifold.

For the proof we refer to the original paper of Ehresmann [Eh1].

Theorem 3.4. The fundamental Schubert cycles of same Ehresmann symbol

$$\begin{bmatrix} b_i & & & \\ b_0 & b_1 & \cdots & b_d \end{bmatrix}$$

w.r.t. various sequences like $\mathcal{S}$ belong to the same algebraic-equivalence class.

Proof. For any two sequences $\mathcal{S},\mathcal{S}'$ there exist some projective transformation of $P^n_{\mathbf{K}}$ which transforms the linear subspaces of $\mathcal{S}$ to those of $\mathcal{S}'$. It transforms therefore the Schubert cycle w.r.t. $\mathcal{S}$ to that of same symbol w.r.t. $\mathcal{S}'$ and hence they are algebraically equivalent. □

Definition 3.7. The *Ehresmann class* of *symbol* (3.7) := The algebraic-equivalence class determined by the fundamental Schubert cycle (3.7).

Notation 3.4. For the sake of simplifying the notations, the algebraic-equivalence class of symbol (3.7) and the Schubert cycle (3.7) will be denoted by the same symbol (3.7) so far no misunderstanding may occur. The operation induced by forming the dual of the class := δ_*.

Theorem 3.5. (W.L.Chow) The Ehresmann classes form an additive basis of the algebraic-equivalence group $Alg_*(GR(n;0,d))$ of the Grassmannian $GR(n;0,d)$.

For the proof we refer to the original papers of Chow [Chow1,2] which contains much more materials of quite general character.

Intersection Ring or Chow Ring of $GR(n;0,d)$.

As $G=GR(n;0,d)$ is non-singular, intersection can be defined among algebraic-equivalence classes so that $ALg_*(G)$ becomes a ring: the *Chow ring* of G.

Theorem 3.6. The Chow ring of $GR(n;0,d)$ in codimension less or equal to d is generated by the classes δ_*P in codimension 1 and δ_*Q_h in codimension h for

$0 \leqslant h \leqslant d$ where

$$\begin{aligned} P &= \begin{bmatrix} 1 \\ (0,d) \end{bmatrix}, \\ Q_h &= \begin{bmatrix} 0 & \\ (0,d-1), & d+h \end{bmatrix}. \end{aligned} \tag{3.11}$$

In particular, we have

$$\delta_* \begin{bmatrix} i & & & \\ (0,i), & b_{i+1}, & \cdots, & b_d \end{bmatrix} = \delta_* \begin{bmatrix} 0 & & & \\ (0,i), & b_{i+1}, & \cdots, & b_d \end{bmatrix} * (\delta_* P)^i, \tag{3.12}$$

$$\delta_* \begin{bmatrix} 0 & & \\ a_0, & \cdots, & a_d \end{bmatrix} * \delta_* \begin{bmatrix} 0 & & \\ b_0, & \cdots, & b_d \end{bmatrix} = \Sigma_c\, \delta_* \begin{bmatrix} 0 & & \\ c_0, & \cdots, & c_d \end{bmatrix}. \tag{3.13}$$

In (3.13) the summation is to be extended over $c = (c_0, \cdots, c_d)$ such that in the ordinary Grassmannian variety $\Omega(n,d)$ we shall have the intersection formula

$$\delta_* [\ a_0, \cdots, a_d\ ] * \delta_* [\ b_0, \cdots, b_d\ ] = \Sigma_c\, \delta_* [\ c_0, \cdots, c_d\ ]. \tag{3.14}$$

Proof. The proof is elementary and will not be given here. For the explicit expression of (3.14) we refer to [H-P2], Chap. XIV, Sect. 6 which can be deduced from the well-known formulae of Pieri and Giambelli.

Algebraic Equivalence Classes of Particular Interest.

Besides the classes P, Q_h and their duals we are particularly interested in the following classes:

Definition 3.8. The *Gamkrelidze-Todd classes* :=

$$GT_{st} = \begin{bmatrix} s-t & \\ (0,d-t), & (d-t+2, d+1) \end{bmatrix} \in Alg_s(GR(n;0,d)), \text{ for } 0 \leqslant t \leqslant s \leqslant d. \tag{3.15}$$

Definition 3.9. The *Chern classes* :=

$$Ch_s = \Sigma_t\ (-1)^t * \binom{d-t+1}{d-s+1} * GT_{st} \in Alg_s(GR(n;0,d)), \text{ for } 0 \leqslant s \leqslant d, \tag{3.16}$$

in which the summation is to be taken over the range $0 \leqslant t \leqslant s$.

Notation 3.5. A partition π of positive integer p :=

$$\pi = (p_1, \cdots, p_s), \text{ where}$$

$$0 < p_1 \leqslant \cdots \leqslant p_s,\ p_1 + \cdots + p_s = p.$$

Definition 3.10. The *Chern class of partition* π :=

$$Ch_\pi = \delta_*(\delta_*\ Ch_{p_1} * \cdots * \delta_*\ Ch_{p_s}) \in Alg_p(GR(n;0,d)). \tag{3.17}$$

From Theorem 3.6 we have now the following relations:

$$
\begin{aligned}
\delta_* Ch_1 &= (d+1)*\delta_* P - \delta_* Q_1. \\
\delta_* Ch_2 &= \binom{d+1}{2}*(\delta_* P)^2 - d*\delta_* P*\delta_* Q_1 + (\delta_* Q_1)^2 - \delta_* Q_2. \\
\delta_* Ch_3 &= \binom{d+1}{3}*(\delta_* P)^3 - \binom{d}{2}*(\delta_* P)^2*\delta_* Q_1 \\
&\quad +(d-1)*\delta_* P*((\delta_* Q_1)^2 - \delta_* Q_2) \\
&\quad -((\delta_* Q_1)^3 - 2*\delta_* Q_1*\delta_* Q_2 + \delta_* Q_3). \\
\delta_* Ch_{11} &= \binom{d+1}{2}^2*(\delta_* P)^2 - 2*(d+1)*\delta_* P*\delta_* Q_1 + (\delta_* Q_1)^2.
\end{aligned} \tag{3.18}
$$

Some Morphisms about an Algebraic Variety.

Consider an irreducible algebraic variety V of dimension d over $\mathbf{K}$ in $P^n_{\mathbf{K}}$. Let Ξ be a generic point of V and T_Ξ the tangent space of V at Ξ which is a linear subspace of dimension d in $P^n_{\mathbf{K}}$ passing through the point Ξ. The pair (Ξ, T_Ξ) is a point of $GR(n;0,d)$ which determines an irreducible subvariety $\hat{V}$ of $GR(n;0,d)$ with (Ξ, T_Ξ) as a generic point:

$$\hat{V} = Spec_{\mathbf{K}}\ (\Xi, T_\Xi) \subset GR(n;0,d). \tag{3.19}$$

Definition 3.11. *Inverse tangential morphism*

$$\tau_*:\ Alg_*(\hat{V}) \longrightarrow Alg_*(V) \tag{3.20}$$

:= The morphism induced by

$$\tau:\ (X,T) \longrightarrow X \tag{3.21}$$

for any point $X \in V$ and $(X,T) \in \hat{V}$.

For any Ehresmann class E of codimension $r \leqslant d$, the intersection of E with $\hat{V}$ is a well-defined algebraic equivalence class of codimension r of $\hat{V}$.

Definition 3.12. *Intersection morphism*

$$\nu_*:\ Alg_{D-r}(GR(n;0,d)) \longrightarrow Alg_{d-r}(\hat{V}),\ 0 \leqslant r \leqslant d \tag{3.22}$$

:= The morphism induced by intersection with $\hat{V}$ in the *non-singular* variety $GR(n;0,d)$.

Chern Classes and Chern Numbers of an Irreducible Variety.

Definition 3.13. For any Ehresmann class E of dimension $r \leqslant d$ as in (3.7) the *Ehresmann class* $E(V)$ of *Ehresmann symbol* E of the variety $V :=$ The class

$$E(V) = \tau_* \nu_* \delta_* E \in Alg_*(V). \tag{3.23}$$

Definition 3.14. For any Chern class Ch_π of partition π of $p \leqslant d$ as in (3.7) the *Chern class* $Ch_\pi(V)$ of partition π of the variety $V :=$ The class

$$Ch_\pi(V) = \tau_* \nu_* \delta_* Ch_\pi \in Alg_p(V). \tag{3.24}$$

Definition 3.15. For any algebraic-equivalence class $c \in Alg_d(GR(n;0,d))$, the *Chern number* or *Chern character* $c(V)$ of the variety $V :=$ The element

$$c(V) = \tau_* \nu_* \delta_* c \in Alg_0(V) \tag{3.25}$$

when $Alg_0(V)$ is identified to $\mathbf{Z}$.

Notation 3.6. For a partition π of d the Chern number corresponding to the class $Ch_\pi \in Alg_*(GR(n;0,d) := ch_\pi(V)$.

Examples in Simple Cases.

Let V_2 be an irreducible algebraic surface in $P^n_{\mathbf{K}}, n \geqslant 3$. Then there are associated to V_2 some classical numerical invariants as follows.

Definition 3.16.

Order $\mu_0(V_2) :=$ The number of points for which a generic linear subspace of dimension $n-2$ in $P^n_{\mathbf{K}}$ meets V_2, or what is the same, the degree of the intersection curve of V_2 by a generic hyperplane of $P^n_{\mathbf{K}}$.

Rank $\mu_1(V_2) :=$ Rank of the intersection curve of V_2 by a generic hyperplane in $P^n_{\mathbf{K}}$.

Class $\mu_2(V_2) :=$ Class of the intersection curve of V_2 by a generic hyperplane in $P^n_{\mathbf{K}}$.

Type $\nu_2(V_2) :=$ The number of simple points of V_2 for which the tangent plane will meet a generic linear subspace of dimension $n-4$ in $P^n_{\mathbf{K}}$.

Remark that type will be defined only in the case $n \geqslant 4$.

Theorem 3.7. For an irreducible algebraic surface in $P^n_{\mathbf{K}}$ we have

$$\begin{aligned}\mu_0(V_2) &= \begin{bmatrix} 2 & & \\ 0 & 1 & 2 \end{bmatrix}(V_2),\\ \mu_1(V_2) &= \begin{bmatrix} 1 & & \\ 0 & 1 & 3 \end{bmatrix}(V_2),\\ \mu_2(V_2) &= \begin{bmatrix} 0 & & \\ 0 & 2 & 3 \end{bmatrix}(V_2),\\ \nu_2(V_2) &= \begin{bmatrix} 0 & & \\ 0 & 1 & 4 \end{bmatrix}(V_2).\end{aligned} \tag{3.26}$$

Proof. This follows directly in comparing the respective definitions. □

Generalized Miyaoka-Yau Inequalities.

For algebraic varieties with arbitrary singularities the Defs. 3.13, 3.14 furnish us a general method of reducing the study of Chern classes and Chern numbers to the study of intersections in Grassmannians. As the intersections in Grassmannians are given explicitely by (3.12)-(3.14), the studies will be ultimately reduced to mere computations. Moreover, in the case of basic field being the complex field, the intersection number of two algebraic subvarieties in complementary dimensions of a non-singular variety like the Grassmannian is always positive. So this will lead to the study of positivity of Chern numbers by means of computations. As a concrete application of the above computational Principle and Method we shall consider the simplest case below.

Miyaoka and Yau (cf. [Miy1],[Yau1,2]) have proved that for a complex algebraic surface without singularities in a complex projective 3-space there holds some remarkable inequality between the Chern numbers, viz.

$$3 * ch_2(V_2) \geqslant ch_{11}(V_2). \tag{3.27}$$

We shall show that this will be true also for any irreducible algebraic hypersurface in $P^3_{\mathbf{K}}$ with arbitrary singularities which follows trivially from the above theory.

Theorem 3.8. For an irreducible complex algebraic surface V_2 in $P^3_{\mathbf{C}}$ with arbitrary singularities we have the Miyaoka-Yau inequality (3.27).

Proof. From (3.18) we have

$$\begin{aligned}ch_2(V_2) &= [\delta_*(3 * (\delta_* \ P)^2 - 2 * \delta_* P * \delta_* Q_1 + (\delta_* Q_1)^2 - \delta_* Q_2)](V_2),\\ ch_{11}(V_2) &= [\delta_*(9 * (\delta_* P)^2 - 6 * \delta_* P * \delta_* Q_1 + (\delta_* Q_1)^2)](V_2).\end{aligned} \tag{3.28}$$

It follows that

$$3 * ch_2(V_2) - ch_{11}(V_2) = [2 * \delta_*(\delta_* Q_1)^2 - 3 * Q_2](V_2). \tag{3.29}$$

By intersection formulae (3.12-14) the right-hand member of the last equality is equal to

$$2 * \begin{bmatrix} 0 & & \\ 0 & 2 & 3 \end{bmatrix} (V_2) - \begin{bmatrix} 0 & & \\ 0 & 1 & 4 \end{bmatrix} (V_2). \tag{3.30}$$

In $P^3_{\mathbf{C}}$ the second term in the last expression is non-existant while the first term is by very definition a non-negative integer. Whence we get from (3.29) the inequality (3.27) as to be proved. □

In a similar way we have also the following

Theorem 3.9. For an irreducible complex hypersurface V_3 of dimension 3 in $P^4_{\mathbf{C}}$ we have for the Chern numbers the inequality

$$4 * ch_{12}(V_3) - 8 * ch_3(V_3) - ch_{111}(V_3) \geqslant 0. \tag{3.31}$$

Proof. By direct computations as in the proof of Theorem 3.8 we have

$$4 * ch_{12}(V_3) - 8 * ch_3(V_3) - ch_{111}(V_3) = 5 * \delta_*(\delta_* Q_1)^3(V_3). \tag{3.32}$$

As $\delta_*(\delta_* Q_1)^3(V_3)$ is a non-negative integer by very definition, so we get (3.31) from (3.32). □

Remark 3.2. Examples show that the Miyaoka-Yau inequality (3.27) will no more be true for irreducible complex algebraic surfaces in $P^4_{\mathbf{C}}$. Cf. e.g. [S-R], p.221.

Remark 3.3. The possible inequalities which can exist among the Chern numbers of an irreducible algebraic hypersurface without singularities in high dimensional projective space were first given by Tai, see e.g. [Tai]. The case of hypersurfaces with arbitrary singularities have then been studied by Shi He by means of the above method. Cf. e.g. [SH1].

Remark 3.4. The Chern classes and Chern numbers have also been introduced for algebraic varieties with arbitrary singularities in various different ways by various authors, notably MacPherson, see e.g. [MP]. None of these methods are however computational in character as the one exhibited here. X.F.Liu has made a comparison of the various definitions due to Nash, Mather, MacPherson, and that of the author, see his paper [LXF].

5.4 A Projection Theorem on Quasi-Varieties.

Parametrization of Geometric Point Sets.

Notation 4.1. Let $\mathbf{K}, X = (x_1, \cdots, x_n)$, etc. be as before.

$T = (t_1, \cdots, t_p) :=$ a set of independent parameters.

$A_{\mathbf{K}}^p(T) :=$ affine space over $\mathbf{K}$ in coordinates T.

$A_{\mathbf{K}}^{n+p}(X,T) := A_{\mathbf{K}}^n(X) \times A_{\mathbf{K}}^p(T)$.

Definition 4.1. A geometric point set $S \subset A_{\mathbf{K}}^n(X)$ is *parametrized* in *parameters* T := There exist rational functions $g_i \in \mathbf{K}(T)$ or pols $g_i \in \mathbf{K}[T]$ for $i = 1, \cdots, n$ such that for each point $X^0 = (x_1^0, \cdots, x_n^0)$ in S, the equations

$$x_i = g_i(t_1, \cdots, t_p),\ i = 1, \cdots, n, \tag{4.1}$$

are solvable for $t_i = t_i^0$ in some extension field of $\mathbf{K}$ with $g_i(t_1^0, \cdots, t_p^0) = x_i^0,\ i = 1, \cdots, n$.

Example 4.1. In the literature one usually says that a circle $C : x_1^2 + x_2^2 = 1$ in the plane $A_{\mathbf{K}}^2(x_1, x_2)$ can be parametrized in a parameter t by means of the rational functions in $\mathbf{K}(x_1, x_2)$ in the form

$$x_1 = \frac{2 * t}{1 + t^2},\ x_2 = \frac{1 - t^2}{1 + t^2}. \tag{4.2}$$

However, this is not quite exact since for the point $X^0 = (x_1^0, x_2^0) = (0, -1)$ on the circle C, the equations (4.2) have no solutions of t at all. This shows that only the point set $C \setminus \{(0, -1)\}$, but not the point set C as a whole, is parametrized in t by the equations (4.2). The parametrization (4.2) of the circle C is thus one only with some *exceptional* points like $X^0 = (0, -1)$.

Remark 4.1. If we do not restrict the notion of parametrization in Def. 4.1 to rational functions or pols, then the circle C in Ex. 4.1 can be parametrized without exceptions by means of trigonometric functions in the usual manner, viz.

$$x_1 = sin\ t,\ x_2 = cos\ t. \tag{4.3}$$

Example 4.2. The elliptic curve

$$x_2^2 = 4 * x_1^3 - g_2 * x - g_3 \tag{4.4}$$

in the complex plane $A_{\mathbf{C}}^2(x_1, x_2)$ has a parametrization in t by means of Weierstrass elliptic functions $\mathcal{P}$ and their derivatives $\mathcal{P}'$, viz.

$$x_1 = \mathcal{P}(t),\ x_2 = \mathcal{P}'(t). \tag{4.5}$$

Remark that modern algebraic geometry shows that the curve (4.4) is impossible to be parametrized by means of rational functions or pols even some exceptional points are allowed.

Implicizitation of Parametrized Point Sets.

Definition 4.2. *Implicitization* of a geometric point set $S \subset A^n_{\mathbf{K}}(X)$ parametrized in $t_1, \cdots, t_p$ by means of (4.1) := A polset $PS \subset \mathbf{K}[X]$ such that S is as much as the same as $Zero(PS)$, or

$$S = Zero(PS) \setminus S' \tag{4.6}$$

where S' is a *negligible* part of S in some sense yet to be precised.

Example 4.2. Given a geometric point set $S \subset A^3_{\mathbf{K}}(x_1, x_2, x_3)$ parametrized in t_1, t_2 by means of equations below:

$$x_1 = t_1 * t_2, \ x_2 = t_1 * t_2^2, \ x_3 = t_1^2. \tag{4.7}$$

First let us show that S is not an algebraic variety. In fact, for independent indeterminates u and v, the point $\Xi = (u * v, v^2, u^4)$ corresponds to $t_1 = u^2, t_2 = \frac{v}{u}$ is in S. For any value $a \neq 0$, the point $A = (0, a, 0)$ is clearly a specialization over $\mathbf{K}$ of Ξ for which no values of t_1, t_2 can exist to give the point A. This shows that S cannot be a variety since otherwise $\Xi \in S$ would imply that $A \in S$ too.

It may be proved that

$$S = Zero(x_2^2 * x_3 - x_1^4 \ / \ x_1 * x_2) + Zero(x_1, x_2). \tag{4.8}$$

It follows that

$$\begin{aligned} S &= Zero(x_2^2 * x_3 - x_1^4) \setminus S', \text{ where} \\ S' &= Zero(x_1, x_3 \ / \ x_2). \end{aligned} \tag{4.9}$$

For $\mathbf{K} = \mathbf{C}$, the part of the point set S in the real space $A^3_{\mathbf{R}}(x_1, x_2, x_3)$ can be visualized to be consisting of a set of parabolas

$$x_1^2 = \pm x_2 * \sqrt{x_3}$$

in planes $x_3 = const. > 0$ shrinking towards the x_2-axis, and also the whole x_3-axis. Remark that the x_2-axis with origin removed consisting of points $(0, a, 0)$ with $a \neq 0$ is wholly not in the point set S.

Implicitization Problem and Geometric-Loci Problem.

Parametrization of a given geometric object is known to be possible only in very rare cases even transcendental functions are used. For this reason we shall leave it out of consideration in the present book. On the other hand we shall consider the implicitization of already parametrized geometric objects in the form of problem below.

Implicitization Problem I. Given a parametrized geometric object S by means of equations (4.1), to determine a polset $PS \subset \mathbf{K}[X]$ such that $Zero(PS)$ is the same as S as much as possible or S differs from $Zero(PS)$ only by a *negligible* part S' as in (4.6).

Remark 4.2. In the literature the equations $PS = 0$, if exist, are usually called the *implicit equations* of the given parametrized object S. The terminology seems to be not fully justified because of the part S' which, in view of (4.9) in Ex. 4.2, is difficult to be made precise. For this reason we shall avoid the use of this terminology in the present book.

An important problem allied to the Implicitization Problem is the following one:

Geometric-Loci Problem L. A moving point P in $A^n_{\mathbf{K}}(X)$ is subject to geometric constraints in the form of

$$h_i(X,T) = 0, \ i = 1, \cdots, r, \tag{4.10}$$

in which $h_i \in \mathbf{K}[X,T]$. To determine the geometric loci of point P as much as possible in the form of equations $PS = 0$ where $PS \subset \mathbf{K}[X]$.

Remark 4.3. The Problem L includes the Problem I as a particular case since (4.1) can be turned into the form of (4.10). Consequently for the Problem L a solution in the form of polynomial equations in $\mathbf{K}[X]$ is usually impossible.

Quasi-Varieties.

In view of (4.8) and (4.9) in Ex. 4.2 the solving of Problem I or L in the form of $Zero(PS), PS \subset \mathbf{K}[X]$, is usually impossible even in very simple cases. However, from (4.8) the solving may become possible if we admit solutions in the form of $Zero(PS/G)$ with $G \in \mathbf{K}[X]$. For this reason we introduce the following defintions.

Definition 4.3. A *quasi-variety* Q in the affine space $A^n_{\mathbf{K}}(X) :=$ A point set Q in the form

$$Q = \bigcup_i Zero(PS_i \ / \ G_i), \tag{4.11}$$

in which $PS_i, i = 1, \cdots, s$, are polsets in $\mathbf{K}[X]$, and $G_i, i = 1, \cdots, s$, are non-zero pols in $\mathbf{K}[X]$.

Remark 4.4. Quasi-varieties in $A^{n+p}_{\mathbf{K}}(X,T)$ or in other affine spaces are defined in the same way.

Projection Problem and Projection Theorem.

We shall look over the Problems I and L from another point of view and put forth the following notations and definitions.

Notation 4.2.

$A^N_{\mathbf{K}}(Z) :=$ Affine space of dimension N over $\mathbf{K}$ with coordinates $Z = (z_1, \cdots, z_N)$.

$K :=$ Subset $\{k_1, \cdots, k_q\}$ of $\{1, \cdots, N\}$ with the k's arranged in increasing order.

$J :=$ The subset of $\{1, \cdots, N\}$ complementary to K.

$A^q_{\mathbf{K}}(Z') :=$ Affine subspace of $A^N_{\mathbf{K}}(Z)$ defined by equations $z_j = 0, j \in J$, with coordinates $Z' = (z_{k_1}, \cdots, z_{k_q})$.

Definition 4.4. *Projection* of an affine space $\mathbf{A}_1 = A^N_{\mathbf{K}}(Z)$ to the subspace $\mathbf{A}_2 = A^q_{\mathbf{K}}(Z')$ as in Notation 4.2 := The map π of $\mathbf{A}_1$ to $\mathbf{A}_2$ defined by $\pi(Z) = Z'$.

Notation 4.3. For $\mathbf{A}_1 = A^{n+p}_{\mathbf{K}}(X, T)$ and $\mathbf{A}_2 = A^n_{\mathbf{K}}(X)$ the projection as defined in Def. 4.4 $:= \pi_{n+p,n}$.

Projection Problem P. Given a variety or a quasi-variety U in the affine space $\mathbf{A}_1 = A^{n+p}_{\mathbf{K}}(X, T)$. To study whether the image of U under the projection map $\pi_{n+p,n}$ is a variety or a quasi-variety or not.

Exs. 4.1, 4.2 show that the answer of Projection Problem P is negative if we restrict U and the images to varieties alone. On the other hand for the case of U and images being both *quasi-varieties* the answer is in the affirmative as given by the following theorem.

Theorem 4.1. Projection Theorem. The image of the projection $\pi_{n+p,n}$ of a quasi-variety in the affine space $\mathbf{A}_1 = A^{n+p}_{\mathbf{K}}(X, T)$ to the affine subspace $\mathbf{A}_2 = A^n_{\mathbf{K}}(X)$ is a quasi-variety too which can be determined explitely in an algorithmic manner.

We shall postpone the proof of this theorem to later paragraphs.

Remark 4.5. It is clear that the Projective Problem P may be considered as an alternative form of Geometric-Loci Problem L.

Remark 4.6. The notion of quasi-variety appears already in usual treatises on algebraic geometry. However, it seems that this notion is rarely studied in the literature. In contrast this notion plays a fundamental role in our theory as seen from the formulation of Well-Ordering Principle and the various Decomposition Theorems in Chap. 3.

Example 4.3. Consider the variety $V = Zero(x_1 * t - 1) \subset A^2_{\mathbf{K}}[x_1, t]$. The projection of V on the subspace $A^1_{\mathbf{K}}(x_1)$ is the quasi-variety $Zero(\varnothing \ / \ x_1)$ which is not a variety in $A^1_{\mathbf{K}}(x_1)$.

Remark 4.7. The varieties, as seen from general formulae (3.2.1) in Sect. 3.2 of Chap. 3, are clearly closed under the operations of union and intersection. The preceding example Ex. 4.3 (and also Exs. 4.1,4.2) shows however that they are not closed under the operation of projection. On the other hand the quasi-varieties are not only closed under the operations of union and intersection, but also closed under the operation of projection, as shown by the Projection Theorem.

Some Lemmas about Quasi-Varieties.

It is clear that the general Projection Theorem will follow by successive applications of the special case for which $p = 1$ or the case of projection of $A^{n+1}_{\mathbf{K}}(X, t)$ to the subspace $A^n_{\mathbf{K}}(X)$. For this reason we shall study in more details quasi-varieties in $A^{n+1}_{\mathbf{K}}(X, t)$ with coordinate t specified. Below we shall arrange the variables involved in the order:

$$x_1 \prec \cdots \prec x_n \prec t. \tag{4.12}$$

Definition 4.5. A *simple quasi-variety* of *type* d in the affine space $A^{n+1}_{\mathbf{K}}(X, t)$:= A quasi-variety Z of the form

$$Z = Zero(XS, F \ / \ G * H), \tag{4.13}$$

in which

$$XS \subset \mathbf{K}[X], \ F, G \in \mathbf{K}[X, t], \ H \in \mathbf{K}[X],$$

and $deg_t \ F = d$ with $d > 0$ while F is lacking for $d = 0$.

The following Lemma will reduce the study of any quasi-variety in $A^{n+1}_{\mathbf{K}}(X, t)$ to that of simple quasi-varieties.

Lemma 4.1. In $A^{n+1}_{\mathbf{K}}(X, t)$ any quasi-variety

$$Zero(PS \ / \ G * H), \tag{4.14}$$

where $PS \subset \mathbf{K}[X, t], G \in \mathbf{K}[X, t]$, and $H \in \mathbf{K}[X]$, is the union of simple quasi-varieties of the form (4.13) for which H is divisible by the initial of F w.r.t. the ordering (4.12), if F is not absent.

Proof. By the Zero-Decomposition Theorems in Sect. 3.4 of Chap. 3 we have for a set of asc-sets AS_k with initial-product IP_k a decomposition of the form

$$Zero(PS) = \bigcup_k Zero(AS_k \ / \ IP_k).$$

According to the ordering (4.12) any AS_k will be of the form

$$AS_k = XS_k + \{F_k\},$$

in which $XS_k \subset \mathbf{K}[X]$ while $F_k \in \mathbf{K}[X,t]$, $deg_t\ F_k = d_k > 0$, if it is not absent. On the other hand we set $d_k = 0$ if F_k is absent. We have then

$$Zero(PS\ /\ G * H) = \bigcup_k\ Zero(XS_k, F_k\ /\ G * H_k),$$
$$H_k = H * IP_k \in \mathbf{K}[X].$$

It is clear that the initial-product IP_k is in $\mathbf{K}[X]$, whether F_k is absent or not. It follows that the members in the last union are simple quasi-varieties of type d_k with IP_k and thus H_k too having the initial of F_k as a factor, if it is not absent. The Lemma is thus proved. □

Let us consider now two particular cases of simple quasi-varieties as in (4.13) for which either F or G is lacking.

Lemma 4.2. For a simple quasi-variety

$$Z_0 = Zero(XS\ /\ G * H)$$

of type 0 in $A^{n+1}_{\mathbf{K}}(X,t)$ as in (4.13) with F lacking the projection $\pi_{n+1,n}(Z_0)$ is a quasi-variety in $A^n_{\mathbf{K}}(X)$.

Proof. If G is lacking then nothing is to be proved. Otherwise let us write G in the canonical form

$$G = B_0 * t^e + \cdots + B_e,\ e = deg_t\ G > 0,\ B_i \in \mathbf{K}[X].$$

Set for $i = 0, 1, \cdots, e$

$$XS_i = XS + \{B_0, \cdots, B_{i-1}\},$$

with the convention $XS_0 = XS$. Then it is easy to see that

$$\pi_{n+1,n}(Z_0) = Zero(XS\ /\ B_e * H) + \bigcup_i\ Zero(XS_i\ /\ B_i * H), \tag{4.15}$$

the union being extended over $i = 0, 1, \cdots, e$. In fact, let (X', t') be any point in Z_0 with $\pi_{n+1,n}(X', t') = X'$. If $t' = 0$ then $G(X', t') = B_e(X') \neq 0$ so that $X' \in Zero(XS/B_e * H)$. Suppose therefore $t' \neq 0$. Then $G(X', t') \neq 0$ so that it is either $B_0(X') \neq 0$ or for some of $i = 1, \cdots, e$ we have $B_0(X') = 0, \cdots, B_{i-1}(X') = 0$, while $B_i(X') \neq 0$. In the former case $X' \in Zero(XS/B_0 * H)$ and in the latter case $X' \in Zero(XS_i\ /\ B_i * H)$. Hence $\pi_{n+1,n}(Z_0)$ is contained in the right-hand side of the identity (4.15). Conversely, let us consider any point X' in the right-hand side of (4.15). If $X' \in Zero(XS/B_e * H)$, then $G(X', 0) = B_e(X') \neq 0$ so that $(X', 0) \in Z_0$

whence $X' = \pi_{n+1,n}(X', 0) \in \pi_{n+1,n}(Z_0)$. If $X' \in Zero(XS/B_0 * H)$, then $B_0(X') \neq 0$. We may then take any t' with $G(X', t') \neq 0$ so that $(X', t') \in Z_0$. If for some of $i = 1, \cdots, e$, we have $X' \in Zero(XS_i \ / \ B_i * H)$, then $B_0(X') = 0, \cdots, B_{i-1}(X') = 0$, while $B_i(X') \neq 0$. In this case we have $G(X', t) = B_i(X') * t^{e-i} + \cdots B_e(X')$ with leading coefficient $B_i(X') \neq 0$. We may then take t' with $G(X', t') \neq 0$ and we have again $(X', t') \in Z_0, X' \in \pi_{n+1,n}(Z_0)$. It follows that the right-hand side of (4.15) is also contained in $\pi_{n+1,n}(Z_0)$. This proves the identity (4.15). As each member in the right-hand side of the identity (4.15) is a quasi-variety in $A^n_{\mathbf{K}}(X)$, the Lemma is proved. □

Lemma 4.3. For a simple quasi-variety

$$Z' = Zero(XS, F \ / \ H)$$

in $A^{n+1}_{\mathbf{K}}(X, t)$ as in (4.13) with G lacking, the projection $\pi_{n+1,n}$ of Z' is a quasi-variety in $A^n_{\mathbf{K}}(X)$.

Proof. Let us write $F \in \mathbf{K}[X, t]$ in the canonical form

$$F = A_0 * t^d + \cdots + A_d, \ A_i \in \mathbf{K}[X], \ d = Deg_t \ F > 0.$$

Set for $i = 0, 1, \cdots, d, d+1$,

$$F_i = A_i * t^{d-i} + \cdots + A_d,$$

$$XS_i = XS + \{A_0, \cdots, A_{i-1}\},$$

with the convention $XS_0 = XS$. Then it is clear that

$$Z' = \bigcup_i \ Zero(XS_i, F_i \ / \ H * A_i),$$

the union being extended over $i = 0, 1, \cdots, d+1$. Remark that the member in the union corresponding to $i = d$ is empty and the member corresponding to $i = d+1$ is already a quasi-variety in $A^n_{\mathbf{K}}(X)$. As in the proof of the previous Lemma it is easy to see that for each $i \leqslant d-1$,

$$\pi_{n+1,n} \ Zero(XS_i, F_i \ / \ H * A_i) = Zero(XS_i \ / \ H * A_i).$$

As each of these is a quasi-variety in $A^n_{\mathbf{K}}(X)$ so that $\pi_{n+1,n}(Z')$ is such a one too. The Lemma is thus proved. □

Proof of the Projection Theorem.

We are now in a position of proving the Projection Theorem for the case of $p = 1$ or the case of projection $\pi_{n+1,n}$.

Proof. By Lemmas 4.1, 4.2, 4.3 it is only necessary to consider quasi-varieties Z as given in the form of (4.13) for which the following conditions (a), (b) and (c) are observed.

(a) $Deg_t\ F = d > 0$.

(b) $Deg_t\ G > 0$.

(c) H is divisible by initial $I \in \mathbf{K}(X)$ of $F \in \mathbf{K}[X, t]$.

Let us form the remainder $R \in \mathbf{K}[X, t]$ of G^d w.r.t. F so that for some non-negative integer s we have

$$I^s * G^d = Q * F + R,\ \ Deg_t\ R < d. \tag{4.16}$$

Set

$$Z' = Zero(XS\ /\ G * H * R). \tag{4.17}$$

Now for any zero $(X', t') \in Z$ of (4.13) we have $G(X', t') \neq 0$ and $H(X') \neq 0$. Hence by (c) we have $I(X') \neq 0$ too. On the other hand $F(X', t') = 0$. Hence by (4.16) $R(X', t') \neq 0$ and so (X', t') belongs to Z'. Therefore Z is contained in Z' so that

$$\pi_{n+1,n}(Z) \subset \pi_{n+1,n}(Z'). \tag{4.18}$$

To see the inclusion relation in the reverse direction let us consider first the case for which R is identically 0. Then Z' is empty so that trivially we shall have $Z' \subset Z$ and so

$$\pi_{n+1,n}(Z') \subset \pi_{n+1,n}(Z). \tag{4.19}$$

Consider therefore the case R not identically 0. Let X' be an arbitrary zero in $\pi_{n+1,n}(Z')$, if there are any. Then there will be some t' with (X', t') belonging to Z'. We have then in particular

$$\begin{aligned} &R(X', t') \neq 0,\\ &G(X', t') \neq 0,\\ &H(X') \neq 0. \end{aligned} \tag{4.20}$$

By (c) we have again

$$I(X') \neq 0.$$

It follows that the degree in t of $F(X',t)$ is same as that of $F(X,t)$, viz. $d>0$. If now any zero t'' of $F(X',t)$ is also a zero of $G(X',t)$ then $F(X',t)$ will be a factor of $G(X',t)^d$. By (4.16) $F(X',t)$ is then also a factor of $R(X',t)$. As degree in t of $R(X',t)$ is less than the degree d of $F(X',t)$ this is only possible if $R(X',t)$ is identically 0 which is in contradiction to $(4.20)_1$. Consequently there is at least one zero t'' of $F(X',t)$ which is not a zero of $G(X',t)$. Then (X',t'') belongs to Z so that $X'=\pi_{n+1,n}(X',t'')$ belongs to $\pi_{n+1,n}(Z)$. This proves again (4.19) and together with (4.18) we have therefore

$$\pi_{n+1,n}(Z)=\pi_{n+1,n}(Z'). \tag{4.21}$$

By Lemma 4.2 $\pi_{n+1,n}(Z')$ is a quasi-variety in $A^n_{\mathbf{K}}(X)$. Hence by (4.21) $\pi_{n+1,n}(Z)$ is also a quasi-variety in $A^n_{\mathbf{K}}(X)$. As the decompositions in the proofs of Lemmas 4.1,4.2,4.3 are all seen to be algorithmic in character, the realization of $\pi_{n+1,n}(Z)$ as quasi-varieties is also algorithmetic in character. This proves completely the Projection Theorem in the case $p=1$ and thus also the Theorem in the general case. □

Remark 4.8. The Projection Theorem is already well-known at least in the case of **K** being the real field **R**, with terminology *semi-algebraic* sets in place of quasi-varieties, cf. e.g. [B-C-R]. The author is ignorant of whether the Projection Theorem is known in the general case. In any way, our proof of the Theorem is *constructive* in character, in contrast to the usual *existential* character treatment prevailing in the modern algebraic geometry. Our proof will thus not only show that the projection of a quasi-variety is still such a one but will also give the precise form of that projection as a quasi-variety.

5.5 Extremal Properties of Real Polynomials.

Extremal Points and Extremal Values.

Notation 5.1.

$X:=(x_1,\cdots,x_n)$.

$O:=$ An open domain in $\mathbf{R}^n(X)$.

$D:=$ A domain in $\mathbf{R}^n(X)$, open or not.

$HS:=\{h_1,\cdots,h_m\}\subset\mathbf{R}[X]$.

$N:=\{1,\cdots,n\}$; $M:=\{1,\cdots,m\}$.

$T:=$ Set of m-tuples $(i_1,\cdots,i_m)$, with $i_k\in N$ and $1\leqslant i_1<\cdots<i_m\leqslant n$.

$X_t := (x_{i_1}, \cdots, x_{i_m})$ for $t = (i_1, \cdots, i_m) \in T$.

$f, g :=$ Pols in $\mathbf{R}[X]$.

Definition 5.1. $X^0 \in \mathbf{R}^n(X)$ is an *extremal point* or *extremal zero* in D of f, or f attains its *local extremum* in D at X^0 under *constraint conditions* $HS = 0$ and $g \neq 0$ $:= X^0 \in DZero_{\mathbf{R}}(HS/g)$ and there is a neighborhood V of X^0 in $\mathbf{R}^n(X) \setminus Zero_{\mathbf{R}}(g)$ such that for $V_D = V \cap D$, either $f(X') \geqslant f(X^0)$ for any $X' \in V_D Zero_{\mathbf{R}}(HS/g)$, or $f(X') \leqslant f(X^0)$ for any $X' \in V_D Zero_{\mathbf{R}}(HS/g)$.

Definition 5.2. A real number r is an *extremal value* of f in D under *constraint conditions* $HS = 0$ and $g \neq 0 := r = f(X^0)$ for some extremal point X^0 of f in D under constraint conditions $HS = 0$ and $g \neq 0$.

Abbreviation 5.1. E-point := Extremal point. E-value := Extremal value. E-zero := Extremal zero.

Notation 5.2.

$EZero_{\mathbf{R}}\{f, g, D, HS\}$:= Set of extremal points in D of f under the conditions $HS = 0$ and $g \neq 0$.

$EVal_{\mathbf{R}}\{f, g, D, HS\}$:= Set of extremal values in D of f under the conditions $HS = 0$ and $g \neq 0$.

$EZero_{\mathbf{R}}\{f, D, HS\}$, or $EZero_{\mathbf{R}}\{f, g, D\}$, or $EZero_{\mathbf{R}}\{f, D\}$:= $EZero_{\mathbf{R}}\{f, g, D, HS\}$ in case g, or HS, or both are absent.

$EVal_{\mathbf{R}}\{f, D, HS\}$, or $EVal_{\mathbf{R}}\{f, g, D\}$, or $EVal_{\mathbf{R}}\{f, D\}$:= $EVal_{\mathbf{R}}\{f, g, D, HS\}$ in case g, or HS, or both are absent.

Remark 5.1. The definitions extend clearly for f, g, h_i all merely *differentiable* functions. The advantages of restricting ourselves to the polynomials will be seen below.

Fundamental Problem of Extremal Point and Extremal Value Determination.

Extremal Problem E or $\mathbf{E}\{f, g, D, HS\}$. Determine $EZero_{\mathbf{R}}\{f, g, D, HS\}$ and $EVal_{\mathbf{R}}\{f, g, D, HS\}$ for given f, g, D, HS.

Example 5.1. Let $n = 2$, $(x_1, x_2) = (x, y)$, $D = \mathbf{R}^2(x, y)$, and $f = x^2 + y^2$, while $HS = \{h_1\}$, $h_1 = (x-1)^3 - y^2$. Then $X^0 = (1, 0)$ is clearly an extremal point of f under the coditions $HS = 0$ with corresponding extremal value $r = 1$. See the accompanying Fig. 5.1.

Remark 5.2. The Problem **E** was one of the problems which have been successfully treated by differential methods since the appearance of calculus. The problem is far from being a trivial one whether from theoretical or practical considerations. For, even in the simplest case of a single function in a single variable in an open interval, the calculus furnishes either only necessary or only sufficient conditions, but no conditions both necesarry and sufficient have ever been given. On the other hand, for the actual determination of the extremal points or extremal values, the calculus applies usually the Lagrange multiplier method which leads to the solving of equations. However, there lacks general successful methods of solving such equations if the functions f, g, h_i are general differentiable ones. We shall see below that the situation becomes very agreeable if the functions are restricted to *polynomial* ones.

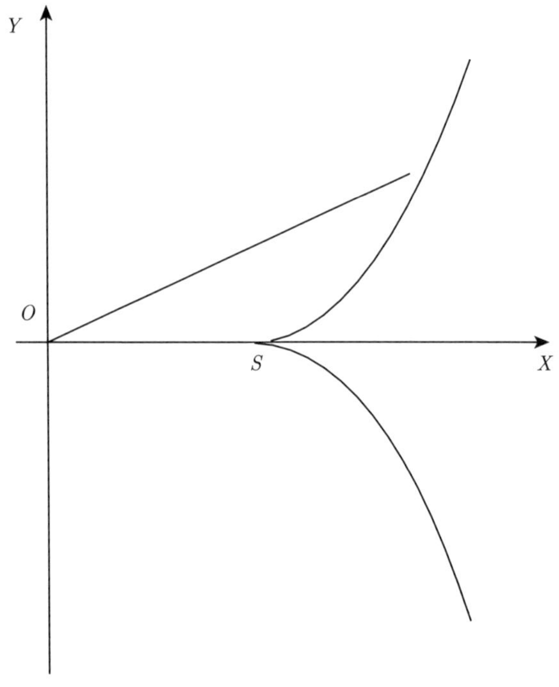

Fig. 5.1

<u>*Lagrange's Multiplier Method.*</u>

We consider below Problem **E** with f, g, D, HS given, for which the domain D is an *open* domain O in $\mathbf{R}^n(X)$.

Definition 5.3. *Lagrangian pol* with *Lagrangian multipliers* $\lambda_j, j \in M$:=

$$L = f + \Sigma_{j \in M}\ \lambda_j * h_j. \tag{5.1}$$

Definition 5.4. *Lagrangian polset* :=

$$LS = \left\{ \frac{\partial L}{\partial x_i}, h_j \mid i \in N, j \in M \right\}. \tag{5.2}$$

Definition 5.5. *Jacobian* of HS for $t = (i_1, \cdots, i_m) \in T$:= The determinant

$$J_t = \frac{\partial(h_1, \cdots, h_m)}{\partial(x_{i_1}, \cdots, x_{i_m})} = \left| \frac{\partial h_j}{\partial x_{i_k}} \right| \tag{5.3}$$

with entry $\frac{\partial h_j}{\partial x_{i_k}}$ in the j-th row and k-th column, for the m-tuple $t = (i_1, \cdots, i_m) \in T$.

Definition 5.6. *Jacobian polset* := The polset

$$JS = \{J_t, h_j \mid t \in T, j \in M\}. \tag{5.4}$$

Notation 5.3.

$\Lambda = (\lambda_1, \cdots, \lambda_m)$:= Set of Lagrangian multipliers.

$\tilde{O} := O \times \mathbf{R}^m(\Lambda) \subset \mathbf{R}^{n+m}(X, \Lambda)$.

Definition 5.7. For $\Lambda = (\lambda_1, \cdots, \lambda_m)$, *Lagrangian projection* := The natural map

$$LProj : \mathbf{R}^{n+m}(X, \Lambda) \longrightarrow \mathbf{R}^n(X), \ with\ LProj(X, \Lambda) = X. \tag{5.5}$$

The following theorem is a classical one, cf. e.g. [Cou], p.198.

Theorem 5.1. For the Problem $\mathbf{E}$ with domain $D = O$ open we have

$$EZero_{\mathbf{R}}\{f, g, O, HS\} \subset LProj\ \tilde{O}Zero_{\mathbf{R}}(LS/g) + OZero_{\mathbf{R}}(JS/g). \tag{5.6}$$

Proof. Suppose that

$$X^0 = (x_1^0, \cdots, x_n^0) \in EZero_{\mathbf{R}}\{f, g, O, HS\} \setminus OZero_{\mathbf{R}}(JS/g). \tag{5.7}$$

Then it is sufficient to prove that

$$X^0 \in LProj\ \tilde{O}Zero_{\mathbf{R}}(LS/g). \tag{5.8}$$

By (5.7) there will be some m-tuple $t = (i_1, \cdots, i_m)$ in T such that

$$J_t \neq 0 \ at\ X^0. \tag{5.9}$$

Let us rewrite $x_{i_k} = y_k$ and the other x's as $u_1, \cdots, u_d$ with $d + m = n$. Then for some permutation $Perm$ we have

$$\begin{aligned} &y_k = x_{i_k},\ k = 1, \cdots m,\ t = (i_1, \cdots, i_m) \in T, \\ &X^0 = (x_1^0, \cdots, x_n^0) = Perm(u_1^0, \cdots, u_d^0, y_1^0, \cdots, y_r^0). \end{aligned} \tag{5.10}$$

By (5.7) we have $X^0 \in O$ and $g \neq 0$ at X^0. Set $U = (u_1, \cdots, u_d)$. Then by implicit function theorem there is some neighborhood $V(\subset O)$ about $U^0 = (u_1^0, \cdots, u_d^0)$ in $\mathbf{R}^d(U)$ and continuously differentiable functions $\phi_j(U)$ in V such that for

$$y_j = \phi_j(u_1, \cdots, u_d),\ j \in M, \tag{5.11}$$

we have

$$y_j^0 = \phi_j(u_1^0, \cdots, u_d^0),\ \text{and} \tag{5.12}$$

$$h_j(Perm(u_1, \cdots, u_d, \phi_1, \cdots, \phi_m)) = 0 \text{ in } V. \tag{5.13}$$

V will be taken to be sufficiently small so that $g \neq 0$ in V. Differentiating (5.13) we get

$$\frac{\partial h_j}{\partial u_k} + \Sigma_{l \in M} \left(\frac{\partial h_j}{\partial y_l} * \frac{\partial \phi_l}{\partial u_k} \right) = 0 \text{ in } V,\ j \in M,\ k = 1, \cdots, d. \tag{5.14}$$

Now as a function in $u_1, \cdots, u_d$,

$$f(Perm(u_1, \cdots, u_d, \phi_1, \cdots, \phi_m)) \tag{5.15}$$

attains clearly its local extremum at $U^0 = (u_1^0, \cdots, u_d^0) \in V$ under no more restraint equational-conditions. It follows that its total derivative w.r.t. each u_k is 0 at U^0, or what is the same,

$$\frac{\partial f}{\partial u_k} + \Sigma_{l \in M} \left(\frac{\partial f}{\partial y_l} * \frac{\partial \phi_l}{\partial u_k} \right) = 0 \text{ at } X^0. \tag{5.16}$$

Differentiating now (5.1) in considering λ_j as functions in X we get

$$\begin{aligned} \frac{\partial L}{\partial y_l} &= \frac{\partial f}{\partial y_l} + \Sigma_{j \in M}\ \lambda_j * \frac{\partial h_j}{\partial y_l} + \Sigma_{j \in M}\ \frac{\partial \lambda_j}{\partial y_l} * h_j,\ l \in M, \\ \frac{\partial L}{\partial u_k} &= \frac{\partial f}{\partial u_k} + \Sigma_{j \in M}\ \lambda_j * \frac{\partial h_j}{\partial u_k} + \Sigma_{j \in M}\ \frac{\partial \lambda_j}{\partial u_k} * h_j,\ k = 1, \cdots, d \end{aligned} \tag{5.17}$$

Now the last summands in $(5.17)_1$, $(5.17)_2$ are both 0 at X^0 since $X^0 \in OZero_{\mathbf{R}}(HS/g)$. On the other hand by (5.9) the coefficient matrix of λ_j in $(5.17)_1$ has its determinant $J_t \neq 0$ at X^0. Hence we may solve the equations

$$\frac{\partial L}{\partial y_l} = 0\ at\ X^0,\ l \in M \tag{5.18}$$

to get $\Lambda = \Lambda^0 = (\lambda_1^0, \cdots, \lambda_m^0)$. Then $(5.17)_1$ gives

$$\frac{\partial L}{\partial y_l} = 0\ at\ (X^0, \Lambda^0),\ l \in M. \tag{5.19}$$

For a fixed $k = 1, \cdots, d$, let us multiply the equations in (5.19) successively by $\frac{\partial \phi_l}{\partial u_k}, l \in M$ and adding, then by (5.14),(5.16), $(5.17)_2$ we get

$$\frac{\partial L}{\partial u_k} = 0 \text{ at } (X^0, \Lambda^0). \tag{5.20}$$

From (5.13),(5.19),(5.20) we see that (X^0, Λ^0) is in $\tilde{O}Zero_{\mathbf{R}}(LS/g)$ and hence X^0 is in $LProj\ \tilde{O}Zero_{\mathbf{R}}(LS/g)$. This proves (5.6) and hence the theorem. □

Remark 5.2. The term involving JS in the right hand side of (5.6) cannot be omitted. In fact, for Ex. 5.1 we have

$$L = x^2 + y^2 + \lambda_1 * ((x-1)^3 - y^2)),$$

$$LS = \{2 * x + 3 * \lambda_1 * (x-1)^2,\ 2 * y - 2 * \lambda_1 * y,\ (x-1)^3 - y^2\},$$

$$JS = \{3 * (x-1)^2,\ -2 * y,\ (x-1)^3 - y^2\}.$$

We see that

$$Zero_{\mathbf{R}}(LS) = \varnothing,\ Zero_{\mathbf{R}}(JS) = \{(1,0)\}$$

Hence the extremal point $X^0 = (1,0)$ is in $Zero_{\mathbf{R}}(JS)$ but not in $LProj\ Zero_{\mathbf{R}}(LS)$.

Remark 5.3. The Theorem 5.1 remains true for functions f, g, h_i to be differentiable ones. It restricts the range of extremal points to be sought for in solving of equations $LS = 0$ and $JS = 0$. Now the solving of such equations is difficult to deal with in the general case, and this is perhaps the reason why the method of Lagrange multipliers is no more in current use as in previous times. However, if we restrict ourselves to the case for which f, g, h_i are all *polynomial* ones, then the equations $LS = 0, JS = 0$ become *polynomial* ones and can be solved by means of our general methods. This is what we shall do below.

Remark 5.4. The Theorem 5.1 furnishes only a necessary condition for a point X^0 to be an extremal one in that it should satisfy the equations $LS = 0, JS = 0$ with some suitable values of Lagrangian multipliers. Thus, even if we are able to solve these equations, there is in general no means to decide whether such solutions will give really extremal values or not. These difficulties will however be resolved in the case of f, g, h_i being polynomial ones as shown below.

Particular Case of HS being an Asc-Set.

Let us consider the particular case for which HS is an asc-set, the domain $D = O$ being open. With some permutation $Perm$ as in (5.10) and $U = (u_1, \cdots, u_d), Y_j =$

$(y_1, \cdots, y_j), y_j = x_{i_j}, j \in M$ we may suppose that $h_j \in HS$ are of the form

$$h_j = I_j * y_j^{d_j} + \text{ lower degree terms in } y_j, \ j \in M, \tag{5.21}$$

the coefficients of y_j-powers being in $\mathbf{R}[U, Y_{j-1}]$.

Lemma 5.1. Suppose that the polset HS is an asc-set with initial-separant-product ISP. Then we have

$$OZero_{\mathbf{R}}(JS/ISP) = \varnothing, \text{ and} \tag{5.22}$$

$$EZero_{\mathbf{R}}\{f, ISP, O, HS\} \subset LProj\ \tilde{O}Zero_{\mathbf{R}}(LS/ISP) \subset OZero_{\mathbf{R}}(HS/ISP). \tag{5.23}$$

Proof. In the Jacobian polset JS there is one given by

$$J_t = \frac{\partial(h_1, \cdots, h_m)}{\partial(y_1, \cdots, y_m)}, \ t = (i_1, \cdots, i_m).$$

By (5.21) we have for $j, k \in M$,

$$\frac{\partial h_j}{\partial y_k} = 0 \ for \ k > j, \text{ and } \frac{\partial h_j}{\partial y_j} = S_j, \tag{5.24}$$

in which S_j is the separant of h_j. It follows that

$$J_t = S_1 * S_2 * \cdots * S_m$$

so that

$$OZero_{\mathbf{R}}(JS/ISP) = \varnothing.$$

This proves (5.22). From Theorem 5.1 it follows also (5.23). □

Lemma 5.2. Let HS and ISP be as in Lemma 5.1. Suppose that

$$f(X) = x_1. \tag{5.25}$$

Let $Proj_1$ be the projection of $\mathbf{R}^{n+r}(X, \Lambda)$ or $\mathbf{R}^n(X)$ to $\mathbf{R}$ defined by

$$\begin{aligned} &Proj_1\ (X, \Lambda) = x_1, \text{ or} \\ &Proj_1\ (X) = x_1. \end{aligned} \tag{5.26}$$

Then there is a *finite* set, eventually empty, of real values

$$K\{x_1, ISP, O, HS\} \subset Proj_1\ \tilde{O}Zero_{\mathbf{R}}(LS/ISP) \subset \mathbf{R}, \tag{5.27}$$

such that

$$EVal_{\mathbf{R}}\{x_1, ISP, O, HS\} \subset K\{x_1, ISP, O, HS\} \subset Proj_1\ OZero_{\mathbf{R}}(HS/ISP). \tag{5.28}$$

Proof. Owing to the chosen ordering of variables x_1 can only be either u_1 or y_1. Consider first the case $x_1 = u_1$. Then by (5.25) we have

$$\frac{\partial f}{\partial y_1} = \frac{\partial f}{\partial y_2} = \cdots = \frac{\partial f}{\partial y_m} = 0, \ \frac{\partial f}{\partial u_1} = 1. \tag{5.29}$$

Let us consider a point

$$\begin{aligned}&(X^0, \Lambda^0) \in \tilde{O}Zero(LS/ISP), \text{ where}\\ &X^0 = (x_1^0, \cdots, x_n^0), \ \Lambda^0 = (\lambda_1^0, \cdots, \lambda_m^0).\end{aligned} \tag{5.30}$$

We have for $j \in M$,

$$\begin{aligned}&\frac{\partial L}{\partial y_j} = 0 \text{ at } (X^0, \Lambda^0) \text{ and}\\ &h_j = 0, \ S_j = \frac{\partial h_j}{\partial y_j} \neq 0 \text{ at } X^0.\end{aligned} \tag{5.31}$$

Differentiating (5.1) w.r.t y_j we get by (5.24),

$$\frac{\partial f}{\partial y_j} + \Sigma_{k \geqslant j}\ \lambda_k * \frac{\partial h_k}{\partial y_j} = 0 \text{ at } (X^0, \Lambda^0), \ j \in M. \tag{5.32}$$

(5.29),(5.31) give then

$$\lambda_j^0 = 0, \ j \in M. \tag{5.33}$$

As $\frac{\partial L}{\partial u_1} = 0$ at (X^0, Λ^0) too this will be in contradiction to the equation

$$\frac{\partial L}{\partial u_1} = \frac{\partial f}{\partial u_1} + \ \Sigma_{j \in M}\ \lambda_j * \frac{\partial h_j}{\partial u_1}$$

which is equal to 1 at (X^0, Λ^0) by (5.33) and (5.29). Hence $\tilde{O}Zero_{\mathbf{R}}(LS/ISP) = \varnothing$ in this case and (5.28) follows trivially from (5.23) in taking

$$K\{x_1, ISP, O, HS\} = \varnothing, \text{ for } x_1 = u_1. \tag{5.34}$$

Consider next the case $x_1 = y_1$. Then by (5.25) we have

$$\frac{\partial f}{\partial y_1} = 1, \ \frac{\partial f}{\partial y_2} = \cdots = \frac{\partial f}{\partial y_r} = 0. \tag*{(5.29)$'$}$$

Let (X^0, Λ^0) be again any zero in $\tilde{O}Zero_{\mathbf{R}}(LS/ISP)$. Then as before (5.25),(5.32) will give us

$$\begin{aligned}&\lambda_j^0 = 0, \ j = 2, \cdots, m, \text{ while}\\ &\lambda_1^0 = -\frac{1}{S_1^0} \neq 0,\end{aligned} \tag*{(5.33)$'$}$$

where S_1^0 is the value of S_1 at X^0. Now $h_j = 0$ at X^0 for any $j \in M$. Owing to our chosen natural ordering of variables X, h_1 will have no variables u_j involved and is a pol in $y_1 = x_1$ alone with constant coefficients. The possible values of x_1^0 for which $h_1 = 0$ is to be satisfied are thus finite in number, being contained in the set of real solutions of $h_1 = 0$ in (5.21). By Lemma 5.1 the set $Proj_1\ \tilde{O}Zero_{\mathbf{R}}(LS/ISP)$ is thus a finite set, eventually empty, of real values which verifies (5.28). The set K in (5.27) may be taken to be

$$\begin{aligned} K\{x_1, ISP, O, HS\} &= Proj_1\ \tilde{O}Zero_{\mathbf{R}}(LS/ISP) \\ &\subset Proj_1\ OZero_{\mathbf{R}}(h_1),\ \text{for } x_1 = y_1. \end{aligned} \tag{5.35}$$

The Lemma 5.2 is now completely proved. □

Finite Kernel Theorem for Open Domain in the General Case.

Consider now the case for which HS is a general polset in variables $x_1, \cdots, x_n$. Let us introduce a new variable x_0 arranged in the order $x_0 \prec x_1$ and take the notations below.

Notation 5.3.

$X^+ := (x_0, x_1, \cdots, x_n)$.

$\mathbf{R}^{n+1} := \mathbf{R}^{n+1}(X^+)$.

$O^+ :=$ The open domain in $\mathbf{R}^{n+1}(X^+)$ defined by $X \in O$ while x_0 is arbitrary.

$\Lambda^+ := (\lambda_0, \lambda_1, \cdots, \lambda_m)$.

$\tilde{O}^+ := O^+ \times \mathbf{R}^{m+1}(\Lambda^+)$.

$h_0 :=$ The new pol

$$h_0(X^+) = f(X) - x_0. \tag{5.36}$$

$HS^+ :=$ The new polset

$$HS^+ = HS + \{h_0\}. \tag{5.37}$$

$g^+ :=$ The pol defined by $g^+(X^+) = g(X)$.

$Proj_0 :=$ The natural projection of $\mathbf{R}^{n+1}(X^+)$ to $\mathbf{R}(x_0)$.

The following lemma is clear.

Lemma 5.3.

$$\{f(X) \mid X \in OZero_{\mathbf{R}}(HS)\} = Proj_0\ O^+Zero_{\mathbf{R}}(HS^+),\ \text{and} \tag{5.38}$$

$$EVal_{\mathbf{R}}\{f, g, O, HS\} = EVal_{\mathbf{R}}\{x_0, g^+, O^+, HS^+\}. \tag{5.39}$$

Theorem 5.2. **(Finite Kernel Theorem** for Open Domain) Let HS be an arbitary polset and f be an arbitrary pol in $\mathbf{R}[X]$. Then there is a *finite* set of real values K such that

$$EVal_{\mathbf{R}}\{f, O, HS\} \subset K \subset \ Proj_0 \ \tilde{O}^+ Zero_{\mathbf{R}}(HS^+). \tag{5.40}$$

Proof. By the Zero-Decomposition Theorem 3.5.2 let us decompose $Zero(HS^+)$ so that for a set of asc-sets $HS_j^+ \subset \mathbf{R}[X^+]$ with initial-separant-product ISP_j we have

$$O^+ Zero_{\mathbf{R}}(HS^+) = \bigcup_j O^+ Zero_{\mathbf{R}}(HS_j^+/ISP_j), \tag{5.41}$$

Let LS_j^+ be the Lagrangian polset formed from HS_j^+ as in Lemmas 5.1, 5.2. By Lemma 5.2 with x_1 replaced by x_0 we have then for each j a finite set of real values K_j such that

$$EVal_{\mathbf{R}}\{x_0, ISP_j, O^+, HS_j^+\} \subset K_j \subset Proj_0 \ \tilde{O}^+ Zero_{\mathbf{R}}(HS_j^+/ISP_j)\}, \text{ where} \tag{5.42}$$

$$K_j = K\{x_0, ISP_j, O^+, HS_j^+\}. \tag{5.43}$$

It is clear from (5.41) that

$$EVal_{\mathbf{R}}\{x_0, O^+, HS^+\} \subset \bigcup_j EVal_{\mathbf{R}}\{x_0, ISP_j, O^+, HS_j^+\}. \tag{5.44}$$

Let us set

$$K = \bigcup_j K_j = \bigcup_j K\{x_0, ISP_j, O^+, HS_j^+\}. \tag{5.45}$$

By (5.39) and (5.41)-(5.45) we have then (5.40) as to be proved. □

Definition 5.8 *Finite Kernel-Set* of the Problem **E** with given $\{f, O, HS\}$:= The finite set K of (5.45).

Notation 5.4. $Ker(f, O, HS) := K$ of (5.45) := The finite kernel-set of Problem E with given $\{f, O, HS\}$

Remark 5.5. Theorem 5.2 remains true with slight modifications if g is present.

Remark 5.6. Theorem 5.2 shows that $EVal_{\mathbf{R}}\{f, O, HS\}$ is a *finite* set though $EZero_{\mathbf{R}}\{f, O, HS\}$ may be *infinite*.

Remark 5.7. The Lagrange's Multiplier Method is applicable for arbitrary *differentiable* functions f, h_j. However, the *finiteness* property implied in Theorem 5.2 is no more true as seen from the simple example of $f = x * sin\frac{1}{x}, O = \{x > 0\} \subset \mathbf{R}^1(x)$. Thus, the *finiteness* of extremal-value set is characteristic for f, h_j being *polynomials*.

Example 5.2 (Example 5.1 revisited). Let us consider Ex. 6.1 again. We have here $O = \mathbf{R}^2(x, y)$. Let us introduce a new variable x_0 and set $h_0 = x^2 + y^2 - x_0$. Then $HS^+ = \{h_0, h_1\}$. For the ordering $x_0 \prec x \prec y$ it is readily found that

$$Zero_{\mathbf{R}}(HS^+) = Zero_{\mathbf{R}}(CS_1/ISP_1) + Zero_{\mathbf{R}}(CS_2/ISP_2) + Zero_{\mathbf{R}}(HS').$$

The polsets and pols in the right-hand side are respectively

$$CS_1 = \{C_{11}, C_{12}\},\ CS_2 = \{C_{21}, C_{22}, C_{23}\},\ HS' = \{h'_1, h'_2, h'_3\},\ \textit{with}$$

$$C_{11} = x^3 - 2 * x^2 + 3 * x - x_0 - 1,\ C_{12} = y^2 + x^2 - x_0.$$

$$C_{21} = x_0 - 1,\ C_{22} = x - 1,\ C_{23} = y.$$

$$h'_1 = 3 * x^2 - 4 * x + 3,\ h'_2 = x^2 + y^2 - x_0,\ h'_3 = (x-1)^3 - y^2.$$

The pols ISP_1, ISP_2 are the initial-separant-products of CS_1, CS_2 respectively. As $h'_1 = 0$ has no real solution so $Zero_{\mathbf{R}}(HS') = \varnothing$. The contributions to the kernel set $K\{f, \mathbf{R}^2, HS\}$ by the two other zero-sets are respectively given by

$$K_1 = \varnothing,\ K_2 = \{1\}.$$

Hence the finite kernel is $K = \{1\}$ which is the projection of the point $(x_0, x, y) = (1, 1, 0)$ on $\mathbf{R}(x_0)$ so the only extremal value $f = 1$ occurs at the only extremal point $(x, y) = (1, 0)$.

$\underline{Finite\ Kernel\ Theorem\ for\ Closed\ Domains.}$

Up to now we have restricted ourselves to the case of an *open* domain $D = O$ for the problem $\mathbf{E}\{f, g, D, HS\}$. Consider now the case of a closed domain D. We shall restrict ourselves to such domains D whose boundary is consisting of pieces of algebraic hypersurfaces in $\mathbf{R}^n(X)$. By introducing new variables and new pols we may usually turn the domain D into a rectangular parallelopiped form in some higher dimensional real space. We may thus suppose without much loss of generality that the closed domain D to be considered is of the form

$$D:\ a_k \leqslant x_k \leqslant b_k,\ k \in N = \{1, \cdots, n\}, \tag{5.46}$$

in which some of the a_k (or b_k) may be $-\infty$ (or $+\infty$).

To reduce the Problem $\mathbf{E}\{f, D, HS\}$ to Problems $\mathbf{E}\{f, O, HS\}$ with O open, let us split the domain D into at most 3^n *open* domains O_τ in some n_τ-dimensional real spaces $\mathbf{R}_\tau^{n_\tau}$ as follows.

Notation 5.5.

$\tau :=$ Any n-tuple $(t_1, \cdots, t_n)$ for which each entry t_k is either $+$, or $-$, or 0.

T := Set of all such n-tuples for which $t_k \neq -$ if $a_k = -\infty$, and $t_k \neq +$ if $b_k = +\infty$.

Below $f \in \mathbf{R}[X], HS \subset \mathbf{R}[X]$, and $\tau = (t_1, \cdots, t_n) \in T$:

n_τ := Number of 0's in τ.

$\mathbf{R}_\tau^{n_\tau}$:= the subspace of $\mathbf{R}^n(X)$ defined by

$$x_k = a_k, \text{ for } t_k = -,$$
$$x_k = b_k, \text{ for } t_k = +.$$

O_τ := Open domain in $\mathbf{R}_\tau^{n_\tau}$ defined by

$$a_k < x_k < b_k, \text{ for } \tau_k = 0.$$

f_τ := Restriction of f on $\mathbf{R}_\tau^{n_\tau}$.

HS_τ := Polset of pols h_τ for $h \in HS$.

The following Lemma is now clear.

Lemma 5.4. If x is a point in $O_\tau \subset \mathbf{R}_\tau^{n_\tau}$ for some $\tau \in T$ and x^τ be the same point considered as a point in $\mathbf{R}^n$, then

$$x^\tau \in EZero_\mathbf{R}\{f, D, HS\} \Longrightarrow x \in EZero_\mathbf{R}\{f_\tau, O_\tau, HS_\tau\}.$$

From the Lemma 5.4 it follows that

$$EZero_\mathbf{R}\{f, D, HS\} \subset \bigcup_{\tau \in T} EZero_\mathbf{R}\{f_\tau, O_\tau, HS_\tau\}. \tag{5.47}$$

Applying now Theorem 5.2 to (5.47), we get readily the following

Theorem 5.3. (Finite Kernel Theorem for Closed Domain) For a Problem $\mathbf{E}\{f, D, HS\}$ with D a closed domain of the type (5.46), there is a *finite* set of real values K such that

$$EVal_\mathbf{R}\{f, D, HS\} \subset K \subset \ Proj_0 \ O^+ Zero_\mathbf{R}(HS^+), \tag{5.48}$$

in which

$$K = \bigcup_{\tau \in T} K\{f_\tau, O_\tau, HS_\tau\}. \tag{5.49}$$

Optimization Problem and Optimization Principle.

Let f, g, D, O, HS be as before.

Definition 5.9. *Optimal* := Either *Least* or *Greatest.*

Definition 5.10. *Optimal point* of a pol f in the domain D under the constraint conditions $HS = 0, g \neq 0$:= A point $X^0 \in DZero_\mathbf{R}(HS/g)$ such that for any point

$X' \in DZero_{\mathbf{R}}(HS/g)$, we have always $f(X') \geqslant f(X^0)$ for optimal = least, or always $f(X') \leqslant f(X^0)$ for optimal = greatest.

Definition 5.11. *Optimal value* of a pol f in the domain D under the constraint conditions $HS = 0, g \neq 0 :=$ The value $f(X^0)$ for X^0 any optimal point of f in domain D under the constraint conditions $HS = 0, g \neq 0$, in supposing such optimal points do exist.

Notation 5.5. $Optm\{f, g, D, HS\} :=$ Optimal value of the pol f in the domain D under the constraint conditions $HS = 0, g \neq 0$, for $Optm = Least$ or $Greatest$, if it exists.

Optimization Problem O. Determine the optimal values $Optm\{f, g, D, HS\}$, if it exists.

Each optimal point, whether least or greatest, is clearly an extremal point for the relevant problem. It follows that we have the following Principle for the actual determination of the optimal values, in supposing such values do exist.

Optimization Principle. The optimal value $Optm\{f, g, D, HS\}$, if it exists, is the optimal value of $EVal_{\mathbf{R}}\{f, g, D, HS\}$.

In case the optimal value, whether least or greatest, does exist, then its determination, by the Optimization Principle, is reduced to the determination of the optimal value of certain more precise set of real values which may even be a finite set. In fact from Theorems 5.1-5.3 we get readily the following theorems as immediate consequences for the solving of Optimization Problem O.

Theorem 5.4. For D an open domain or a closed domain of type (5.46) we have, in supposing that the optimal value exists,

$$Optm\{f, D, HS\} = Optm\ Ker\{f, D, HS\}.$$

Theorem 5.5. For D an open domain O we have, in supposing that the optimal value exists,

$$Optm\{f, g, O, HS\} = Optm_{X \in O}\ \{f(X) \mid X \in LJZ\}, \text{ where}$$

$$LJZ = LProj\ \tilde{O}Zero_{\mathbf{R}}(LS/g) + OZero_{\mathbf{R}}(JS/g)\}.$$

In the last expression LS, JS are respectively the Lagrangian polset and the Jacobian polset of the problem $E\{f, g, O, HS\}$.

Chapter 6

Applications to Polynomial Equations-Solving

6.1 Basic Principles of Polynomial Equations-Solving: The Char-Set Method.

Recall of Descartes Program from Sect. 1.3.

In Sect. 1.3 of Chap. 1 we have cited the *Program* of Descartes about problem-solving which may be schematized as in the following diagram:

$$\begin{aligned} \textit{Solving of any problem} &\longrightarrow \textit{Solving of a problem in mathematics} \\ &\longrightarrow \textit{Solving of a problem in algebra} \\ &\longrightarrow \textit{Solving of a system of algebraic equations} \\ &\longrightarrow \textit{Solving of a single algebraic equation.} \end{aligned}$$

Though Descartes scheme is absurd in apparence, the importance of Descartes thought is undiscutable. In fact, in recent years that polynomial equations solving may play an important role in all kinds of problems has been gradually well-regconized by the scientific world. Diverse methods have appeared to deal with the problem of solving a system of polynomial equations $PS = 0$ with $PS \subset \mathbf{K}[X], X = (x_1, \cdots, x_n)$ and $\mathbf{K}$ a field of characteristic 0, most frequently the rational field $\mathbf{Q}$.

Numerical Methods and Symbolic Methods of Polynomial Equations-Solving.

There are two main kinds of methods for *polynomial equations-solving*: the numerical ones and the symbolic ones. For the numerical methods we may cite the generalized Newton-Raphson method and its diverse variants, the homotopy continuation method, etc. These methods start directly from the given numerical datas in the equations and apply some limiting process in order to converge toward some solution of the given polynomial equations. In general these methods will give only one solution among the set of all possible solutions and are thus *local* in character in the main. The homotopy continuation method, based on some topological considerations, under circumstances may give the set of all possible solutions and is thus sometimes *global* in character. This is particularly true in the case of Eigenvalue determination and T.Y.Li, one of the pioneers of the method has shown how this may be true in some general cases. In any way all these methods give only solutions in approximate values and suffer from error control as well as stability control. In contrast to the numerical methods the symbolic ones treat at the outset all the approximate numerical datas as indeterminates and only symbolic computations are to be used during the procedure. The solutions will be given in some acceptable algebraic form and all the possible solutions may be expressed in this way so that the methods are *global* in character. Among such symbolic methods we may cite in particular the Buchberger method based on Groebner basis construction of an ideal, cf. e.g. [Bu1]. See also e.g. [Kal], [Koba], [Laz1,2]. The method has been proved to be quite successful in the 0-dimensional case, i.e. the case for which only a *finite* number of solutions can exist. To the knowledge of the author, the more general case is yet under way of serious study.

For us we shall adopt the symbolic method but follow a way entirely different from those of Buchberger or any others. Our method, under the name of *char-set method*, had in fact its root in the development of *Polynomial Equations-Solving* in ancient China as reviewed below.

Recall of Polynomial Equations-Solving in Ancient China from Sect. 1.2.

Equations-solving was the main topic of studies throughout the historical development of Chinese ancient mathematics. It is no exageration to say that our development of ancient mathematics was unconciously along the line of thought of Descartes and may be considered as an actual realization of Descartes Program. In fact, in the earliest classic *Nine Chapters of Arithmetic* in 2c. B.C. there had already appeared the method of solving a system of linear equations together with the introduction of negative numbers which is essentially the same as the present-day Gaussian elimination method. A concrete example is given in Sect. 1.2. This method, when expressed in modern way, consists in reducing the given set of equations to a form which in modern notations is in general equivalent to a set of equations as shown below:

$$
\begin{aligned}
c_{11} * x_1 \ + \ d_1 \ &= \ 0, \\
c_{22} * x_2 \ + \ c_{21} * x_1 + \ d_2 \ &= \ 0, \\
\cdots\cdots \\
c_{nn} * x_n \ + \ \cdots \ + \ c_{n1} * x_1 + \ d_n \ &= \ 0,
\end{aligned} \tag{1.1)y}
$$

In the last set of equations (eventually some may be lacking) all the non-missing leading coefficients c_{ii} are non-zero and the solutions are then readily found (eventually in terms of some parameters corresponding to the leading variables x_i of the missing equations). Later on, in the classic *Jade Mirror of Four Elements* of Zhu Shijie in +1303 of Yuan Dynasty there appeared a method which indicates a way of solving in principle general non-linear system of equations. The method consists in reducing by elimination the given system to some normalized form which will be easy to solve. A concrete example of Zhu's method with some comments has also been given in Sect. 1.2. Thus, a geometrical problem was led to a set of three equations in three unknowns which in modern notations may be written in the form of (1.2.11), the unknowns being z, x, y. By Zhu's method this polynomial set is reduced to a system, (with x assumed to be non-zero) of the form (1.2.13) for which the unknowns z, x, y are entered in succession. For the convenience of the readers we reproduce the equations (1.2.13) as (1.2) given below.

$$
\begin{aligned}
&z^4 - 6 * z^3 + 4 * z^2 + 6 * z - 5 \ = \ 0, \\
&x * (7 + 3 * z - z * * 2) - (6 + 7 * z + 3 * z^2 - z^3) = \ 0, \\
&y - (x * z - x^2 - z + x) \ = \ 0.
\end{aligned} \tag{1.2}
$$

Solving now the equations (1.2) successively in single unknowns z, x, and y we get one of the solutions to be $(z, x, y) = (5, 3, 4)$. In general Qin Jiushao's method in +1247 of Song Dynasty may be applied to get all possible real solutions if required. We see that, though the method of Zhu is very incomplete and has many flaws as seen from Exs. 3.4.1 and 3.5.1, it furnishes nevertheless clues for the solving of general polynomial equations.

Suggestions from Chinese Ancient Achievements.

The equations (1.1) and (1.2) suggest the introduction of the notions of *triangular sets* and *asc-sets* as in Sect. 3.3. We restate the definitions in slightly different form below:

Let us consider a polset $AS \subset \mathbf{K}[X]$ of some particular form. Suppose that the indeterminates $x_1, \cdots, x_n$ can be rearranged as $u_1, \cdots, u_d, y_1, \cdots, y_r$ $(r + d = n)$

such that the pols in AS can be arranged as $A_1, \cdots, A_r$ in canonical forms below:

$$\begin{aligned} A_1 &= I_1 * y_1^{d_1} + \mathit{lower\ degree\ terms\ in}\ y_1, \\ A_2 &= I_2 * y_2^{d_2} + \mathit{lower\ degree\ terms\ in}\ y_2, \\ &\cdots\cdots \\ A_r &= I_r * y_r^{d_r} + \mathit{lower\ degree\ terms\ in}\ y_r. \end{aligned} \tag{1.3}$$

In (1.3) for each i the coefficients of y_i-powers in A_i are pols in $\mathbf{K}[u_1, \cdots, u_d, y_1, \cdots, y_{i-1}]$. Then the polset AS is a *triangular set.* If furthermore each pol I_i has a degree in y_j less than $deg(A_j) = d_j$ for any pair (i, j) with $j < i$, then the polset AS is an *ascending set* or in abbreviation an *asc-set.* The pols I_i are called the *initials* and $IP = \Pi\ I_i$ the *initial-product* of AS.

General Principle for Polynomial Equations-Solving.

Clearly a triangular set or an asc-set defined by (1.3) is just a generalization of the polsets occuring in the left-sides of (1.1) and (1.2). We remark the following simple principle for *polynomial equations-solving*:

Principle PES. A set of polynomial equations of the form $AS = 0$ in which AS is a triangular set or an asc-set as in (1.3) may be considered as solved for $y_1, \cdots, y_r$ as definite algebraic functions of $u_1, \cdots, u_d$, so far the initials are all non-zero. If definite values of u's are known, then definite values of y's will also be determined for the solutions, whenever the initials will not be turned into 0. In a word, *$Zero(AS/IP)$ and more generally $Zero(AS/IP * G)$ for any pol $G \in \mathbf{K}[X]$ may be considered as a well-determined set.*

The above simple Principle furnishes us the clue for the solving of arbitrary polynomial equations $PS = 0$ in reducing it to the following problems:

Problem PES1. Determine a triangular set or an asc-set AS from given polset PS such that the zero-set of AS is closely related to that of PS which will give partial determination of an important part of $Zero(PS)$ and serves to lead to a complete determination of $Zero(PS)$.

Problem PES2. Determine a set of asc-sets (eventually triangular sets) AS_i from given polset PS such that the zero-set of PS will be determined through the determination of the zero-sets of the asc-sets (or the triangular sets) AS_i.

The Char-Set Method for Polynomial Equations-Solving.

Along the lines of thought furnished by our ancestors we have introduced concepts and techniques from modern mathematics, mainly borrowed from the two classical texts of J.F.Ritt, [Ritt1,2], in order to reformulate and to make precise and rigorous the procedure originated from Zhu Shijie's work in year 1303. The outcome is a general *char-set method* described in the form of the Well-Ordering Principle of Sect. 3.4 and Zero-Decomposition Theorems of Sect. 3.5. They permit to solve completely the two Problems PES1 and PES2 at least in the theoretical sense and are reproduced below:

Well-Ordering Principle. There are algorithms which permit to determine for a given polset PS in a finite number of steps a particular asc-set CS called a *char-set* of PS such that

$$Zero(PS) = Zero(CS/IP) + \bigcup_i Zero(PS_i), \tag{1.4}$$

in which

$$PS_i = PS + CS + I_i \text{ or } PS + I_i, \tag{1.5}$$

and IP is the initial-product or the product of all initials I_i of CS.

Zero-Decomposition Theorem. There are algorithms which permit to determine for a given polset PS in a finite number of steps a finite set of asc-sets (or triangular sets) AS_j such that

$$Zero(PS) = \bigcup_j Zero(AS_j/IP_j), \tag{1.6}$$

in which each IP_j is the initial-product of the corresponding asc-set (eventually a triangular set) AS_j.

There are also variants of the Zero-Decomposition Theorems which we refer to Chap. 3.

With the two theorems above the problem of *polynomial-equations solving* is seen to be completely settled at least in the theoretical sense.

An Example of General Quadratic System.

To illustrate the use of the char-set method for polynomial equations-solving let us consider the example below:

Example 1.1. Let PS be the polset consisting of three *generic* quadratic pols P_j in x_1, x_2, x_3 :

$$P_j \ = \ \Sigma\ c^j_{ghk} * x_3^g * x_2^h * x_1^k,\ j = 1, 2, 3. \tag{1.7}$$

In each P_j Σ is running over non-negative integers g, h, k such that $g + h + k \leqslant 3$. The c^j_{ghk} are all constants to be treated as indeterminates. Applying our algorithm according to diagram (S) with $PS_i = BS_{i-1} + RS_{i-1}$ as in Sect. 3.4 to $PS = PS_0$ we get successively polsets PS_i with index-sets as follows.

$$PS_0 : [10\ x_3\ 2],\ [10\ x_3\ 2],\ [10\ x_3\ 2],$$
$$PS_1 : [18\ x_3\ 1],\ [18\ x_3\ 1],\ [10\ x_3\ 2],$$
$$PS_2 : [108\ x_2\ 3],\ [258\ x_2\ 4],\ [18\ x_3\ 1],$$
$$PS_3 : [16211\ x_2\ 2],\ [108\ x_2\ 3],\ [18\ x_3\ 1].$$

In the next step we have to form for the two pols of index-sets $[16211\ x_2\ 2], [108\ x_2\ 3]$ in PS_3 the remainder of one w.r.t. the other. This is however out of reach of our computer used, a SUN3/140. Even a much bigger computer seems to be difficult to carry out such computations up to the end.

Terms- and Degree-Explosion in the Char-Set Method.

The above example shows that, for the pols occuring during the procedure as given in scheme (S) of Sect. 3.4, the number of terms may increase rapidly to so high to become out of control. In fact, for two pols $P_1, P_2 \subset \mathbf{K}[X]$ with number of terms t_1, t_2 respectively, the number of terms of the product $P_1 * P_2$ may rise up to the order of $t_1 * t_2$. Furthermore, remainder formation is a basic operation to be used constantly in the procedure which is itself composed of a series of more fundamental operations of the following type: Consider two pols P_1, P_2 of same class $c > 0$ with degrees d_1, d_2 respectively. Write $P_i, (i = 1, 2)$, in the form $P_i = I_i * x_c^{d_i} + P_i'$ and suppose $d_1 \geqslant d_2$, I_i being the initial of P_i so that P_i' is either of class $< c$ or of class c but of degree $< d_i$. Then the fundamental operation consists of forming the pol $P_3 = I_2 * P_1' - I_1 * P_2' * x_c^{d_1 - d_2}$. Let t_{i1}, t_{i2}, t_1, t_2 be the number of terms of I_1, I_2, P_1, P_2 respectively. The number of terms of P_3 may then rise up to the order of $t_{i2} * (t_1 - t_{i1}) + t_{i1} * (t_2 - t_{i2})$. What is more grave is that the degree of P_3 in a variable $x_j, j < c$, may also rise up to $d_{j1} + d_{j2}'$ or $d_{j1}' + d_{j2}$ if $d_{j1}, d_{j2}, d_{j1}', d_{j2}'$ are the degrees in x_j of I_1, I_2, P_1', P_2' respectively. This rise up of degrees will cause immense troubles in the subsequent procedure.

Collins-Li Phenomena in the Char-Set Method.

There are also other shortcomings of the method. For example, let us consider two pols $P_1, P_2 \subset \mathbf{K}[X]$ of same class c and degrees $d_1 \geqslant d_2$. Let us form a series of pols

$$P_1, P_2, \cdots, P_k, \cdots$$

with each P_i, $i \geqslant 3$, the *remainder* of P_{i-2} w.r.t. P_{i-1}, considered all as pols of x_c. Collins has proved that for remainders in the sense of pseudo-remainders, the latter pols in the above sequence from some point onwards will inevitably contain initials of previous pols as factors. This remains true if the remainders are taken to be in the Ritt sense. Z.M.Li has proved, as a generalization of the above phenomenon discovered by Collins, that there will occur inevitably during the procedure in accordance to the scheme (S) of Sect. 3.4, undesirable redundant factors which are initials of pols in various basic sets BS previously formed. The occurence of such factors renders the pols appearing in the procedure too large to make computations too complicate than what it ought to be so. There may happen also other kinds of unexpected factors of unclear origin. As a partial remedy of the terms- and degree-explosion due to such phenomena we have to remove all such undesirable initial factors by simple division or split the pols occuring in the procedure into factors whenever possible. The influence of such operations on the solutions are governed by the formulae below:

$$\begin{aligned} &Zero(PS) = Zero(PS/F) + Zero(PS + \{F\}),\\ &Zero(PS + \{F_1 * F_2\}) = Zero(PS + \{F_1\}) + Zero(PS + \{F_2\}). \end{aligned} \tag{1.8}$$

Based on the above observations a package due to D.K.Wang named *wsolve* has been implemented on some SPARC2 station. It turns out that the package is quite effective for polynomial equations-solving and permit to solve a number of interesting problems as we shall see in the sequel.

$\underline{Systematic\ Condensed\ Form\ of\ Coefficients.}$

An alternative partial remedy to the terms-explosion is a *method of condensed coefficients*. For this purpose we shall first introduce some systematic notations for the pols and coefficients involved as follows.

Let the given polset $PS \subset \mathbf{K}[X]$ be consisting of pols say $P_1, \cdots, P_k$. Starting from PS let us form according to algorithms described in Sects. 3.4,3.5 by remainder-formation or Replacement Rules successively pols P_{k+1}, P_{k+2}, etc. In a pol P_j of the sequence we shall denote by c_t^j the coefficient of a monomial

$$X^t = x_n^{d_n} * ... * x_1^{d_1},$$

with $t = (d_n, \cdots, d_1)$. We may also group some terms in the pol P_j with same part $X^u, u = (d_n, ..., d_k)$, or $(d_n, \cdots, d_k, 0, \cdots, 0)$, with $k-1$ $0's$, where $k > 1$, into a single term. The corresponding coefficient of X^u will then be denoted as c_u^j, so that

$$c_u^j = \Sigma_v\ c_{uv}^j * X'^v,$$

in which $X' = (x_{k-1}, ..., x_1)$ and the summation is to be extended over integer tuples $v = (d_{k-1}, ..., d_1)$. The coefficients like c_u^j or c_{uv}^j will henceforth be called *condensed coefficients* of the respective pols. Thus, to stress the roles of x_3 in P_j of Ex. 1.1, we may write

$$P_j = c_2^j * x_3^2 + c_1^j * x_3 + c_0^j,\ j = 1, 2, 3, \tag{1.9}$$

in which

$$\begin{aligned}
&c_2^j = c_{200}^j,\ c_1^j = c_{11}^j * x_2 + c_{10}^j,\ c_0^j = c_{020}^j * x_2^2 + c_{01}^j * x_2 + c_{00}^j,\\
&c_{11}^j = c_{110}^j,\ c_{10}^j = c_{101}^j * x_1 + c_{100}^j,\ c_{01}^j = c_{011}^j * x_1 + c_{010}^j,\\
&c_{00}^j = c_{002}^j * x_1^2 + c_{001}^j * x_1 + c_{000}^j.
\end{aligned} \tag{1.10}$$

An Example for the Method of Condensed Coefficients.

Example 1.2. Let us consider Ex. 1.1 again and assume that the coefficients c^j's for $j = 1, 2, 3$ are independent constant indeterminates. By simple manipulations the polset $PS = PS_0 = \{P_1, P_2, P_3\}$ is seen to be 0-equivalent to the system $PS_1 = \{P_4, P_5, P_6\}$ with P_4, etc. given in condensed form below:

$$\begin{aligned}
P_4 &= x_3^2 + c_{10}^4 * x_3 + c_{02}^4 * x_2^2 + c_{01}^4 * x_2 + c_{00}^4,\\
P_5 &= x_3 * x_2 + c_{10}^5 * x_3 + c_{02}^5 * x_2^2 + c_{01}^5 * x_2 + c_{00}^5,\\
P_6 &= c_{10}^6 * x_3 + c_{02}^6 * x_2^2 + c_{01}^6 * x_2 + c_{00}^6.
\end{aligned} \tag{1.11}$$

In (1.11) we have for $i = 4, 5, 6$,

$$\begin{aligned}
&c_{10}^i = c_{101}^i * x_1 + c_{100}^i,\\
&c_{02}^i = c_{020}^i,\ c_{01}^i = c_{011}^i * x_1 + c_{010}^i,\\
&c_{00}^i = c_{002}^i * x_1^2 + c_{001}^i * x_1 + c_{000}^i.
\end{aligned} \tag{1.12}$$

Remark that all c^i's, $i = 4, 5, 6$, are definite rational functions of the c^j's for $j = 1, 2, 3$. For the sake of simplicity we shall consider the c_{ghk}^i's for $i = 4, 5, 6$ as independent indeterminates in what follows.

Form now on, the remainders of P_4, P_5 w.r.t. P_6, all considered as pols in x_3. We get then two further pols

$$\begin{aligned}
P_7 &= c_{030}^7 * x_2^3 + c_{02}^7 * x_2^2 + c_{01}^7 * x_2 + c_{00}^7,\\
P_8 &= c_{040}^8 * x_2^4 + c_{03}^8 * x_2^3 + c_{02}^8 * x_2^2 + c_{01}^8 * x_2 + c_{00}^8.
\end{aligned} \tag{1.13}$$

In (1.13) we have

$$\begin{aligned}
c_{030}^7 &= -\ c_{02}^6,\\
c_{02}^7 &= c_{10}^6 * c_{02}^5 - c_{02}^6 * c_{10}^5 - c_{01}^6,\\
c_{01}^7 &= c_{10}^6 * c_{01}^5 - c_{01}^6 * c_{10}^5 - c_{00}^6,\\
c_{00}^7 &= c_{10}^6 * c_{00}^5 - c_{00}^6 * c_{10}^5.
\end{aligned} \tag{1.14}$$

$$\begin{aligned}
c_{040}^8 &= (c_{02}^6)^2,\\
c_{03}^8 &= 2 * c_{02}^6 * c_{01}^6,\\
c_{02}^8 &= (c_{10}^6)^2 * c_{02}^4 - c_{10}^6 * c_{02}^6 * c_{10}^4 + 2 * c_{02}^6 * c_{00}^6 + (c_{01}^6)^2,\\
c_{01}^8 &= (c_{10}^6)^2 * c_{01}^4 - c_{10}^6 * c_{01}^6 * c_{10}^4 + 2 * c_{01}^6 * c_{00}^6,\\
c_{00}^8 &= (c_{10}^6)^2 * c_{00}^4 - c_{10}^6 * c_{00}^6 * c_{10}^4 + (c_{00}^6)^2.
\end{aligned} \tag{1.15}$$

The equations like (1.14) and (1.15) which express the condensed coefficients of pols in terms of those of preceding pols, will be henceforth called *auxiliary equations.* Applying now Replacement Rules of Sect. 3.5 to P_7, P_8 we get an eliminant P_9 of class 2 and degree 1, and a resultant P_{10} deprived of x_2. Express all the c's in P_9, P_{10} in terms of c^i's for $i = 4, 5, 6$ by means of the auxiliary equations (1.14), (1.15), it turns out that both P_9 and P_{10} have factors of the form c^6_{02} and

$$c^6_{10} = c^6_{101} * x_1 + c^6_{100}.$$

Remove now these factors and denote the quotient pols thus obtained still by P_9, P_{10}. If we further expand P_9, P_{10} by means of (1.12), then P_9, P_{10} will be pols of index-sets [1503 x_2 1] and [34411 x_1 8] respectively. By verification we see that all remainders of P_4, P_5, P_6 w.r.t. the asc-set $CS = \{P_{10}, P_9, P_6\}$ are all 0. It follows that CS is a char-set of $PS_1 = \{P_4, P_5, P_6\}$ and hence also one of $PS = \{P_1, P_2, P_3\}$ so far the removed factors c^6_{02} and c^6_{10} are considered as non-zero. Remark that if we express the coefficients of pols in CS in terms of those of the original coefficients of P_1, P_2, P_3 and clear of denominators, then the number of terms would be of order of milliards.

Remark that during the procedure we have removed the factors c^6_{02}, c^6_{10}. The removal of first factor as a non-zero constant is immaterial, but the removal of the second factor shows that the Well-Ordering Principle and (1.8) will give us an identity of the form

$$Zero(PS) = Zero(CS/IP * c^6_{10}) + Zero(PS + \{IP\} \ / \ c^6_{10}). + Zero(PS + \{c^6_{10}\}).$$

In assuming that the c^j's for $j = 4, 5, 6$ are independent indeterminates it is now easily verified that both P_9 and P_{10} are irreducible. Simple arguments will show that the above identity will be reduced to the following one:

$$Zero(PS) = Zero(CS). \tag{1.16}$$

A Further Example of Czapor and Geddes.

Example 1.3. As a further example let us consider the solving of the polynomial equations due to Czapor and Geddes as in Ex. 3.4.2 and Ex. 4.4.2, viz. $PS = 0$ in unknowns x, y, z where PS is consisting of 3 pols given by:

$$\begin{aligned} P_1 &= x^2 + a * y * z + d * x + g, \\ P_2 &= y^2 + b * z * x + e * y + h, \\ P_3 &= z^2 + c * x * y + f * z + k. \end{aligned} \tag{1.17}$$

The entities $a, \cdots, k$ will be treated as independent constant indeterminates and the unknowns x, y, z are just previous x_1, x_2, x_3 but we shall preserve the present notations. We see that P_1, P_2, P_3 here in (1.17) are already in the form of P_5, P_6, P_4 in (1.11) if we set

$$x_1 = x,\ x_2 = y,\ x_3 = z,\ \text{and}$$

$$\begin{aligned} &c^4_{11} = 0,\ c^4_{10} = f,\ c^4_{02} = 0,\ c^4_{01} = c * x,\ c^4_{00} = k; \\ &c^5_{10} = 0,\ c^5_{02} = 0,\ c^5_{01} = 0,\ c^5_{00} = (x^2 + d * x + g)/a; \\ &c^6_{10} = b * x,\ c^6_{02} = 1,\ c^6_{01} = e,\ c^6_{00} = h. \end{aligned} \tag{1.18}$$

With these values in (1.15) we see that the pols P_9, P_{10} in Ex. 1.2 become pols of number of terms reduced to 137 and 35 respectively. The polset $PS = \{P_1, P_2, P_3\}$ in (1.17) has thus a char-set CS of index-sets [137 x 8], [35 y 1], [4 z 1] and $Zero(PS) = Zero(CS)$ as in Ex. 3.4.2.

Some Concluding Remarks.

Remark 1.1. Let us denote by *maxt* the maximum number of terms of pols occuring in the procedure due to the char-set method or some others. Then all pols during the whole procedure of above Ex. 1.3 are quite small ones with $maxt = 324$ and computations are quite simple too. The underlying reason is that not only we have avoided the appearance of all redundant factors, but also we have expressed all pols in condensed form with coefficients of pols in successive polsets in terms of those of previous polsets by means of auxiliary equations, in accordance to the Propositions 1, 2 and 5 of Sect. 3.5.

Remark 1.2. A great shortcoming of the above method of condensed coefficients is the following one: A pol in its ordinary non-condensed form may have some factors which may be detected by means of the present-day software and may be removed if required. However, when the pol is put in the form with condensed coefficients, it will be in general no more factorizable and there may be no way of detecting such possible factors.

Remark 1.3. Some particular techniques or strategies may be taken to limit the size of pols by mere inspecting in adopting the man-machine interactive manner for the computations. For example, we may replace a pair of pols P, Q in a polset PS by some 0-equivalent pair P', Q' which are linear combinations of P, Q so that the new polset PS' thus gotten may become easier for the determination of its zero-set.

Remark 1.4. Even if we remove all the redundant factors whenever possible, and if some particular strategies are taken, the pols occuring in the procedure may still be too large to permit the carrying out of further computations. This is almost unavoidable owing to the general occurence of parameters in symbolic method. It seems that the only issue is to apply a *hybrid* method in combining together the symbolic as well as the numerical method, especially for practical applications. We shall however leave this topic to next section .

6.2 A Hybrid Method of Polynomial Equations-Solving

The Necessity of a Hybrid Method of Polynomial Equations-Solving.

As we have remarked in last section, there are two kinds of general methods for polynomial equations-solving, the symbolic one and the numerical one, the basic field **K** being understood to be **R** or **C** unless otherwise stated. Both methods have their own merits and demerits. Unlike the numerical method, the symbolic method has the advantage of being *exact* in character. Consider however the simple problem of solving a system of three quadratic equations in three variables x, y, z. Whether the coefficients of these equations are parametric or approximate numerical ones, for the symbolic method we have to replace these numerical values by indeterminates. As shown in examples Ex. 1.1 and Ex. 1.2, it is found that the final equations, say the one in x alone, will be either difficult to be determined or will have a number of terms too large to have any interest or of any usefulness. The symbolic method has therefore the disadvantage of having very likely too big pols during the procedure. However, if we replace the intermediate coefficients of the given equations by their original numerical values, then the final pol, being of degree 8 in x, may have 9 terms only in the worst case. This shows that the symbolic method may be quite unpracticable, especially for problems arising from practice. On the other hand the usual numerical methods can hardly give the totality of all possible solutions. It seems thus one of the issues is to use a hybrid method, in keeping the advantages of both methods, symbolic as well as numerical ones, while avoiding the disadvantages of both methods.

Difficulties for a Hybrid Method.

Remark that if we treat by symbolic method equations with numerical coefficients, then there will be accumulated errors of coefficients in computations which will render the final results unreliable. Remark also that factorization is almost inevitable in symbolic methods, while factorization of pols with approximate numerical coefficients lacks exact meaning. Consequently in order to give a satisfactory hybrid method of polynomial equations-solving, we have to settle first the two following problems:

Problem H1. The error estimation and error control during the hybrid computations in the procedure.

Problem H2. The factorization of pols with approximate numerical coefficients.

These two problems will be treated separately below.

Estimation of Roots of a Numerical Algebraic Equation.

It is well-known that roots of a polynomial equation in a single unknown are continuous in the coefficients and are highly sensitive to even slight variation of coefficients. To be precise, let us cite below a theorem due to Ostrowski:

Theorem 2.1. (Ostrowski) Consider two pols

$$f(z) = a_0 * z^n + \cdots + a_n, \tag{2.1}$$

$$g(z) = b_0 * z^n + \cdots + b_n. \tag{2.2}$$

Assume that $a_0 * a_n \neq 0$ and let the roots of $f(z) = 0$ be $z_1, \cdots, z_n$. Suppose that for a certain positive τ with

$$4 * n * \tau^{\frac{1}{n}} \leqslant 1, \tag{2.3}$$

we have for $i = 1, \cdots, n$,

$$|b_i - a_i| \leqslant |a_i| * \tau. \tag{2.4}$$

Then the n roots $y_1, \cdots, y_n$ of $g(z) = 0$ can be ordered in such a way that we have for each $i = 1, \cdots, n$,

$$|\frac{y_i}{z_i} - 1| < 8 * n * \tau^{\frac{1}{n}}. \tag{2.5}$$

The theorem shows that for an equation say of degree 4, in order to get roots accurate up to two decimal places, it requires that coefficients of the equation should be accurate up to perhaps 8 or more decimal places. Examples show that it is impossible to improve such estimation of errors. It seems thus the evaluation of errors becomes practically useless. In spite of this there are no reasons to be discouraged. Cf. the discussions by Ostrowski in [O2], p.111 and p.221.

Exact Representation of Approximate Numbers.

In the sequel we shall restrict ourselves to the case of $\mathbf{K} = \mathbf{R}$, the case of $\mathbf{K} = \mathbf{C}$ being entirely similar.

For our purposes we shall treat the problem in the following way. For an approximate numerical value say a, let us write it in the form $a_0 + t_0$, where a_0 is a *known exact rational* number close to a, while t_0 is a sufficiently small number, *unknown* but *exact*. For example, we may write π in the form $\pi = \dfrac{31416}{10000} + t$ in which $\dfrac{31416}{10000}$ and t are both *exact* numbers, t being an unknown number with its magnitude $|t| < 10^{-4}$.

Definition 2.1. *Exact Representation* of an approximate number $a := a_0 + t_0, a_0$ being a known exact rational number, and t_0 an unknown exact number with $|t_0| < 10^{-N}, N$ a positive integer.

Definition 2.2. For exact representation of $a = a_0 + t_0$ in above definition, $a_0 :=$ *Essential part* of a, $t_0 :=$ *Tiny part* of a, and $N :=$ *Precision degree* of the representation.

Exact Representation of Pols with Approximate Coefficients.

Let us consider a system of polynomial equations $PS = 0$ with $PS \subset \mathbf{R}[x_1, \cdots, x_n]$.

Definition 2.3. *Exact representation to precision degree* N of a polset $PS :=$ The polset gotten from PS in replacing each approximate numerical coefficient c_i in any pol of PS by its exact representation $c_{i0} + t_i$ to same prescribed precision degree N, where c_{i0} are the essential parts and t_i are the tiny parts with all $|t_i| < 10^{-N}$.

Definition 2.4. For the exact representation of PS in the above definition, t_i := *Tinies*, N := *Precision degree* of the representation, and any linear combination of products of k tinies with definite rational coefficients := A *tiny* of *order k*.

Hybrid Method of Polynomial Equations-Solving.

Let us solve now $PS = 0$ with approximate numerical coefficients in the following way. First we fix a precision degree N and a positive integer s. Transform now the given polset into its exact representation to be still denoted by PS. Solve now $PS = 0$ by some symbolic method, say by the char-set method. To determine such a char-set CS of PS let us proceed in the usual way by the char-set algorithm. During the procedure we shall however set any tiny of order greater than s to be 0. Let the first pol of the char-set CS thus gotten be

$$C_1 = (a_0 + T_0) * x_1^n + \cdots + (a_n + T_n), \tag{2.6}$$

in which T_k are tinies of order less than or equal to s and a_k are exact known rational numbers. Similarly for the other char-pols C_i.

Definition 2.5. The polset $\{C_i\}$ obtained above := *Approximate char-set* to *order* s of the given polset PS.

In the same way in using other methods as in Sects. 3.4,3.5, we may get *approximate* char-series, *approximate* zero-decomposition of $Zero(PS)$, etc. which we shall not enter.

Suppose that in (2.6) $a_0 * a_n \neq 0$. Let C_{10} be the pol given by

$$C_{10} = a_0 * x_1^n + \cdots + a_n. \tag{2.7}$$

We treat the other pols C_i in the same way to get pols C_{i0}. Let us solve the equation $C_{10} = 0$ of (2.7) instead of the equation $C_1 = 0$ of (2.6), and then the other pols

$C_{i0} = 0$ in succession. The theorem of Ostrowski, in case it may be applied, will give a rough estimate of the errors in the solutions. Thus the above give us hybrid methods of polynomial equations-solving with error estimates roughly governed by Ostrowski Theorem.

Remark 2.1. To arrive at (2.7) we have to replace each numerical data coefficient c_i of the given equations $PS = 0$ by an *exact* pol $c_{i0} + t_i$ of two terms. The pols occuring during the procedure of arriving at the char-set CS may thus be very big so that the computations may become too complicate to be carried out. However, for each pol occuring in the procedure, we keep only such terms which have degrees $\leqslant s$ in t_i. Thus, if the t_i's are m in number, then the number of terms in C_{10} will be at most $(n+1)*(m+1)$ for $s=1$ and $(n+1)*\left(1+m*\dfrac{m+3}{2}\right)$ for $s=2$. Similarly for other pols occuring in the procedure. In this way the polynomials will be controlled to have a reasonable number of terms and the computations may be carried out up to the end without much difficulties.

Remark 2.2. As already remarked, to get low degree of accuracy in solving $C_{10} = 0$ instead of $C_1 = 0$ and others, it requires to have high precision of the coefficients a_i and hence also high precision of the original data coefficients c_i. This means that the precision degree N should be chosen to be sufficiently high at the outset. This disadvantage becomes however just the advantage in applying the above method. For if t_i have magnitudes $< 10^{-N}$, then products of the form $t_i * t_j$ and $t_i * t_j * t_k$ will have magnitudes $< 10^{-2*N}$ and 10^{-3*N} respectively. Hence this will give us safety in neglecting terms of degree $\geqslant 3$ or $\geqslant 2$ in t_i, so far N is chosen sufficiently large.

Remark 2.3. Suppose that none of the essential parts of any initial become 0 during the procedure, then the above hybrid methods will raise no theoretical difficulties and the Ostrowski Theorem will furnish means of error estimate and control, at least theoretically so. It is however not clear under what situations some initial in the procedure may have its essential part 0. Nevertheless it will be shown below that in practical problems it will rarely be so.

Remark 2.4. It may also happen that for some number $a + \Sigma_k\, b_k * T_k$ with T_k product of k tinies which occur in the procedure some of the coefficients b_k of the tiny parts T_k may be sufficiently large in comparing with a so that $b_k * T_k$ will be at first sight no more a *tiny* number. This apparent defect may however be remedied by sufficiently raising the precision degree. In contrast it shows that the usual method of using directly the approximate numerical values in computation which amounts to set all t_i to 0 is theoretically unreliable.

Healthy and Pathological Approximation of Numerical Data Coefficients.

We shall show below how to settle the difficulties arisen from exact representation of numerical data coefficients in the char-set method as pointed out in Remark 2.3.

For this purpose let us consider a polset $PS \subset \mathbf{R}[x_1, \cdots, x_n]$ whose zero-set is to be determined. Let us try to do this in determining first the char-set of PS by the char-set method say in accordance to the scheme (S) in Sect. 3.3. Treat first the approximate numerical datas occuring in the pols of PS as independent indeterminates. Proceeding as in scheme (S) of Sect. 3.3 let us arrange the pols of the given polset PS as well as the successively introduced non-zero remainders in a definite order according to its order of appearance, say

$$P_1, P_2, \cdots, P_M. \tag{2.8}$$

Let us fix now a precision degree N and replace each approximate numerical data coefficients c_i by its exact representation $c_{i0} + t_i$ to the precision degree $N' > N$ with c_{i0} the essential part and t_i the corresponding tiny. Proceed in the same way as before in treating the tinies t_i as independent indeterminates and arrange the pols of PS as well as the successive new non-zero remainders in the definite order according to their order of appearance we will get a sequence of pols below which are clearly in 1-1-correspondence with the preceding one:

$$P_1', P_2', \cdots, P_M'. \tag{2.8$'$}$$

In fact, it is clear that each P_k' is gotten from the corresponding P_k in replacing each indeterminate c_i in P_k by its exact representation $c_{i0} + t_i$.

Definition 2.6. An exact representation $c_i = c_{i0} + t_i$ is *pathological* at stage k := For some monomial in x_i's in P_k' of the sequence (2.8)$'$ its coefficient as a pol in the tinies has its part not involving any tiny equal to 0. In this case such a monomial := A *pathological monomial.*

Definition 2.7. An exact representation $c_i = c_{i0} + t_i$ is *healthy* := It is pathological at no stage in the sequence (2.8)$'$.

Suppose that for the exact representation $c_i = c_{i0} + t_i$ the sequence (2.8)$'$ is pathological first time at stage k. If we set any tiny of certain order to be 0 and we proceed as before, then there may be dangers that the subsequent pols may be a sequence entirely different in form as in (2.8) or (2.8)$'$. The final char-set thus obtained may be *qualitatively* different from that gotten from (2.8) or (2.8)$'$ and will be entirely unreliable. On the other hand if the exact representation is healthy then even if we set all the tinies to be 0 during the procedure, we may proceed as before and we will get a sequence in the same form as (2.8) or $(2.8)'$ and the approximate char-set is of the same form as that gotten from (2.8) or $(2.8)'$.

Our problem is thus reduced to the seeking of some *healthy* representation of the approximate numerical data coefficients c_i of given precision degree N of the given polset PS. For this we require the following Lemma.

Lemma 2.1. Given a finite collection $\mathcal{P}$ of pols in $t_i, i = 1, \cdots, r$ and sufficiently small $\epsilon > 0$, there are points $(d_1, \cdots, d_r)$ in the neighborhood $V = \{|t_i| < \epsilon, i = 1, \cdots, r\}$ of the origin $(0, \cdots, 0)$ in the real space $\mathbf{R}^r(t_1, \cdots, t_r)$ such that for $t_i = d_i + t_i', i = 1, \cdots, r$, any pol in the collection $\mathcal{P}$ when expressed as pol in t_i' will have its term not involving any t_i' unequal to 0. Moreover, the points in V verifying the above condition will form a set open and dense in V.

Proof. For each pol P in $\mathcal{P}$ $Zero_{\mathbf{R}}(P)$ defines a *real* algebraic variety V_P of topological dimension atmost $r-1$ in the space $\mathbf{R}^r(t_1, \cdots, t_r)$ of real dimension r. See Th. 5.1.15. In any neighborhood of the origin $(t_1, \cdots, t_r) = (0, \cdots, 0)$ the points $(d_1, \cdots, d_r)$ for which none of the pols in the collection $\mathcal{P}$ is zero form then a set open and dense in that neighborhood. Any such point in this open and dense set may be served as a point $(d_1, \cdots, d_r)$ as required. □

Theorem 2.2. Let $c_i = c_{i0} + t_i$ be any exact representation of the approximate numerical data coefficients of pols in the given polset $PS \subset \mathbf{R}[x_1, \cdots, x_n]$ with a precision order N. Then for $\epsilon > 0$ sufficiently small there are healthy representations $c_{i0}' + t_i'$ with essential parts c_{i0}' and tinies t_i' of precision degree N such that $|c_{i0}' - c_{i0}| < \epsilon$.

Proof. We may suppose without loss of generality that the given approximation $c_i = c_{i0} + t_i$ is of precision degree $N' > N$. If the approximation is not healthy let $\mathcal{P}$ be the collection of all pols in t_i which are coefficients of pathological monomials of such pathological pols P_k' in $(2.8)'$. In applying Lemma 2.1 we may take $\epsilon>0$ sufficiently small and take some $(d_1, \cdots, d_r)$ in the neighborhood $|t_i| < \epsilon$ verifying the conditions in the Lemma. Then the exact representation $c_i = c_{i0}' + t_i'$ with $c_{i0}' = c_{i0} + d_i$, $t_i' = t_i - d_i$ will be a healthy one of precision degree N as required. □

Remark 2.5. As the points $(d_1, \cdots, d_r)$ may be arbitrarily chosen from a subset open and dense in some sufficiently small neighborhood of the origin in the space $\mathbf{R}^r(t_1, \cdots, t_r)$, it is almost sure that in randomely choosing such a point $(d_1, \cdots, d_r)$ with d_i sufficiently small, the modified exact representation $c_i = c_{i0}' + t_i'$ with $c_{i0}' = c_{i0} + d_i, t_i' = t_i - d_i$ will be a healthy one. It follows that the above Theorem 2.2 actually gives a solution of Problem H1 in the practice.

Approximate Factorization of Multivariate Polynomials.

We come now to the solving of Problem H2.

For symbolic method factorization of multivariate pols is actually unavoidable. However, if the coefficients of a pol are in approximate numerical values, then it is difficult to decide whether the given pol is factorizable or not. For example, if the coefficient 3.0 in the pol $f = x^3 + y^3 + z^3 - 3.0 * x * y * z$ is only an approximate numerical value, then it is somewhat illegal to say that f has a factor $g = x + y + z$. What we can say atmost is only that f will have *approximately* the factor g. Such a notion of approximate factorization occured already in some writings of Zassenhaus and recently it appears also in some papers of Sasaki et al, cf. e.g. [S-S-K-S]. Below we shall introduce the notion of *approximate factorization* in the following way.

Let us consider a pol in exact representation $f \in \mathbf{Q}[x, Y, t_1, \cdots, t_r]$ in which $Y = (y_1, \cdots, y_l)$ is a set of parameters and t_i are tinies treated as parameters too. First let us consider the case for which Y is absent while the basic field is a general field $\mathbf{K}$ of characteristic 0.

Definition 2.6. An *approximate factorization* of $f \in K[x, t_1, \cdots, t_r]$ up to *order* s as a pol in x := A product of the form

$$f = g * h \ mod \ T^{s+1}, \tag{2.9}$$

in which g, h are pols in $\mathbf{K}[x, t_1, \cdots, t_r]$ of positive degree in x, and $mod \ T^{s+1}$ means that (2.9) is an equality up to some tiny of order greater than s.

In order to approximately factorize f let us first remark that, by atmost a preliminary linear transformation in x and t_i, we may without loss of generality lay down the following supposition

(S) The leading coefficient in x of $f \in \mathbf{K}[x, t_1, \cdots, t_r]$ is free of t_i.

Under the supposition (S) the pol f can be written in the form

$$f = f_0 + f_1 + \cdots + f_m, \tag{2.10}$$

in which for each j, f_j is homogeneous of degree j in $t_1, \cdots, t_r$.

Let us introduce a new parameter u, replace each t_i by $u * t_i$ and consider the pol thus obtained from a pol $P \in \mathbf{K}[x, t_1, \cdots, t_r]$ as a pol $P^* \in \mathbf{K}[x, t_1, \cdots, t_r, u]$ Then (2.10) can be rewritten in the form

$$f^* = f_0 + u * f_1 + \cdots + u^m * f_m. \tag{2.10)*$$

We recover f from f^* by just setting u equal to 1.

Lemma 2.2. To factorize f as in (2.9) is equivalent to factorize f^* in the form below:

$$f^* = g^* * h^* \ mod \ u^{s+1}, \tag{2.11}$$

in which g^*, h^* are pols in $\mathbf{K}[x, t_1, \cdots, t_r, u]$ of degree in x both greater than 1, and $mod \ u^{s+1}$ means that (2.11) is an equality up to terms each having a factor in u of degree greater than s, or $f^* - g^* * h^*$ is a multiple of u^{s+1}.

Proof. If (2.11) is true, then setting $u = 1$ we will get an approximate factorization of f in the form of (2.9) with pols $g, h \in \mathbf{K}[x, t_1, \cdots, t_r]$. Conversely, if we have an approximate factorization of f in the form of (2.9) with pols $g, h \in \mathbf{K}[x, t_1, \cdots, t_r]$, then in replacing each t_i by $u * t_i$ we will get from (2.9) an approximate factorization of f^* in the form of (2.11). □

As usual, we base our method of the above approximate factorization in the form of (2.11) on Hensel construction as exhibited in [VdW2], Sect. 14.

Theorem 2.3. Suppose that in (2.10) f_0 satisfies (S) and g_0, h_0 in $\mathbf{K}[x]$ are known with

$$f_0 = g_0 * h_0. \tag{2.12}$$

Suppose that the degrees in x of f_0, g_0, h_0 be respectively n, p, q all > 0 with $n = p+q$. Suppose further that g_0, h_0 have a common factor D in x. Then for a given s there is some algorithm which permits to decide whether f as a pol in $\mathbf{K}[x, t_1, \cdots, t_r]$ is approximately factorizable up to order s, and in the affirmative to determine such an approximate factorization as in (2.9). Moreover, if g_0, h_0 are prime to each other, then such an approximate factorization is always possible which will be directly given by the algorithm.

Proof. Let us consider pols f_0, g_0, h_0 all as pols in $\mathbf{K}[x]$ and the pols g, h to be determined as pols in $\mathbf{K}[x, t_1, \cdots, t_r]$. With this understood let us write now f^*, g^* and h^* in the form below:

$$\begin{aligned} f^* &= f_0 + u * f_1 + \cdots + u^s * f_s \ mod \ u^{s+1}, \\ g^* &= g_0 + u * g_1 + \cdots + u^s * g_s \ mod \ u^{s+1}, \\ h^* &= h_0 + u * h_1 + \cdots + u^s * h_s \ mod \ u^{s+1}, \end{aligned} \tag{2.13}$$

in which for each k all f_k, g_k, h_k in $\mathbf{K}[x, t_1, \cdots, t_r]$ are homogeneous of degree k in $t_1, \cdots, t_r$. Our method consists in determining the g_k, h_k successively for $k = 1, 2$, etc. from f and g_0, h_0 verifying (2.12).

Let the degree in x of D be $d \geqslant 0$. If g_0, h_0 are prime to each other then we may take $D = 1$ and $d = 0$ in what follows. In any way we have

$$g_0 = D * g_0', \ h_0 = D * h_0'$$

so that g_0', h_0' considered as pols in $\mathbf{K}[x]$ with degrees in x equal to $p' = p-d, q' = q-d$ are prime to each other. We have then

$$\alpha * g_0' + \beta * h_0' = 1, \tag{2.14}$$

in which $\alpha, \beta \in \mathbf{K}[x]$ have degrees in x less than or equal to $q'-1, p'-1$ respectively.

For each $0 < k \leqslant s$, let

$$\begin{aligned} F_k &= f_0 + u * f_1 + \cdots + u^k * f_k, \\ G_k &= g_0 + u * g_1 + \cdots + u^k * g_k, \\ H_k &= h_0 + u * h_1 + \cdots + u^k * h_k. \end{aligned} \tag{2.15}$$

in which g_j, h_j all in $\mathbf{K}[x, t_1, \cdots, t_r]$ are homogeneous in degree j of $t_1, \cdots, t_r$ and are yet to be determined. In order to prove the theorem we shall try to determine g_k, h_k of degrees in x less than p, q respectively for $k = 1, 2, \cdots, s$ in succession to satisfy the equations below:

$$F_k = G_k * H_k \ mod \ u^{k+1}. \tag{2.16}$$

For this purpose let us set for each $0 < k \leqslant s$,

$$f_k^0 = f_k - \Sigma_k \ g_i * h_j, \ k \geqslant 1, \tag{2.17}$$

in which Σ_k means summation over pairs (i, j) with $i > 0, j > 0, i+j = k$. Owing to the fact that $f_0 = g_0 * h_0$ the equations (2.16) are equivalent to a series of equations below:

$$g_0' * h_k + h_0' * g_k = \frac{f_k^0}{D}, \ k \geqslant 1. \tag{2.18}$$

in which f_k^0 is given by (2.17). We shall determine $g_k, h_k \in \mathbf{K}[x, t_1, \cdots, t_r]$ recursively from (2.18). It is clear that whenever f_k^0 is not divisible by D then there will exist no such solutions g_k, h_k and the approximate factorization in question is impossible. Suppose the contrary so that $f_k'^0 = \dfrac{f_k^0}{D}$ is a *pol* in $\mathbf{K}[x, t_1, \cdots, t_r]$ whose degree in x is then less than $n' = n - d = p + q' = p' + q$.

Owing to (2.14) we see that an evident solution of (2.18) is given by

$$(g_k, h_k) = (\beta * f_k'^0, \alpha * f_k'^0). \tag{2.19}$$

The pols g_k, h_k are however too high in degree of x to meet the purpose. To get solutions (g_k, h_k) with degrees less than or equal to $p-1, q-1$ respectively let us first remark that each of the equations (2.18) is of the form

$$g_0' * H + h_0' * G = F, \tag{2.20}$$

in which g_0', h_0', F are known pols in $\mathbf{K}[x]$ of degree in x equal to p', q' and less than n' respectively, while G, H are pols in $\mathbf{K}[x]$ to be determined with degrees in x less than p', q' respectively. By supposition (S) we can therefore write g_0', h_0' in the form $(a_i, b_i \in \mathbf{K})$

$$\begin{aligned} g_0' &= a_{p'} * x^{p'} + \cdots + a_0, \\ h_0' &= b_{q'} * x^{q'} + \cdots + b_0. \end{aligned} \tag{2.21}$$

Let F, G, H in (2.20) be respectively of the form $(e_i, c_i, d_i \in \mathbf{K}[t_1, \cdots, t_r])$

$$\begin{aligned} F &= e_{n'-1} * x^{n'-1} + \cdots + e_0, \\ G &= c_{p'-1} * x^{p'-1} + \cdots + c_0, \\ H &= d_{q'-1} * x^{q'-1} + \cdots + d_0. \end{aligned} \tag{2.22}$$

Then (2.20) gives a set of equations

$$\Sigma_i(c_i * b_{r-i} + d_i * a_{r-i}) - e_r = 0, \tag{2.23$_r$}$$

for $r = 0, 1, \cdots, n' - 1$. In $(2.23)_r$ the summation Σ_i runs over i from 0 to r, and $a_i = 0$ for $i > p'$, $b_i = 0$ for $i > q'$, $c_i = 0$ for $i > p' - 1$, and $d_i = 0$ for $i > q' - 1$. Let M be the usual Sylvester matrix of the two pols g_0', h_0'. Then (2.23) is equivalent to the following matrix equation

$$(d_{q'-1}, \cdots, d_0, c_{p'-1}, \cdots, c_0) \ * \ M \ = (e_{n'-1}, \cdots, e_0). \tag{2.24}$$

Since g_0', h_0' have no common factor in x, the determinant of M is non-zero so that the inverse matrix M^{-1} with elements in $\mathbf{K}$ is well-defined. From (2.24) we get therefore

$$(d_{q'-1}, \cdots, d_0, c_{p'-1}, \cdots, c_0) \ = (e_{n'-1}, \cdots, e_0) \ * \ M^{-1}. \tag{2.25}$$

Denote the column vectors of M^{-1} by $V_1, \cdots, V_n$ and the vector $(e_{n'-1}, \cdots, e_0)$ by E. Then we have explicit formulas for G and H below.

$$\begin{aligned} G &= \Sigma_{p'} \ (E.V_{n'-i}) * x^i, \\ H &= \Sigma_{q'} \ (E.V_{q'-i}) * x^i, \end{aligned} \tag{2.26}$$

in which $\Sigma_{p'}, \Sigma_{q'}$ means summation over i from 0 to $p' - 1$ and 0 to $q' - 1$ respectively. It is clear that the solution-pair (G, H) of (2.20) is unique.

Starting now from $k = 1$ for (2.18) we determine in succession (g_k, h_k) for $k = 1, 2, \cdots, s$, so far f_k^0 is divisible by D. The corresponding approximate factors g, h of the original pol f will then be readily gotten from (2.13) by setting $u = 1$ with however $g_k, h_k \in \mathbf{K}[x, t_1, \cdots, t_r]$. If at some stage k the pol f_k^0 is not divisible by D, then the algorithm stops and the pol f is not approximately factorizable already to the order of $k \leqslant s$. This proves the theorem. □

Approximate Factorization of Pols with Parameters.

Consider now the problem of approximate factorization of pols in exact representation $f \in \mathbf{Q}[x, Y, t_1, \cdots, t_r]$ involving parameters $Y = (y_1, \cdots, y_l)$ with tinies $(t_1, \cdots, t_r)$ as before. By atmost a linear transformation if necessary we may suppose without loss of generality that the leading coefficient of f as a pol of x is free of Y and $t_1, \cdots, t_r$. Let us introduce the extension field $\mathbf{K} = \mathbf{Q}(Y)$ and consider f as a pol in $\mathbf{K}[x, t_1, \cdots, t_r]$. Then f satisfies the supposition (S) and Theorem 2.3 may be applied. If we turn back $\mathbf{K}$ to $\mathbf{Q}(Y)$ and write the possible approximately factorized pols in the form of rational functions in $x, Y, t_1, \cdots, t_r$, then there will appear in the denominators some pols in Y alone. By clearing of denominators we get then as a corollary of Theorem 2.3 the following theorem:

Theorem 2.4. Suppose that for $f \in \mathbf{Q}[x, Y, t_1, \cdots, t_r]$ in (2.10) each f_k is in $\mathbf{Q}[x, Y, t_1, \cdots, t_r]$ and is homogeneous of degree k in $t_1, \cdots, t_r$, while $f_0 \in \mathbf{Q}[x]$ is free of Y and $t_1, \cdots, t_r$. Suppose that f_0 satisfies (2.12) in which g_0, h_0 in $\mathbf{Q}[x]$ are known pols with degrees in x both positive. Then there is some algorithm which permits to decide whether there exist pols $I, I_1, I_2 \in \mathbf{Q}[Y]$ and pols $g, h \in \mathbf{Q}[x, Y, t_1, \cdots, t_r]$ so that for given s we have

$$I * f = g * h \ mod \ T^{s+1}, \qquad (2.9)'$$

in which the parts in g, h not involving any $t_1, \cdots, t_r$ are $I_1 * g_0, I_2 * h_0$ and $I_1 * I_2 = I$.

Remark. 2.6. For applications of the above theorem to actual char-set computations of a polset $PS \subset \mathbf{R}[X], X = (x_1, \cdots, x_n)$ we may encounter the problem of approximate factorization of some pol f during the procedure. The variable x in f will then be some x_m while the Y will be a part of the variables $x_1, \cdots, x_{m-1}$ preceding x_m in the natural ordering of variables. It follows that the pol I in (2.9)′ may be absorbed in the initial of f which causes thus no trouble for further computations. It follows that Theorems 2.3 and 2.4 furnish actually a solution to Problem H2.

Remark 2.7. For some improvements of approximate factorization and some instructive examples on hybrid computation we refer to the doctoral thesis of Zhi Li-hong. Cf. [Zhi].

Some Examples.

Example 2.1. Consider the pol

$$f = x^3 + y^3 + z^3 - 3.0 * x * y * z$$

with 3.0 an approximate numerical coefficient. Let us take the exact representation $3+t$ for 3.0 with t a tiny. We shall consider y, z as Y and try to approximately factorize f up to order 2. With notations as in the proof of Theorems 2.3, 2.4 we see that

$$\begin{aligned}
&f_0 = x^3 - 3*x*y*z + (y^3+z^3) = g_0 * h_0,\\
&f_1 = -t*x*y*z, \text{ with}\\
&g_0 = x+(y+z),\ h_0 = x^2 - x*(y+z) + (y^2+z^2-y*z).
\end{aligned}$$

Successively we get

$$M = \begin{bmatrix} 1 & y+z & 0 \\ 0 & 1 & y+z \\ 1 & -(y+z) & y^2+z^2-y*z \end{bmatrix},$$

$$Det\ M = 3*(y^2+z^2+y*z),$$

$$M^{-1} = \begin{bmatrix} 2*(y^2+z^2)+y*z & (y+z)*(y^2+z^2-y*z) & (y+z)^2 \\ y+z & y^2+z^2-y*z & -(y+z) \\ -1 & 2*(y+z) & 1 \end{bmatrix} / \ Det\ M,$$

$$g_1 = \frac{(y+z)*y*z*t}{Det\ M},$$

$$h_1 = -\frac{(y+z)*y*z*t*x-(y^2+z^2-y*z)*y*z*t}{Det\ M},$$

$$g = g_0 + g_1,\ h = h_0 + h_1,$$

$$f = g*h \ mod\ t^2.$$

By clearing of denominators we get finally the required approximate factorization of f of order 2.

The method of approximate factorization works naturally also for exact factorization, in taking s sufficiently large. It furnishes thus an alternative method of factorization of multivariate pols in ordinary sense, under the restriction imposed by supposition (S).

Example 2.2. Let us consider Ex. 4.5.3 again which is taken from [W-R], Ex. D17, viz. the factorization of the pol f below:

$$\begin{aligned}
f = {} & x^4 + (-z+3)*x^3 + (z^3+(y-3)*z-y^2-13)*x^2\\
&+(-z^4+(y^2+3*y+15)*z+6)*x\\
&+y*z^4+2*z^3+(-y^3-15*y)*z-2*y^2-30.
\end{aligned}$$

Set

$$y = u * t_1, z = u * t_2.$$

Then f becomes

$$\begin{aligned}
f &= f_0 + u * f_1 + \cdots + u^5 * f_5, \text{ where} \\
f_0 &= x^4 + 3 * x^3 - 13 * x^2 + 6 * x - 30, \\
f_1 &= x * t_2 * (-x^2 - 3 * x + 15), \\
f_2 &= t_1 * (x^2 * t_2 - x^2 * t_1 + 3 * x * t_2 - 15 * t_2 - 2 * t_1), \\
f_3 &= t_2 * (x^2 * t_2^2 + x * t_1^2 + 2 * t_2^2), \\
f_4 &= -t_2 * (x * t_2^3 + t_1^3), \\
f_5 &= t_2^4 * t_1.
\end{aligned}$$

We have

$$f_0 = g_0 * h_0, \text{ where}$$

$$g_0 = x^2 + 2, \; h_0 = x^2 + 3 * x - 15$$

are co-prime. We find successively

$$\begin{aligned}
&\alpha = \frac{1}{307} * (3 * x + 26), \; \beta = -\frac{1}{307} * (3 * x + 17), \\
&(g_1, h_1) = (-x * t_2, 0), \; (g_2, h_2) = (t_2 * t_1, -t_1^2), \\
&(g_3, h_3) = (0, t_2^3), \; (g_4, h_4) = (0, 0), \; (g_5, h_5) = (0, 0), \\
&g = x^2 + 2 - u * x * t_2 + u^2 * t_2 * t_1, \\
&h = x^2 + 3 * x - 15 + u^2 * t_1^2 + u^3 * t_2^3.
\end{aligned}$$

It is found that $f - g * h = 0$. By setting $u = 1, t_1 = y, t_2 = z$, we find that f is factored as

$$f = (x^2 - x * z + z * y + 2) * (x^2 + 3 * x + z^3 - y^2 - 15).$$

6.3 Solving of Problems in Enumerative Geometry.

Problems of General Enumerative Geometry.

As illustrations of our general method of polynomial equations-solving let us consider the problems of enumerative geometry. Such problems can be formulated in the following general form:

Problem EG. Find number of geometrical entities of certain type verifying some prescribed geometrical conditions.

In general, by introducing suitable coordinates, the geometrical entities in question can be expressed by points in some affine spaces $\mathbf{K}^n(X), X = (x_1, \cdots, x_n), \mathbf{K}$

a field of characteristic 0, with coordinates X satisfying some polynomial equations. The prescribed geometrical conditions in question will also in general be expressed in the form of polynomial equations in the coordinates X, together with eventually some other parameters. In any way, we shall restrict ourselves to Problems EG of above characterization. The solving of such problems leads thus to the solving of some polynomial equations which may then be treated by our general methods of polynomial equations-solving.

Classical Enumerative Geometry as a Chapter of Algebraic Geometry.

The classical enumerative geometry forms a fascinating chapter of modern algebraic geometry. The solving of the corresponding problems is based on the Counting Principle of the theory of algebraic correspondence as described in Sect. 5.2. Using the method to solve the counting problems has some restrictions below.

First, the geometrical entities are restricted to ones in *projective* spaces over algebraically closed fields of characteristic 0, mainly the complex field. The geometrical conditions are also restricted to be of a projective character.

Secondly, the number of solutions to be determined holds only in the *generic* cases and the number has to be counted based on the notion of *multiplicity*. In non-generic cases there is no general way of treatment.

A Simple Example : Circles Touching Three Lines in the Plane.

Our method has no above restrictions. The geometrical entities and conditions may be affine, euclidean, projective, or others, so far they may be expressed in polynomial forms under suitable coordinates. The number of solutions is to be counted not by multiplicities, but in an *exact* way (i.e. with multiplicities not taking into acount). Moreover, all possible cases may be considered, whether *generic* or not. Furthermore, our method will not only give the *number* of solutions, but also the solutions themselves. As illustration let us consider the simple example below.

Example 3.1. Find the number of circles touching three given lines in the ordinary plane.

The example, though a trivial one, is nevertheless quite illuminating. For the study let us take one of the lines L_0 as the X-axis. To avoid unnecessary complications let us suppose that the other two lines L_1, L_2 both intersect L_0 in different points of which one will be taken to be the origin. In coordinates (X, Y) the two lines L_1, L_2 will have then equations of the form

$$\begin{aligned} L_1 &: \ X + v_1 * Y = 0, \\ L_2 &: \ X + v_2 * Y - w_2 = 0, \ w_2 \neq 0. \end{aligned} \tag{3.1}$$

Consider a circle Z touching the line L_0 for which the equation is of the form

$$Z:\ (X-x)^2+(Y-y)^2=y^2. \tag{3.2}$$

For Z to touch L_1, L_2 the conditions will be $PS=0$ where $PS=\{P_1,P_2\}$ with

$$\begin{aligned} P_1 &= (x+v_1*y)^2-(1+v_1^2)*y^2,\\ P_2 &= (x+v_2*y-w_2)^2-(1+v_2^2)*y^2. \end{aligned} \tag{3.3}$$

Let the ordering of variables be taken as $y \prec x$. Then the char-set is $CS=\{C_1,C_2\}$ with

$$\begin{aligned} C_1 =&\ 4*y^4*(v_2-v_1)^2-8*y^3*w_2*(v_2-v_1)*(v_1*v_2+1)\\ &-4*y^2*w_2^2*((v_2-v_1)^2-(v_2*v_1+1))+4*y*w_2^3*(v_2-v_1)-w_2^4,\\ C_2 =&\ 2*x*(y*(v_2-v_1)-w_2)-2*y*w_2*v_2+w_2^2. \end{aligned} \tag{3.4}$$

By Well-Ordering Principle we have then

$$\begin{aligned} &Zero(PS\ /\ w_2)\\ =&\ Zero(CS\ /\ (v_2-v_1)*(y*(v_2-v_1)-w_2)*w_2)\\ &+Zero(PS+\{v_2-v_1\}\ /\ w_2)\\ &+Zero(PS+\{y*(v_2-v_1)-w_2\}\ /\ (v_2-v_1)*w_2). \end{aligned} \tag{3.5}$$

It is readily verified that the last zero-set is empty unless $v_2=-v_1$. We have therefore 3 cases to consider according to $v_2^2\neq v_1^2, v_2=v_1$, or $v_2=-v_1\neq 0$. The zero-sets in these cases are found to be respectively the following ones whose union is $Zero(PS\ /\ w_2)$:

$$\begin{aligned} &Zero(CS\ /\ w_2)\ for\ v_2^2\neq v_1^2,\ v_1,v_2\neq 0,\\ &Zero(CS^1\ /\ w_2)\ for\ v_2=v_1\neq 0,\\ &Zero(CS^2\ /\ w_2)+Zero(CS^3\ /\ w_2),\ for\ v_2=-v_1\neq 0. \end{aligned} \tag{3.6}$$

The char-sets $CS^i=\{C_i^1, C_2^i\}$, $i=1,2,3$ are given respectively as follows.

$$\begin{aligned} C_1^1 &= 4*y^2*(v_1^2+1)-w_2^2,\\ C_2^1 &= 2*x+2*y*v_1-w_2; \end{aligned} \tag{3.7}$$

$$\begin{aligned} C_1^2 &= 2*v_1*y+w_2,\\ C_2^2 &= 4*v_1^2*x^2-4*v_1^2*w_2*x-w_2^2; \end{aligned} \tag{3.8}$$

$$\begin{aligned} C_1^3 &= 4*y^2-4*v_1*w_1*y-w_2^2,\\ C_2^3 &= 2*x-w_2. \end{aligned} \tag{3.9}$$

The geometrical meaning of the various zero-sets are quite evident. In fact, $(3.6)_1$ and $(3.6)_3$ show that in the generic case for which the 3 lines in question form a proper triangle there are 4 circles touching the 3 lines which correspond to the 4 inscribed and escribed circles of the triangle. Remark that when the given lines are all real it is also easily verified that all the 4 touching circles are all real too. On the other hand $(3.6)_2$ shows that there will be only two circles in case L_1, L_2 are parallel non-coincident lines which intersect L_0 in two different points.

Further Examples : Appolonius and Malfatti Problems.

As further illustrations let us cite the two classical problems below.

Appolonius Problem A. Determine the number of circles touching three given circles in the ordinary plane.

Malfatti Problem M. Given a triangle in the ordinary plane determine the number of triples of circles touching each other and each touching two sides of the triangle.

Before giving in details the solving of the above problems let us first make some remarks.

In the real ordinary plane a line or a circle will separate the plane into two parts. A circle tangent to a line will touch the line either in this side or in that side of the line. Similarly, two circles tangent to each other will be either in inner contact or in outer contact. However, if we are dealing with ordinary planes over a non-real field of characteristic 0, e.g. the complex field, then the above notions will be devoid of meaning. In the real case we shall distinguish the two manners of tangency of lines or circles with circles in a different way by introducing the notions of *oriented* lines and circles as follows.

Oriented Coordinates of Oriented Lines.

In what follows we consider the ordinary plane over the field $\mathbf{R}$ with (X, Y) as the running coordinates.

Definition 3.1. The lines

$$L: \ u * X + v * Y = w, \quad L': \ u' * X + v' * Y = w'$$

are *positively equivalent* := There is a non-zero element $k \in \mathbf{R}$ such that

$$u' = k^2 * u, \ v' = k^2 * v, \ w' = k^2 * w.$$

Definition 3.2. An *oriented line* := An equivalence class of lines under positive equivalence.

Remark 3.1. For abus of language an oriented line will be identified to any line in the equivalence class if no confusion can arise.

Definition 3.3. The *oppositely oriented* line of an oriented line $u*X+v*Y=w$:= The oriented line $-u*X-v*Y=-w$.

Notation 3.1. The oppositely oriented line of an oriented line $+L$ (respectively $-L$) := $-L$ (respectively $+L$).

Definition 3.4. The *oriented coordinates* of an oriented line $+L : u*X+v*Y=w := (u:v:w)^+$.

Definition 3.5. The *normal form* of an oriented line := The unique line $u*X+v*Y=w$ in the equivalence class for which $u^2+v^2=1$.

Definition 3.6. The *normalized* coordinates of an oriented line := The oriented coordinates of the line in normal form in the equivalence class.

Remark 3.2. As a simple example, we may attribute the positively oriented X-axis the equation $+Y=0$ or the normalized oriented coordinates $(0:1:0)^+$. Then the oppositely oriented Y-axis will be $-Y=0$ with normalized oriented coordinates $(0:-1:0)^+$.

Oriented Coordinates of Oriented Circles.

Any circle in the real plane with center (x,y) and radius $|r| \geqslant 0$ has an equation of the form

$$(X-x)^2+(Y-y)^2=r^2. \tag{3.10}$$

In case $r \neq 0$ we shall associate to it two *oriented* circles according to whether we take $+r$ or $-r$ from r^2.

Definition 3.7. The triples (a,b,r) and $(a,b,-r)$:= The *oriented coordinates* of the two *opposively oriented* circles associated to the circle (3.9).

Definition 3.8. The *oriented radius* of the oriented circle of oriented coordinates $(a,b,r) := r$.

Notation 3.2. The oppositely oriented circle of an oriented circle $+Z$ (respectively $-Z$) := $-Z$ (respectively $+Z$).

Remark 3.3. The notions of oriented lines and circles may be naturally extended to *oriented planes* and *oriented spheres* in ordinary real space.

Oriented Tangency of Oriented Lines and Oriented Circles.

Definition 3.9. An oriented line with normalized oriented coordinates (u,v,w) is *tangent in the same sense* or *in the opposite sense* to an oriented circle of oriented coordinates (x,y,r) :=

$$\begin{aligned} &u*x+v*y=w+r, \text{ or} \\ &u*x+v*y=w-r. \end{aligned} \tag{3.11}$$

Definition 3.10. Two oriented circles of oriented coordinates (x_1, y_1, r_1), (x_2, y_2, r_2) are *tangent* to each other *in the same sense* or *in the opposite sense* :=

$$\begin{aligned}(x_1-x_2)^2+(y_1-y_2)^2&=(r_1-r_2)^2, \text{ or}\\(x_1-x_2)^2+(y_1-y_2)^2&=(r_1+r_2)^2.\end{aligned} \tag{3.12}$$

Remark 3.4. As a simple example, the oriented circle of oriented coordinates (x, y, y) is tangent to the positively oriented X-axis $+Y = 0$ in the same sense.

Solving of Appolonius Problem.

Example 3.2. Appolonius Problem A.

Let the 3 given circles $Z_i, i = 1, 2, 3$ be oriented with oriented coordinates (a_i, b_i, c_i), and the circle Z to be found be also oriented with oriented coordinates (x, y, r). Let us introduce e_i with $e_i^2 = 1$ to indicate the manner of tangency of Z with Z_i. Then the geometrical conditions that Z touches each of Z_i will be expressed in the form of equations

$$P_i \equiv (x-a_i)^2+(y-b_i)^2-(r-e_i*c_i)^2=0,\ i=1,2,3. \tag{3.13}$$

With a_i, b_i, c_i, e_i considered as parametric constants and variables r, x, y in the ordering $r \prec x \prec y$ the char-set of $PS = \{P_1, P_2, P_3\}$ is readily found to be of the form $CS = \{C_1, C_2, C_3\}$ with index-sets of C_i given as

$$[593\ r\ 2],\ [30\ x\ 1],\ [30\ y\ 1]. \tag{3.14}$$

Let us consider the *generic* case alone for which the initials of C_i are all non-zero. Then for each set of fixed values of triples $(e_1, e_2, e_3) = (\pm1, \pm1, \pm1)$, there are two oriented circles as solutions. The totality of oriented circles touching the 3 given oriented circles in various manners is thus apparently of number 16. However, the pol C_1 is seen to involve r in such a manner that C_1 will remain unchanged if the quadruple (r, e_1, e_2, e_3) is changed to $(-r, -e_1, -e_2, -e_3)$. Hence the actual number of circles touching 3 given circles in the *generic* case is 8. Furthermore, the equation in r is of degree 2 so complex solutions may occur. We leave aside however the question of reality conditions and the exact number of *real* solutions which can be determined in each concrete case.

Solving of Malfatti Problem M.

Example 3.3. Malfatti Problem M.

Let the triangle be $A_1A_2A_3$ and the centers of the 3 circles Z_i to be found be O_i, $i = 1, 2, 3$. It is clear that A_1O_1, A_2O_2, A_3O_3 should be the bisectors of the respective angles of the triangle. There are now two cases to be distinguished according as whether the 3 bisectors are concurrent or not. Let us consider first the case that these bisectors are concurrent at I, which is the incenter or any one of the 3 escribed centers of the triangle.

Let us take the coordinate system so that

$$A_1 = (a_1, 0), \ A_2 = (a_2, 0), \ I = (0, c). \tag{3.15}$$

Then the vertex A_3 is given by

$$A_3 = (a_3, b_3), \text{ where} \tag{3.16}$$

$$a_3 = \frac{c^2 * (a_1 + a_2)}{c^2 + a_1 * a_2}, \quad b_3 = \frac{2 * c * a_1 * a_2}{c^2 + a_1 * a_2}. \tag{3.17}$$

Let the centers O_i of the circles Z_i be given by (x_i, y_i), $i = 1, 2, 3$. Then the coordinates x_i, y_i are connected by equations $P_i = 0$, $i = 1, 2, 3$, where

$$\begin{aligned}
P_1 &= y_1 * a_1 + x_1 * c - c * a_1; \\
P_2 &= y_2 * a_2 + x_2 * c - c * a_2; \\
P_3 &= y_3 * c * (a_2 + a_1) + x_3 * (c^2 - a_2 * a_1) - c^2 * (a_2 + a_1).
\end{aligned} \tag{3.18}$$

On the other hand the radii of Z_i will be $|y_1|, |y_2|, |r_3|$ respectively with

$$r_3 = \frac{c * (a_3 - x_3)}{a_3} = \frac{c * (b_3 - y_3)}{b_3 - c}. \tag{3.19}$$

Now let us orient the circles Z_i in such a way that the oriented circles $+Z_i$ will have respectively the oriented coordinates

$$+Z_1 = (x_1, y_1, +y_1), \ +Z_2 = (x_2, y_2, +y_2), +Z_3 = (x_3, y_3, +r_3). \tag{3.20}$$

The geometrical conditions that $Z_i, i = 1, 2, 3$ will touch each other will be expressed by the equations $P_j = 0$, $j = 4, 5, 6$, where

$$\begin{aligned}
P_4 =& \ 2 * y_2 * y_1 * (e_3 - 1) + (x_2 - x_1)^2; \\
P_5 =& -4 * y_3^2 * c^2 * a_2 * a_1 \\
&+2 * y_3 * y_1 * e_2 * (c^4 - a_2^2 * a_1^2) - 2 * y_3 * y_1 * (c^2 - a_2 * a_1)^2 \\
&+4 * y_3 * c * a_2 * a_1 * (c^2 + a_2 * a_1) - 4 * y_1 * e_2 * c * a_1 * a_2 * (c^2 - a_2 * a_1)^2 \\
&+(x_3 - x_1)^2 * (c^2 - a_2 * a_1)^2 - 4 * c^2 * a_2^2 * a_1^2; \\
P_6 =& -4 * y_3^2 * c^2 * a_2 * a_1 \\
&+2 * y_3 * y_2 * e_1 * (c^4 - a_2^2 * a_1^2) - 2 * y_3 * y_2 * (c^2 - a_2 * a_1)^2 \\
&+4 * y_3 * c * a_2 * a_1 * (c^2 + a_2 * a_1) - 4 * y_2 * e_1 * c * a_1 * a_2 * (c^2 - a_2 * a_1)^2 \\
&+(x_3 - x_2)^2 * (c^2 - a_2 * a_1)^2 - 4 * c^2 * a_2^2 * a_1^2.
\end{aligned} \tag{3.21}$$

In (3.21) we have introduced e_i with $e_i^2 = 1,\ i = 1,2,3$ to indicate the manner of tangency between the circles.

Let the variables be now arranged in the ordering below:

$$x_1 \prec x_2 \prec x_3 \prec y_1 \prec y_2 \prec y_3. \tag{3.22}$$

Set $PS = \{P_1, \cdots, P_6\}$. Let us write e.g. $Zero^{(+-+)}$ for the zero-set with $e_1 = +1, e_2 = -1, e_3 = +1$. Then the package *wsolve* readily gives

$$\begin{aligned}
&Zero^{(s)}(PS) = Zero(CS^{(s)}/IP^{(s)}),\\
&\qquad\qquad \text{for } (s) = (+++), (++-), (+-+), (-++),\\
&Zero^{(s)}(PS) = \textstyle\bigcup_{k=1,2} Zero(CS_k^{(s)}/IP_k^{(s)}),\\
&\qquad\qquad \text{for } (s) = (+--), (-+-), (--+), (---).
\end{aligned} \tag{3.23}$$

In (3.23) IP denotes naturally the corresponding initial-product.

For $(s) = (+++)$ the corresponding char-set is $CS^{(+++)} = \{x_1, x_2, x_3, y_1 - c, y_2 - c, y_3 - c\}$ so that the 3 circles Z_i degenerate into the same in- or es-cribed circle with center I.

For (s) different from $(---)$ and $(+++)$ the solution circles have also some trivialities. For example, for $(s) = (-++)$ there are two solution sets of 3 circles Z_i both with $x_1 = x_2$ so that Z_1, Z_2 touch each other and the X-axis at the same point.

For $(s) = (---)$ the corresponding char-series $CS_1^{(s)}$ and $CS_2^{(s)}$ each has pols of index-sets as shown below:

$$[49\ x_1\ 4], [78\ x_2\ 1], [212\ x_3\ 1], [3\ y_1\ 1], [76\ y_2\ 1], [200\ y_3\ 1]. \tag{3.24}$$

Hence we get for this case of (s) for given in- or es-center I 8 sets of 3 circles Z_i. As there are 4 possibilities for the choice of I, we get in all 32 such sets of 3 circles as solutions. These are the *generic* solutions given by Steiner, cf. e.g. [St],[Pet].

The case that the 3 bisectors are non-concurrent may be treated in the same manner and all possible solution sets of 3 circles, trivial or generic, have also been completely determined. We remark only that this case seems to have been neglected by Steiner and other geometers working on this problem. We leave aside again the reality considerations.

Lines on Cubic Surface in Projective Space.

A cubic surface S_3 in a projective space $P_{\mathbf{K}}^3(Z)$ with $Z = (z_0 : z_1 : z_2 : z_3)$ is given by an equation $S = 0$ with

$$\begin{aligned} S= & a_1 * z_0^3 + a_2 * z_1^3 + a_3 * z_2^3 + a_4 * z_3^3 \\ & + a_5 * z_0^2 * z_1 + a_6 * z_0^2 * z_2 + a_7 * z_0^2 * z_3 \\ & + a_8 * z_0 * z_1^2 + a_9 * z_1^2 * z_2 + a_{10} * z_1^2 * z_3 \\ & + a_{11} * z_0 * z_2^2 + a_{12} * z_1 * z_2^2 + a_{13} * z_2^2 * z_3 \\ & + a_{14} * z_0 * z_3^2 + a_{15} * z_1 * z_3^2 + a_{16} * z_2 * z_3^2 \\ & + a_{17} * z_0 * z_1 * z_2 + a_{18} * z_0 * z_1 * z_3 + a_{19} * z_0 * z_2 * z_3 + a_{20} * z_1 * z_2 * z_3. \end{aligned} \tag{3.25}$$

Example 3.4. Problem. Determine the number of lines lying on the cubic surface S_3 supposed to have only a finite number of lines on it.

It can be shown by means of the Counting Principle of the theory of algebraic correspondence that the number of lines lying on a cubic surface, if not infinite, is always 27, with multiplicities taken into account, see e.g. [VdW2]. We shall apply on the other hand our general method to this problem which permits to determine the *exact* number (i.e. with multiplicities not taking into account) of such lines on an arbitrary cubic surface and also the lines themselves.

For this purpose let us remark that the cubic surface S_3 should not be a cone and should have no plane as a component since otherwise there will be an infinity of lines lying on S_3. This being understood we shall take a coordinate tetrahedron $U_0U_1U_2U_3$ as follows. First we may take a point U_3 on S_3 such that the tangent plane T_3 at U_3 will intersect S_3 in a non-degenerate cubic curve C_3. Next we may take a point U_2 on C_3 such that the tangent plane T_2 of S_3 at U_2 is different from T_3 which intersects C_3 in a further point U_1 besides U_2. The tangent plane T_1 of S_3 at U_1 can contain neither U_2 nor U_3 since otherwise the line U_1U_2 (or U_1U_3) will meet the cubic surface S_3 in 4 points, two at U_1, and two at U_2 (or U_3), which is absurd. The tangent plane T_1 of S_3 at U_1 will intersect then T_2 in a line which will intersect S_3 in 3 points, two being at U_1, and the third one will be taken to be U_0.

From the construction T_3 is the plane $U_1U_2U_3$, T_2 the plane $U_0U_1U_2$ and T_1 contains the line U_0U_1. It follows respectively

$$a_{15} = a_{16} = 0, \ a_{11} = a_{12} = 0, \ \text{and } a_8 = 0. \tag{3.26}$$

Consider now an arbitrary line L with Pluecker coordinates $(p_{01} : p_{02} : p_{03} : p_{12} : p_{13} : p_{23})$. The line L, if not in any one of the coordinate planes $z_i = 0, i = 0, 1, 2, 3$, will intersect these planes in the points $P_0 = (0 : p_{01} : p_{02} : p_{03})$, $P_1 = (-p_{01} : 0 : p_{12} : p_{13})$, $P_2 = (-p_{02} : -p_{12} : 0 : p_{23})$, $P_3 = (-p_{03} : -p_{13} : -p_{23} : 0)$. We shall suppose that L lies wholly on S_3. By construction the line U_2U_3 is not on S_3 so that $p_{01} \neq 0$. Then L is on S_3 if and only if for any u the point $Q_u = P_0 + u.P_1$ is on S_3. It follows that for L lying wholly on S_3 it is necessary and sufficient that $P_{ui} = 0, i = 0,1,2,3$ with the pols P_{ui} given below:

$$\begin{aligned}
P_{u0} = & p_{03} * p_{02}^2 * a_{13} + p_{03} * p_{02} * p_{01} * a_{20} \\
& + p_{03} * p_{01}^2 * a_{10} + p_{02} * p_{01}^2 * a_9, \\
P_{u1} = & p_{13} * p_{02}^2 * a_{13} + p_{13} * p_{02} * p_{01} * a_{20} \\
& + p_{13} * p_{01}^2 * a_{10} + 2 * p_{12} * p_{03} * p_{02} * a_{13} \\
& + p_{12} * p_{03} * p_{01} * a_{20} + p_{12} * p_{01}^2 * a_9 \\
& - p_{03}^2 * p_{01} * a_{14} - p_{03} * p_{02} * p_{01} * a_{19} \\
& - p_{03} * p_{01}^2 * a_{18} - p_{02} * p_{01}^2 * a_{17}, \\
P_{u2} = & 2 * p_{13} * p_{12} * p_{02} * a_{13} + p_{13} * p_{12} * p_{01} * a_{20} \\
& - 2 * p_{13} * p_{03} * p_{01} * a_{14} - p_{13} * p_{02} * p_{01} * a_{19} \\
& - p_{13} * p_{01}^2 * a_{18} + p_{12}^2 * p_{03} * a_{13} \\
& - p_{12} * p_{03} * p_{01} * a_{19} - p_{12} * p_{01}^2 * a_{17} \\
& + p_{03} * p_{01}^2 * a_7 + p_{02} * p_{01}^2 * a_6 + p_{01}^3 * a_5, \\
P_{u3} = & - p_{13}^2 * p_{01} * a_{14} + p_{13} * p_{12}^2 * a_{13} \\
& - p_{13} * p_{12} * p_{01} * a_{19} + p_{13} * p_{01}^2 * a_7 \\
& + p_{12} * p_{01}^2 * a_6.
\end{aligned}$$

Besides we have the relation $PL = 0$ for the Pluecker coordinates where

$$PL = p_{23} * p_{01} - p_{13} * p_{02} + p_{12} * p_{03}. \tag{3.28}$$

We have then a polset $PS = \{P_{u0}, P_{u1}, P_{u2}, P_{u3}, PL\}$ and the lines on S_3 will be given by the set $Zero(PS)$.

Let us arrange the variables $p_{ij}, i < j$ in the order

$$p_{01} \prec p_{02} \prec p_{03} \prec p_{12} \prec p_{13} \prec p_{23}.$$

Then the char-set CS is found to be consisting of 5 pols $C_i,\ i = 1, \cdots, 5$ with C_1 irreducible and with index-sets given below:

$$[15895\ p_{02}\ 27],\ [4\ p_{03}\ 1],\ [1245\ p_{12}\ 1], [10\ p_{13}\ 1],\ [3\ p_{23}\ 1].$$

The initials are $I_i, i = 1, \cdots, 5$ with

$$I_1 = a_5^2 * a_6 * a_9 * a_{13} * a_{15}.$$

It follows that in the generic case for which $I_i \neq 0,\ i = 1, \cdots, 5$ and discriminant of C_1 w.r.t. p_{02} not equal to 0 too the number of lines on S_3 will be exactly 27 and the lines themselves are given by $Zero(PS)$. The non-generic case may also be studied in the same way if we like by setting some of I_i and/or the discriminant of C_1 to be 0.

6.4 Central Configurations in Planet Motions and Vortex Motions

Central Configurations in Celestial Mechanics.

Notation 4.1.

$J_n = \{1, \cdots, n\}$;

$J_n^2 = \{(i, j) \mid i, j \in J_n, i \neq j\}$;

$m_1, \cdots, m_n :=$ Masses of n particles moving under mutual Newtonian gravitational attractions.

$\mathbf{r}_1, \cdots, \mathbf{r}_n :=$ Positions of these masses at a certain moment, with $\mathbf{r}_i \neq \mathbf{r}_j$ for $i \neq j$.

$\begin{bmatrix} m_1 & m_2 & \cdots & m_n \\ \mathbf{r}_1 & \mathbf{r}_2 & \cdots & \mathbf{r}_n \end{bmatrix}$:= A *configuration* formed of masses m_i at positions $\mathbf{r}_i$, $i \in J_n$.

Definition 4.1. The configuration $\begin{bmatrix} m_1 & m_2 & \cdots & m_n \\ \mathbf{r}_1 & \mathbf{r}_2 & \cdots & \mathbf{r}_n \end{bmatrix}$ is a *rigid configuration* (respectively a *central configuration*) w.r.t. the masses $m_1, \cdots, m_n :=$ There are initial velocities of the masses m_i such that under the Newtonian gravitational attractions the configurations formed by the masses during the motion will remain *congruent* (respectively *similar*) to the initial one.

As a direct consequence of Newtonian mechanics for a rigid or a central configuration $\begin{bmatrix} m_1 & m_2 & \cdots & m_n \\ \mathbf{r}_1 & \mathbf{r}_2 & \cdots & \mathbf{r}_n \end{bmatrix}$ the center of mass of the masses $m_1, \cdots, m_n$ may be considered to be fixed during the motion.

Definition 4.2. An *inertial coordinate system* associated to a rigid or a central configuration $\begin{bmatrix} m_1 & m_2 & \cdots & m_n \\ \mathbf{r}_1 & \mathbf{r}_2 & \cdots & \mathbf{r}_n \end{bmatrix}$:= A cartesian coordinate system for which the origin is at the fixed center of mass of the masses $m_1, \cdots, m_n$.

Again as a direct consequence of Newtonian mechanics, we have the following proposition:

Proposition 4.1. For a rigid or a central configuration $\begin{bmatrix} m_1 & m_2 & \cdots & m_n \\ \mathbf{r}_1 & \mathbf{r}_2 & \cdots & \mathbf{r}_n \end{bmatrix}$

we have in any associated inertial coordinate system

$$\Sigma_i\ m_i * \mathbf{r}_i = 0. \tag{4.1}$$

Definition 4.3. Two rigid configurations w.r.t. same masses $m_1, \cdots, m_n$ belong to the same *configuration class* := They have same center of mass and there exists an orientation-preserving similarity transformation which keeps the center of mass fixed and transforms one configuration to the other.

Some Historical Account about Central Configurations.

Clearly a rigid configuration is also a central configuration with same set of masses. The converse is known to be true which we omit however the proof. See e.g. [Wint]. It follows that a central configuration determines a class of rigid configurations and vice versa.

In what follows only inertial coordinate systems are considered.

Notation 4.2.

$q_3(n) = q_3(n; m_1, \cdots, m_n)$:= Number of classes of rigid configurations with given masses $m_1, \cdots, m_n$.

$q_2(n) = q_2(n; m_1, \cdots, m_n)$:= Number of classes of rigid configurations with given masses $m_1, \cdots, m_n$ for which the masses m_i are situated in the same plane.

$q_1(n) = q_1(n; m_1, \cdots, m_n)$:= Number of classes of rigid configurations with given masses $m_1, \cdots, m_n$ for which the masses m_i are situated in the same line.

Remark 4.1. Our notations differ from those of Wintner in his book [Wint] in that q_2, q_3 of that book correspond to $q_2 - q_1, q_3 - q_2$ here.

Results known. For given masses $m_1, \cdots, m_n$ we have:

$q_1(2) = q_2(2) = q_3(2) = 1$;

$q_1(3) = 3$ (Euler 1767);

$q_2(3) = q_3(3) = 5$ (Lagrange 1772);

$q_1(n) = \dfrac{n!}{2}$ (Moulton 1910, [Mou]);

$q_3(n) > q_2(n)$ for each $n > 4$ with some particular sets of $m_1, \cdots, m_n$ (Waldvogel 1972, [Wald]).

It is clear that

$$q_1(n; m_1, \cdots, m_n) \leqslant q_2(n; m_1, \cdots, m_n) \leqslant q_3(n; m_1, \cdots, m_n). \tag{4.2}$$

It is known that $q_1(n; m_1, \cdots, m_n)$ is finite for all n and masses $m_1, \cdots, m_n$ owing to the results of Moulton.

Wintner Conjecture. $q_3(n; m_1, \cdots, m_n)$ and hence also $q_2(n; m_1, \cdots, m_n)$ are finite for all n and masses $m_1, \cdots, m_n$.

In recent years Smale and his followers have studied central configurations with topological methods via Morse critical point theory. See e.g. [Sm1,2]. It seems that Wintner's conjecture remains open. In this section we shall restrict ourselves to the actual determination of rigid or central configurations in treating it as a problem of polynomial equations-solving. For this purpose we shall adopt an inertial coordinate system associated to the rigid or central configuration $\begin{bmatrix} m_1 & m_2 & \cdots & m_n \\ \mathbf{r}_1 & \mathbf{r}_2 & \cdots & \mathbf{r}_n \end{bmatrix}$ and the following notations will be used:

Notation 4.3.

$\mathbf{r}_i = (x_i, y_i), i \in J_n;$

$x_{ij} = x_i - x_j,\ y_{ij} = y_i - y_j,\ (i,j) \in J_n^2;$

$r_{ij} = \sqrt{x_{ij}^2 + y_{ij}^2},\ (i,j) \in J_n^2.$

Fundamental Equations for Rigid Configurations.

For a rigid configuration $\begin{bmatrix} m_1 & m_2 & \cdots & m_n \\ \mathbf{r}_1 & \mathbf{r}_2 & \cdots & \mathbf{r}_n \end{bmatrix}$ in a plane it is known that during the motion the plane will be fixed to keep the configuration congruent to the original one. Take now planar inertial coordinate system (x, y) with origin O at the center of mass. Then the configuration will move around O with an angular velocity ω such that

$$\left.\begin{aligned} \omega^2 * x_i &= \Sigma_{j \neq i}\, m_j * \frac{x_{ij}}{r_{ij}^3}, \\ \omega^2 * y_i &= \Sigma_{j \neq i}\, m_j * \frac{y_{ij}}{r_{ij}^3}, \end{aligned}\right\} \quad i \in J_n. \qquad (4.3)_n$$

Up to non-zero constant factors the right-hand sides represent the total Newtonian attractive forces exerted on the masses m_i by all the other masses, while the left-hand side represent the corresponding centrifugal forces. The equations (4.3) show in particular that ω is a constant. Furthermore, the equations are invariant under orientation-preserving similarity transformations with origin keeping fixed. However the constant ω may then be changed.

We have assumed $r_{ij} \neq 0, (i,j) \in J_n^2$, which means that collisions are out of consideration. The determination of rigid configurations and hence also central configurations amounts then to the solving of polynomial equations obtained from (4.3) by clearing of denominators. To achieve this we shall introduce some preliminary transformations of (4.3) to make easier the solving.

Let us take the complex number field $\mathbf{C}$ as the basic field and introduce complex variables u_i, v_i, z_{ij}, w_{ij} as follows ($\mathbf{i}$ stands for $\sqrt{-1}$).

$$\begin{aligned} u_i &= x_i + \mathbf{i} * y_i,\ v_i = x_i - \mathbf{i} * y_i,\ i \in J_n, \\ z_{ij} &= \frac{1}{(u_i - u_j) * r_{ij}},\ w_{ij} = \frac{1}{(v_i - v_j) * r_{ij}},\ (i,j) \in J_n^2, \\ m_0 &= -\,\omega^2. \end{aligned} \tag{4.4}_n$$

From (4.3) we have now the following set of polynomial equations:

$$\left.\begin{aligned} m_0 * u_i + \Sigma_{j\neq i}\ m_j * w_{ij} = 0, \\ m_0 * v_i + \Sigma_{j\neq i}\ m_j * z_{ij} = 0. \end{aligned}\right. \quad i \in J_n; \tag{4.5}_n$$

$$\left.\begin{aligned} &(u_i - u_j) * z_{ij} * r_{ij} - 1 = 0, \\ &(v_i - v_j) * w_{ij} * r_{ij} - 1 = 0, \\ &r_{ij}^2 - (u_i - u_j) * (v_i - v_j) = 0. \end{aligned}\right. \quad (i,j) \in J_n^2; \tag{4.6}_n$$

$$r_{ij} - r_{ji} = 0,\ z_{ij} + z_{ji} = 0,\ w_{ij} + w_{ji} = 0,\ \ (i,j) \in J_n^2. \tag{4.7}_n$$

To these we may also add

$$\begin{aligned} \Sigma_i\ m_i * u_i = 0, \\ \Sigma_i\ m_i * v_i = 0, \end{aligned} \tag{4.8}_n$$

which follows from (4.1) or (4.3).

Remark 4.2. In the above equations Σ is to be extended either on $i \in J_n$ or $j \in J_n$ as is evident from the context.

Remark 4.3. In the above equations u_i, v_i, etc. should satisfy the following conditions to meet the reality of the situation:

$m_0, m_1, \cdots, m_n$ are all real numbers, with $m_0 < 0, m_i > 0,\ i \in J_n$.

All $r_{ij}, (i,j) \in J_n^2$, are real and positive.

u_i, v_i for $i \in J_n$, are conjugate complex numbers, $\mathbf{i}$ being $\sqrt{-1}$.

$z_{ij} \neq 0, w_{ij} \neq 0$, for $(i,j) \in J_n^2$.

Definition 4.4. For a rigid or central configuration $\begin{bmatrix} m_1 & m_2 & \cdots & m_n \\ \mathbf{r}_1 & \mathbf{r}_2 & \cdots & \mathbf{r}_n \end{bmatrix}$ *reality conditions* := The conditions in Remark 4.3.

Determination of Rigid or Central Configurations for $n = 3$.

The determination of rigid or central configurations for $n = 2$ is trivial. Let us proceed to the case $n = 3$ which amounts to the problem in the example below.

Example 4.1. To solve the system of equations $(4.5)_3$-$(4.8)_3$ for the variables $u_i, v_i,\ m_0, r_{ij}$ with $i \in J_3, (i,j) \in J_3^2$ in terms of the parameters $m_i,\ i \in J_3$.

For the solving let us first replace the variables u_2, u_3, v_2, v_3 by $u_{12}, u_{13}, v_{12}, v_{13}$ in setting

$$\begin{aligned} u_{12} &= u_1 - u_2,\ u_{13} = u_1 - u_3, \\ v_{12} &= v_1 - v_2,\ v_{13} = v_1 - v_3, \end{aligned} \tag{4.9}$$

so that

$$u_2 - u_3 = u_{13} - u_{12}, \quad v_2 - v_3 = v_{13} - v_{12}. \tag{4.10}$$

To solve $(4.5)_3$-$(4.8)_3$ is then easily seen to be equivalent to the solving of the system $QS = 0$ below for variables $u_1, u_{12}, u_{13}, v_1, v_{12}, v_{13}, m_0, r_{12}, r_{13}, r_{23}$ in terms of known parameters m_1, m_2, m_3. Here $QS = \{Q_1, \cdots, Q_9\}$ with

$$\begin{aligned}
Q_1 =& u_1 * (m_3 + m_2 + m_1) - u_{13} * m_3 - u_{12} * m_2, \\
Q_2 =& (v_{13} - v_{12}) * v_{13} * r_{23} * r_{13} * (r_{12}^3 * m_0 + m_2 + m_1) \\
& + v_{13} * v_{12} * r_{12} * m_3 * (r_{23} - r_{13}) - v_{12}^2 * r_{23} * r_{12} * m_3, \\
Q_3 =& v_{13}^2 * r_{23} * r_{13} * m_2 + v_{13} * v_{12} * r_{23} * r_{13} * (r_{13}^2 * r_{12} * m_0 - m_2) \\
& + v_{13} * v_{12} * r_{12} * (r_{23} * m_3 + r_{23} * m_1 + r_{13} * m_2) \\
& - v_{12}^2 * r_{23} * r_{12} * (r_{13}^3 * m_0 + m_3 + m_1), \\
Q_4 =& v_1 * u_{13} * u_{12} * r_{13} * r_{12} * m_0 + u_{13} * r_{13} * m_2 + u_{12} * r_{12} * m_3, \\
Q_5 =& (u_{13} - u_{12}) * u_{13} * r_{23} * r_{13} * (r_{12}^3 * m_0 + m_2 + m_1) \\
& + u_{13} * u_{12} * r_{12} * m_3 * (r_{23} - r_{13}) - u_{12}^2 * r_{23} * r_{12} * m_3, \\
Q_6 =& u_{13}^2 * r_{23} * r_{13} * m_2 + u_{13} * u_{12} * r_{23} * r_{13} * (r_{13}^2 * r_{12} * m_0 - m_2) \\
& + u_{13} * u_{12} * r_{12} * (r_{23} * m_3 + r_{23} * m_1 + r_{13} * m_2) \\
& - u_{12}^2 * r_{23} * r_{12} * (r_{13}^3 * m_0 + m_3 + m_1), \\
Q_7 =& v_{12} * u_{12} - r_{12}^2, \quad Q_8 = v_{13} * u_{13} - r_{13}^2, \\
Q_9 =& (v_{13} - v_{12}) * (u_{13} - u_{12}) - r_{23}^2.
\end{aligned} \tag{4.11}$$

Remark that the reality conditions have to be satisfied. Moreover $u_{12}, u_{13}, v_{12}, v_{13}$, $r_{12}, r_{13}, r_{23}, m_0, m_1, m_2, m_3$ are all non-zero. Let NZ be the product of all these non-zero variables and parameters. Then the problem is reduced to the determination of

$$Zero_{(rc)}(QS/NZ) \subset Zero_{\mathbf{C}}(QS/NZ), \tag{4.12}$$

in which (rc) means that the zeros are restricted to those for which the reality conditions are satisfied.

To determine (4.12) let us first arrange the variables and parameters involved in the following order:

$$m_1 \prec m_2 \prec m_3 \prec r_{12} \prec r_{13} \prec r_{23} \prec m_0 \prec u_{12} \prec u_{13} \prec v_{12} \prec v_{13} \prec u_1 \prec v_1. \tag{4.13}$$

Let us apply now the Replacement Rules in replacing Q_2, Q_3 by the remainder Q_{10} of Q_2 w.r.t. Q_3 and the resultant Q_{11} of Q_2 and Q_3 w.r.t. v_{13}. The resultant Q_{11} has an index set [57 m_0 3] while Q_{10} is of the form

$$Q_{10} = v_{13} * g_0 + v_{12} * h_0, \tag{4.14}$$

where g_0, h_0 have index sets [12 m_0 2] and [8 m_0 2] respectively. Similarly, let us replace Q_5, Q_6 by the remainder Q_{12} of Q_5 w.r.t. Q_6 and the resultant Q_{13} of Q_5 and Q_6 w.r.t. u_{13}. It turns out that Q_{13} is the same as Q_{11} so that it may be removed while Q_{12} is of the form

$$Q_{12} = u_{13} * g_0 + u_{12} * h_0, \tag{4.15}$$

Forming remainders of Q_8 w.r.t. Q_{10}, Q_{12}, Q_7 in eliminating v_{13}, u_{13}, v_{12} successively we get a pol Q_{14} which may be factored as

$$\begin{aligned} Q_{14} &= f_a * f_b, \\ f_a &= -r_{13} * g_0 + r_{12} * h_0, \\ f_b &= +r_{13} * g_0 + r_{12} * h_0. \end{aligned} \tag{4.16}$$

It follows from $Q_{14} = 0$ that either $f_a = 0$ or $f_b = 0$. From the remainders of Q_9 w.r.t. Q_8, Q_7, Q_{10}, Q_{12} in succession we get a pol Q_{15} of the form

$$Q_{15} = (-r_{23}^2 + r_{13}^2 + r_{12}^2) * g_0 + 2 * r_{12}^2 * h_0. \tag{4.17}$$

Consider first the case $f_a = 0$. Then after reduction it turns out that Q_{15} will be factored into three factors $f_{ai},\ i = 1, 2, 3$ as shown below:

$$\begin{aligned} f_{a1} &= g_0, \\ f_{a2} &= r_{23} - r_{13} - r_{12}, \\ f_{a3} &= r_{23} + r_{13} + r_{12}. \end{aligned} \tag{4.18}$$

Consider next the case $f_b = 0$. Then Q_{15} after reduction will be factored into factors $f_{bi},\ i = 1, 2, 3$ with

$$\begin{aligned} f_{b1} &= g_0, \\ f_{b2} &= r_{23} - r_{13} + r_{12}, \\ f_{b3} &= r_{23} + r_{13} - r_{12}. \end{aligned} \tag{4.19}$$

From the above it follows easily that

$$Zero_{(rc)}(QS/NZ) = \bigcup_k Zero_{(rc)}(AS_k/NZ) + \bigcup_k Zero_{(rc)}(BS_k/NZ). \tag{4.20}$$

In (4.20) the union is to be extended over $k = 1, 2, 3$ and

$$\begin{aligned} AS_k &= QS + \{Q_{10}, Q_{11}, Q_{12}\} + \{f_a, f_{ak}\}, \\ BS_k &= QS + \{Q_{10}, Q_{11}, Q_{12}\} + \{f_b, f_{bk}\}, \end{aligned} \quad k = 1, 2, 3. \tag{4.21}$$

Remark that owing to $u_{12} \neq 0, v_{12} \neq 0$, the sets $\{f_a, f_{a1}\}$ and $\{f_b, f_{b1}\}$ are by (4.15), $(4.18)_1, (4.19)_1$ both 0-equivalent to $\{g_0, h_0\}$. Owing to the reality conditions we have also $f_{a3} \neq 0$ so that

$$ZA_3 = Zero_{(rc)}(AS_3/NZ) = \varnothing. \tag{4.22}$$

On the other hand by our general method we have

$$\begin{aligned} ZA_1 &= Zero_{(rc)}(AS_1/NZ) = Zero_{(rc)}(CS_{a1}/NZ), \\ ZA_2 &= Zero_{(rc)}(AS_2/NZ) = Zero_{(rc)}(CS_{a2}/NZ). \end{aligned} \tag{4.23}$$

In (4.23) $CS_{ak} = \{C_{ak1}, \cdots, C_{ak8}\}, k = 1, 2$ are triangular sets or asc-sets with pols $C_{aki}, i = 1, \cdots, 8$ given below:

$$\begin{aligned} C_{a11} &= r_{13} - r_{12}, \\ C_{a12} &= r_{23} - r_{12}, \\ C_{a13} &= m_0 * r_{12}^3 + (m_3 + m_2 + m_1), \\ C_{a14} &= u_{13}^2 - u_{13} * u_{12} + u_{12}^2, \\ C_{a15} &= v_{12} * u_{12} - r_{12}^2, \\ C_{a16} &= v_{13} * u_{13} - r_{12}^2, \\ C_{a17} &= u_1 * (m_3 + m_2 + m_1) - u_{13} * m_3 - u_{12} * m_2, \\ C_{a18} &= v_1 * u_{13} * u_{12} * r_{12} * m_0 + u_{13} * m_2 + u_{12} * m_3; \end{aligned} \tag{4.24}$$

$$\begin{aligned} C_{a21} &= r_{13}^5 * (m_2 + m_1) + r_{13}^4 * r_{12} * (3 * m_2 + 2 * m_1) \\ &\quad + r_{13}^3 * r_{12}^2 * (3 * m_2 + m_1) - r_{13}^2 * r_{12}^3 * (3 * m_3 + m_1) \\ &\quad - r_{13} * r_{12}^4 * (3 * m_3 + 2 * m_1) - r_{12}^5 * (m_3 + m_1), \\ C_{a22} &= r_{23} - r_{13} - r_{12}, \\ C_{a23} &= m_0 * r_{13}^2 * r_{12}^3 * (r_{13} + r_{12})^2 \\ &\quad + r_{13}^2 * (r_{13} + r_{12})^2 * (m_2 + m_1) - (2 * r_{13} + r_{12}) * r_{12}^3 * m_3, \\ C_{a24} &= u_{13} * r_{12} + u_{12} * r_{13}, \end{aligned} \tag{4.25}$$

and C_{a2k} same as C_{a1k}, for $k = 5, 6, 7, 8$. From C_{a11}, C_{a12} it is readily seen that ZA_1 gives the central configuration of Lagrange while ZA_2 gives that of Euler for which the particles m_1, m_2, m_3 are collinear with m_1 lying between m_2 and m_3. In particular, C_{a21} gives a quintic equation in $\dfrac{r_{13}}{r_{12}}$ which has one and only one positive real solution owing to the reality conditions as originally given by Euler.

The sets $ZB_1 = Zero_{(rc)}(BS_1/NZ)$ coincides with $ZA_1 = Zero_{(rc)}(AS_1/NZ)$ and the sets $ZB_k = Zero_{(rc)}(BS_k/NZ)$, $k = 2, 3$ are similar to $ZA_2 = Zero_{(rc)}$ (AS_2/NZ) which correspond to the other two collinear central configurations of Euler.

The set of all possible central configurations in case of $n = 3$ is thus completely determined as formulated in the form of following theorem:

Theorem 4.1. The only possible central configurations of three non-vanishing masses are the classical ones due to Euler and Lagrange but no others.

Rigid Configurations in Vortex Motions.

Consider now vortex movements in an incompressible and nonviscous fluid extending to infinity.

Notation 4.4.

$F_1, \cdots, F_n :=$ n parallel rectilinear vortex filaments moving under their own influences.

$k_1, \cdots, k_n :=$ Strengths of the n vortex filaments F_i.

$P :=$ A fixed plane perpendicular to all the n filaments F_i with a coordinate system (x, y).

$\mathbf{r}_1, \cdots, \mathbf{r}_n :=$ Traces of vortex filaments F_i on the plane P, or vectors from the origin O to that trace.

$\mathbf{e} :=$ Unit vector orthogonal to the fixed plane P.

$r_{ij} :=$ Distance between the parallel filaments F_i and F_j.

$\begin{bmatrix} k_1 & k_2 & \cdots & k_n \\ \mathbf{r}_1 & \mathbf{r}_2 & \cdots & \mathbf{r}_n \end{bmatrix} :=$ A configuration of n parallel vortex filaments F_i with strengths k_i and traces $\mathbf{r}_i$ on the fixed plane P.

Hypothesis K.

$$\Sigma_i\ k_i \neq 0. \tag{4.26}$$

Under Hypothesis K it is known that there will exist a *center* in the finite part of the plane P such that, if the center is taken as the origin O, then we will have

$$\Sigma_i\ k_i * \mathbf{r}_i = 0, \text{ or}$$

$$\begin{aligned} \Sigma_i\ k_i * x_i &= 0, \\ \Sigma_i\ k_i * y_i &= 0. \end{aligned} \tag{4.27}$$

Consider now a vortex configuration $\begin{bmatrix} k_1 & k_2 & \cdots & k_n \\ \mathbf{r}_1 & \mathbf{r}_2 & \cdots & \mathbf{r}_n \end{bmatrix}$.

Definition 4.5.
$\begin{bmatrix} k_1 & k_2 & \cdots & k_n \\ \mathbf{r}_1 & \mathbf{r}_2 & \cdots & \mathbf{r}_n \end{bmatrix}$ is a *fixed configuration* := The configuration remains fixed during the motion under its own influences.

Definition 4.6.
$\begin{bmatrix} k_1 & k_2 & \cdots & k_n \\ \mathbf{r}_1 & \mathbf{r}_2 & \cdots & \mathbf{r}_n \end{bmatrix}$ is a *rigid configuration* := The configuration remains congruent to the original one during the motion under its own influences.

Remark 4.4. There is no analogue of fixed configurations for particles moving under mutual Newtonian attractions.

Some Elementary Properties of Rigid Vortex Configurations.

Below $\begin{bmatrix} k_1 & k_2 & \cdots & k_n \\ \mathbf{r}_1 & \mathbf{r}_2 & \cdots & \mathbf{r}_n \end{bmatrix}$ will be a vortex configuration verifying Hypothesis K so that the center of the system exists which is taken to be the origin of the coordinate system.

Proposition 4.1. For a rigid vortex configuration which verifies Hypothesis K the whole configuration will rotate about the fixed center with a uniform angular velocity.

Proof. Let the origin O be at the fixed center. Let us denote by $\dfrac{c(t)}{2*\pi}$ the angular velocity about O of the whole rigid configuration at time t. Then the velocity $\mathbf{v}_i(t)$ of i-th filament at time t will be

$$\mathbf{v}_i(t) = \frac{c(t)}{2*\pi} * (\mathbf{e} \times \mathbf{r}_i(t)).$$

Now the velocity of filament F_i due to filament F_j at time t is given by

$$\mathbf{v}_{ji}(t) = \frac{1}{2*\pi} * k_j * \frac{\mathbf{e} \times (\mathbf{r}_i(t) - \mathbf{r}_j(t))}{r_{ij}^2}.$$

Hence the velocity of F_i at time t is

$$\mathbf{v}_i(t) = \frac{1}{2*\pi} * \Sigma_{j\neq i}\, k_j * \frac{\mathbf{e} \times (\mathbf{r}_i(t) - \mathbf{r}_j(t))}{r_{ij}^2}.$$

Comparing the two expressions of $\mathbf{v}_i(t)$, we get

$$\Sigma_{j\neq i}\, k_j * \frac{\mathbf{r}_i(t) - \mathbf{r}_j(t)}{r_{ij}^2} \quad = \quad c(t) * \mathbf{r}_i(t).$$

As $\mathbf{r}_i(t)$ differs from $\mathbf{r}_i(0) = \mathbf{r}_i$ by a rotation independent of i, we see that $c(t)$ is independent of t. Hence $\frac{c(t)}{2*\pi} = \frac{c}{2*\pi}$ is the uniform angular velocity of the whole rigid configuration as to be proved. □

Remark 4.5. From the proof we have

$$\Sigma_{j\neq i}\ k_j * \frac{\mathbf{r}_i - \mathbf{r}_j}{r_{ij}^2} \ = \ c * \mathbf{r}_i. \tag{4.28}$$

Moreover we see that $c = 0$ corresponds to the case of a fixed vortex configuration.

Proposition 4.2. For a fixed vortex configuration it is necessary that

$$\Sigma_{i\neq j}\ k_i * k_j \ = \ 0. \tag{4.29}$$

Proof. From (4.28) we get

$$c * \Sigma_i\ k_i * \mathbf{r}_i^2 = \Sigma_{i\neq j}\ k_i * k_j.$$

As $c = 0$ for a fixed vortex configuration, we get (4.29). □

Fundamental Equations of Rigid Vortex Configurations.

From (4.28) we get

$$\left.\begin{aligned} \Sigma_{j\neq i}\ k_j * \frac{x_{ij}}{r_{ij}^2} &= c * x_i, \\ \Sigma_{j\neq i}\ k_j * \frac{y_{ij}}{r_{ij}^2} &= c * y_i. \end{aligned}\right\} \quad i \in J_n. \tag*{$(4.30)_n$}$$

Let us introduce u_i, v_i, z_{ij}, w_{ij} as given below:

$$\begin{aligned} &u_i = x_i + \mathbf{i} * y_i,\ v_i = x_i - \mathbf{i} * y_i,\ i \in J_n, \\ &z_{ij} = \frac{1}{u_i - u_j},\ w_{ij} = \frac{1}{v_i - v_j},\ (i,j) \in J_n^2, \\ &k_0 \ = \ -c. \end{aligned} \tag*{$(4.31)_n$}$$

Then (4.30) will be transformed into the equations below:

$$\left.\begin{aligned} k_0 * u_i + \Sigma_{j\neq i}\ k_j * w_{ij} = 0, \\ k_0 * v_i + \Sigma_{j\neq i}\ k_j * z_{ij} = 0, \end{aligned}\right. \quad i \in J_n; \tag*{$(4.32)_n$}$$

$$\left.\begin{aligned} &(u_i - u_j) * z_{ij} - 1 = 0, \\ &(v_i - v_j) * w_{ij} - 1 = 0, \\ &r_{ij}^2 - (u_i - u_j) * (v_i - v_j) = 0, \end{aligned}\right. \quad (i,j) \in J_n^2; \tag*{$(4.33)_n$}$$

$$r_{ij} - r_{ji} = 0,\ z_{ij} + z_{ji} = 0,\ w_{ij} + w_{ji} = 0,\ \ (i,j) \in J_n^2. \tag*{$(4.34)_n$}$$

To these we may also add

$$\begin{aligned}\Sigma_i\ k_i * u_i &= 0,\\ \Sigma_i\ k_i * v_i &= 0.\end{aligned} \qquad (4.35)_n$$

The variables and parameters involved in $(4.32)_n - (4.35)_n$ are subject to some *reality conditions*, viz.

$k_1, \cdots, k_n$ are real and non-zero, and k_0 is real.

u_i, v_i, are complex conjugates for $i \in J_n$.

r_{ij} are real and positive, for $(i,j) \in J_n^2$.

z_{ij}, w_{ij} are all non-zero, for $(i,j) \in J_n^2$.

r is actually equal to 1.

Determination of Rigid Vortex Configurations in Case of $n = 3$.

In comparing $(4.31)_n$-$(4.35)_n$ with $(4.3)_n$-$(4.8)_n$ we see that the determination of rigid vortex configurations will be analogous to that of rigid planet configurations. Consider in particular the case $n = 3$ or the problem below:

Example 4.2. To solve the system of equations $(4.32)_3$-$(4.35)_3$ for variables u_i, v_i, k_0, r_{ij} with $i \in J_3$, $(i,j) \in J_3^2$ in terms of the parameters k_i, $i \in J_3$.

For fixed configurations we have $k_0 = -c = 0$ and the solving is quite easy. So in what follows we shall consider only the case $k_0 \neq 0$.

For this purpose let us replace as before u_2, u_3, v_2, v_3 by $u_{12}, u_{13}, v_{12}, v_{13}$ by means of (4.9), (4.10). Then the solving of $(4.32)_3 - (4.35)_3$ is equivalent to the solving of $QS = 0$ under the relevant reality conditions where $QS = \{Q_1, \cdots, Q_9\}$ with Q_i given below:

$$\begin{aligned}
Q_1 &= u_1 * (k_3 + k_2 + k_1) - u_{13} * k_3 - u_{12} * k_2,\\
Q_2 &= (v_{13} - v_{12}) * v_{13} * (k_0 * r_{12}^2 + k_2 + k_1) - v_{12}^2 * k_3,\\
Q_3 &= v_{13}^2 * k_2 + (v_{13} - v_{12}) * v_{12} * (k_0 * r_{13}^2 + k_3 + k_1),\\
Q_4 &= v_1 * u_{13} * u_{12} * k_0 + u_{13} * k_2 + u_{12} * k_3,\\
Q_5 &= (u_{13} - u_{12}) * u_{13} * (k_0 * r_{12}^2 + k_2 + k_1) - u_{12}^2 * k_3,\\
Q_6 &= u_{13}^2 * k_2 + (u_{13} - u_{12}) * u_{12} * (k_0 * r_{13}^2 + k_3 + k_1),\\
Q_7 &= v_{12} * u_{12} - r_{12}^2,\\
Q_8 &= v_{13} * u_{13} - r_{13}^2,\\
Q_9 &= (v_{13} - v_{12}) * (u_{13} - u_{12}) - r_{23}^2.
\end{aligned} \qquad (4.36)$$

We have to determine

$$Zero_{(rc)}(QS/NZ) \subset Zero_{\mathbf{C}}(QS/NZ), \qquad (4.37)$$

in which (rc) means that the zeros should be chosen among those verifying the relevant reality conditions and NZ is the product of the non-zero variables and parameters $k_0, k_1, k_2, k_3, r_{12}, r_{13}, r_{23}$.

Proceed now as in Example 4.1 with the same ordering of variables (m_i replaced by k_i) we get finally the result below:

$$Zero_{(rc)}(QS/NZ) = \bigcup_{k=1,2,3,4} Zero_{(rc)}(CS_k/NZ), \tag{4.38}$$

In (4.38) we have $CS_k = \{C_{k1}, \cdots, C_{k8}\}$ for each k with C_{ki} given below:

$$\begin{aligned}
C_{11} &= r_{13} - r_{12},\\
C_{12} &= r_{23} - r_{12},\\
C_{13} &= k_0 * r_{12}^2 + (k_3 + k_2 + k_1),\\
C_{14} &= u_{13}^2 - u_{13} * u_{12} + u_{12}^2,\\
C_{15} &= v_{12} * u_{12} - r_{12}^2,\\
C_{16} &= v_{13} * u_{13} - r_{12}^2,\\
C_{17} &= u_1 * (k_3 + k_2 + k_1) - u_{13} * k_3 - u_{12} * k_2,\\
C_{18} &= v_1 * u_{13} * u_{12} * k_0 + u_{13} * k_2 + u_{12} * k_3,
\end{aligned} \tag{4.39}$$

$$\begin{aligned}
C_{21} &= r_{13}^3 * (k_2 + k_1) + r_{13}^2 * r_{12} * (2 * k_2 + k_1)\\
&\quad - r_{13} * r_{12}^2 * (2 * k_3 + k_1) - r_{12}^3 * (k_3 + k_1),\\
C_{22} &= r_{23} - r_{13} - r_{12},\\
C_{23} &= k_0 * r_{13} * r_{12}^2 * (r_{13} + r_{12}) + r_{13}^2 * (k_2 + k_1)\\
&\quad + r_{13} * r_{12} * (k_2 + k_1) - r_{12}^2 * k_3,\\
C_{24} &= u_{13} * r_{12} + u_{12} * r_{13},
\end{aligned} \tag{4.40}$$

and C_{2i} same as C_{1i}, for $i = 5, 6, 7, 8$. The polsets CS_3, CS_4 are similar to CS_2. The zero-set of CS_1 is analogous to the Lagrange case while those of CS_2, CS_3, CS_4 are analogous to the Euler cases of planet motions. We summarize the results in the form of the following theorem:

Theorem 4.2. The only possible rigid configurations of three parallel vortex filaments in an incompressible and non-viscous fluid extending to infinity for which the strengths are non-zero and verify hypothesis K are either of Lagrange equilateral type or of Euler rectilinear type but are no others.

Remark 4.6. As k_1, k_2, k_3, though non-zero, may be either positive or negative, so from $C_{21} = 0$ we will get either none, or 1, or 2, or 3 positive real roots for the corresponding rectinear vortex motions, in comparing with the single one rectilinear planet motion in that case.

6.5 Solving of Inverse Kinematic Equations in Robotics.

Description of a Robot.

We collect below a number of terminologies used in robotics.

Definition 5.1. An *n-robot* or an *n-manipulator* := An open linkage consisting of *n links* successively connected by *joints*.

Prismatic or *revolutory joint* of a robot := Connection of two successive links moveable along or around the axis.

Hand or *end effector* of a robot := The end of the open linkage.

6*R-robot* := A robot with 6 revolutory axis. Similarly for others, e.g. *RRRRPR*-robot := A robot with 4 revolutory, then one prismatic, and finally a revolutory joint.

To fix the ideas, only 6*R*-robots will be considered so that below by a *6-robot* will be meant a 6R-robot. The case involving prismatic axis can be dealt with by due modifications.

Geometric Entities associated to a robot.

Notation 5.1. (See Fig.5.1.) Suppose that the successive $(i-1)$-th and i-th, also i-th and $(i+1)$-th joint axis do not intersect. Then:

J_{i-1}, J_i, J_{i+1} := $(i-1)$-th, i-th, and $(i+1)$-th joint axis all supposed definitely oriented.

H_iO_i := Common perpendicular of J_i, J_{i+1} with H_i on J_i and O_i on J_{i+1}, oriented in the direction from H_i to O_i. Similarly for $H_{i-1}O_{i-1}$.

a_i := Directed length of H_iO_i, measured in the direction from H_i to O_i.

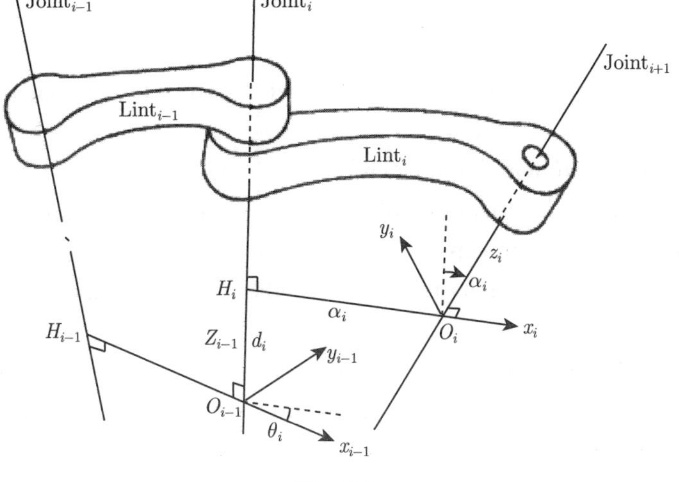

Fig. 5.1.

$d_i :=$ Oriented distance from O_{i-1} to H_i along the oriented axis J_i.

$\alpha_i :=$ Oriented angle from the oriented parallel of the oriented axis J_i at H_i to the oriented parallel of the oriented axis J_{i+1} at H_i.

$\theta_i :=$ Oriented angle from the oriented line at O_{i-1} in the direction from H_{i-1} to O_{i-1} and the oriented parallel at O_{i-1} in the direction from H_i to O_i.

Remark 5.1. The angles α_i, θ_i are determined only up to multples of $2 * \pi$.

Remark 5.2. When J_i and J_{i+1} are parallel to each other, there will be an infinity of possible common perpendiculars and any one of them may be taken as H_iO_i with H_i on J_i and O_i on J_{i+1}.

Remark 5.3. When J_i and J_{i+1} intersect in a point, this point will be taken to be $H_i = O_i$ and the common perpendicular H_iO_i will be understood to be one through the point $H_i = O_i$ and orthogonal to the plane determined by J_i and J_{i+1} and will be definitely oriented in some manner.

Definition 5.2.

$d_i, a_i, \alpha_i :=$ *Geometric constraints* of the robot which are known intrinsic parameters

$\theta_i :=$ *Control parameters* of the robot.

Geometric Spaces and Maps associated to a robot.

Consider an n-robot below:

Definition 5.3. *Configuration Space* $\mathbf{H} :=$ Space of possible positions of the robot hand.

Definition 5.4. *Control Space* $\boldsymbol{\Theta} :=$ Space of control parameters $\theta_1, \cdots, \theta_n$.

Definition 5.5. *Robot map* $:=$ The correspondence

$$\phi_n : \boldsymbol{\Theta}_n \longrightarrow \mathbf{H}, \tag{5.1}$$

with $\phi_n(\theta_1, \cdots, \theta_n)$ the hand position uniquely determined by the control parameters $\theta_1, \cdots, \theta_n$.

Definition 5.6. *Range* of the robot $:=$ The image of the map ϕ_n.

Remark 5.4. Θ_n is a torus of real dimension n, and $\mathbf{H}$ is the space of the group of rigid motions in the real 3-space and is of real dimension 6.

Remark 5.5. ϕ_n is a many to one map. The range is a subset of $\mathbf{H}$ usually of dimension n.

Remark 5.6. For a point in the range, the inverse image is in general a finite subset in $\boldsymbol{\Theta}_n$ for $n = 6$. It has been first time proved by C.G.Liang that for a *generic* n-robot with $n = 6$, the number of points in the inverse image is *generically* 16 if the number is not infinite. Cf. e.g. [Liang1,2].

Let $n = 6$. We shall write simply ϕ instead of ϕ_6 and similarly for others. When a point in $\boldsymbol{\Theta}$ varies along a path, the image point will in general move along a path in $\mathbf{H}$ too. But there may occur exceptions. So we lay down the following definitions.

Definition 5.7. A *hand-position* of a 6-robot := A point $P = \phi(\theta)$ in the range of the robot for some θ in Θ.

Definition 5.8. A *generic* hand-position of a 6-robot := A hand-position P_0 such that there is only a finite number of different points $\theta_0^{(j)}$ in Θ with $\phi(\theta_0^{(j)}) = P_0$ verifying the following condition: The robot map ϕ is a local homomorphsim for each j of a neighborhood N_0^j of $\theta_0^{(j)}$ on a neighborhood N of P_0 and the neighborhoods $N_0^{(j)}$ are disjoint from each other.

Remark 5.7. It follows from the definition that when the hand moves along a path of hand-positions P_t through a generic hand-position P_0 for $t \in (-t_0, +t_0)$, $t_0 > 0$ sufficiently small, each $\theta_0^{(j)}$ of $\phi^{-1}(P_0)$ will undergo a path $\theta_t^{(j)}$ disjoint from each other.

Definition 5.9. A *singular* hand-position of a 6-robot := A hand-position of the robot which is not generic.

Among the singular hand-positions of a 6-robot we may cite the following ones.

Definition 5.10. A *bifurcation* hand-position of a 6-robot := A hand-position P_0 of the robot such that there exists a path P_t and paths $\theta'_t, \theta''_t, t \in (-t_0, 0], t_0 > 0$ with $\phi(\theta'_t) = \phi(\theta''_t) = P_t$ for $t \in (-t_0, 0]$, such that $\theta'_t \neq \theta''_t$ for $t \in (t_0, 0)$ while $\theta'_0 = \theta''_0$.

Definition 5.11. A *dead-position* of a 6-robot := A hand-position P_0 of the robot such that there exist some paths P_t and $\theta_t, t \in (-t_0, 0], t_0 > 0$ with $\phi(\theta_t) = P_t$, which cannot be extended to *smooth* paths θ_t and $P_t, t \in (-t_0, +t_1), t_1 > 0$, with $\phi(\theta_t) = P_t$.

Definition 5.12. A *degeneracy-position* of a 6-robot := A hand-position P_0 of the robot for which ϕ^{-1} is an infinite set.

Clearly it is important to know for a 6-robot not only its range but also all the possible singular hand-positions.

Some General Problems in Robotics.

Problem R1. Determine the range in the configuration space $\mathbf{H}$.

Problem R2. Given a point in the range of the configuration space $\mathbf{H}$, determine its inverse images in $\boldsymbol{\Theta}$.

Problem R3. Determine the set of all singular positions. In particular, determine the sets of bifurcation positions, dead positions, and degeneracy positions separately.

Denavit-Hartenberg Coordinate Systems.

Notation 5.2. (See Fig.5.1.)

$F_0 = OX_0Y_0Z_0 :=$ Fixed coordinate system or frame attached to the robot with $J_0 = J_1$ the oriented OZ_0 axis.

$F_i = OX_iY_iZ_i$ $(0 < i \leqslant 6) := i$-th coordinate system or frame with O_iX_i along the direction from H_i to O_i, O_iZ_i along the direction of oriented axis J_{i+1}, and F_i in similar orientation as F_0, the frame F_6 being attached to the hand.

$A_{i,j} :=$ Matrix of j-th frame F_j w.r.t. the i-th frame F_i.

Definition 5.13. *Denavit-Hartenberg coordinate systems* or *frames* for an n-robot := The coordinate systems or frames $OX_iY_iZ_i = F_i, i = 0, 1, \cdots, n$.

Following the usage in robotics we use a (4×4)-matrix to denote a spatial transformation, with the last row $(0, 0, 0, 1)$, the (3×3)-matrix at the left-upper corner the rotation part, and the last column without the last element 1 as the vector of translation. Then we have:

Theorem 5.1.

$$A_{i-1,i} = \begin{bmatrix} cos\ \theta_i & -sin\ \theta_i * cos\ \alpha_i & sin\ \theta_i * sin\ \alpha_i & a_i * cos\ \alpha_i \\ sin\ \theta_i & cos\ \theta_i * cos\ \alpha_i & -cos\ \theta_i * sin\ \alpha_i & a_i * sin\ \alpha_i \\ 0 & sin\ \alpha_i & cos\ \alpha_i & d_i \\ 0 & 0 & 0 & 1 \end{bmatrix}, \tag{5.2}$$

$$A_{i,i-1} = \begin{bmatrix} cos\ \theta_i & sin\ \theta_i & 0 & -a_i \\ -sin\ \theta_i * cos\ \alpha_i & cos\ \theta_i * cos\ \alpha_i & sin\ \alpha_i & -d_i * sin\ \alpha_i \\ sin\ \theta_i * sin\ \alpha_i & -cos\ \theta_i * sin\ \alpha_i & cos\ \alpha_i & -d_i * cos\ \alpha_i \\ 0 & 0 & 0 & 1 \end{bmatrix}. \tag{5.3}$$

Proof. This follows from

$$A_{i-1,i} = Rot(OZ_i, \theta_i).Trans(0, 0, d_i).T(a_i, 0, 0).Rot(OX_i, \alpha_i), \tag{5.4}$$

in which $Rot(J, \lambda)$ means a rotation about an oriented axis J around an oriented angle λ, and $Trans(v)$ means a translation of a vector v.

Inverse Kinematic Equations.

Along with the Problems R1-3 we may state the following much simpler problem:

Problem R4. Suppose the hand position is known. To determine the control parameters θ_i so that the hand is at the given position.

Let the hand position be given by the matrix

$$H = A_{0,6} = \begin{bmatrix} h_{11} & h_{21} & h_{31} & h_{41} \\ h_{12} & h_{22} & h_{32} & h_{42} \\ h_{13} & h_{23} & h_{33} & h_{43} \\ 0 & 0 & 0 & 1 \end{bmatrix}. \tag{5.5}$$

Then the problem is equivalent to the solving for θ_i of the following equation in terms of h_{ij} supposed known:

$$A_{01}.A_{12}.\cdots.A_{56} = H. \tag{5.6}$$

Definition 5.14. *Inverse kinematic equations* of an n-robot := Equations (5.6) in θ_i with H known.

With this definition the Problem R4 may be restated as

Problem R4′. Solve the inverse kinematic equations (5.6) for θ_i in terms of h_{ij}.

Remark 5.8. With (5.5) the configuration space of a robot may be considered as a 6-dimensional subspace of the 12-dimensional real space $\mathbf{R}^{12}$ with coordinates $h_{ij}, i = 1, \cdots, 4,\ j = 1, 2, 3$, which verify the conditions for the (3×3)-matrix $(h_{ij})_{i,j=1,2,3}$ to be an orthogonal one.

Robot of Puma Type.

The solving of inverse kinematic equations for a general 6-robot is extremely complicate so we shall restrict ourselves to some special type of robots.

Definition 5.15. A 6R-robot is of *Puma type* := The successive pairs of the 6 joint axis are either parallel or perpendicular to each other.

There are in all 2^5 different Puma type 6R-robot. To fix the idea let us consider the Puma type robot for which

$$J_1 \perp J_2 \parallel J_3 \perp J_4 \perp J_5 \perp J_6. \tag{5.7}$$

Definition 5.16. A *Puma R-robot* := A $6R$-robot verifying the conditions (5.7), see Fig. 5.2.

Inverse Kinematic Equations of Puma R-Robot.

For a Puma R-robot let us first orient the joint axis and the O_iH_i directions in such a way that all right angles α_i will be $+\frac{\pi}{2}$. Thus, the successive constraint angles α_i will be:

$$(\alpha_1, \alpha_2, \cdots, \alpha_6) = \left(+\frac{\pi}{2}, 0, +\frac{\pi}{2}, +\frac{\pi}{2}, +\frac{\pi}{2}, 0\right). \tag{5.8}$$

Besides we shall choose the Denavit-Hartenberg frames F_0 and F_6 such that O_0Z_0 lies on and in the direction of the oriented axis J_1, and $O_6 = H_6$ coincides with O_5 and O_6Z_6 with O_5Z_5 in the same direction. This will induce the simplification

$$d_6 = 0. \tag{5.9}$$

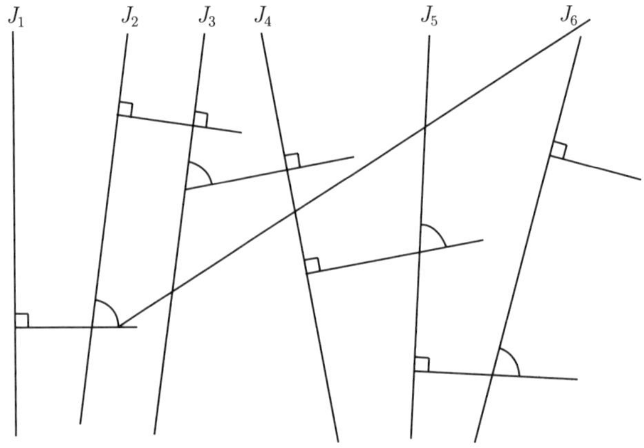

Fig. 5.2.

Set now

$$\begin{aligned} c_i &= cos\ \theta_i, \quad s_i = sin\ \theta_i, \quad i = 1,3,4,5,6, \\ c_{23} &= cos\ \theta_{23}, \quad s_{23} = sin\ \theta_{23}, \\ \theta_{23} &= \theta_2 + \theta_3, \ d_{23} = d_2 + d_3. \end{aligned} \tag{5.10}$$

Then to solve Problem R4 for Puma R-robot is equivalent to solve for $c_i, s_i, (i \neq 2)$, and c_{23}, s_{23} in terms of a_i, d_i and h_{ij} considered as known parameters. For this purpose let us replace the equation (5.6) by the equivalent equation

$$A_{34}.A_{45}.A_{56} = A_{32}.A_{21}.A_{10}.H. \tag{5.11}$$

Introduce for simplicity the pols f_{ij} by setting

$$\begin{aligned} &f_{i1} := s_1 * h_{i2} + c_1 * h_{i1} - \delta_{4i} * a_1, \ i = 1,2,3,4; \\ &f_{i2} := -s_1 * h_{i1} + c_1 * h_{i2} + \delta_{4i} * d_{23}, \ i = 1,2,3,4; \\ &f_{43} := h_{43} - d_1; \end{aligned} \tag{5.12}$$

Then (5.10) will give rise to a system of polynomial equations $P_i = 0, i = 1, \cdots, 12$, with

$$
\begin{aligned}
&P_1 := s_6 * s_4 + c_6 * c_5 * c_4 - s_{23} * h_{13} - c_{23} * f_{11};\\
&P_2 := s_6 * c_4 - c_6 * c_5 * s_4 - f_{12};\\
&P_3 := c_6 * s_5 - s_{23} * f_{11} + c_{23} * h_{13};\\
&P_4 := s_6 * c_5 * c_4 - c_6 * s_4 + s_{23} * h_{23} + c_{23} * f_{21};\\
&P_5 := s_6 * c_5 * s_4 + c_6 * c_4 - f_{22};\\
&P_6 := s_6 * s_5 + s_{23} * f_{21} - c_{23} * h_{23};\\
&P_7 := s_5 * c_4 - s_{23} * h_{33} - c_{23} * f_{31};\\
&P_8 := s_5 * s_4 + f_{32};\\
&P_9 := c_5 + s_{23} * f_{31} - c_{23} * h_{33};\\
&P_{10} := c_5 * c_4 * a_5 + s_4 * d_5 + c_4 * a_4 + c_3 * a_2 - s_{23} * f_{43} - c_{23} * f_{41} + a_3;\\
&P_{11} := c_5 * s_4 * a_5 + s_4 * a_4 - c_4 * d_5 + f_{42};\\
&P_{12} := s_5 * a_5 + s_3 * a_2 - s_{23} * f_{41} + c_{23} * f_{43} + d_4.
\end{aligned} \tag{5.13}
$$

Besides these we have also the trigonometrical identities $P_i = 0, i = 13, \cdots, 18$ for the trigonometric functions with P_i given below:

$$
\begin{aligned}
&P_i = c_{i-12}^2 + s_{i-12}^2 - 1, \ i = 13, 15, 16, 17, 18;\\
&P_{14} = c_{23}^2 + s_{23}^2 - 1.
\end{aligned} \tag{5.14}
$$

The Problem R4 for Puma R-robot is then equivalent to the determination of $Zero_{\mathbf{R}}(PS)$ for

$$PS = \{P_1, \cdots, P_{18}\}. \tag{5.15}$$

Preliminary Simplification of Zero(PS) for Puma R-Robot.

To solve $PS = 0$ or to determine $Zero(PS)$ let us first make some preliminary transformations for the sake of simplifying the computations. Thus, with the aid of the equation $P_{16} = 0$ the pair P_1, P_2 is clearly 0-equivalent to the pair P_1', P_2' where

$$P_1' = s_4 * P_1 + c_4 * P_2, \ P_2' = c_4 * P_1 - s_4 * P_2. \tag{5.16}$$

Similarly, the pairs P_4, P_5 and P_7, P_8 are 0-equivalent to the pairs P_4', P_5' and P_7', P_8' respectively where

$$
\begin{aligned}
&P_4' = -s_4 * P_4 + c_4 * P_5, \ P_5' = c_4 * P_4 + s_4 * P_6;\\
&P_7' = s_4 * P_7 - c_4 * P_8, \ P_8' = c_4 * P_7 + s_4 * P_8.
\end{aligned} \tag{5.17}
$$

Furthermore, the pairs P_2', P_3 and P_5', P_6 are 0-equivalent to the pair P_2'', P_3' and P_5'', P_6' respectively where

$$
\begin{aligned}
&P_2'' = s_5 * P_2' - c_5 * P_3, \ P_3' = c_5 * P_2' + s_5 * P_3.\\
&P_5'' = s_5 * P_5' - c_5 * P_6, \ P_6' = c_5 * P_5' + s_5 * P_6.
\end{aligned} \tag{5.18}
$$

Write now P_i as P_i' for $i \geqslant 9$ and rewrite P_2'', P_5'' as P_2', P_5', then PS is 0-equivalent to the polset

$$PS' = \{P_1', \cdots, P_{18}'\}, \text{ where} \tag{5.19}$$

$$\begin{aligned}
P_1' &:= s_6 - c_4 * f_{12} - s_4 * c_{23} * f_{11} - s_4 * s_{23} * h_{13};\\
P_2' &:= c_6 - c_5 * c_4 * (c_{23} * f_{11} - s_{23} * h_{13})\\
&\quad + c_5 * s_4 * f_{12} + s_5 * c_{23} * h_{13} - s_5 * s_{23} * f_{11};\\
P_3' &:= -c_5 * c_{23} * h_{13} + c_5 * s_{23} * f_{11}\\
&\quad - s_5 * c_4 * c_{23} * f_{11} - s_5 * c_4 * s_{23} * h_{13} + s_5 * s_4 * f_{12};\\
P_4' &:= c_6 - c_4 * f_{22} - s_4 * c_{23} * f_{21} - s_4 * s_{23} * h_{23};\\
P_5' &:= s_6 + c_5 * c_4 * c_{23} * f_{21} + c_5 * c_4 * s_{23} * h_{23}\\
&\quad - c_5 * s_4 * f_{22} - s_5 * c_{23} * h_{23} + s_5 * s_{23} * f_{21};\\
P_6' &:= c_5 * c_{23} * h_{23} - c_5 * s_{23} * f_{21}\\
&\quad + s_5 * c_4 * c_{23} * f_{21} + s_5 * c_4 * s_{23} * h_{23} - s_5 * s_4 * f_{22};\\
P_7' &:= c_4 * f_{32} + s_4 * c_{23} * f_{31} + s_4 * s_{23} * h_{33};\\
P_8' &:= s_5 - c_4 * c_{23} * f_{31} - c_4 * s_{23} * h_{33} + s_4 * f_{32};\\
P_i' &:= P_i, \ i = 9, \cdots, 18.
\end{aligned} \tag{5.20}$$

The determination of $Zero_{\mathbf{R}}(PS)$ is thus equivalent to the determination of $Zero_{\mathbf{R}}(PS')$.

A Particular Puma R-Robot : Puma560.

A further particular Puma type robot is furnished by the Puma560 in current use which may be defined as follows.

Definition 5.17. A *Puma560 robot* := A Puma 6R-robot with joint axis J_i satisfying the further relations below:

$$J_4, J_5, J_6 \ \textit{concurrent and no two coincident}. \tag{5.21}$$

The condition (5.21) as well as the natural geometry situation at robot-hand imply that

$$a_4 = a_5 = d_5 = 0. \tag{5.22}$$

On the other hand we shall suppose that

$$a_2, \ a_3, \ d_{23}, \ d_4, \ \textit{all} \ \neq 0, \tag{5.23}$$

The pols $P_{10}', P_{11}' P_{12}'$ in (5.19) will then be simplified to the following form:

$$\begin{aligned}
P_{10}' &:= c_3 * a_2 - s_{23} * f_{43} - c_{23} * f_{41} + a_3;\\
P_{11}' &:= s_1 * h_{41} - c_1 * h_{42} - d_{23};\\
P_{12}' &:= s_3 * a_2 - s_{23} * f_{41} + c_{23} * f_{43} + d_4.
\end{aligned} \tag{5.24}$$

Char-Set of PS or PS′ for Puma560.

Let us introduce an ordering of the variables $c's$ and $s's$ as follows:

$$s_1 \prec c_1 \prec s_{23} \prec c_{23} \prec s_4 \prec c_4 \prec s_5 \prec c_5 \prec s_6 \prec c_6 \prec s_3 \prec c_3. \tag{5.25}$$

Then our char-set algorithm gives readily a triangular set (which can be readily turned into a char-set if we want)

$$CS = \{C_1, \cdots, C_{12}\} \tag{5.26}$$

with C_i given below:

$$\begin{aligned}
C_1 :&= s_1^2 * (h_{42}^2 + h_{41}^2) - 2 * s_1 * h_{41} * d_{23} - h_{42}^2 + d_{23}^2;\\
C_2 :&= c_1 * h_{42} - s_1 * h_{41} + d_{23};\\
C_3 :&= 4 * s_{23}^2 * (f_{43}^2 + f_{41}^2) * (d_4^2 + a_3^2)\\
&\quad -4 * s_{23} * (f_{43} * a_3 + f_{41} * d_4) * (f_{43}^2 + f_{41}^2 + d_4^2 + a_3^2 - a_2^2)\\
&\quad +(f_{43}^2 + f_{41}^2 + d_4^2 + a_3^2 - a_2^2)^2\\
&\quad -4 * (f_{43} * d_4 - f_{41} * a_3)^2;\\
C_4 :&= 2 * c_{23} * (f_{43} * d_4 - f_{41} * a_3) - 2 * s_{23} * (f_{43} * a_3 + f_{41} * d_4)\\
&\quad +f_{43}^2 + f_{41}^2 + d_4^2 + a_3^2 - a_2^2;\\
C_5 :&= s_4^2 * (c_{23} * f_{31} + s_{23} * h_{33})^2\\
&\quad +s_4^2 * f_{32}^2 - f_{32}^2;\\
C_6 :&= c_4 * f_{32} + s_4 * c_{23} * f_{31} + s_4 * s_{23} * h_{33};\\
C_7 :&= s_5 - c_4 * c_{23} * f_{31} - c_4 * s_{23} * h_{33} + s_4 * f_{32};\\
C_8 :&= c_5 - c_{23} * h_{33} + s_{23} * f_{31};\\
C_9 :&= s_6 - c_4 * f_{12} - s_4 * c_{23} * f_{11} - s_4 * s_{23} * h_{13};\\
C_{10} :&= c_6 - c_4 * f_{22} - s_4 * c_{23} * f_{21} - s_4 * s_{23} * h_{23};\\
C_{11} :&= s_3 * a_2 + c_{23} * f_{43} - s_{23} * f_{41} + d_4;\\
C_{12} :&= c_3 * a_2 - c_{23} * f_{41} - s_{23} * f_{43} + a_3.
\end{aligned} \tag{5.27}$$

The non-trivial initials of C_i with some non-zero factors removed are given by

$$\begin{aligned}
I_1 &:= h_{42}^2 + h_{41}^2;\\
I_2 &:= h_{42};\\
I_3 &:= f_{43}^2 + f_{41}^2;\\
I_4 &:= f_{43} * d_4 - f_{41} * a_3;\\
I_5 &:= (c_{23} * f_{31} + s_{23} * h_{33})^2 + f_{32}^2;\\
I_6 &:= f_{32}.
\end{aligned} \tag{5.28}$$

Note that for the *real* solutions $I_1 \neq 0$ owing to $C_2 = 0$ with $d_{23} \neq 0$. Furthermore $I_6 \neq 0$ would imply $I_5 \neq 0$ and $I_3 = 0$ would imply $f_{43} = f_{41} = 0$ so that $I_4 = 0$. Hence the Well-Ordering Principle gives us the following theorem:

Theorem 5.1. For the Puma560 the char-set CS of the polset PS in (5.13) with $PS = 0$ the corresponding inverse kinematic equations is given by (5.26) and (5.27) and we have

$$Zero_{\mathbf{R}}(PS) = Zero_{\mathbf{R}}(CS/f_{32} * h_{42} * I_4) + Zero_{\mathbf{R}}(PS + \{h_{42}\}) + Zero_{\mathbf{R}}(PS + \{f_{32}\}) + Zero_{\mathbf{R}}(PS + \{I_4\}). \tag{5.29}$$

We shall leave aside the further determination of the last three zero-sets in (5.29) which can be carried out by our general method. We remark only that from the form of char-set there are generically atmost 8 possible real solutions for a given hand-position and the singularities of the Puma560 robot can be studied via the zeros of either the initials or the discriminants of the pols of CS.

Degenerate Case of Vanishing Initials for Puma560.

It would be very cumbersome to determine the last three zero-sets in the right-hand side of (5.29). So instead of doing this we shall draw some conclusions directly from the expressions (5.27) of CS. For this purpose we lay down first some elementary Lemmas as follows. Remark that we are interested in *real* solutions only.

Lemma 5.1. Let A, B, C be in $\mathbf{R}$ with A, B not both 0 and

$$A * cos\ \theta + B * sin\ \theta + C = 0. \tag{5.30}$$

For (5.30) to have real solutions in θ it is necessary and sufficient that

$$A^2 * (A^2 + B^2 - C^2) \geqslant 0. \tag{5.31}$$

In that case we have

$$\begin{aligned} sin\ \theta &= \frac{-B * C + \varepsilon * A * \sqrt{A^2 + B^2 - C^2}}{A^2 + B^2}, \\ cos\ \theta &= \frac{-A * C - \varepsilon * B * \sqrt{A^2 + B^2 - C^2}}{A^2 + B^2}, \end{aligned} \tag{5.32}$$

in which $\varepsilon = \pm 1$.

The proof is elementary and will be omitted.

From (5.31) we see that it is reasonable to lay down the following definition:

Definition 5.18. *Discriminant* of the left-hand side of (5.30) := The expression $A^2 + B^2 - C^2$.

Lemma 5.2. If $f_{32} = 0$ then

$$N_1 \equiv (h_{32} * h_{41} - h_{31} * h_{42})^2 - (h_{31}^2 + h_{32}^2) * d_{23}^2 = 0. \tag{5.33}$$

Proof. We have by (5.12)

$$f_{32} = -s_1 * h_{31} + c_1 * h_{32} = 0.$$

As $Zero(PS) = Zero(PS') \subset Zero(CS)$, by (5.27) we have also

$$C_2 = c_1 * h_{42} - s_1 * h_{41} + d_{23} = 0$$

too. Whence

$$c_1 * (h_{42} * h_{31} - h_{41} * h_{32}) = -h_{31} * d_{23},$$
$$s_1 * (h_{42} * h_{31} - h_{41} * h_{32}) = -h_{32} * d_{23}.$$

We get then readily (5.33). □

Lemma 5.3. If $I_4 = f_{43} * d_4 - f_{41} * a_3 = 0$, then either $h_{41} = h_{42} = 0$ or

$$N_2 \equiv (h_{42}^2 + h_{41}^2) - (f_{43} * d_4 + a_1 * a_3)^2 - d_{23}^2 = 0. \quad (5.34)$$

Proof. By $(5.12)_2$ and $I_4 = 0$, we have

$$c_1 * h_{41} + s_1 * h_{42} - (f_{43} * d_4 + a_1 * a_3) = 0.$$

Together with $C_2 = 0$ we get

$$c_1 * (h_{42}^2 + h_{41}^2) = (f_{43} * d_4 + a_1 * a_3) * h_{41}) - h_{42} * d_{23},$$
$$s_1 * (h_{42}^2 + h_{41}^2) = (f_{43} * d_4 + a_1 * a_3) * h_{42}) - h_{41} * d_{23}.$$

Whence we get

$$(h_{42}^2 + h_{41}^2) * N_2 = 0$$

and the Lemma is readily proved. □

Remark 5.8. If $N_2 = 0$ then h_{41}, h_{42} cannot be both 0 since d_{23} is assumed to be non-zero.

Theorem 5.2. If h_{42}, N_1, N_2 are all non-zero then

$$Zero_{\mathbf{R}}(PS) = Zero_{\mathbf{R}}(CS), \quad (5.35)$$

where CS is given by (5.26) and (5.27).

Proof. From the Lemmas we should have $f_{32} \neq 0, I_4 \neq 0$. So the theorem follows from (5.29). □

*Range and Singularities of Puma*560.

We shall restrict ourselves to the *generic* case of Puma560 for which none of the initials in (5.28) is equal to 0. By Theorem 5.2 this is particularly the case when

$h_{42} * N_1 * N_2 \neq 0$. In that case we get directly from Lemma 5.1 the following Lemma and Theorem:

Lemma 5.4. The discriminants of C_2, C_4, C_6 in the case none of the initials in (5.28) are zero are given by

$$\begin{aligned} &D_2 = h_{41}^2 + h_{42}^2 - d_{23}^2 \geqslant 0, \\ &D_4 = 4 * a_2^2 * (d_4^2 + a_3^2) - (f_{43}^2 + f_{41}^2 - d_4^2 - a_3^2 - a_2^2)^2 \geqslant 0, \\ &D_6 = (c_{23} * f_{31} + s_{23} * h_{33})^2 + f_{32}^2 \geqslant 0, \end{aligned} \tag{5.36}$$

in which ($\varepsilon_2 = \pm 1$)

$$\begin{aligned} &f_{41} = \varepsilon_2 * D_2 - a_1, \\ &f_{31} = \frac{(h_{41} * h_{32} - h_{42} * h_{31} * d_{23} + \varepsilon_2 * (h_{42} * h_{32} + h_{41} * h_{31}) * \sqrt{D_2}}{h_{41}^2 + h_{42}^2}. \end{aligned} \tag{5.37}$$

Moreover, the angles $\theta_1, \theta_{23}, \theta_4$ will be determined from $C_2 = 0, C_4 = 0, C_6 = 0$ by (5.31),(5.32),(5.36), and (5.37).

Theorem 5.3. The part of range of Puma560 for which $h_{42} * N_1 * N_2 \neq 0$ is given by

$$\begin{aligned} &h_{41}^2 + h_{42}^2 \geqslant d_{23}^2, \\ &4 * a_2^2 * (d_4^2 + a_3^2) \geqslant (f_{43}^2 + f_{41}^2 - d_4^2 - a_3^2 - a_2^2)^2. \end{aligned}$$

There are in general 8 possible sets of control parameters for a given hand position in the above range and bifurcations will occur when one of the above inequalities becomes an equality.

Chapter 7

Appicaltions to Geometry Theorem-Proving

7.1 Basic Principles of Mechanical Geometry Theorem-Proving.

Notations. $\mathbf{K}, X = (x_1, \cdots, x_n)$, etc. are as before.

Historical Development of Geometry Theorem-Proving Reviewed.

We have described in Chap. 2 the historical development of geometry theorem-proving into four stages. The turning point from the first stage of Euclidean fashion proving to Descartesian fashion proving occured in 17C owing to the appearance of analytic geometry of Descartes. This turned the theorem-proving by logical deductions based on axiomatic systems into theorem-proving by algebraic computations based on coordinate systems. However, the jump from axiomatic systems to coordinate systems is not a trivial one. It was Hilbert who first time in his classic *Grundlagen der Geometrie* filled this gap as in the diagram below:

$$\textit{Axiomatic System} \longrightarrow_{(1)} \textit{Number System} \longrightarrow_{(2)} \textit{Coordinate System}. \qquad (*)$$

While the Step (2) seems nowadays somewhat routine in character the Step (1) is quite involved. We discuss this below in more details.

Hilbert's Axiomatic System of Euclidean Geometry.

Let us restrict ourselves to the planar case. In order to lay down a rigorous foundation of Euclidean geometry, Hilbert had divided the axioms into 5 groups:

H-I. Axioms of incidence.

H-II. Axioms of order.

H-III. Axioms of congruence.

H-IV. Axioms of parallels.

H-V. Axioms of continuity.

As pointed out by P.K.Rashevsky in his preface to the Russian translation of Hilbert's classic, the 5th group of axioms on continuity plays actually no role in the founding of Euclidean geometry, a fact of fundamental importance for the *mechanization* of proving of geometry theorems. On the other hand the other four groups of axioms are intermingled together. For example, the axioms of order have to be used to define concepts about congruence. With such an axiomatic system Hilbert showed how a number system which is actually a field of characteric 0 may be introduced and then coordinate system may be introduced in the usual way as shown in the diagram $(*)$ above.

Roles of Desargues Theorem and Pascal Theorem.

In Hilbert's classic it was shown that two theorems play important roles in the introduction of number system which is intrinsically associated to the geometry in consideration as defined by a certain set of axioms among the above 5 groups. One theorem is the Desargues Theorem as described in Ex. 2.1.4$'$ which is actually a particular case of the general Desargues theorem in current terminology. Hilbert showed how a number system, though not necessary a field of characteristic 0, may still be introduced in the geometry without axioms of congruence but admitting the Desargues Theorem as an axiom. Moreover, under such an axiom the plane geometry may be extended to a space geometry which shows the particular geometrical significance of the Desargues Theorem. Another theorem playing a particular role is the Pascal Theorem, actually a particular case of Pappus theorem which in turn is a particular case of the Pascal theorem in current terminology. Again as shown by Hilbert, if we deprive of the axioms of congruence, but admit the Pascal Theorem as an axiom, then the number system introduced will obey the commutative law of multiplication and is a field of characteric 0. This shows thus the particular role enjoyed by the Pascal Theorem or Pascal Axiom: its close connection with the multiplication commutativity.

Desarguesian Geometry and Pascalian Geometry.

In the book [WWT5] of the present author we suggested a modified version of axiomatic system of Hilbert. The modified system considers first a system consisting of the axioms of incidence and parallelism with two further axioms. One is the *Axiom of Infinity* to deprive of the case corresponding to the usual finite geometry, and the other one is the Desargues Axiom corresponding to the two Desargues Theorems as in Ex. 2.1.4′. Without going into details which we refer to the book [WWT5] let us cite the following definitions and theorems.

Definition 1.1. A *Desarguesian axiomatic system* := The system of Axioms H-I, H-IV, the Axiom of Infinity, and the Desargues Axiom.

Definition 1.2. A (plane) *unordered Desarguesian geometry* := The geometry of points and lines verifying the Desarguesian axiom system.

Definition 1.3. A *Desarguesian number system* := A number system which possesses two binary operations + and $*$ obeying the usual laws of addition and multiplication except the commutative law of multiplication and has a subset identified to the usual field of rational numbers.

Theorem 1.1. To a (plane) unordered Desarguesian geometry may be intrinsically associated a unique Desarguesian number system.

Theorem 1.2. In a (plane) unordered Desarguesian geometry coordinate systems may be introduced in which a line will be represented by an equation in the form $a * x + b * y = c$, in which (x, y) are the coordinates, and a, b, c are in the intrinsically associated Desargues number system with a, b not both 0.

Remark 1.1. In a (plane) unordered Desarguesian geometry the equation $x * a + y * b = c$ is in general different from the equation $a * x + b * y = c$ and may not be a line.

Definition 1.4. A *Pascalian axiomatic system* := The axiomatic system consisting of the axioms in the Desarguesian axiom system as well as the Pascal Axiom.

Definition 1.5. A (plane) *unordered Pascalian geometry* := The geometry of points and lines verifying the Pascalian axiomatic system.

Theorem 1.3. To a (plane) unordered Pascalian geometry may be intrinsically associated a unique field of characteristic 0.

Theorem 1.4. In a (plane) unordered Pascalian geometry coordinate systems may be introduced in which a line will be represented by an equation in the form $a*x+b*y = c$, in which (x, y) are the coordinates, and a, b, c are in the intrinsically associated field with a, b not both 0.

Metric Geometry, Ordinary Geometry, and Euclidean Geometry.

By adding further axioms of Hilbert's axiomatic system to the Pascalian axiomatic system we get a hierarchy of (plane) geometries among which the most important are the following ones.

Definition 1.6. A (plane) *(unordered) metric geometry* := The geometry of points and lines verifying the Pascalian axiomatic system as well as Hilbert's axioms of congruence.

Definition 1.7. A (plane) *ordinary geometry* := The geometry of points and lines verifying the Pascalian axiomatic system as well as Hilbert's axioms of order, and axioms of congruence.

Definition 1.8. A (plane) *Euclidean geometry* := The geometry of points and lines verifying the Pascalian axiomatic system as well as Hilbert's axioms of order, axioms of congruence, and axioms of continuity.

Algebrization of Geometry Concepts.

In all the above geometries coordinate systems may be introduced so that various geometrical entities and geometrical relations will be expressed in the form of algebraic expressions or algebraic relations. If the geometry in question verifies also the Pascal Axiom so that multiplication is commutative, then it has been proved that the following theorem is true.

Theorem 1.5. In the hierarchy of geometries between the unordered Pascalian geometry and Euclidean geometry any geometry entities may be expressed in the form of polynomials and any geometry relations may be expressed in the form of polynomial equations, inequations, eventually inequalities, connected by logical connectives $\wedge, \vee, \neg$.

Remark 1.2. The algebraic expressions or relations involved in the above theorem may be quite complicated. For example, let $P_i = (a_i, b_i), i = 1, \cdots, 4$, be four points such that the line P_1P_2 is *geometrically* parallel to another line P_3P_4 in the usual sense. Then first of all, for P_1P_2 and P_3P_4 to be actual *geometrical* lines, the points P_1, P_2 and also the points P_3, P_4 should be different. Algebraically this means that the following inequations should be verified:

$$\neg(a_2 = a_1 \wedge b_2 = b_1), \ \neg(a_4 = a_3 \wedge b_4 = b_3). \tag{1.1}$$

Only in this case the geometric lines P_1P_2 and P_3P_4 will be defined respectively by the equations below:

$$\begin{vmatrix} x & y & 1 \\ a_1 & b_1 & 1 \\ a_2 & b_2 & 1 \end{vmatrix} = 0, \quad \begin{vmatrix} x & y & 1 \\ a_3 & b_3 & 1 \\ a_4 & b_4 & 1 \end{vmatrix} = 0. \tag{1.2}$$

Secondly, for the two geometric lines P_1P_2, P_3P_4 to be *geometrically* parallel in the usual sense they should be different from each other or P_3, P_4 should not be on the *geometric* line P_1P_2. Algebraically we should have the following inequations verified, viz.

$$\begin{vmatrix} a_1 & b_1 & 1 \\ a_2 & b_2 & 1 \\ a_3 & b_3 & 1 \end{vmatrix} \neq 0, \quad \begin{vmatrix} a_1 & b_1 & 1 \\ a_2 & b_2 & 1 \\ a_4 & b_4 & 1 \end{vmatrix} \neq 0. \tag{1.3}$$

Only in this case *geometrical* parallelism of the two *different geometric* lines P_1P_2, P_3P_4 may be defined in the usual *geometrical* sense and may be turned into the algebraic relation below:

$$(a_2 - a_1) * (b_4 - b_3) - (a_4 - a_3) * (b_2 - b_1) = 0. \tag{1.4}$$

It follows that in the usual geometric sense the parallelism of two lines P_1P_2, P_3P_4 is actualy algebraically equivalent to a set of equations and inequations (1.1)-(1.4). If the geometrical concepts are turned into algebraic expressions and relations in such a way, it will be clearly very difficult to study geometry by algebraic means. To avoid such undesirable complexities we adopt the following measure.

We say that the first one of equations in (1.2) represents an *algebraic* line joining the two points P_1 and P_2, whether the two points P_1, P_2 be different or not or whether the corresponding inequation in (1.1) be verified or not. In the former case P_1P_2 is an ordinary *geometric* line while in the latter case we shall say that P_1P_2 is a *degenerate* line or a line *in degenerate sense.* A *degenerate* line may therefore be a point set consisting of all points in the plane. Furthermore, we shall say that two *algebraic* lines P_1P_2 and P_3P_4 represented by the equations (1.2) are *algebraically* parallel if (1.4) is verified. Only in case that both (1.1) and (1.3) are verified P_1P_2 and P_3P_4 will be true geometrical lines different from each other and are *geometrically* parallel to each other in the usual sense when (1.4) is true. On the other hand, we see that the two *algebraic* lines will be *algebraically* parallel if either or both are *degenerate* or if the two lines coincide when both are non-degenerate. In such cases we shall say that the two algebraic lines are *degenerately* parallel or parallel *in degenerate sense.*

Instead of saying an *algebraic* line or two *algebraic* lines being *algebraically* parallel we shall say by abuse of language a line or two parallel lines in droping away the modifier *algebraic.* For *non-degenerate* lines or *non-degenerately* parallel lines the corresponding algebraic relations (1.1) or (1.3) will then be called *non-degeneracy conditions.* Other *algebraic* geometrical entities and relations as well as *degenerate* ones may be introduced in a similar way. In what follows geometric entities and relations will all be understood to be in such *algebraic* sense in the above manner.

Remarks on Degeneracy of Geometry Theorems.

In usual geometry theorems the geometric entities and geometric relations are understood to be in the usual exact geometrical sense so that *degeneracy* are implicitely discarded. Otherwise the theorem may not be true or even devoid of meaning.

Example 1.1. Theorem. For a parallelogram $ABCD$ the two diagonals AC and BD will intersect at their common mid-point (or have a common mid-point).

For this simple theorem to be true it has been implicitely assumed that the sides AB, BC, CD, DA are all true geometric lines or none of the point-pairs A, B; B, C; C, D; D, A will coincide. Moreover, it is also implicitely assumed that the pairs of opposite sides, assumed to be true geometric lines, are truly geometrically parallel to each other in the usual sense. In fact, the theorem will no more be true or even meaningless if these implicit assumptions are not observed. For example, if the four sides are true geometrical lines but the sides AB, CD are only degenerately parallel to each other so that AB, CD degenerate into the same line with C, D on the line AB. Then the mid-points of AC and BD will in general no more the same point. Moreover, intersection here will be devoid of meaning.

Example 1.2. Let us consider Ex. 2.1.4′ again. We see already that for the case there considered the theorem will no more be true if the corresponding sides $BC, B'C'$ of the two triangles $ABC, A'B'C'$ are *degenerately* parallel. Similarly, for the converse part of the Desargues Theorem in Ex. 2.1.4′, the theorem will no more be true if the triangles $ABC, A'B'C'$ are *degenerate* so that they are not truely triangles in the usual sense but collapsed lines. See the accompanying Fig. 1.1 below. See also the Figs. 2.1.4-6 in Chap. 2 and Fig. 3.6.6 in Chap. 3.

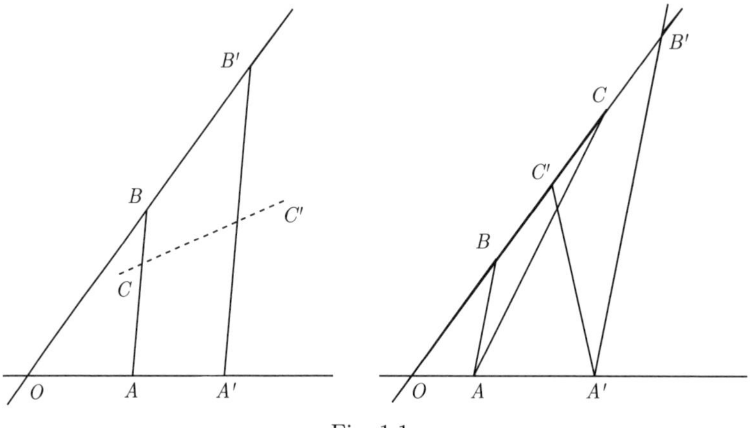

Fig. 1.1.

Notions of Elementary Geometries.

All the geometries (planar or spatial) described above may be said to be *subsidiary* to the ordinary geometry or Euclidean geometry in certain sense. There are however variety of geometries not of such character, e.g. projective gemetries, line geometry, non-Euclidean geometries, circle geometries, etc. In [WWT5] it was shown how these geometries may be defined in basing on an axiomatic system and how a number system and then a coordinate system may be introduced as in the case of ordinary geometry or geometries subsidiary to it. It was also shown that for the above geometries the number system intrinsically associated to such geometries is actually a field of characteristic 0 so far some axiom analogous to the preceding Pascal theorem in Hilbert's terminology is adjoined to the defining axiomatic system of the relevant geometry. For such geometries we may then introduce coordinate systems, the algebrization of geometric concepts, and then treat geometry theorem-proving in the Descartesian fashion without further reference to its origin — the underlying axiomatic system. We shall even adopt the following extremist measure:

We shall utterly forsake the Euclidean way of basing geometries on axiomatic systems. On the contrary we shall treat geometries directly from number systems and coordinate systems in the Descartesian way. In fact, in mathematical researches on geometries we have already followed this way of treatment. Thus, to study the planar projective geometry we may start from the field of real or complex numbers and introduce homogeneous coordinates with points as the ratios of triplets $(a : b : c)$, where a, b, c being numbers in the field not all 0. We may then define lines, the collinearity of points, the concurrency of lines, the cross ratios of four points on a line, etc. by means of algebraic equations or expressions. The study of projective geometry will then be done by means of algebraic manipulations of such algebraic expressions and equations. In the same way we may develop various kinds of geometries without any reference to the possible underlying axiomatic system. If we follow the Kleinian viewpoint of geometries, then the notion of an axiomatic system is clearly superfluous at all. Consider furthermore the case of differential geometries. For 3-dimensional case we define usually a *surface* or a *curve* by a single equation or a pair of equations in the coordinates verifying some differentiability and regularity conditions. The same may also be defined by means of parametric equations. In any way we shall define properties of geometrical significance of these curves or surfaces directly from these equations. The geometrical studies of curves and surfaces are then developed by means of purely algebraic and differential manipulations. No axiomatic systems will be involved. In fact, the present author is ignorant of any axiomatic systems for such differential geometries. Even the possibility of introducing such axiomatic systems seems to be doubtful.

It is for these reasons that we shall study the geometries in starting directly from a number field (mainly a field of characteristic 0) and coordinate systems in this number field without consideration of step (1) and even step (2) in the diagram $(*)$.

For differential geometries some operations of differentiation should be introduced in the basic field. In the present book we shall restrict ourselves to geometries for which differentiation is not involved. We shall call vaguely all such geometries *elementary* geometries, to distinguish from *differential* geometries in the usual sense.

Geometry Theorems and Geometry Theorem-Proving in Elementary Geometries.

Consider now any kind of elementary geometry as described above. We suppose that the basic field, or the number field intrinsically associated to the geometry is a field $\mathbf{K}$ of characteristic 0 with a coordinate system $X = (x_1, \cdots, x_n)$. We suppose further that geometric entities or geometric relations may be expressed in terms of polynomials in $\mathbf{K}[X]$. Geometry theorems may then in general be put in the form $T = \{H_T, C_T\}$, with H_T the hypothesis in the form of polynomial or rational equations, inequations, or inequalities, connected by $\wedge, \vee, \neg$. Similarly for the conclusion part C_T. In any way only geometry theorems which may be put in such forms will be considered in the present book. With this understood the domain of all such theorems will be denoted by $\mathbf{T}$ as defined below:

Definition 1.9. *Tarski domain* of geometry theorems := Domain of geometry theorem in the form of $T = \{H_T, C_T\}$, in which H_T := *Hypothesis* and C_T := *Conclusion* both in the form of sets of polynomial or rational equations, inequations, inequalities, connected by logical connectives.

Theorem 1.6. Tarski Theorem. There is an algorithm which permits to decide in a finite number of steps whether a theorem $T = \{H_T, C_T\}$ in the domain $\mathbf{T}$ is true or not, or whether C_T is a consequence of H_T or not.

Remark 1.3. Not every well-known theorem in usual ordinary geometry is in the domain $\mathbf{T}$. For example, the Jordan theorem stating that any planar simple closed polygon will divide the plane in two parts is such a theorem.

Particular Cases : *Domains* **H** *and* **E**

Definition 1.10. The *Hilbertian Domain* of geometry theorems **H** := The domain of theorems $T = \{HYP, CONC\}$ for which $HYP = \{P_1, \cdots, P_r\}$ with $P_k = x_k - R_k, R_k \in \mathbf{K}(X_{k-1}, U)$ and $CONC \in \mathbf{K}(X, U)$, U being a finite set of parameters. The *hypothesis* and the *conclusion* of the theorem T := The equations $HYP = 0$ and $CONC = 0$ respectively.

Definition 1.11. The *Equality Domain* of geometry theorems **E** := The domain of theorems $T = \{HYP, CONC\}$ for which $HYP \subset \mathbf{K}[X]$ and $CONC \in \mathbf{K}[X]$. The *hypothesis* and the *conclusion* of the theorem T := The equations $HYP = 0$ and $CONC = 0$ respectively.

We shall leave the study of geometry theorem-proving in the domain **H** to the next section. Below we shall lay down some general principles for the proving of theorems in the domain **E**, leaving actual proving of theorems in the domain **E** to Sect. 7.3.

Basic Principle for Proving of Theorems in **E** :

Each zero X^0 in $Zero(HYP)$ is a geometrical configuration in eventually an extended space $\mathbf{K}'^n, \mathbf{K}'$ being an extension field of **K**, verifying hypothesis of the theorem T. Conclusion is true for such a geometrical configuration if $CONC = 0$ at the corresponding zero X^0. To prove a geometrical theorem thus amounts to seek for a significant part P of $Zero(HYP)$ such that $CONC = 0$ on that part P. The theorems on zero-set structures of $Zero(HYP)$ will serve to offer means of finding such parts P and to verify that $CONC = 0$ on P.

Abbreviation. MTP := Mechanical geometry theorem-proving.

Well-Ordering Principle $\longrightarrow$

MTP-Principle I. For a given theorem $T = \{HYP, CONC\}$ in the domain **E** let CS be a char-set of the hypothesis polset HYP and IP be the initial-product of CS. Then

$$Remdr(CONC/CS) = 0 \tag{1.5}$$

is a sufficient condition for

$$Zero(HYP/IP) \subset Zero(CONC). \tag{1.6}$$

In other words (1.5) is a sufficient condition for the theorem T to be true so far

$$IP \neq 0. \tag{1.7}$$

Moreover, the condition (1.5) is also a necessary one for the above to be true in case CS is an irreducible asc-set.

Proof. Let us put

$$R = Remdr(CONC/CS). \tag{1.8}$$

By the remainder formula (3.8) in Chap.3 we have for some pols Q_i and some power-product I_{CONC} of initials of pols C_i in CS, a formula of the form below:

$$I_{CONC} * CONC = \Sigma_i \ Q_i * C_i + R. \tag{1.9}$$

By the Well-Ordering Principle Theorem 4.1 in Chap.3 we have also

$$Zero(CS/IP) = Zero(HYP/IP) \subset Zero(HYP) \subset Zero(CS). \tag{1.10}$$

Suppose first that $R = 0$. Consider any geometrical configuration verifying the hypothesis in question, i.e. some zero X^0 of HYP. By second part of (1.10) X^0 is also a zero of all pols C_i. If X^0 is not a zero of IP or not a zero of any initial of C_i, then X^0 is not a zero of I_{CONC}. It follows then from (1.9) with $R = 0$ that X^0 is a zero of $CONC$, or (1.6) is true. This proves the first part of the Principle.

Suppose next that CS is irreducible and (1.6) is true so that theorem T is true for any geometrical configuration corresponding to $X = X^0$ for which none of the initials of CS is 0. If in particular $X = \Xi^0$ is the extended zero defined by the irreducible asc-set CS, then by Theorem 3.6.2 none of the initials of CS at Ξ^0 is 0 so that $\Xi^0 \in Zero(CS/IP) = Zero(HYP/IP)$. This Ξ^0 will be a zero of all pols C_i and by the asumption (1.6) it is also a zero of $CONC$. By (1.9) Ξ^0 is then a zero of R. As R is reduced w.r.t. the irreducible asc-set CS, we have again $R = 0$ by Theorem 3.6.2. This proves the second part of the Principle. □

Zero Decomposition Theorem $\longrightarrow$

MTP-Principle II. For a given theorem $T = \{HYP, CONC\}$ in the domain **E** let AS_i be a set of asc-sets determined from the hypothesis polset HYP such that

$$Zero(HYP) = \bigcup_i Zero(AS_i/IP_i) \tag{1.11}$$

in which each IP_i is the initial-product of the corresponding asc-set AS_i. If

$$Remdr(CONC/AS_i) = 0, \tag{1.12}$$

then the theorem T will be true on that part $Zero(AS_i/IP_i)$ of $Zero(HYP)$ or

$$Zero(AS_i/IP_i) \subset Zero(CONC). \tag{1.13}$$

Conversely, if AS_i is irreducible and the theorem T is true on the part $Zero(AS_i/IP_i)$ or (1.13) is true, then we have (1.12).

Proof. The proof is similar to that of Principle I in replacing CS by AS_i. □

Variety Decomposition Theorem $\longrightarrow$

MTP-Principle III. For a given theorem $T = \{HYP, CONC\}$ in the domain **E** let IRR_k be a set of irreducible asc-sets determined from the hypothesis polset HYP such that

$$Zero(HYP) = \bigcup_k Var[IRR_k]. \tag{1.14}$$

Then the theorem T will be true on that irreducible component $Var[IRR_k]$ if and only if

$$Remdr(CONC/IRR_k) = 0. \tag{1.15}$$

Proof. By MTP-Principle II (1.15) is true if and only if

$$Zero(IRR_k/IP_k) \subset Zero(CONC), \tag{1.16}$$

in which IP_k is the initial-product of the irreducible asc-set IRR_k. From (3.6.11) in Chap. 3 we have

$$Zero(IRR_k/IP_k) \subset Var[IRR_k]. \tag{1.17}$$

Hence from

$$Var[IRR_k] \subset Zero(CONC), \tag{1.18}$$

it would follow (1.16) and hence (1.15). Conversely suppose that (1.15) is true so that (1.16) is true too or the theorem T is true on the part $Zero(IRR_k/IP_k)$. Let Ξ_k be a generic point of $Var[IRR_k]$. Then by Theorem 3.6.1 we have

$$\Xi_k \in Zero(IRR_k/IP_k). \tag{1.19}$$

Hence from (1.16) it would follow $\Xi_k \in Zero(CONC)$. As all points of $Var[IRR_k]$ are specializations over **K** of Ξ_k we have then (1.18). Hence (1.18) is equivalent to (1.16) or (1.15) as to be proved. □

Some Remarks.

Remark 1.4. From MTP-Principle I we see that (1.7) actually gives the *non-degeneracy conditions* for the theorem in question $T = \{HYP, CONC\}$ to be true if (1.5) is verified. In that case we shall say that the theorem T is *generically* true under the non-degeneracy conditions (1.7).

Remark 1.5. For theorems in the domain **E** the MTP-Principles I-III give three different methods of geometry theorem-proving. Principle I is the simplest one to apply but somewhat incomplete in that theorems will only be proved in the *generic* sense. The method in Principle III is the most complete one in that no non-degeneracy conditions will be involved but is not so easy to apply.

Remark 1.6. The algorithm of Tarski for domain **T** in which inequalities are really involved is so complicated that it is very difficult to prove any theorems interesting. However we shall later consider some special cases with entirely different methods which will become feasible and effective.

Remark 1.7. In latter sections we shall give concrete examples to illustrate the various methods of geometry theorem-proving.

Remark 1.8. Throughout the whole book we have restricted ourselves to the case of the basic field being of characteristic 0. However, so far theorem-proving is concerned, the proving of theorems in *finite* geometries for which the basic field is of prime characteristic is of high interest in itself. We shall nontheless leave this topic in referring only to some interesting original papers of Lin and Liu, see e.g. [L-L1,2].

7.2 Mechanical Proving of Geometry Theorems of Hilbertian Type.

In what follows we shall suppose that the intrinsically associated number system of a geometry is a field of charateristic 0 which is usually the field of rational numbers **Q**. The field will be denoted by **K** and the notations in previous chapters will be hencefoth used.

Explicit Constructibility and Theorems of Hilbertian Type.

Along the general line of thought about geometry theorem-proving as described in Sect. 7.1, we adopt some coordinate system to turn geometry entities and geometry relations into algebraic expressions involving variables $X = (x_1, \cdots, x_n)$ arranged in the usual definite order which may be coordinates of points, coefficients of line equations, ratios, distances, etc.

Definition 2.1. A geometric entity is *explicitly constructible* from some already constructed entities := The variables appeared in that entity are in the field of rational functions $\mathbf{K}(X')$, where X' are variables appeared in the already constructed entities as well as some newly introduced parameters.

Definition 2.2. A geometry theorem is of *explicitly constructible type* or of *Hilbertian type* := The geometry entities involved in the theorem can be entered in succession such that each geometry entity is either arbitrary in terms of arbitrary parameters or is explicitly constructible from the previous ones.

Hilbert Mechanization Theorem.

As a consequence of the above definitions for a geometry theorem $T = \{HYP, CONC\}$ of Hilbertian type or one in domain $\mathbf{H}$ the variables corresponding to the geometry entities involved can be arranged in a definite order say

$$x_1 \prec x_2 \prec \cdots \prec x_n$$

such that for the hypothesis we have

$$x_{i_k} = \frac{P_{i_k}}{Q_{i_k}}, \quad P_{i_k}, Q_{i_k} \in \mathbf{K}[X_{i_k-1}], \tag{2.1}_k$$

in which i_k runs over a subset $N' = \{i_1, \cdots, i_r\} \subset N = \{1, 2, \cdots, n\}$ with $i_1 < \cdots < i_r$. The equations $(2.1)_k$ form hypothesis set HYP consisting of rational functions $x_{i_k} = \frac{P_{i_k}}{Q_{i_k}}$. Cf. Def. 1.10. To verify whether the theorem is true, it is only necessary to eliminate x_{i_k} successively in reverse order by means of the equations (2.1). More precisely, let us set U as the set consisting of variables x_i with $i \in N \setminus N'$ while for $s > 0, X_s^+ = U + X_s$ and form the scheme below:

$$CONC = R_r(X) \longrightarrow_{(r)} R_{r-1}(X_{i_r-1}^+) \longrightarrow_{(r-1)} \cdots \longrightarrow_{(1)} R_0(X_{i_1-1}^+) = R_0(U). \quad (*)$$

In the scheme $(*)$ each step (k) consists of replacing x_{i_k} in R_k by $\frac{P_{i_k}}{Q_{i_k}}$ of (2.1). Then the theorem in question is *generically true* if and only if the final result R_0 of elimination is 0. Here generically true means that the theorem is true under the *non-degeneracy conditions* $Q_0 \neq 0$ and $Q_{i_k} \neq 0, i_k \in N'$ where Q_0 is the denominator of $CONC$.

To facilitate the computations we may also modify the scheme $(*)$ to $(**)$ below:

$$CONC = R_r(X) \longrightarrow_{(r)} N_{r-1}(X_{i_r-1}^+) \longrightarrow_{(r-1)} \cdots \longrightarrow_{(1)} N_0(X_{i_1-1}^+) = N_0(U). \quad (**)$$

In $(**)$ each N_k is the numerator of the rational function obtained from the previous N_{k+1} in replacing x_{i_k} by $\dfrac{P_{i_k}}{Q_{i_k}}$ of $(2.1)_k$, the N_r being the numerator of $CONC = R_r(X)$. As an alternative procedure, we may also first reduce each equation $(2.1)_k$ to the form

$$x_{i_k} = \frac{P'_{i_k}}{Q'_{i_k}}, \ P'_{i_k}, Q'_{i_k} \in \mathbf{K}[U], \tag{$(2.1)'_k$}$$

by means of $(2.1)_j, j < k$, and then apply the procedures $(*)$ or $(**)$.

As the procedures $(*)$ and $(**)$ are mechanical ones, the above gives us the following theorem. See also Sect. 2 of Chap. 2.

Theorem 2.1. (Hilbert Mechanization Theorem) The class of geometry theorems of Hilbertian type or the class of theorems in domain $\mathbf{H}$ is mechanizable.

Examples of Explicit Constructibility.

According to Hilbert's original formulation, the explicit constructibility is restricted to line-intersection, line-parallelism, and points-collineation, in a word, constructions of a *linear* character. It was D.M.Wang who first pointed out that the applicability of Hilbert's idea of explicit constructibility may be much extended beyond the geometrical constructions of a linear character, e.g. to perpendicularity. Others like X.S. Gao, X.L.Lu, et al, had further greatly extended the domain of such explicit constructions. The following are some of such examples restricted to ordinary geometry. The coordinate system in the ordinary plane or space will be (Y_1, Y_2) and (Y_1, Y_2, Y_3) respectively in what follows.

Construction 2.1. The point of intersection of two lines is explicitly constructible. Similarly the line joining two points is explicitly constructible.

In fact, let the two given lines be of the form

$$\begin{aligned} L_1: \ u_1 * Y_1 + u_2 * Y_2 = u_3, \\ L_2: \ u_4 * Y_1 + u_5 * Y_2 = u_6, \end{aligned} \tag{2.2}$$

in which u_i are new parameters or variables already introduced. The point of intersection will then be given by (z_1, z_2) with

$$z_1 = \frac{u_3 * u_5 - u_6 * u_2}{u_1 * u_5 - u_2 * u_4}, \ z_2 = \frac{u_1 * u_6 - u_3 * u_4}{u_1 * u_5 - u_2 * u_4}. \tag{2.3}$$

These equations show that the point of intersection is explicitly constructible. The non-degeneracy condition is seen to be given by

$$u_1 * u_5 - u_2 * u_4 \neq 0, \tag{2.4}$$

which means that the two lines are not (algebraically) parallel. Similarly, for two given points $P_1 = (u_1, u_2), P_2 = (u_3, u_4)$ the line P_1P_2 will be given by

$$z_1 * Y_1 + z_2 * Y_2 + z_3 = 0, \text{ with}$$

$$z_1 = u_2 - u_4,\ z_2 = u_3 - u_1,\ z_3 = u_1 * u_4 - u_2 * u_3. \tag{2.3$'$}$$

The non-degeneracy conditions are then

$$u_2 \neq u_4, \text{ or } u_1 \neq u_3. \tag{2.4$'$}$$

Construction 2.2. The distance-square of two points is explicitly constructible, but the distance itself is not so.

For the two points $(u_1, u_2), (u_3, u_4)$ given the distance-square is given by $(u_3 - u_1)^2 + (u_4 - u_2)^2$ and is thus explicitly constructible. On the other hand the distance itself cannot be expressed as a rational function of the u_i's and is thus non-explicitly constructible.

We list below a few further explicit constructions for reference for which the equations expressing the newly introduced variables in terms of those already known are easily determined.

Construction 2.3. The foot of the perpendicular from a point to a line is explicitly constructible, but the length of the perpendicular is not so.

Construction 2.4. The reflection of a point or a line w.r.t. a line is explicitly constructible, but the bisector of an angle is not so.

Construction 2.5. The points of intersection of a line through a point on a circle with that circle is explicitly constructible, but the points of intersection of a line with a circle is not so.

Construction 2.6. The circumcenter of a triangle is explicitly constructible, but the in- or es-centers of a triangle is not so. The radius of the circumscribed circle as well as any of the inscribed or escribed circles are non-explicitly constructible too. The case is similar for spheres circumscribing or inscribing a tetrahedron in ordinary space.

Construction 2.7. The tangent lines to a circle from a point not on that circle is not explicitly constructible, but the tangent line at a point on the circle is explicitly constructible.

The list of explicit and non-explicit constructibility can clearly be multiplied almost indefinitely, the more so when geometries beyond the ordinary or the Euclidean ones are considered.

Examples of Theorems of Hilbertian Type.

Even for the class of explicitly constructible theorems of linear character considered originally by Hilbert, it englobes already a lot of quite non-trivial interesting theorems. As examples we may mention, besides the Pappus line theorem and the Pascal line theorem, also the theorems about 6 Pappus lines of triads of points on two lines, and the Pascal conic theorem, all discovered and proved by the present author. With the class enlarged as in the previous constructions, let us give below for illustrations some such theorems involving circles or metric properties which are all of *non-linear* character and at first sight hardly seems to be explicitly constructible. For each theorem we have to deliberately well-arrange the geometric entities in some order in order to render the theorem explicitly constructible.

Example 2.1. (Simson Line Theorem) Let D be a point on the circum-circle of a triangle ABC. Then the feet of perpendiculars E, F, G from D to the sides AB, AC, BC are collinear.

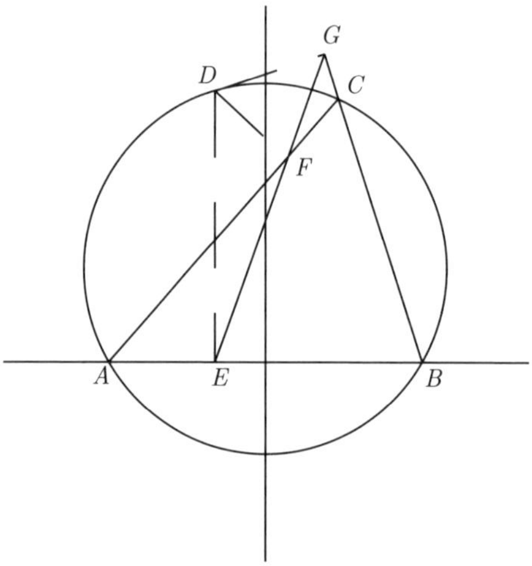

Fig. 2.1.

Proof. To see that the theorem is of Hilbertian type let us take coordinates such that A, B and the circum-center O of the triangle ABC are given as $A = (x_1, 0)$, $B = (-x_1, 0)$, $O = (0, x_2)$. To determine C, D let us consider the lines AC, AD to be

$$\begin{aligned} AC: &\quad x_4 * (Y_1 - x_1) = x_5 * Y_2, \\ AD: &\quad x_8 * (X - x_1) = x_9 * Y, \end{aligned}$$

with x_4, x_5, x_8, x_9 as arbitrary parameters. Then $C = (x_6, x_7), D = (x_{10}, x_{11})$ are explicitly constructible by Construction 2.5 and are given by

$$x_6 = \frac{x_1 * (x_4^2 - x_5^2) + 2 * x_2 * x_4 * x_5}{x_4^2 + x_5^2}, \quad x_7 = \frac{2 * x_4 * (x_2 * x_4 - x_1 * x_5)}{x_4^2 + x_5^2}, \tag{2.5}$$

$$x_{10} = \frac{x_1 * (x_8^2 - x_9^2) + 2 * x_2 * x_8 * x_9}{x_8^2 + x_9^2}, \quad x_{11} = \frac{2 * x_8 * (x_2 * x_8 - x_1 * x_9)}{x_8^2 + x_9^2}. \tag{2.6}$$

Finally the feet of perpendiculars $G = (x_{12}, x_{13})$, $F = (x_{14}, x_{15})$, $E = (x_{16}, x_{17})$ are all explicitly constructible by Construction 2.3 and their coordinates are given by

$$\begin{aligned} x_{12} &= \frac{x_{10} * (x_6 + x_1)^2 + x_7 * x_{11} * (x_6 + x_1) - x_1 * x_7^2}{(x_6 + x_1)^2 + x_7^2}, \\ x_{13} &= \frac{x_7 * ((x_6 + x_1) * (x_{10} + x_1) + x_{11})}{(x_6 + x_1)^2 + x_7^2}; \end{aligned} \tag{2.7}$$

$$\begin{aligned} x_{14} &= \frac{x_1 * x_4^2 + x_4 * x_5 * x_{11} + x_{10} * x_5^2}{x_4^2 + x_5^2}, \\ x_{15} &= \frac{x_4 * (-x_1 * x_5 + x_4 * x_{11} + x_{10} * x_5)}{x_4^2 + x_5^2}; \end{aligned} \tag{2.8}$$

$$x_{16} = x_{10}, \quad x_{17} = 0. \tag{2.9}$$

The rational functions corresponding to the equations (2.5)-(2.9) form the hypothesis set HYP. The conclusion that E, F, G are collinear is equivalent to say that

$$CONC \equiv \begin{vmatrix} x_{12} & x_{13} & 1 \\ x_{14} & x_{15} & 1 \\ x_{16} & x_{17} & 1 \end{vmatrix} = 0. \tag{2.10}$$

The proof that $CONC = 0$ is readily done by successive elimination of x_{17}, x_{16}, x_{15}, x_{14}, x_{13}, x_{12}, x_{11}, x_{10}, x_7, x_6 by means of (2.9),(2.8), (2.7),(2.6),(2.5) in succession. In accordance to scheme (**) the successive index-sets of the pols N_k are as follows:

$$CONC: [6\ x_{17}\ 1] \longrightarrow [4\ x_{15}\ 1] \longrightarrow [9\ x_{13}\ 1] \longrightarrow [12\ x_{11}\ 2] \longrightarrow [12\ x_9\ 1] \longrightarrow 0.$$

It follows that the theorem is *generically* true under the non-degeneracy conditions

$$x_4^2 + x_5^2 \neq 0, \ x_8^2 + x_9^2 \neq 0, \ (x_6 + x_1)^2 + x_7^2 \neq 0,$$

which show that the lines AC, AD are to be true geometric lines and the point C is not to be coincident to the point B in case the field $\mathbf{K}$ is the real field. □

Example 2.2. Feuerbach Theorem (1822). For a triangle the nine point circle is tangent to the inscribed as well as the three escribed circles.

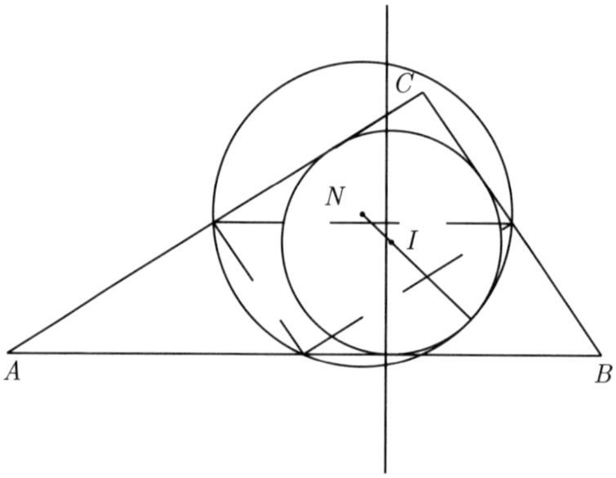

Fig. 2.2.

Proof. In order to make the theorem to be of explicitly constructible type let us take the coordinate system and arrange the construction of geometric entities so that vertices of a triangle ABC and the center I, either the in-center or any one of the es-centers, be given as $I = (0, x_1)$, $A = (x_2, 0)$, $B = (x_3, 0)$, $C = (x_4, x_5)$. By Construction 2.4 the lines AC, BC, as reflections of AB w.r.t. AI, BI respectively, are both explicitly constructible, so is their intersection C. By Construction 2.1 the coordinates (x_4, x_5) of C are then given by

$$x_4 = \frac{x_1^2 * (x_2 + x_3)}{x_1^2 + x_2 * x_3}, \quad x_5 = \frac{2 * x_1 * x_2 * x_3}{x_1^2 + x_2 * x_3}. \tag{2.11}$$

Now the mid-points of the sides AB, AC, BC are clearly explicitly constructible. By Construction 2.5 the circum-center of the mid-points triangle, i.e. the center of the nine-point circle $N = (x_6, x_7)$ is explicitly constructible too and the coordinates x_6, x_7 are given by

$$x_6 = \frac{x_2 + x_3 + 2 * x_4}{4}, \quad x_7 = \frac{x_5^2 + (x_3 - x_4) * (4 * x_6 - 2 * x_2 - x_3 - x_4)}{4 * x_5}. \tag{2.12}$$

The radius-square x_8 of the 9-point circle and the distance-square x_9 between the centers of the 9-point circle and the in- or es-cribed circle are by Construction 2.2 both explicitly constructible with

$$x_8 = x_7^2 + \left(x_6 - \frac{x_2 + x_3}{2}\right)^2, \quad x_9 = (x_7 - x_1)^2 + x_6^2. \tag{2.13}$$

The conclusion, irrespective of the manner of tangency, is equivalent to say that

$$CONC \equiv x_9^2 + x_8^2 - 2 * x_9 * x_8 - 2 * x_9 * x_1^2 - 2 * x_8 * x_1^2 + x_1^4 = 0. \tag{2.14}$$

Irrespective of whether I is an in-center or an es-center and whether the tangency asserted by the theorem is an inner or an outer contact, the conclusion is seen to be true by simple verification of (2.14) by means of scheme (**) with successive index-sets given below:

$$CONC : [6\ x_9\ 2] \longrightarrow [11\ x_8\ 2] \longrightarrow [23\ x_7\ 1] \longrightarrow [45\ x_6\ 2]$$
$$\longrightarrow [18\ x_5\ 1] \longrightarrow [24\ x_4\ 3] \longrightarrow 0.$$

The non-degeneracy conditions are readily seen to be

$$x_1^2 + x_2 * x_3 \neq 0, \ \ x_5 \neq 0.$$

The geometrical meaning of $x_1^2 + x_2 * x_3 \neq 0$ is that AC and BC are not parallel to each other. That of $x_5 \neq 0$ is that C does not lie on the side AB. Both non-degeneracy conditions are thus trivially observed. □

Explicit Constructions Involving Angles.

In Book III of Euclid's *Elements*, there are two propositions Props. 21, 22 which assert for four points A, B, C, D different from each other lying on the same circle, the *angles* $<(ACB)$ and $<(ADB)$ may be either equal or complementary to each other. Such ambiguity would cause much troubles for the proving of geometry theorems in Euclidean fashion: Each time when the propositions are to be applied, we have to verify whether the angles are situated in the same segment or in opposite segments, a fact not easy to ascertain. To avoid such troubles we suggest the following notion of *oriented angles.*

Definition 2.3. Given in the ordinary plane two true geometric lines L_1, L_2. The *oriented angle* from L_1 to L_2 := The angle mod π for rotating about their intersecting point in counterclockwise sense from L_1 to L_2 if L_1, L_2 intersect. Otherwise := 0 mod π.

Notation 2.1. The oriented angle from L_1 to $L_2 := <(L_1, L_2) = -<(L_2, L_1)$.

Remark 2.1. The oriented angle is defined up to multiples of π so that its tangent value is uniquely determined.

Remark 2.2. To avoid the possible infinite value of the tangent of an angle, we shall express the tangent in the form of a ratio $t_1 : t_2$. The usual *slope* of a line in the ordinary plane which is the tangent of the oriented angle from the horizontal axis to that line, will also be expressed as a ratio.

Proposition 2.1. For any three true geometric line L_1, L_2, L_3 in arbitrary position in the ordinary plane, we have $<(L_1, L_3) = <(L_1, L_2) + <(L_2, L_3)$.

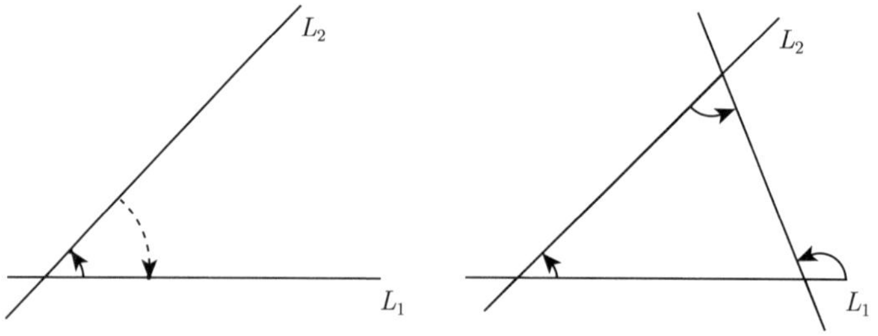

Fig. 2.3.

Remark 2.3. The above equalities as well as similar ones in the sequel are all to be understood as congruences mod π unless otherwise stated.

The Props. 21,22 in Euclid's *Elements* can now be unified to the following one:

Proposition 2.2. In the ordinary plane four points A, B, C, D different from each other will lie on the same circle if and only if $<(AC, BC) = <(AD, BD)$.

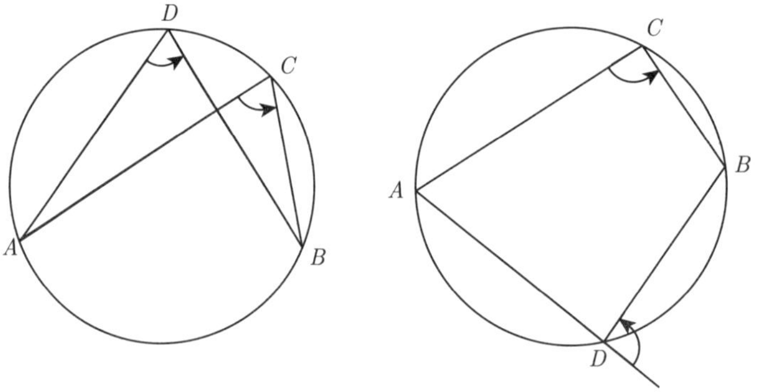

Fig. 2.4.

Construction 2.8. The tangent value of an oriented angle $<(L_1, L_2)$ is explicitly constructible.

Construction 2.9. Given a point P and a line L in the ordinary plane, the line L' through P with given tangent value of $<(L, L')$ is explicitly constructible.

Definition 2.4. For an oriented angle $<(L_1, L_2)$ of intersersecting lines L_1, L_2 any line L through the intersecting point is a *bisector* of the orinted angle $:= <(L_1, L_2) = 2* <(L_1, L)$.

Definition 2.5. For an oriented angle $<(L_1, L_2)$ of intersecting lines L_1, L_2 any line L through the intersecting point is a *trisector* of the oriented angle := $<(L_1, L_2) = 3* <(L_1, L)$.

Remark 2.4. For two intersecting lines L_1, L_2 the oriented angle $<(L_1, L_2)$ has two bisectors and three trisectors.

Defintion 2.6. For any trisector L of an oriented angle $<(L_1, L_2)$ with intersecting lines L_1, L_2 the reflection L' of L_1 w.r.t. L := The *secondary trisector* associated to the trisector L.

Geometry Theorems of Hilbertian Type Involving Angles.

By means of the notion of oriented angles geometry theorems which are difficult to deal with may be turned into theorems of Hilbertian type and become thus easily proved. We cite a few examples below as illustrations.

Example 2.3. Steiner Theorem. Over the three sides of a triangle ABC let us erect all inwards or all outwards equilateral triangles ABC', ACB', BCA'. Then the three lines AA', BB', CC' are concurrent.

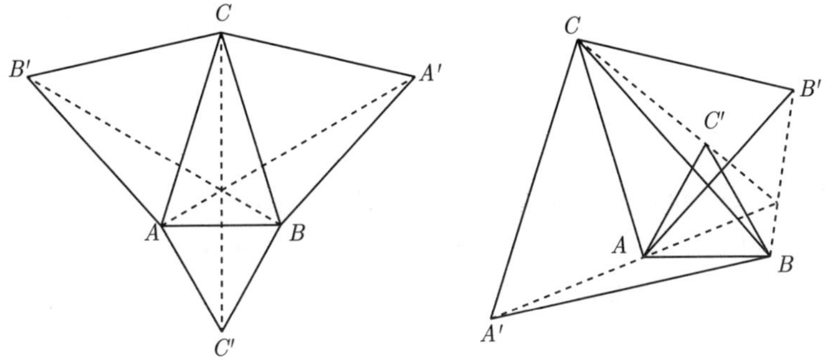

Fig. 2.5.

Proof. Take $A = (x_1, 0), B = (x_2, 0), C = (0, x_3)$. Let $\mathbf{Q}(m)$ be the extension field of $\mathbf{Q}$ in adjoining m for which $m^2 = 3$ (i.e. $m = \pm\, tan\frac{\pi}{3}$). Let $AB', AC', BA', BC', CA', CB'$ be the lines through A, B, C such that

$$\begin{aligned} tan < (AC, AB') = \ tan < (BA, BC') = \ tan < (CB, CA') = m, \\ tan < (AB, AC') = \ tan < (BC, BA') = \ tan < (CA, CB') = -m. \end{aligned} \tag{2.15}$$

By Constructions 2.8 and 2.1 the points of intersection $A' = (x_4, x_5), B' = (x_6, x_7)$, $C' = (x_8, x_9)$ of these lines will be given by

$$x_4 = \frac{m * x_3 + x_2}{2}, \quad x_5 = \frac{m * x_2 + x_3}{2},$$
$$x_6 = \frac{-m * x_3 + x_1}{2}, \quad x_7 = \frac{x_3 - m * x_1}{2}, \tag{2.16}$$
$$x_8 = \frac{x_2 + x_1}{2}, \quad x_9 = \frac{m * (-x_2 + x_1)}{2}.$$

The conclusion that AA', BB', CC' are concurrent is

$$CONC \equiv \begin{vmatrix} x_5 & x_1 - x_4 & x_1 * x_5 \\ x_7 & x_2 - x_6 & x_2 * x_7 \\ x_3 - x_9 & x_8 & x_3 * x_8 \end{vmatrix} = 0. \tag{2.17}$$

The index-sets of the proof procedure according to scheme (**) runs now as follows:

$$CONC : [14\ x_9\ 1] \longrightarrow [18\ x_7\ 1] \longrightarrow [22\ x_5\ 1] \longrightarrow 0.$$ □

One of the most celebrated theorems in ordinary geometry is the Morley Theorem stated below which involves the trisectors of a triangle.

Example 2.4. Morley Theorem. For a triangle ABC let L_A, L_B, L_C be each one of the trisectors of the oriented angles $<(AB, AC), <(BC, BA), <(CA, CB)$. Let L'_A, L'_B, L'_C be their respective associated secondary trisectors. Let D, E, F be the intersection points of the respective pairs of lines $(L_A, L'_B), (L_B, L'_C), (L_C, L'_A)$. Then the triangle DEF is an equilateral one if

$$<(L_A, L'_A) + \ <(L_B, L'_B) + \ <(L_C, L'_C) = \pm\frac{\pi}{3}. \tag{2.18}$$

Proof. Let us take the coordinates with $A = (x_1, 0),\ B = (x_2, 0),\ D = (0, x_3)$. Let the slopes of L_A, L'_A, AC and those of L'_B, L_B, BC be respectively x_4, x_5, x_6 and x_7, x_8, x_9. See Fig. 2.6 below.

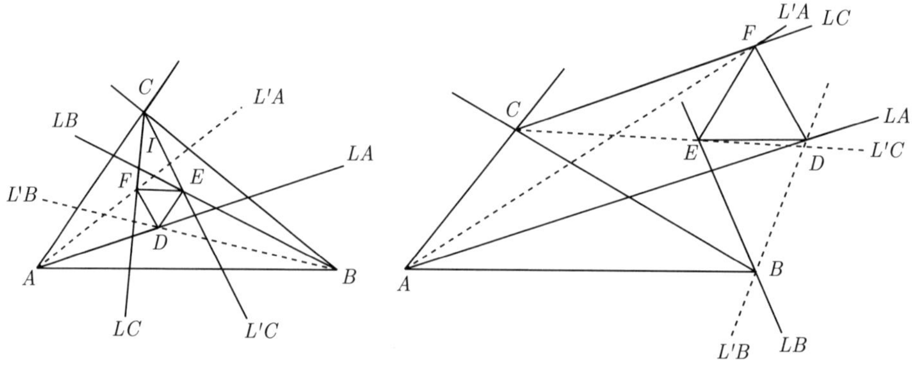

Fig. 2.6.

By Constructions 2.1 and 2.4 we have

$$\begin{aligned} x_4 &= -\frac{x_3}{x_1}, \quad x_5 = \frac{2*x_4}{1-x_4^2}, \quad x_6 = \frac{3*x_4 - x_4^3}{1-3*x_4^2}; \\ x_7 &= -\frac{x_3}{x_2}, \quad x_8 = \frac{2*x_7}{1-x_7^2}, \quad x_9 = \frac{3*x_7 - x_7^3}{1-3*x_7^2}. \end{aligned} \tag{2.19}$$

The point of intersection of lines AC, BC will be $C = (x_{10}, x_{11})$ with

$$x_{10} = \frac{x_9 * x_2 - x_6 * x_1}{x_9 - x_6}, \quad x_{11} = x_6 * (x_{10} - x_1). \tag{2.20}$$

Let $tan\left(\pm\frac{\pi}{3}\right) = a$ so that $a^2 = 3$. Set now $tan < (L_C, L'_C) = x_{12}$ and slopes of $L_C, L'_C = x_{13}, x_{14}$. Then we have

$$\begin{aligned} x_{12} &= \frac{-x_4 + x_7 + a + x_4 * x_7 * a}{1 + x_4 * x_7 + x_4 * a - x_7 * a}, \\ x_{13} &= \frac{x_{12} + x_6}{1 - x_{12} * x_6}, \quad x_{14} = \frac{x_{12} + x_{13}}{1 - x_{12} * x_{13}}. \end{aligned} \tag{2.21}$$

Finally, the points of intersections $E = (x_{15}, x_{16}), F = (x_{17}, x_{18})$ will be given by

$$\begin{aligned} x_{15} &= \frac{x_{14} * x_{10} - x_8 * x_2 - x_{11}}{x_{14} - x_8}, \quad x_{16} = x_8 * (x_{15} - x_2), \\ x_{17} &= \frac{x_{13} * x_{10} - x_5 * x_1 - x_{11}}{x_{13} - x_5}, \quad x_{18} = x_5 * (x_{17} - x_1). \end{aligned} \tag{2.22}$$

The equations (2.22)-(2.25) form the hypothesis system. The conclusions are $|DE|^2 = |DF|^2 = |EF|^2$. By symmetry it is sufficient to prove that $|DE|^2 = |DF|^2$, or

$$CONC \equiv x_{17}^2 + (x_{18} - x_3)^2 - x_{15}^2 - (x_{16} - x_3)^2 = 0.$$

If we reduce first each of $x_i, i = 8, \cdots, 19$ to the form $\frac{P'_i}{Q'_i}$ with P'_i, Q'_i in x_1, x_2, x_3 alone as in $(2.1)'$. Then the index-sets of the proof procedure according to scheme $(**)$ runs as follows:

$$CONC : [8\ x_{18}\ 2] \longrightarrow [125\ x_{16}\ 2] \longrightarrow 0. \qquad \square$$

Remark 2.5. If we proceed by scheme $(**)$ directly from (2.1), then the pols occuring in the procedure arise rapidly to thousands of terms and the procedure fails.

Remark 2.6. Each of the oriented angles $< (AB, AC), < (BC, BA), < (CA, CB)$ has three trisectors, so there may be formed in all 27 triangles DEF. However, under the restriction (2.21) only 18 of them have been proved to be equilateral ones. In fact, the other 9 triangles are in general non-equilateral ones.

A 6-Pole Theorem.

As a further example of geometry theorem proving involving angles let us consider a plane E moving over a fixed plane E_0. The motion of E from one position to another will be either a translation or a rotation.

Definition 2.7. For the rotation of a moving plane over a fixed plane from one position to another the center of rotation := *Pole* of the rotation.

Notation 2.2. For a rotation of a moving plane over a fixed plane from position E_i to E_j the pole and the angle of rotation := P_{ij} and θ_{ij} respectively. Clearly $P_{ij} = P_{ji}$ and θ_{ij} is determined only up to multiples of $2 * \pi$ with $\theta_{ij} = -\theta_{ji}$ (naturally $mod\ 2 * \pi$).

The following proposition is well-known and is fundamental in the kinematic analysis of mechanism studies.

Proposition 2.3. For three positions E_i, E_j, E_k of the moving plane over a fixed plane suppose that all of the motions from one position to another one are all rotations. Then we have

$$< (P_{ij}P_{ik}, P_{ij}P_{jk}) = \frac{\theta_{ij}}{2}. \tag{2.23}$$

The equality is understood to be congruence $mod\ \pi$.

Example 2.5. A 6-Pole Theorem. Suppose a moving plane over a fixed plane take up four different positions $E_i, i = 1, 2, 3, 4$, with each of the motions from one position E_i to E_j a rotation with pole P_{ij} and angle of rotation θ_{ij}. Then the three mid-points $M_{12,34}, M_{13,24}, M_{14,23}$ of the three pairs of poles $(P_{12}, P_{34}), (P_{13}, P_{24}), (P_{14}, P_{23})$ are collinear.

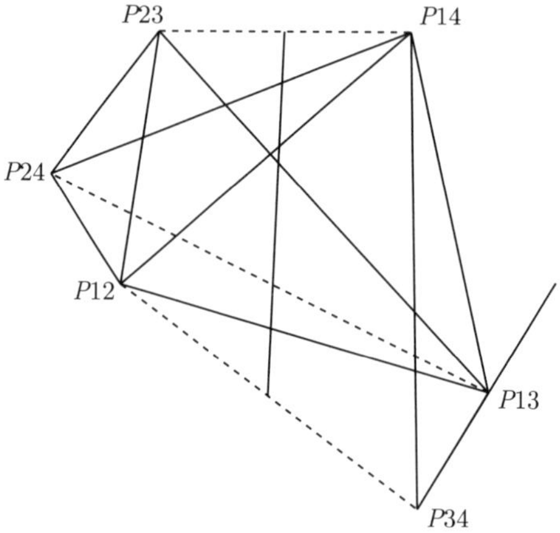

Fig. 2.7.

Proof. The points of $P_{12}, P_{13}, P_{23}, P_{14}$ may be arbitrarily chosen so let us take coordinates such that

$$P_{12} = (0,0),\ \ P_{13} = (x_1, 0), P_{23} = (x_2, x_3),\ \ P_{14} = (x_4, x_5).$$

The point P_{24} is now no more arbitrary but is constrained according to Proposition 2.3 by the relation

$$< (P_{12}P_{13}, P_{12}P_{23}) =< (P_{12}P_{14}, P_{12}P_{24}) \ \left(= \frac{\theta_{12}}{2}\right).$$

The line $P_{12}P_{24}$ is thus of the form

$$(x_4 * x_2 - x_5 * x_3) * Y_1 = (x_4 * x_3 + x_5 * x_2) * Y_2.$$

We may thus take $P_{24} = (x_6, x_7)$ arbitrarily on this line with x_0 an arbitrary constant and

$$x_6 = (x_4 * x_2 - x_5 * x_3) * x_0,\ \ x_7 = (x_4 * x_3 + x_5 * x_2) * x_0. \tag{2.24}$$

By Proposition 2.3 the lines $P_{13}P_{34}$ and $P_{14}P_{34}$ are constrained by the relations

$$< (P_{13}P_{14}, P_{13}P_{34}) =< (P_{13}P_{12}, P_{13}P_{23}) \ \left(= \frac{\theta_{13}}{2}\right),$$

of which the tangent value is $x_3 : (x_2 - x_1)$. By Construction 2.9 the line $P_{13}P_{34}$ has slope $x_8 : x_9$ with

$$\begin{aligned} x_8 &= -(x_4 - x_1) * x_3 - x_5 * (x_2 - x_1), \\ x_9 &= -(x_4 - x_1) * (x_2 - x_1) + x_5 * x_3. \end{aligned} \tag{2.25}$$

Similarly we have

$$< (P_{14}P_{13}, P_{14}P_{34}) =< (P_{14}P_{12}, P_{14}P_{24}) \ \left(= \frac{\theta_{13}}{2}\right)$$

of which the tangent value is $x_{10} : x_{11}$ with

$$\begin{aligned} x_{10} &= -\ x_5 * (x_6 - x_4) + x_4 * (x_7 - x_5), \\ x_{11} &= x_4 * (x_6 - x_4) + x_5 * (x_7 - x_5). \end{aligned} \tag{2.26}$$

It follows that the line $P_{14}P_{34}$ has slopes $x_{12} : x_{13}$ with

$$\begin{aligned} x_{12} &= -(x_4 - x_1) * x_{10} - x_5 * x_{11}, \\ x_{13} &= -(x_4 - x_1) * x_{11} + x_5 * x_{10}. \end{aligned} \tag{2.27}$$

The point $P_{34}=(x_{14},x_{15})$ is then by Construction 2.1 given by

$$\begin{aligned} x_{14}&=\frac{x_{13}*x_9*x_5+x_{13}*x_8*x_1-x_{12}*x_9*x_4}{x_{13}*x_8-x_{12}*x_9},\\ x_{15}&=\frac{x_8*(x_{13}*x_5-x_{12}*x_4+x_{12}*x_1)}{x_{13}*x_8-x_{12}*x_9}. \end{aligned}\tag{2.28}$$

The mid-points are then

$$M_{12,34}=\left(\frac{x_{14}}{2},\frac{x_{15}}{2}\right),\ M_{13,24}=\left(\frac{x_6+x_1}{2},\frac{x_7}{2}\right),\ M_{14,23}=\left(\frac{x_4+x_2}{2},\frac{x_5+x_3}{2}\right).\tag{2.29}$$

The conclusion may be put in the form

$$CONC\equiv\begin{vmatrix} x_{14} & x_{15} & 1\\ x_6+x_1 & x_7 & 1\\ x_4+x_2 & x_5+x_3 & 1 \end{vmatrix}=0.$$

The index-sets of the proof procedure according to scheme (**) runs now as follows:

$$CONC:[13\,x_{15}\,1]\longrightarrow[24\,x_{13}\,1]\longrightarrow[46\,x_{11}\,1]\longrightarrow[54\,x_9\,1]\longrightarrow[83\,x_7\,2]\longrightarrow 0.\quad\square$$

A Theorem about Cubic Curve.

The explicit construction and the corresponding method of proving geometry theorems are not restricted to ordinary geometry or its alike alone. Thus, we have e.g. the following

Construction 2.10. Give in a projective plane a cubic curve and two points on it. The third point of intersection of the joining line of the two given points with the cubic is explicitly contructible.

Example 2.6. Cubic Syzygy Theorem. Let $P_i,\ i=1,\cdots,8$ be 8 points on a cubic C in the projective plane such that the triads of points $P_1,P_2,P_5;P_3,P_4,P_6;P_1,P_3,P_7;\ P_2,P_4,P_8;$ are all collinear. Then the point of intersection P_9 of the lines P_5P_6,P_7P_8 are also on the cubic.

Proof. Let us take the homogeneous coordinates $(Y_1:Y_2:Y_3)$ such that

$$P_1=(1:0:0),\ P_2=(0:1:0),\ P_3=(0:0:1),\ P_4=(1:1:1).$$

Then the cubic C will have an equation of the form

$$\begin{aligned} &x_1*Y_1*Y_2*(Y_1-Y_3)+x_2*Y_1*Y_3*(Y_1-Y_2)\\ &+x_3*Y_1*Y_2*(Y_2-Y_3)+x_4*Y_2*Y_3*(Y_2-Y_1)\\ &+x_5*Y_3*Y_1*(Y_3-Y_2)+x_6*Y_2*Y_3*(Y_3-Y_1)=0. \end{aligned}\tag{2.30}$$

By Construction 2.10 the points $P_j, j = 5, 6, 7, 8$ will be given by

$$P_5 = (x_3 : -x_1 : 0),\ P_6 = (x_5 + x_6 : x_5 + x_6 : x_1 + x_3),$$
$$P_7 = (x_5 : 0 : -x_2),\ P_8 = (x_3 + x_4 : x_2 + x_5 : x_3 + x_4).$$

The point $P_9 = (x_7 : x_8 : x_9)$ is by analogue of Construction 2.1 given by

$$\begin{aligned} x_7 &= x_6 * x_4 + x_6 * x_3 + x_5 * x_4, \\ x_8 &= x_6 * x_2 + x_5 * x_2 + x_5 * x_1, \\ x_9 &= x_4 * x_1 + x_3 * x_2 + x_3 * x_1. \end{aligned} \tag{2.31}$$

The conclusion to be proved is that P_9 lies on the cubic or

$$\begin{aligned} CONC \equiv\, & x_1 * x_7 * x_8 * (x_7 - x_9) + x_2 * x_7 * x_9 * (x_7 - x_8) \\ & + x_3 * x_7 * x_8 * (x_8 - x_9) + x_4 * x_8 * x_9 * (x_8 - x_7) \\ & + x_5 * x_9 * x_7 * (x_9 - x_8) + x_6 * x_8 * x_9 * (x_9 - x_7) = 0. \end{aligned}$$

The theorem is now proved with index-sets of the proof procedure according to scheme (**) running as follows:

$$CONC : [12\ x_9\ 2] \longrightarrow [36\ x_8\ 2] \longrightarrow [60\ x_7\ 2] \longrightarrow 0. \qquad \square$$

7.3 Mechanical Proving of Geometry Theorems involving Equalities Alone.

In the previous section we have seen that many geometry theorems may be unexpectedly so arranged to become theorems of Hilbertian type by deliberate considerations. There are also theorems for which very likely such arrangements are actually impossible. In order to avoid such somewhat artificial arrangements eventually possible and to proceed along more natural lines let us restrict ourselves to the class of theorems **E** which involves equalities but not inequalities. Then there are three MTP-Principles which may be applied. Let us consider first the simplest case of MTP-Principle I.

Examples of MTP-Principle I based on Well-Ordering Principle.

For convenience of the readers let us reproduce the MTP-Principle I below:

MTP-Principle I. For a given theorem $T = \{HYP, CONC\}$ in the domain **E** let CS be a char-set of the hypothesis polset HYP. Then

$$R \equiv Remdr(CONC/CS) = 0 \tag{3.1}$$

is a sufficient condition for theorem T to be *generically* true under the *non-degeneracy condition* $IP = 0$, in which IP is the initial-product of the char-set CS. Moreover, the condition $R=0$ is also a necessary one for the above to be true in case CS is an irreducible asc-set.

Example 3.1. 8-Triads Theorem. From 8 mutually different points $A_i, i=1,\cdots,8$ in circular order with $A_{i+8}=A_i$, there may be formed 8 triads $T_i=A_iA_{i+1}A_{i+3}, i=1,\cdots,8$. If all of these triads are collinear, then *generically* the triad $A_1A_2A_3$ is collinear too and in fact *generically* all the 8 points will lie on the same line.

Proof. Let us call the configuration consistng of the 8 points $A_i=(x_i,y_i), i=1,\cdots,8,(A_{i+8}=A_i)$ different from each other for which the 8 triads $T_i=A_iA_{i+1}A_{i+3}, i=1,\cdots,8$, are collinear a *8-triad configuration.* The configuration will be said to be *non-trivial* if the 8 points A_i are not on the same line.

For the given 8-triad configuration take now coordinates such that

$$A_1=(0,0),\ A_2=(x_2,0),\ A_i=(x_i,y_i),\ i=3,\cdots,8. \tag{3.2}$$

Then $T_1=A_1A_2A_4$ collinear implies $y_4=0$ and $T_i,\ i=2,\cdots,8$, collinear give us a hypothesis polset $HYP=\{P_1,\cdots,P_7\}$ with P_i given below:

$$\begin{aligned}
P_1&=y_5*(x_3-x_2)-y_3*(x_5-x_2),\\
P_2&=y_6*(x_4-x_3)+y_3*(x_6-x_4),\\
P_3&=y_7*(x_5-x_4)-y_5*(x_7-x_4),\\
P_4&=y_8*(x_6-x_5)-y_6*(x_8-x_5)+y_5*(x_8-x_6),\\
P_5&=y_7*x_6-y_6*x_7,\\
P_6&=y_8*(x_7-x_2)-y_7*(x_8-x_2),\\
P_7&=y_8*x_3-y_3*x_8.
\end{aligned} \tag{3.3}$$

To prove the theorem let us first remark that the collinearity of $A_1A_2A_3$ implies already that the configuration is a trivial one. In fact, suppose that $A_1A_2A_3$ is collinear so that $y_3=0$, then we have

$$\begin{aligned}
y_3=0&\Longrightarrow_{(A_3\neq A_2)}\ x_3\neq x_2\ \Longrightarrow_{(P_1=0)}\ y_5=0\\
&\Longrightarrow_{(A_5\neq A_4)}\ x_5\neq x_4\ \Longrightarrow_{(P_3=0)}\ y_7=0\\
&\Longrightarrow_{(A_4\neq A_3)}\ x_4\neq x_3\ \Longrightarrow_{(P_2=0)}\ y_6=0\\
&\Longrightarrow_{(A_6\neq A_5)}\ x_6\neq x_5\ \Longrightarrow_{(P_4=0)}\ y_8=0.
\end{aligned}$$

It follows then all the 8 points A_i are collinear and the configuration becomes a trivial one.

Take now the ordering

$$x_2\prec x_3\prec x_4\prec x_5\prec x_6\prec x_7\prec x_8\prec y_3\prec y_5\prec y_6\prec y_7\prec y_8. \tag{3.4}$$

The char-set of HYP is easily found to be $CS = \{C_1, \cdots, C_5\}$ with

$$C_1 = y_3 * I_1, \; C_i = P_{i-1}, \; i = 2, \cdots, 5,$$

in which I_1 has an index set $[12 \; x_7 \; 1]$. The initial-product of CS is

$$IP = I_1 * (x_3 - x_2) * (x_4 - x_3) * (x_5 - x_4) * (x_6 - x_5).$$

It follows that the theorem is *generically* true under the non-degeneracy conditions

$$I_1 \neq 0, \; x_3 \neq x_2, \; x_4 \neq x_3, \; x_5 \neq x_4, \; x_6 \neq x_5.$$

If we adjoin I_1 to HYP, it will be found that the new hypothesis polset HYP' will have a char-set $CS' = \{I_1, y_3 * I_2, C_2, \cdots, C_5\}$ with $C_i, i = 2, \cdots, 5$ as before and

$$I_2 = (x_4 - x_2) * I_2',$$

in which I_2' has an index set $[24 \; x_8 \; 1]$. Consequently if we have $I_1 = 0$ but $x_4 \neq x_2$ and $I_2' \neq 0$ then the theorem is still *generically* true. If we adjoin both I_1, I_2' to HYP and proceed as before the new hypothesis polset will have a char-set $CS'' = \{I_1, I_2', y_3 * I_3, C_2, \cdots, C_5\}$ in which

$$I_3 = (x_6 - x_5) * (x_5 - x_4) * (x_4 - x_3) * (x_3 - x_2) * I_3'',$$

I_3'' being of an index set $[20 \; x_6 \; 2]$. Combining these together we conclude that the theorem is *generically true* under the non-degeneracy conditions $x_i \neq 0, x_i \neq x_j$ for all $i, j = 2, \cdots, 8, i \neq j$ and any one of I_1, I_2', I_3'' is $\neq 0$. □

Remark 3.1. From Ex. 3.1 we see that the non-degeneracy conditions may be so intricate to have any evident geometrical meaning and to be foreseen. However, our method permits to discover such non-degeneracy conditions in an automatic way.

Remark 3.2. From the proof we see that the 8-triads theorem will no more be true if the $x_i, i = 1, \cdots, 8$, are mutually different and all I_1, I_2', I_3'' are 0. In fact, there are ∞^8 configurations in some extended affine space (e.g. the complex space) such that all the points A_i are different from each other and all the 8 triads $A_i A_{i+1} A_{i+3}$ are collinear but $A_1 A_2 A_3$ is not collinear. The situation will however be entirely different if we restrict the configurations to be in the *real* affine space, cf. Ex. 4.4 in the next section.

Example 3.2. Desargues Theorem. Let $ABC, A'B'C'$ be two triangles such that the corresponding sides are parallel to each other. Suppose that AA', BB' intersect in a point O. Then CC' will also pass through O.

Proof. This is one part of the theorem considered in Ex. 2.1.4′ and Ex. 3.6.3. It is easy to express the theorem in an explicitly constructible form and then prove it as in the last section. However, we shall prove it in a different way for the sake of illustrating our more general method.

To prove the theorem let us take coordinates as in Ex. 3.6.3, viz.

$$A = (x_1, 0), \quad B = (0, x_3), \quad C = (x_4, x_5),$$
$$A' = (x_2, 0), \quad B' = (0, x_6), \quad C' = (x_7, x_8).$$

The hypothesis polset is then $HYP = \{H_1, H_2, H_3\}$ as in Ex. 3.6.3 which we reproduce below:

$$\begin{aligned} H_1 &= x_1 * x_6 - x_2 * x_3, \\ H_2 &= x_4 * (x_8 - x_6) - x_7 * (x_5 - x_3), \\ H_3 &= (x_4 - x_1) * x_8 - x_5 * (x_7 - x_2). \end{aligned} \tag{3.5}$$

The conclusion pol is given by

$$CONC \ = \ x_4 * x_8 - x_5 * x_7. \tag{3.6}$$

The polset $HYP = \{H_1, H_2, H_3\}$ is not in the form of an asc-set. It is readily found that the char-set is found to be $CS = \{C_1, C_2, C_3\}$ with

$$\begin{aligned} C_1 &= H_1, \\ C_2 &= x_7 * (x_1 * x_3 - x_1 * x_5 - x_3 * x_4) + x_4^2 * x_6 + x_2 * x_4 * x_5 - x_2 * x_3 * x_4, \\ C_3 &= H_3. \end{aligned} \tag{3.7}$$

It turns out that (3.1) is verified. Hence the theorem is true generically under the non-degeneracy conditions

$$I_1 \equiv x_1 \neq 0, \ I_2 \equiv x_1 * x_3 - x_1 * x_5 - x_3 * x_4 \neq 0, \ I_3 \equiv x_4 - x_1 \neq 0,$$

the I_1, I_2, I_3 being the initials of C_1, C_2, C_3 respectively.

Remark 3.3. The condition $I_2 \neq 0$ means that the triangle ABC does not degenerate to a line which is indispensable as we see from the diagram Fig. 3.6.2 or 3.6.6 in Chap. 3, Sect. 6. In fact, let us adjoin I_2 to HYP and determine the char-set of the enlarged hypothesis polset. It will be found that in this case the remainder of the conclusion pol is no more 0 so that the theorem is no more true if the triangle ABC degenerates to a line.

Remark 3.4. The non-degeneracy condition $I_3 \neq 0$ means that AC is not parallel to the Y-axis. To see whether this condition is really necessary for the truth of the theorem we may proceed further in adjoining to HYP the new hypothesis pol I_3, determining the char-set of the enlarged hypothesis polset, and then the remainder of the conclusion pol w.r.t. this new char-set. It is readily found that the remainder is 0 so that the theorem remains true under that non-degeneracy condition. See the diagram Fig. 3.6.5 in Sect. 6 of Chap. 3. We may continue to do the same for any non-degeneracy condition if it is considered worthy of so doing. Such a measure may be applied to any theorem in the class **E**.

Remark 3.5. The degeneracy condition $I_1 = x_1 = 0$ means that A_1 coincides with O. To see whether the theorem is still true in some *degenerate* sense we may adjoin the pol x_1 to HYP to form a new hypothesis set and proceed as before to see what happens about the conclusion in this case. The procedures are complicate, tedious and geometrically uninteresting so it is better to leave it away.

Remark 3.6. Suppose we formulate the truth of a theorem $T = \{HYP, CONC\}$ in class **E** to be

$$HYP = 0 \longrightarrow CONC = 0. \tag{3.8}$$

By Hilbert's Zero Theorem this is equivalent to

$$CONC^k \in Ideal(HYP) \tag{3.9}$$

for some positive integer k. However, for the above Desargues Theorem with HYP, $CONC$ given by (3.5),(3.6) we see readily that (3.9) cannot be true. For example, the point $(x_1, \cdots, x_8) = (2, 4, 1, 1, 2, 4, 3, 1)$ satisfies the equations $Ideal(HYP) = 0$ or $HYP = 0$ but not $CONC = 0$. Cf. Ex. 3.6.3 and Fig. 3.6.6 in Chap. 3, Sect. 6. This shows that the treatment by ideal theory is inadequate for geometry theorem-proving. The underlying reason is that the occurence of non-degeneracy conditions usually in the form of inequalities is difficult to handle by means of ideals. This is one of the reasons that we have forsaken the ideal-theoretical treatment current in present-day researches while adopting the zero-set treatment throughout the whole book.

Examples of Geometry Theorem-Proving based on MTP-Principle II.

The MTP-Principle I for proving of theorems in class **E** is simple to apply but also very incomplete. To get more information we may apply the MTP-Principle II based on the Zero-Decomposition Theorem which we reproduce below:

MTP-Principle II. For a given theorem $T = \{HYP, CONC\}$ in the domain **E** let AS_i be a set of asc-sets determined from the hypothesis polset HYP such that

$$Zero(HYP) = \bigcup_i Zero(AS_i/IP_i) \tag{3.10}$$

in which each IP_i is the initial-product of the corresponding asc-set AS_i. Then the theorem T will be true on that part $Zero(AS_i/IP_i)$ of $Zero(HYP)$ or

$$Zero(AS_i/IP_i) \subset Zero(CONC) \tag{3.11}$$

if

$$Remdr(CONC/AS_i) = 0. \tag{3.12}$$

Example 3.2a (Ex. 3.2 revisited.)

Let us consider the Desargues Theorem in Ex. 3.2 or Ex. 3.6.3 again. As in Ex. 3.6.3 let us consider x_1, x_2, x_3 as non-zero independent constant parameters. Then we have (3.6.24) or

$$Zero(HYP) = \bigcup_i Zero(AS_i/IP_i), \tag{3.13}$$

in which i runs from 1 to 5 and AS_i are asc-sets given by (3.6.25) with IP_i the corresponding initial-products. It is readily found that

$$\begin{aligned} &Remdr(CONC/AS_i) = 0,\ i = 1, 5, \\ &Remdr(CONC/AS_i) \neq 0,\ i = 2, 3, 4. \end{aligned} \tag{3.14}$$

By MTP-Principle II the theorem is therefore true on the first and last parts of configurations but not so on the three other parts, in accord with the geometrical situations. See the diagrams Figs. 3.6.1-5 in Chap. 3, Sect. 6.

Remark 3.7 The decomposition of $Zero(HYP)$ as in (3.13), $(3.13)'$ is usually not unique. The members in the decomposition may rise up to quite high number even if we impose at the outset some non-zero conditions to deprive off uninteresting or undesirable cases. In spite of these the MTP-Principle II will give us much more complete information about the truth of a theorem than the Principle I.

Geometry Theorem-Proving based on MTP-Principle III.

We recall the MTP-Principle III below:

MTP-Principle III. For a given theorem $T = \{HYP, CONC\}$ in the domain $\mathbf{E}$ let IRR_k be a set of irreducible asc-sets determined from the hypothesis polset HYP such that

$$Zero(HYP) = \bigcup_k Var[IRR_k]. \tag{3.15}$$

Then the theorem T will be true on that irreducible component $Var[IRR_k]$ if and only if

$$Remdr(CONC/IRR_k) = 0. \tag{3.16}$$

Example 3.2b. (Ex. 3.2, Ex. 3.2a revisited.)

Let us consider the Desargues Theorem again. It happens that all the asc-sets AS_i in (3.13) are irreducible asc-sets. Rename AS_i by IRR_i, then the decomposition of the variety $Zero(HYP)$ in (3.10) with HYP given by (3.5) in considering x_1, x_2, x_3 as non-zero constants, will be sharpened to (3.6.28), viz.

$$Zero(HYP) = Var[IRR_1] + Var[IRR_2]. \tag{3.17}$$

It turns out that

$$Remdr(CONC/IRR_1) = 0, \text{ while}$$
$$Remdr(CONC/IRR_2) \neq 0.$$

By MTP-Principle III it follows that the Desargues Theorem is true on the irreducible component $Var[IRR_1]$, but not on the component $Var[IRR_2]$. The geometrical reason is quite clear. See Remark 3.3.

Remark 3.8. The decomposition in (3.17) with no redundant members is, unlike that in (3.10), unique up to an order. In fact, the MTP-Principle III furnishes us the most complete information about the truth of a geometrical theorem in comparing with the other two Principles I and II. However, to arrive at (3.15) the factorization of pols in algebraic function field is required which is not an easy task. It seems that so far geometry theorem-proving is concerned, the Principle I, incomplete as it is, is preferred since usually the *generic truth* of a theorem *alone* will be sufficient for our purpose. See Remark 3.4 for this point.

Example 3.3. In- or Ex-Center Theorem. The bisectors of the three angles of a triangle intersect three by three in 4 points.

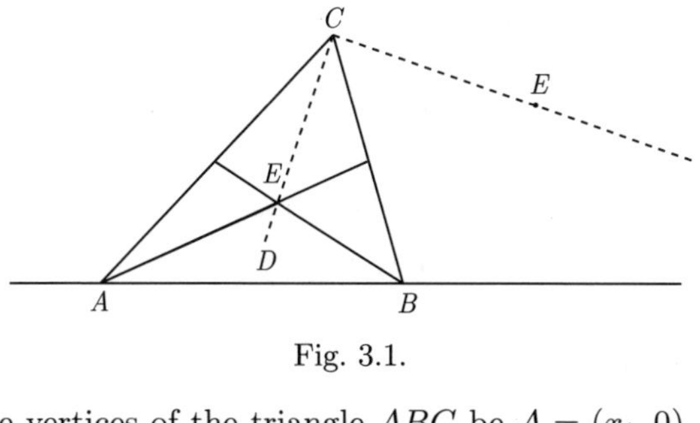

Fig. 3.1.

Proof. Let the vertices of the triangle ABC be $A = (x_1, 0)$, $B = (x_2, 0)$, $C = (0, x_3)$, Let the bisectors of the angles A, B, C be AD, BD, CE. Let $D = (x_4, x_5)$, and slope of CE be $x_7 : x_6$. Then the hypothesis polset will be $HYP = \{H_1, \cdots, H_5\}$

with H_i given below:

$$\begin{aligned}H_1 &= x_5^2 * x_3 - 2 * x_5 * x_1 * (x_4 - x_1) - x_3 * (x_4 - x_1)^2,\\ H_2 &= x_5^2 * x_3 - 2 * x_5 * x_2 * (x_4 - x_2) - x_3 * (x_4 - x_2)^2,\\ H_3 &= x_7^2 * x_3 * (x_2 + x_1) + 2 * x_7 * x_6 * (x_3^2 - x_1 * x_2) - x_6^2 * x_3 * (x_2 + x_1).\end{aligned} \tag{3.18}$$

The conclusion pol corresponding to the concurrency of bisectors AD, BD, CF or D on CF is given by

$$CONC = x_4 * x_7 - x_6 * (x_5 - x_3). \tag{3.19}$$

The char-set $CS = \{C_1, C_2, C_3\}$ of HS is given by:

$$\begin{aligned}C_1 =& -4 * x_4^4 + 8 * x_4^3 * (x_1 + x_2)\\ & -4 * x_4^2 * (x_1^2 + 3 * x_1 * x_2 + x_2^2 - x_3^2)\\ & -4 * x_4 * (x_1 + x_2) * (x_3^2 - x_1 * x_2) + x_3^2 * (x_1 + x_2)^2,\\ C_2 =& 2 * x_5 * (x_4 - x_1 - x_2) - x_3 * (2 * x_4 - x_1 - x_2),\\ C_3 =& x_7^2 * x_3 * (x_1 + x_2) + 2 * x_7 * x_6 * (x_3^2 - x_1 * x_2) - x_6^2 * x_3 * (x_1 + x_2).\end{aligned} \tag{3.20}$$

We see that (3.1) is not verified so that the theorem is apparently not true, as is geometrically evident. See Fig. 3.1 above. In fact, the char-set CS is not an irreducible one. If we apply MTP-Principle III, then we get with x_1, x_2, x_3 considered as independent parameters:

$$Zero(HYP) = Var[IRR_1] + Var[IRR_2]. \tag{3.21}$$

In (3.21) the irreducible asc-sets are $IRR_i = \{C_1, C_2, C_{i3}\},\ i = 1, 2$, with

$$\begin{aligned}C_{13} &= 2 * x_7 * x_4 * (x_4 - x_1 - x_2) - x_6 * x_3 * (x_1 + x_2),\\ C_{23} &= x_7 * x_3 * (x_1 + x_2) + 2 * x_6 * x_4 * (x_4 - x_1 - x_2).\end{aligned} \tag{3.22}$$

It is seen that (3.16) is verified on the first component but not on the second one of (3.21) which corresponds to the geometric fact that one of the two bisectors of $< C$ will pass through D while the other is not. The triads of concurrent bisectors are seen to be of number 4, corresponding to the 4 possible in- or ex-centers D given by $C_1 = 0$ and $C_2 = 0$.

Ambiguities in Ordinary Geometry.

In geometry there are many ambiguities corresponding to non-explicit constructibilities which may cause great troubles in applying the above MTP-Principles. As such ambiguities in ordinary geometry let us just cite the following few ones.

Ambiguity 1. A line will meet a circle in two points undistinguishable from each other.

Ambiguity 2. From a point two tangents may be drawn to a circle not containing that point which are undistinguishable from each other.

Ambiguity 3. Given two intersecting lines there are two bisectors undistinguishable from each other.

Ambiguity 4. There are four common tangents of two circles undistinguishable from each other.

An Example of Secant Theorem.

As an example how troubles occur when there are ambiguities let us consider the simple Secant Theorem below which has raised some controversies in the literature.

Example 3.4. Secant Theorem. Through a point S two secants are drawn to cut a circle in points A_1, A_1' and A_2, A_2' respectively. Then the products on the secants of the segments from S to the points of intersection are equal, i.e.

$$|SA_1| * |SA_1'| = |SA_2| * |SA_2'|. \tag{3.23}$$

See the diagram below.

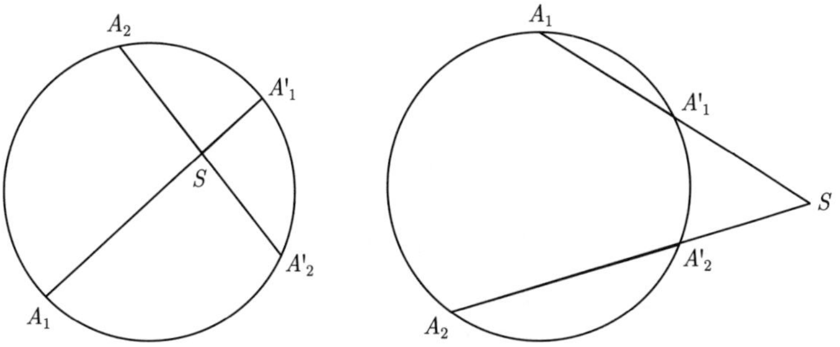

Fig. 3.2.

Let us try to prove the theorem in applying MTP-Principle I in the following *false* way. Let the point S be (x_0, y_0), while the circle C and the lines L_1, L_2 through S be respectively

$$\begin{aligned} C:\ X^2+Y^2 &= 1, \\ L_i:\ a_i*(X-x_i) &= b_i*(Y-b_i),\ i=1,2. \end{aligned} \tag{3.24}$$

In (3.24) X, Y are the running coordinates and a_i, b_i are to be treated as indeterminate constants. Let the points A_i, A_i' be given by

$$A_i = (x_i, y_i),\ A_i' = (x_i', y_i'),\ i = 1, 2, \tag{3.25}$$

and the distances SA_1, SA_1', SA_2, SA_2' be respectively z_1, z_1', z_2, z_2'. Then we have the hypothesis polset $HYP = \{H_1, \cdots, H_{12}\}$ with H_i given below:

$$
\begin{aligned}
H_1 &= (a_1^2 + b_1^2) * x_1^2 + 2 * a_1 * x_1 * (b_1 * y_0 - a_1 * x_0) + (b_1 * y_0 - a_1 * x_0)^2 - b_1^2,\\
H_2 &= b_1 * (y_1 - y_0) - a_1 * (x_1 - x_0),\\
H_3 &= (a_1^2 + b_1^2) * x_1'^2 + 2 * a_1 * x_1' * (b_1 * y_0 - a_1 * x_0) + (b_1 * y_0 - a_1 * x_0)^2 - b_1^2,\\
H_4 &= b_1 * (y_1' - y_0) - a_1 * (x_1' - x_0),\\
H_5 &= (a_1^2 + b_1^2) * x_2^2 + 2 * a_1 * x_2 * (b_1 * y_0 - a_1 * x_0) + (b_1 * y_0 - a_1 * x_0)^2 - b_1^2,\\
H_6 &= b_1 * (y_2 - y_0) - a_1 * (x_2 - x_0),\\
H_7 &= (a_1^2 + b_1^2) * x_2'^2 + 2 * a_1 * x_2' * (b_1 * y_0 - a_1 * x_0) + (b_1 * y_0 - a_1 * x_0)^2 - b_1^2,\\
H_8 &= b_1 * (y_2' - y_0) - a_1 * (x_2' - x_0),\\
H_9 &= z_1^2 - (x_1 - x_0)^2 - (y_1 - y_0)^2,\\
H_{10} &= z_1'^2 - (x_1' - x_0)^2 - (y_1' - y_0)^2,\\
H_{11} &= z_2^2 - (x_2 - x_0)^2 - (y_2 - y_0)^2,\\
H_{12} &= z_2'^2 - (x_2' - x_0)^2 - (y_2' - y_0)^2.
\end{aligned}
\tag{3.26}
$$

The conclusion pol is given by

$$CONC = z_2^2 * z_2'^2 - z_1^2 * z_1'^2. \tag{3.27}$$

Let us take the ordering of variables as

$$x_0 \prec y_0 \prec x_1 \prec y_1 \prec x_1' \prec y_1' \prec x_2 \prec y_2 \prec x_2' \prec y_2' \prec z_1 \prec z_1' \prec z_2 \prec z_2'. \tag{3.28}$$

Then the hypothesis polset HYP is already in the form of an asc-set. It is readily verified that

$$Remdr(CONC/HYP) \neq 0. \tag{3.29}$$

It follows from our MTP-Principle I that no conclusion can be drawn whether the secant theorem is true or not by the above method. We cannot, however, to draw uncautiosly the conclusion that the *secant theorem* which is so well known to be a real *true theorem* is *not true*. This seemingly paradoxical conclusion is sometimes due to a misunderstanding of the MTP-Principle I. In fact, the Principle gives for the vanishing of the remainder of the conclusion pol only a *sufficient* condition for the *generic* truth of the theorem but not a necassary one. The condition becomes both necassary and sufficient for the *generic* truth only when the asc-set in question is an *irreducible* one. However, the above asc-set HYP is actually not an irreducible one and this causes the failure of the above procedure.

The reason of the reducibility of HYP lies in the ambiguity of intersection points of the lines with a circle so that points in the pairs A_1, A_1' and A_2, A_2' are undistinguishable from each other. To proceed on the right way we may either apply the technique used in the preceding section or first resolve the asc-set HYP into irreducible ones as follows.

Subtract H_1 from H_3 we get a new pol

$$H_3' = H_{31}' * H_{32}', \text{ where}$$

$$H_{31}' = x_1' - x_1, \; H_{32}' = (a_1^2 + b_1^2) * (x_1' + x_1) + 2 * a_1 * (b_1 * y_0 - a_1 * x_0).$$

Similarly in subtracting H_5 from H_7 we get a new pol

$$H_7' = H_{71}' * H_{72}', \text{ where}$$

$$H_{71}' = x_2' - x_2, \; H_{72}' = (a_1^2 + b_1^2) * (x_2' + x_2) + 2 * a_1 * (b_1 * y_0 - a_1 * x_0).$$

It is clear that

$$Zero(HYP) = \bigcup_{i=1,\cdots,4} Zero(HYP_i),$$

in which HYP_i are all irreducible asc-sets as given below:

$$\begin{aligned} HYP_1 &= \{H_1, H_2, H_{31}', H_4, H_5, H_6, H_{71}', H_8, H_9, H_{10}, H_{11}, H_{12}\}, \\ HYP_2 &= \{H_1, H_2, H_{31}', H_4, H_5, H_6, H_{72}', H_8, H_9, H_{10}, H_{11}, H_{12}\}, \\ HYP_3 &= \{H_1, H_2, H_{32}', H_4, H_5, H_6, H_{71}', H_8, H_9, H_{10}, H_{11}, H_{12}\}, \\ HYP_4 &= \{H_1, H_2, H_{32}', H_4, H_5, H_6, H_{72}', H_8, H_9, H_{10}, H_{11}, H_{12}\}. \end{aligned} \tag{3.30}$$

It is now readily verified that

$$Remdr(CONC/HYP_i) \neq 0, \; i = 1, 2, 3, \text{ while } Remdr(CONC/HYP_4) = 0.$$

Let IP_i be the initial-product of HYP_i, $i = 1, \cdots, 4$ which are all irreducible asc-sets. Then it follows from the MTP-Principle II that the Secant Theorem is true only on the part $Zero(HYP_4/IP_4)$ but not on the other 3 parts $Zero(HYP_i/IP_i)$, $i = 1, 2, 3$. We may also apply MTP-Principle III to see that the Secant Theorem is true on the whole variety $Var[HYP_4]$ but not true for the other three varieties. The geometrical meaning of this conclusion is quite evident.

Remark 3.9. By deliberate arrangements of the geometric entities the Secant Theorem may be turned into one of Hilbertian type and then be proved as in Sect. 7.2.

Remark 3.10. There are various special devices of dealing with theorems with ambiguities of ordinary geometry. Not only various theorems in last section, viz. the theorems of Feuerbach, Morley, Steiner may be proved and analyzed in details by means of the MTP-Principles described above, but also such delicate theorems like that of Thébault-Taylor-Chou may be treated in the same way. However, as we are mainly interested in the general Principles of proving geometry theorems, rather than in the theorems themselves, we leave the topic to original papers due to Chou, Gao, Wang, Wu, *et al.*

7.4 Mechanical Proving of Geometry Theorems involving Inequalities.

Inequations and Inequalities in a Real Field.

In this section the basic field **K** will be the real field **R**. Though most of our condiderations may be extended to the general *real closed field*, we shall restrict ourselves to the field **R**, to avoid any superficial generalizations.

According to Seidenberg, we may reduce an inequation or an inequality in **R** to the existence of solutions of certain pols in an unknown y as shown in the lemma and the examples below.

Lemma 4.1. Let a, b, c be constants and y an unknown variable in **R**. Then:

$$\begin{array}{ll} a > 0 & \Longleftrightarrow Zero_{\mathbf{R}}(a * y^2 - 1) \neq \varnothing, \\ a < 0 & \Longleftrightarrow Zero_{\mathbf{R}}(a * y^2 + 1) \neq \varnothing, \\ a \geqslant 0 & \Longleftrightarrow Zero_{\mathbf{R}}(y^2 - a) \neq \varnothing, \\ a \leqslant 0 & \Longleftrightarrow Zero_{\mathbf{R}}(y^2 + a) \neq \varnothing, \\ a \neq 0 & \Longleftrightarrow Zero_{\mathbf{R}}(a * y - 1) \neq \varnothing, \\ a, b \ same\ sign & \Longleftrightarrow Zero_{\mathbf{R}}(a * b * y^2 - 1) \neq \varnothing, \\ a, b \ opposite\ sign & \Longleftrightarrow Zero_{\mathbf{R}}(a * b * y^2 + 1) \neq \varnothing, \\ c \ between\ a, b & \Longleftrightarrow Zero_{\mathbf{R}}(b - c) * (c - a) * y^2 - 1) \neq \varnothing, \ (a \neq b), \\ c \ not\ between\ a, b & \Longleftrightarrow Zero_{\mathbf{R}}(b - c) * (c - a) * y^2 + 1) \neq \varnothing, \ (a \neq b). \end{array}$$

Example 4.1. For three mutually different points

$$A_1 = (a_1, a_4), \ A_2 = (a_2, a_5), \ A_3 = (a_3, a_6)$$

lying on the same line in the real plane the order relation that A_2 lies between A_1, A_3 is equivalent to

$$Zero_{\mathbf{R}}((a_2 - a_1) * (a_2 - a_3) * y^2 + 1) + Zero_{\mathbf{R}}((a_5 - a_4) * (a_5 - a_6) * y^2 + 1) \ \neq \ \varnothing.$$

Example 4.2. That two points $A = (a_1, a_2)$, $A' = (a'_1, a'_2)$ lie on the same side or different sides of a line in the real plane $\mathbf{R}^2(x_1, x_2)$

$$L(x_1, x_2) \equiv c_1 * x_1 + c_2 * x_2 + c_0 = 0,$$

c_1, c_2 not both 0, is respectively equivalent to

$$Zero_{\mathbf{R}}(L(a_1, a_2) * L(a'_1, a'_2) * y^2 - 1) \neq \varnothing, \text{ or}$$
$$Zero_{\mathbf{R}}(L(a_1, a_2) * L(a'_1, a'_2) * y^2 + 1) \neq \varnothing.$$

Geometry Theorem-Proving in Domain **T**.

Let $X = (x_1, \cdots, x_n)$ be as before. Consider a theorem T in the Tarski domain **T**. By considering geometric concepts in the usual exact sense, both the hypothesis and the conclusion of T will be an aggregate of polynomial equations, inequations, and inequalities in the variables X, connected by logical connectives $\wedge, \vee, \neg$. By means of the method of quantifier elimination Tarski has given a universal procedure of proving or disproving any such theorem in the domain **T**, see the Tarski Theorem in Sect. 7.1. We may also reformulate the theorem in the following form:

Tarski Theorem. The domain **T** of geometry theorems is *mechanizable*.

The algorithm of Tarski is however so complicated that it is very difficult to prove geometrically interesting theorems. Tarski's method has been greately improved by later investigators. Among these improvements the most notable and successful one is due to Collins by his CAD method, or more precisely Cylindrical Algebraic Decomposition method. So far the CAD method has produced a number of results of geometrical or mathematical interest and it is on the way of further improvements.

The high complexities of the algorithm of Tarski or Collins arise from two sources. First, the problems considered by them are of very general character, viz. the decidability in a general real closed field. Secondly, so far geometry theorems are concerned, the geometrical concepts are in true geometric sense for which the algebraic equivalence will be expressed in a very complicated form, see the preceding Sect. 7.1. In order to lessen such complexities and thus to raise the efficiency of the procedures, we have adopted the following means, with the sacrifice of generality and completeness.

Naturally we shall restrict ourselves to narrower domain of applications. For geometries we shall modify the geometry concepts, so that their algebraic equivalence will be in quite simple form as explained in Sect. 7.1. With this understood let us consider now geometry theorems T in the domain **T** for which the hypothesis and conclusion are both aggregates of polynomial equations, inequations, and inequalities connected by $\vee, \wedge, \neg$, the polynomials being in $\mathbf{R}[X]$. Applying the Seidenberg technique as in Lemma 4.1 and adjoining new variables $Y = (y_1, \cdots, y_m)$ to X if necessary, we can turn the hypothesis into the form $HYP = 0$, with HYP a single polset in $\mathbf{R}[X, Y]$, as well as eventually some simple inequations or inequalities. We shall further restrict ourselves, without much loss of generality, to the case that the conclusion is of the form $CONC = 0, \neq 0, > 0, < 0, \geqslant 0$, or $\leqslant 0$, for some pol $CONC \in \mathbf{R}[X, Y]$. Remark that the *existence* of the new variables Y occuring in HYP are already implied by the very hypothesis itself. Let us denote the theoorem as $T = \{HYP, CONC, \mapsto\}$ in which $\mapsto$ stands for $=, \neq, >, <, \geqslant$, or $\leqslant$, meaning that the conclusion of the theorem T is $CONC \mapsto 0$. Now we can transform by Lemma 4.1, in adjoining further some new variable y_0 if necessary, the conclusion into the form of $CONC' = 0$, with $CONC' \in \mathbf{R}[X, Y, y_0]$. However, unlike the case of hypothesis, the *existence* of such new variable y_0 is no more implied by the conclusion or the theorem itself, but is to be *proved.* There is thus no advantage at all in turning the conclusion into a single equation. For this sake we shall refrain from ourselves to introduce any new variables further. In rewriting X, Y as X, we shall formulate the geometry theorem-proving in the restricted domain $\mathbf{T}'$ of $\mathbf{T}$ in the following form:

Definition 4.1. $\mathbf{T}'$:= The domain of geometry theorems $T = \{HYP, CONC \mapsto \}$, with $HYP \subset \mathbf{R}[X]$, $CONC \in \mathbf{R}[X]$, and $\mapsto$ stands for $=, \neq, >, <, \geqslant$, or $\leqslant$. HYP := *Hypothesis polset* of T, $CONC$:= *Conclusion pol* of T.

Definition 4.2. For theorem $T = \{HYP, CONC, \mapsto\} \in \mathbf{T}'$, *hypothesis* of theorem $T := HYP = 0$, and *conclusion* of $T := CONC \mapsto 0$.

Problem T′. Find universal procedure which permits to decide in a finite number of steps whether for any theorem $T = \{HYP, CONC, \mapsto\} \in \mathbf{T}'$, $CONC \mapsto 0$ is a consequence of $HYP = 0$ or not.

Remark 4.1. For theorems in $\mathbf{T}'$, it is sometimes convenient to reserve simple inequations or inequalities instead of introducing new variables y_i. This amounts to disregard some uninteresting or unreasonable degeneracies at the outset.

Geometry Theorem-Proving in the restricted Domain $\mathbf{T}'$.

Consider a theorem $T = \{HYP, CONC, \mapsto\}$ in the domain $\mathbf{T}'$. If $\mapsto$ stands for $=$, then the theorem T is also in the domain $\mathbf{E}$ which may be dealt with by any of the MTP-Principles as described in Sect. 7.2. Hence in what follows we shall consider only theorems $T = \{HYP, CONC, \mapsto\} \in \mathbf{T}' \setminus \mathbf{E}$, for which $\mapsto$ stands for $\neq, >, <, \geqslant$ or $\leqslant$, but not for $=$. The MTP-Principles may still be applied in a modified form as follows.

Consider e.g. the MTP-Principle I. Let CS be a char-set of $HYP \subset \mathbf{R}[X]$ and I_i be the initials of pols C_i in CS. By Sect. 3.3 we may form the squared-remainder R of $CONC$ w.r.t. CS such that

$$(\Pi_i\ I_i^{2*s_i}) * CONC = \Sigma_i\ Q_i * C_i + R, \tag{4.1}$$

in which $Q_i, R \in \mathbf{R}[X]$ with R reduced w.r.t. CS, and s_i are some non-negative integers. To prove $CONC \mapsto 0$ amounts then to prove

$$R \mapsto 0$$

under the conditions $C_i = 0$. The latter is sometimes much easier to be done than the original requirement of proving the conclusion $CONC \mapsto 0$ directly from the hypothesis $HYP = 0$. The other two MTP-Principles may be applied in the same manner. This furnishes us thus a method of proving theorems in $\mathbf{T}'$ with partial success, though incomplete at all.

Example 4.3. The Pasch Theorem. Given a triangle ABC and a line L passing through none of the vertices and intersecting the 3 sides BC, AC, AB in the points D, E, F. Then either none or just two of D, E, F are inside the segments BC, AC, AB.

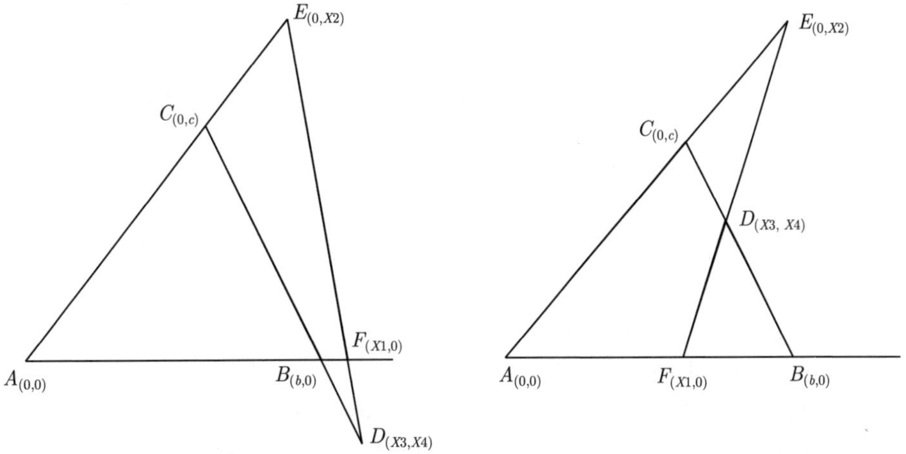

Fig. 4.1.

Proof. Let us suppose that E, F are either both inside or both outside the segments AC, AB respectively. Then we have to prove that D is outside the segment BC. The other cases of the theorem may be deduced from this case by *reductio absurdo*.

For this purpose let us take oblique coordinates so that

$$A=(0,0),\ B=(b,0),\ C=(0,c),\ F=(x_1,0),\ E=(0,x_2),\ D=(x_3,x_4). \quad (4.2)$$

in which $b\neq 0, c\neq 0$. By Lemma 4.1 we have then for some y_1, y_2 the hypothesis polset $HYP=\{H_1,H_2,H_3,H_4\}$ with

$$\begin{aligned} H_1 &= y_1^2 * x_1 * (b-x_1)+e,\\ H_2 &= y_2^2 * x_2 * (c-x_2)+e,\\ H_3 &= x_3 * (b*x_2-c*x_1)-b*x_1*(x_2-c),\\ H_4 &= x_4 * (b*x_2-c*x_1)+c*x_2*(x_1-b). \end{aligned} \quad (4.3)$$

In (4.3) e is either -1 or +1, corresponding respectively to the cases of E, F being both inside or both outside the segments AC, AB. The conclusion pol is

$$CONC = x_3 * (b-x_3),$$

and we have to prove that

$$x_3 * (b-x_3) < 0.$$

Take the ordering

$$y_1 \prec y_2 \prec x_1 \prec x_2 \prec x_3 \prec x_4.$$

Then HYP is already in the form of an asc-set and is itself a char-set of HYP. The squared-remainder formula for $CONC$ gives us

$$y_1^4 * y_2^4 * (b*x_2-c*x_1)^2 * CONC \equiv -\ b^2 * y_1^2 * y_2^2 \quad mod\ HYP.$$

The squared-remainder is thus

$$-\ b^2 * y_1^2 * y_2^2 < 0$$

so that $CONC < 0$ too as to be proved. □

Real 8-Triad Configurations.

Example 4.4. (Example 7.3.1 revisited).

From the proof of Ex. 7.3.1 we have seen that there are ∞^8 non-trivial 8-triad configurations in the *complex* affine space. On the contrary S.C.Chou and X.S.Gao have discovered and proved the following beautiful theorem. Cf. [C-G2].

Theorem 4.1 (Chou-Gao). In the *real* affinc space there exists no non-trivial 8-triad configurations.

Proof. Suppose that $\{A_1,\cdots,A_8\}$ is a *non-trivial* 8-triad configuration with A_i given by (7.3.2) and satisfying $P_j=0, j=1,\cdots,7$ in (7.3.3). From the proof of the 8-triad theorem there we have then $y_3 \neq 0$. We should have further $y_5 \neq 0$ since otherwise from $P_1 = 0$ we would have $x_5 = x_2$ or $A_5 = A_2$. Similarly from $P_2 = 0, P_3 = 0, P_7 = 0$, we should have respectively $y_6 \neq 0, y_7 \neq 0$, and $y_8 \neq 0$. Besides we should have $x_2 \neq 0, x_4 \neq 0$ and $x_4 \neq x_2$ in order that $A_4 \neq A_2$. It follows that

$$NZ \equiv y_3 * y_5 * y_6 * y_7 * y_8 * x_4 * x_2 * (x_4 - x_2) \neq 0.$$

Let us take an ordering of variables reverse to that in (7.3.4). Then by *wsolve* we get a decomposition of $Zero(HYP/NZ)$ into 10 components of asc-sets $AS_j, j = 1, \cdots, 10$ with initial-products IP_j such that

$$Zero(HYP/NZ) = \bigcup_j Zero(AS_j/IP_j * NZ).$$

Among the 10 asc-sets 4 of them will lead to the identity of some pairs of the 8 points A_i and should be discarded. For the other 6 asc-sets each of them is consisting of some pols linear in the leading variables and a single pol quadratic in the leading variable. For example, one of such quadratic pol is of the form

$$P = y_3^2 * (3 * y_7^2 - 3 * y_7 * y_8 + y_8^2) - y_3 * y_7 * y_8 * (3 * y_7 - y_8) + y_7^2 * y_8^2.$$

The discriminant of P w.r.t. y_3 is

$$-3 * y_7^2 * y_8^2 * (y_7 - y_8)^2 < 0,$$

since $y_7 \neq y_8$ otherwise $P_6 = 0$ would imply $x_7 = x_8$ so that $A_7 = A_8$. Hence the corresponding asc-set has no *real* zeros and similarly for the other five asc-sets. This proves that HYP has no real zeros so far $NZ \neq 0$. The theorem is thus proved. □

Triangles with Equal Bisectors.

A sophiscated problem about triangles with equal bisectors has perplexed geometers in 19th century. It is trivial to prove that an isosceles triangle ABC with sides $|AC| = |BC|$ have corresponding *internal* bisectors of the angles A and B equal to each other. However, the converse, though it is true, is far from being trivial to prove in the usual Euclidean fashion. Moreover, if at least one of the two bisectors is an *external* one, then it may happen that the two bisectors of the angles A and B may be equal with the corresponding sides $|AC|$ and $|BC|$ unequal to each other. To see how this may happen let us consider e.g. the case that both bisectors of the angles A, B, say AD and BE are *external* ones. Let us take the vertices A, B as well as the line $l = BC$ to be fixed with the vertex C moving on the fixed line l. The *external* bisector of angle B will then also be a fixed line m. Let us draw now a line through

A and parallel to m to meet l in point C_-. Then for $C = C_-$, the external bisector BE_- of angle B will have the end E_- receding to infinity so that $|BE_-| = +\infty$. So for the external bisector AD_- of the angle A we have $|AD_-| < |BE_-|$. Draw now through A a line parallel to l and form the reflection of the line AB w.r.t. this line which meets l in the point C_+. Then for the triangle ABC_+ the external bisector AD_+ of angle A is the line parallel to l with D_+ on l receding to infinity. Hence $|AD_+| = +\infty$ so that $|AD_+| > |BE_+|$ for the two external bisectors in this case of $C = C_+$. It follows that, when the vertex C moves along the line l from C_- to C_+, there will be some position C_0 for which the two external bisectors AD_0, BE_0 of the angles A, B will have their lengths equal to each other. For such a triangle we have then equal *external bisectors* without being isosceles. As the line l may be varied with vertex B fixed, there will be an infinity of such non-isosceles triangles with two equal *external* bisectors, see Fig. 4.2 above.

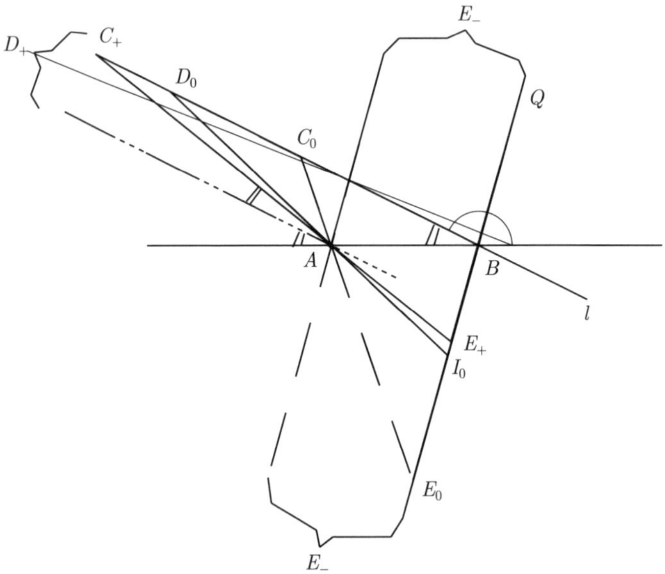

Fig. 4.2.

Similar situation occurs when the two bisectors are one internal and the other *external*. The case becomes however entirely different when both bisectors are *internal* ones. We have in fact the following theorem.

Example 4.5. Equi-Bisector Theorem. A triangle with two equal *internal* bisectors of its two angles is isosceles with the opposite sides of the two angles equal to each other. On the other hand, there are infinities of non-isosceles triangles for which the bisectors not both *internal* of the opposite angles of the unequal sides, are equal to each other.

Proof. Consider a triangle ABC with the bisectors AD, BE, internal or external, of the angles A, B intersecting at the in- or es-center I and having equal lengths. Let us take coordinates such that

$$\begin{aligned} A &= (-x_1, 0),\ B = (+x_1, 0),\ I = (x_2, x_3), \\ C &= (x_4, x_5),\ D = (x_6, x_7),\ E = (x_8, x_9). \end{aligned} \tag{4.4}$$

We have then a hypothesis set $HYP = \{H_i \mid i = 1, \cdots, 7\}$ with

$$\begin{aligned} H_1 &= x_4 * (x_3^2 + x_2^2 - x_1^2) + 2 * x_2 * (x_3^2 - x_2^2 + x_1^2), \\ H_2 &= x_5 * (x_3^2 + x_2^2 - x_1^2) - 2 * x_3 * (x_2^2 - x_1^2), \text{ etc.} \end{aligned} \tag{4.5}$$

The polset HYP will have a char-set $CS = \{C_i \mid i = 1, \cdots, 7\}$ for which the first pol C_1 is given by

$$\begin{aligned} C_1 &= x_2 * C_1', \text{ where} \\ C_1' &= (x_3^2 + x_2^2 - x_1^2)^2 * (x_3^2 + 2 * (x_2^2 - x_1^2)) \\ &\quad -4 * x_1^2 * (x_2^2 - x_1^2)^2. \end{aligned} \tag{4.6}$$

To see whether from $C_1 = 0$ which is a consequence of $HYP = 0$ we should have $x_2 = 0$ or $C_1' = 0$ let us remark that up to now we have not yet taken into account whether the bisectors AD and BE are *internal* ones or *external* ones. To fix the ideas let us assume that $x_1 > 0, x_5 > 0$ so that A lies on the left side of B on the X-axis while C is above the Y-axis. Now the bisector AD will be an *internal* or an *external* one according as the slope $\dfrac{x_3}{x_2 + x_1} = e_1$ of AI is > 0 or < 0. Similarly the bisector BE will be an *internal* or an *external* one according as the slope $\dfrac{x_3}{x_2 - x_1} = -e_2$ of BE is < 0 or > 0. It follows in particular that $x_2^2 - x_1^2$ is of same sign as $-e_1 * e_2$ and we have

$$\frac{x_3^2}{x_2^2 - x_1^2} = -e_1 * e_2. \tag{4.7}$$

If $e_1 * e_2 < 0$ then the two bisectors will be one *internal* and the other *external*. On the other hand let $e_1 * e_2 > 0$. Then AD, BE are either both *internal* or both *external* according as I and C are on the same side or on the opposite side of the X-axis, or according as $x_3 * x_5 > 0$ or $x_3 * x_5 < 0$. By H_2 in (4.5) we have

$$x_3 * x_5 = \frac{2 * x_3^2}{1 - e_1 * e_2}.$$

It follows that the bisectors AI, BI will be both *internal* or both *external* according as $0 < e_1 * e_2 < 1$ or $e_1 * e_2 > 1$.

Consider first the case $e_1 * e_2 > 0$. Then by (4.6) and (4.7) we have

$$C_1' = (x_2^2 - x_1^2)^3 * \left((1 - e_1 * e_2)^2 * (2 - e_1 * e_2) - \frac{4 * x_1^2}{x_2^2 - x_1^2}\right). \qquad (4.6)'$$

Suppose first that $0 < e_1 * e_2 < 1$ so that both bisectors are *internal* ones. Then $x_2^2 - x_1^2 < 0$ and we see from (4.6)′ that C_1' is always negative. Hence $C_1' \neq 0$ and from $C_1 = 0$ we have necessarily $x_2 = 0$. This means that the triangle is isosceles which proves the first part of the theorem.

Suppose next that $e_1 * e_2 > 1$ so that both bisectors are *external* ones. Then we see from (4.6) there will be an infinity of possible non-isosceles triangles with equal *external* bisectors. In fact, in the running coordinates (X, Y) the es-center $I = (x_2, x_3)$ will undergo an oval curve of degree 6 defined by

$$Y^6 + 4 * Y^4 * (X^2 - a^2) + 5 * Y^2 * (X^2 - a^2)^2 + 2 * (X^2 - a^2)^3 = 4 * a^2 * (X^2 - a^2)^2,$$

in which $x_1 = a$ is a positive constant. The oval is easily seen to touch the vertical lines through A, B with a height slightly less than the height of an equilateral triangle over AB. See Fig. 4.3 below.

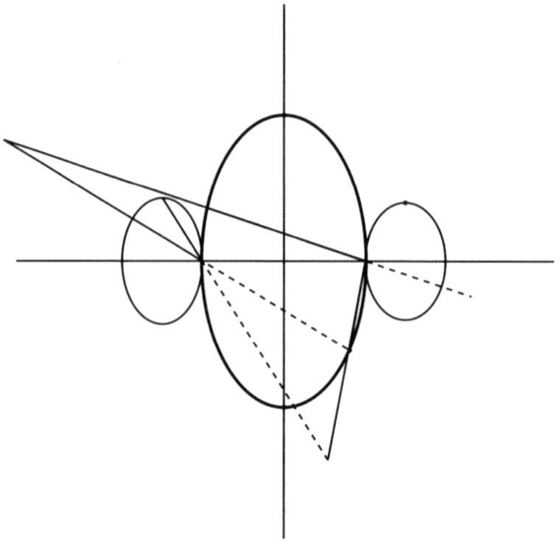

Fig. 4.3.

The case $e_1 * e_2 < 0$ for which the two bisectors are one *internal* and the other *external* may be similarly treated. We see thus there are infinities of non-isosceles triangles for which the bisectors one *internal* and one *external* are equal to each other. The corresponding es-center $I = (x_2, x_3)$ will undergo two oval curves of degree 6 which touch the two vertical lines through A, B respectively and also the two vertical lines through the points $(+\sqrt{3} * x_1, 0)$ and $(-\sqrt{3} * x_1, 0)$ respectively.

The heights of the ovals are slightly less than the widths of the ovals. See again Fig. 4.3.

The theorem is now completely proved. □

Remark 4.2. The above result is a consequence of discussions between S.C.Chou, X.L.Lu, and the present author. In particular, the diagram Fig. 4.3 is due to X.L.Lu.

An Alternative Method for Proving of Theorems in Class $\mathbf{T}'$.

The finiteness theorem in Sect. 5.5 may be applied to the proof of theorems in class $\mathbf{T}'$. Some examples given below may be served to illustrate such applications.

Example 4.3a. Pasch Theorem revisited.

For the triangle ABC with a line intersecting the sides BC, AC, AB in the points D, E, F different from the vertices let us suppose that D is inside the segment BC and E is outside the segment AC in the order of ACE. Let us prove that F is inside the segment AB.

For this purpose let us take oblique coordinates as in (4.2) with

$$b > 0,\ c > 0,\ x_2 > c,\ \text{and}$$

$$0 < x_3 < b,\ 0 < x_4 < c. \tag{4.8}$$

Now instead of considering D as a fixed point in the open segment BC, let us take D to be a point varying on the closed segment BC including the two end points B, C. Instead of (4.8) we shall have x_3, x_4 varying in the closed domain

$$\bar{D} : 0 \leqslant x_3 \leqslant b, \quad (\text{or } 0 \leqslant x_4 \leqslant c). \tag*{(4.8)$'$}$$

From collinearity of D, E, F we have $H_1 = 0, H_2 = 0$ where

$$\begin{aligned} H_1 &= x_3 * (b * x_2 - c * x_1) - b * x_1 * (x_2 - c), \\ H_2 &= x_4 * (b * x_2 - c * x_1) + c * x_2 * (x_1 - b). \end{aligned} \tag{4.9}$$

We shall take the ordering of variables to be

$$x_1 \prec x_3 \prec x_4,$$

noting that $x_2 (> c)$ is fixed. The problem is now to determine the least and greatest values of x_1 clearly exist for x_3, x_4 varying in the closed domain $\bar{D}$ under the constraint equation $H_1 = 0, H_2 = 0$. The variable x_1 plays thus the role of x_0 in the Finite Kernel Theorem in Sect. 5.5. For the above purpose let us split now the domain $\bar{D}$ into 3 open ones O_1, O_2, O_3 in respective affine spaces with O_1 defined by (4.8) and O_2, O_3 defined by $x_3 = 0$ and $x_3 = b$ respectively. For O_2 we have $x_1 = 0$

and for O_3 we have $x_1 = b$, being the projections on x_1 of the points $(x_3, x_4) = (0, c)$ and $(b, 0)$ respectively. The kernel sets for O_2 and O_3 are thus

$$K_2 = \{0\}, K_3 = \{b\}. \tag{4.10}$$

On the other hand the corresponding kernel set for O_1 is easily found to be

$$K_1 = \varnothing. \tag{4.11}$$

(4.10) and (4.11) give now the kernel set of our problem as

$$K = K_1 + K_2 + K_3 = \{0, b\} \subset \mathbf{R}.$$

It follows that the least and greatest values of x_1 are 0 and b corresponding to the positions $F = A, D = C$ and $F = B, D = B$ respectively. For other positions of $D \neq C, \neq B$, i.e. for D inside the segment BC, we have then necessarily $0 < x_1 < b$ or F inside the segment AB, which proves the theorem. □

Remark 4.3. The proof seems to be awkwardly too involved. However, it serves as a simple illustration of the general principle to follow in reducing the proving of theorems in domain $\mathbf{T}'$ to an optimization problem for which the method in Sect. 5.5 may then be applied. In fact, quite difficult theorems in domain $\mathbf{T}'$ have been proved in this way. Cf. the relevant papers of the present author on inequalities. We shall satisfy ourselves in giving some further examples below in order to illustrate this principle.

Example 4.6. Quadrilateral Convexity Theorem. Let $ABCD$ be a convex quadrilateral with points O, P, Q, R on the inside of the sides AD, AB, BC, and CD respectively. Then $OPQR$ is also a convex quadrilateral.

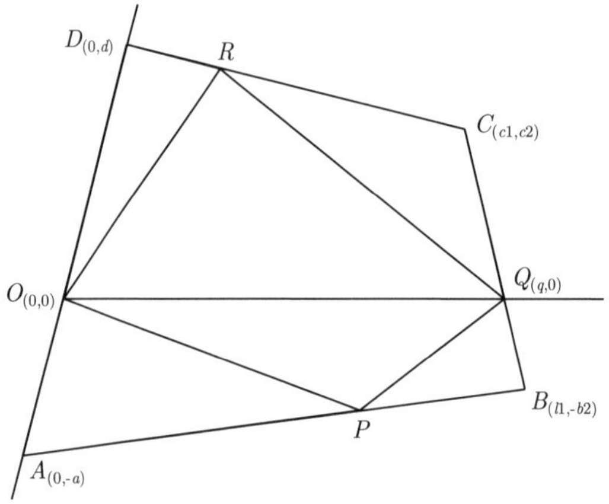

Fig. 4.4.

Proof. Owing to convexity of $ABCD$ oblique coordinates can be so taken that

$$O=(0,0),\ A=(0,-a),\ B=(b_1,-b_2),\ C=(c_1,c_2),\ D=(0,d),$$

$$Q=(q,0),\ q=\frac{b_1*c_2+b_2*c_1}{b_2+c_2},$$

$$P=r_0.B+(1-r_0).A=(r_0*b_1,-r_0*b_2-(1-r_0)*a),\ 0<r_0<1,$$

$$R=r.C+(1-r).D=(r*c_1,r*c_2+(1-r)*d),\ 0<r<1.$$

The constants a,b_1,b_2,c_1,c_2,d will all be positive so that q is also positive. Let us set

$$k=\frac{Oriented\ Area(\triangle OQR)}{Oriented\ Area(\triangle OQP)}$$

so that $H=0$ where

$$H=k*(r_0*b_2+(1-r_0)*a)+(r*c_2+(1-r)*d).$$

We shall prove that $k<0$ or R,P are on opposite sides of the line OQ. To do this we shall, as in the case of Ex. 4.3a, instead of considering R as a fixed point inside the open segment CD with r constant and $0<r<1$, take R as a point varying on the closed segment CD with r varying in the closed domain

$$\bar{D}:0\leqslant r\leqslant 1.$$

Let us treat r_0 as a constant and take ordering of variables to be

$$k\prec r.$$

We come then to the problem of optimizing k for the domain $\bar{D}$ under the constraint equation $H=0$. Here k will play again the role of x_0 as in the Finite Kernel Theorem of Sect. 5.5. Split now the domain $\bar{D}$ into open ones O_1,O_2,O_3 in the respective affine spaces corresponding to the cases $0<r<1, r=0$, and $r=1$ respectively. Suppose first that $c_2\neq d$ or CD is not parallel to OQ. It is then readily found that the kernel sets for the open domains O_i and for $\bar{D}$ are respectively ($Proj_k$ means projection on coordinate k)

$$\begin{aligned}
K_1&=\varnothing,\\
K_2&=\{k_2\}=Proj_k\ \{(k,r)=(k_2,0)\},\\
K_3&=\{k_3\}=Proj_k\ \{(k,r)=(k_3,1)\},\\
K\ &=\ K_1+K_2+K_3=\{k_2,k_3\},
\end{aligned}$$

in which

$$k_2 = -\frac{d}{r_0 * b_2 + (1 - r_0) * a} < 0,$$
$$k_3 = -\frac{c_2}{r_0 * b_2 + (1 - r_0) * a} < 0.$$

We see therefore both the least and the greatest value of k are negative. If $c_2 = d$ then the least and greatest value of k are both equal to the same negative value $k_2 = k_3$. In any way we have $k < 0$ so that R, P are on opposite sides of the line OQ. In the similar way we prove that O, Q are on the opposite sides of the line PR. It follows that the quadrilateral $OPQR$ is convex and the theorem is proved. □

An Alternative Proof of Equi-Bisector Theorem.

Example 4.5a. Equi-Bisector Theorem revisited.

Let us consider the equi-bisector theorem again. We shall prove below the first part of the theorem by means of the alternative method based on the Finite Kernel Theorem of Sect. 5.5. As before we shall assume in present case $x_5 > 0, x_1 = a_1 > 0$ a positive constant, and $e_1 > 0, e_2 > 0$ so that AD, BE are both *internal* bisectors. Take $x_3 = a_3 > 0$ to be a positive constant too. Then it is necessary that $a_1 > a_3$ to ensure the validity of the above assumptions.

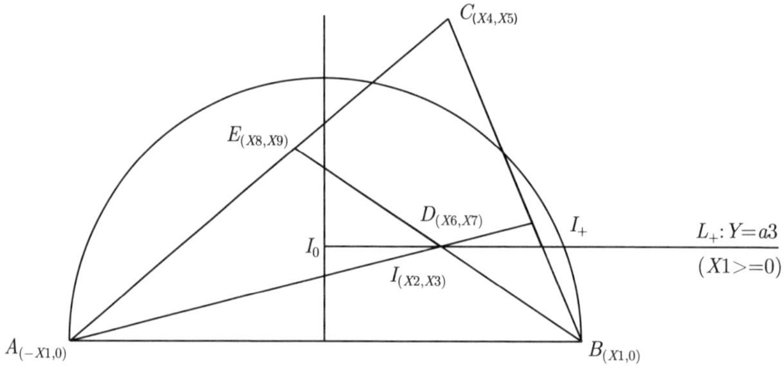

Fig. 4.5.

For the proof let us consider I not as a fixed point but a point moving on the half-lines

$$L_+: \ X \geqslant 0, \ Y = a_3,$$
$$L_-: \ X \leqslant 0, \ Y = a_3.$$

Let us draw a semi-circle on AB as diameter in the upper half plane which intersects the half-lines L_+ and L_- in the points $I_+ = (a_2, a_3)$ and $I_- = (-a_2, a_3)$ respectively, where $a_2 = +\sqrt{a_1^2 - a_3^2} > 0$. To ensure I to be the in-center of the triangle we shall restrict I to move on the closed segments I_0I_+ and I_0I_- where $I_0 = (0, a_3)$, see the figure Fig. 4.5. Remark that for $I = I_+$ or $I = I_-$ the sides AC, BC will be parallel to each other so that C goes to infinity and ABC will no longer be a triangle properly speaking. This does not have any affect on our method, however. For, we may neglect the vertex C and restrict our configuration to be consisting of the points A, B, I, D, E only.

Consider now the hypothesis pols H_i in (4.5) again. For our purpose let us introduce a new variable x_0, replace H_7 by $H_7' = H_7 - x_0$ and the polset HYP by the new one $HYP' = \{H_3, \cdots, H_6, H_7'\}$. As a first step of the proof we come to the determination of the least value of x_0 under the constraint conditions $HYP' = 0$, when I runs over the closed interval I_0I_+. As in the preceding examples, we shall divide the closed interval into three parts: the open interval $0 < x_2 < a_2$, and the two points I_0 and I_+. For the first part it will be found that the kernel set is again $K_1 = \varnothing$, while for the other two parts the kernel sets are respectively $K_2 = \{0\}$ and $K_3 = \{16 * a_1 * a_2\}$ for which $a_1 * a_2 > 0$. It follows that the least value of x_0 when I moves on the closed interval I_0I^+ is 0 which occurs when and only when $I = I_0$, for which the triangle becomes isosceles. Similarly, when the point I moves over the closed interval I_0I_-, then the greatest value of x_0 will be 0 which occurs when and only when $I = I_0$, i.e. when and only when the triangle ABC is isosceles. It follows that the least value of $|x_0|$ when I moves over the closed interval I_-I_+ will occur when and only when the triangle ABC is isosceles.

The same is true for arbitraly fixed $x_3 = a_3 > 0$. It follows that the least value of $|x_0|$ when I moves over the upper plane in keeping I as an in-center is 0 which occurs when and only when the triangle ABC is isosceles. This proves the theorem.

□

Chapter 8

Diverse Applications

8.1 Applications to Automated Discovering of Unknown Relations and Automated Determination of Geometric Loci.

Notations. $\mathbf{K}, X = (x_1, \cdots, x_n)$, etc. as before.

Automated Discovering of Unknown Relations.

Consider a triangle with three sides known. Then the triangle will be completely determined up to a rigid motion. In particular the area of the triangle will be completely determined. It follows that the area should be a certain expression in terms of the three sides. Moreover, if the orientation of the triangle is taken into consideration, then in terms of the known sides there will be two values of the area in opposite signs. So it is legitimate to guess that the area-square of a triangle is some possibly *rational* expression in the three sides. The problem is now to determine the *exact* form of this expression which should exist according to the above analysis. Of course such exact expression was already known in the early history of mathematics, viz. in the form of Heron's formula in ancient Europe and in the form of Qin's formula in ancient China, see Ex. 2.1.2 and formula (2.3.5) in Chap. 2.

The derivations of Qin's and Heron's formulae, actually equivalent to each other, are however, quite ingenious and even quite intricate. The problem thus arises of whether there are methods of arriving at these formulae in a simpler and more natural manner. As there are a lot of problems of the similar kind in the determination of unknown relations supposed exist or already known to exist, we formulate a general problem below:

Problem R. To discover unknown relations possibly existing between some geometrical (or algebraic) entities.

One of the simplest examples is to discover the relation, *supposed unknown*, between the area of a triangle and its three sides. To solve the Problem R let us formulate the general method and underlying principle as follows.

Principle and **Method**. Let HYP be the polset whose vanishing expresses the underlying geometrical conditions of the problem. The variables x_i will be so chosen and so arranged that the first few x_i are just the geometrical (or algebraic) entities of which the connecting relation is to be sought for. Let CS be a char-set of HYP with C_i as its pols. By the Well-Ordering Principle we have

$$Zero(HYP) \subset Zero(CS). \tag{1.1}$$

Hence we have always $C_1 = 0$ which is in general just the relation to be sought for.

Examples about Distances, Areas, and Volumns :

Example 1.1. Qin-Heron Formula for Area of a Triangle.

For a triangle $A_0A_1A_2$, let x_1, x_2, x_3, x_4 be the three sides $|A_0A_1|, |A_0A_2|, |A_1A_2|$, and the area respectively whose yet unknown relation is to be sought for. Choose coordinates so that

$$A_0 = (0,0),\ A_1 = (x_5, 0),\ A_2 = (x_6, x_7).$$

Remark that the variables x_i occuring in the coordinates of A_0, A_1, A_2 have been so chosen to be successive to $x_1, \cdots, x_4$ in the natural order. The geometrical conditions are now in the form of equations $HYP = 0$ where $HYP = \{h_1, h_2, h_3, h_4\}$ with

$$\begin{aligned} h_1 &= 2 * x_4 - x_5 * x_7, \\ h_2 &= x_5 - x_1, \\ h_3 &= x_2^2 - x_6^2 - x_7^2, \\ h_4 &= x_3^2 - (x_6 - x_5)^2 - x_7^2. \end{aligned} \tag{1.2}$$

The char-set is readily found to be consisting of 4 pols of which the first one is given by

$$C_1 = 2 * x_1^2 * x_2^2 + 2 * x_1^2 * x_3^2 + 2 * x_2^2 * x_3^2 - x_1^4 - x_2^4 - x_3^4 + 16 * x_4^2. \tag{1.3}$$

The relation $C_1 = 0$ is just the Qin-Heron Formula to be sought for.

Example 1.2. Relation between Volume and Edges of a Tetrahedron.

For the tetrahedron $A_0A_1A_2A_3$ let the volume be x_7 and the six edges be

$$\begin{aligned} &|A_0A_1| = x_1,\ |A_0A_2| = x_2,\ |A_0A_3| = x_3, \\ &|A_1A_2| = x_4,\ |A_1A_3| = x_5,\ |A_2A_3| = x_6. \end{aligned} \tag{1.4}$$

Let us assign now x_i with $i > 7$ to the coordinates of the four vertices such that

$$A_0 = (0,0,0),\ A_1 = (x_8, 0, 0),\ A_2 = (x_9, x_{10}, 0),\ A_3 = (x_{11}, x_{12}, x_{13}). \tag{1.5}$$

Then we get a system of polynomial equations $h_i = 0,\ i = 1, \cdots, 7$, which give the volume and edges in terms of the variables $x_1, \cdots, x_{13}$ with h_i given below:

$$\begin{aligned}
h_1 &= 6 * x_7 + x_8 * x_{10} * x_{13}, \\
h_2 &= x_8 + x_1, \\
h_3 &= x_2^2 - x_9^2 - x_{10}^2, \\
h_4 &= x_3^2 - x_{11}^2 - x_{12}^2 - x_{13}^2, \\
h_5 &= x_4^2 - (x_9 - x_8)^2 - x_{10}^2, \\
h_6 &= x_5^2 - (x_{11} - x_8)^2 - x_{12}^2 - x_{13}^2, \\
h_7 &= x_6^2 - (x_{11} - x_9)^2 - (x_{10} - x_{13})^2.
\end{aligned} \tag{1.6}$$

Form now the char-set CS of the polset $HYP = \{h_1, \cdots, h_7\}$ it is readily found that for the first char-pol C_1 the equation $C_1 = 0$ will give us automatically the expression of the volume x_7 in terms of the edges $x_1, \cdots, x_6$ to be found as follows:

$$144 * x_7^2 = -\Sigma_1\ (x_1^2 * x_6^4 + x_6^2 * x_1^4) + \Sigma_2\ (x_2^2 * x_1^2 * x_5^2) - \Sigma_3\ (x_1^2 * x_2^2 * x_4^2). \tag{1.7}$$

In the formula the $\Sigma' s$ are summations to be extended over respectively ranges as shown below:

Σ_1 over 3 pairs of opposite edges $(x_1, x_6), (x_2, x_5), (x_3, x_4)$,

Σ_2 over 12 triples of non-closed edges (x_2, x_1, x_5),

Σ_3 over 4 triples of edges forming a triangle like (x_1, x_2, x_4).

Consider the particular case that the tetrahedron collapses into a planar one so that the volume is 0. Then the equation (1.7) becomes

$$-\Sigma_1 + \Sigma_2 - \Sigma_3 = 0, \tag{1.8}$$

which gives a relation between the six mutual distances between the four points lying in the same plane. Note that the equation (1.8) is homogeneous in the 6 edges $x_i,\ i = 1, \cdots, 6$, and is of degree 4 in each edge, which is evident directly from geometric considerations.

Remark 1.1. We may ask the same question as in Exs. 1.1, 1.2 about the volume of a 4-simplex in a 4-dimensional Euclidean space $\mathbf{R}^4$, of a 5-simplex in $\mathbf{R}^5$, and the relation between the mutual distances of 5 points in an $\mathbf{R}^3$, or that of any 6 points in $\mathbf{R}^4$. We may ask even those for a Euclidean space of arbitrary but *fixed* dimension. One can solve these problems in just the same manner as above so far the dimension of the Euclidean space is fixed in advance and so far the limitations of the memory space and running time of a computer permit to carry out such a computation. On the other hand, the determination of such a formula for a general dimension n is entirely out of reach of our mechanization method. Remark that such formulae have already been discovered by Cayley *et al* by ingenious manipulations of determinants. This shows that, although the use of computers furnishes us a powerful tool in mathematical studies, we cannot solely rely upon it without resort to usual methods of mathematics.

A Pentagon Theorem of Gauss.

Example 1.3. Gauss Pentagon Theorem. In a short note appeared in [Gau2] of C.F.Gauss, it was solved an interesting problem below:

> Beliebige 5 Punkte A, B, C, D, E einer Ebene sind, je zwei, durch gerade Linien verbunden. Man kennt die somit entstehenden 5 Dreiecke EAB, ABC, BCD, CDE, DEA ihrem Inhalte nach, und verlangt daraus den Inhalt des Fuenfecks $ABCDE$.

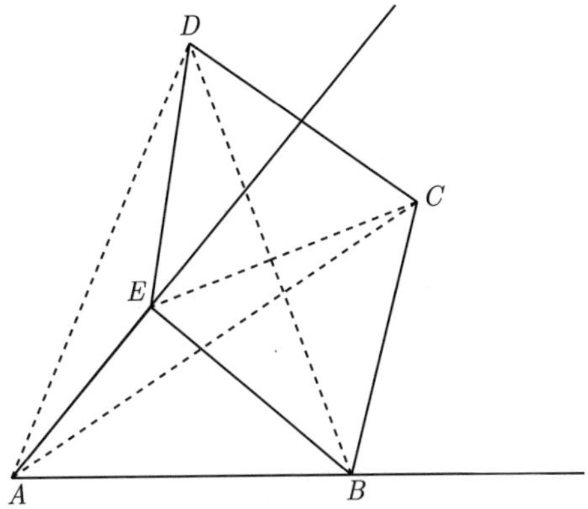

Fig. 1.1.

Gauss solved the problem in an ingenious manner, seemingly in supposing that the pentagon $ABCDE$ is *convex* in form. With our general method we solve the problem as follows.

Suppose that A, B, E are not on a line and let us take the coordinate system with AB, AE as the X- and Y-axis. Following our underlying Principle let us denote the areas of the 5 triangles ABC, BCD, CDE, DEA, EAB and the pentagon $ABCDE$ by $x_1, \cdots, x_6$ respectively. Note that all the areas will be considered as oriented ones in accordance to the indicated ordering of vertices. Irrespective whether the pentagon is convex or not we have also here

$$Area(ABCDE) = Area(ABC) + Area(ACD) + Area(ADE).$$

Let the vertices be now

$$A = (0,0),\ B = (x_7, 0),\ C = (x_8, x_9),\ D = (x_{10}, x_{11}),\ E = (0, x_{12}).$$

Remark that in oblique coordinates the oriented area of a triangle with vertices in the order of $(x_i, y_i), (x_j, y_j), (x_k, y_k)$ will be given by

$$\frac{1}{2} * \begin{vmatrix} x_i & y_i & 1 \\ x_j & y_j & 1 \\ x_k & y_k & 1 \end{vmatrix}$$

up to some definite non-zero value D which depends only on the coordinate system. We have thus a hypothesis polset $HYP = \{h_1, \cdots, h_6\}$ with h_i given below, the non-zero definite value D having been dropped away:

$$\begin{aligned}
h_1 &= 2 * x_1 - x_7 * x_9, \\
h_2 &= 2 * x_2 - x_7 * x_9 - x_8 * x_{11} + x_9 * x_{10} + x_7 * x_{11}, \\
h_3 &= 2 * x_3 - x_8 * x_{11} - x_{10} * x_{12} + x_8 * x_{12} + x_9 * x_{10}, \\
h_4 &= 2 * x_4 - x_{10} * x_{12}, \\
h_5 &= 2 * x_5 - x_7 * x_{12}, \\
h_6 &= 2 * x_6 - 2 * x_1 - x_8 * x_{11} + x_9 * x_{10} - 2 * x_4.
\end{aligned} \tag{1.9}$$

The char-set CS of HYP is found to be $CS = \{C_1, \cdots, C_6\}$ with the first char-pol C_i given below:

$$C_1 = x_6^2 - (x_1 + x_2 + x_3 + x_4 + x_5) * x_6 + (x_1 * x_2 + x_2 * x_3 + x_3 * x_4 + x_4 * x_5 + x_5 * x_1). \tag{1.10}$$

The relation $C_1 = 0$ is just the beautiful formula given by Gauss. Remark that here no presupposition about convexity is required. It is however required that the points A, B, E are not in the same line for the choice of the coordinate axis to be possible. Otherwise it is entirely arbitrary. If required we may add the condition that A, B, E are collinear to the hypothesis to see what happens in this degenerate case. As it is of little geometrical interest we may satisfy ourselves to stop here.

Some Further Geometry Examples.

Example 1.4. Determine the ex-radius of the circumscribing sphere S of a tetrahedron in terms of its 6 edges.

Let us take the coordinates so that the 4 vertices and the 6 edges of the tetrahedron will be given as in (1.4) and (1.5). Let the ex-center of the sphere S be (x_{14}, x_{15}, x_{16}) and the ex-radius be x_7'. Then we have for the hypothesis a polset HYP' consisting of the pols $h_2, \cdots, h_7$ in (1.6) as well as the pols h_1', h_8, h_9, h_{10} given below:

$$\begin{aligned}
h_1' &= x_7'^2 - x_{14}^2 - x_{15}^2 - x_{16}^2,\\
h_8 &= x_7'^2 - (x_{14} - x_8)^2 - x_{15}^2 - x_{16}^2,\\
h_9 &= x_7'^2 - (x_{14} - x_9)^2 - (x_{15} - x_{10})^2 - x_{16}^2,\\
h_{10} &= x_7'^2 - (x_{14} - x_{11})^2 - (x_{15} - x_{12})^2 - (x_{16} - x_{13})^2.
\end{aligned} \tag{1.11}$$

The first pol C_1' of the char-set CS' of HYP' is easily determined. In comparing with the char-pol C_1 in Ex. 1.2 and the sums Σ_i there we find readily the following formula:

$$9 * Vol^2 * Diam^2 = \sigma * (\sigma - \alpha) * (\sigma - \beta) * (\sigma - \gamma), \tag{1.12}$$

in which $Vol, Diam$ stand for the volume of the tetrahedron and the diameter of the sphere S respectively, while

$$\sigma = \frac{\alpha + \beta + \gamma}{2}, \text{ with} \tag{1.13}$$

$$\alpha = x_1 * x_6,\ \beta = x_2 * x_5,\ \gamma = x_3 * x_4 \tag{1.14}$$

corresponding to the 3 pairs of opposite sides x_1, x_6; x_2, x_5; x_3, x_4 respectively.

Remark 1.2. The author is at a loss where to find the formula (1.12) in the literature of pasttimes which is here discovered in an automatic manner. Note the connection of the right-hand side of (1.12) with the Ptolemy formula about 4 points on the same circle for which $Vol = 0$.

Distances and Areas in Hyperbolic Plane.

Our method makes no difference in whatever kind of geometry it may be. To fix the ideas, let us consider the case of plane hyperbolic geometry, for which Beltrami coordinates will be used in what follows. We remark that in studies of non-Euclidean geometries transcendental functions are intensively used. On the other hand our method has to deal solely with polynomials of purely *algebraic* character. This is however not an unsurmountable barrier to the application of our method since we are dealing actually with *algebraic* relations between the *transcendental* functions, but not the transcendental functions themselves. It has indeed been explained in the book [WWT5] and the earliest paper [WWT1] about mechanical theorem-proving that there are already some examples illustrating how we can deal with such problems involving transcendental functions. In the examples below one can see again how this is done.

Example 1.5. Find the relation between the 6 distances of 4 points in a hyperbolic plane.

For the 6 distances A_iA_j between the 4 given points A_0, A_1, A_2, A_3 let us set

$$\begin{aligned} x_1 &= cosh\ A_0A_1,\ x_2 = cosh\ A_0A_2,\ x_3 = cosh\ A_0A_3, \\ x_4 &= cosh\ A_1A_2,\ x_5 = cosh\ A_1A_3,\ x_6 = cosh\ A_2A_3. \end{aligned} \tag{1.15}$$

Take now the coordinate system such that the 4 points A_i are given in Beltrami coordinates by

$$A_0 = (0,0),\ A_1 = (x_7, 0),\ A_2 = (x_8, x_9),\ A_3 = (x_{10}, x_{11}).$$

We have then a hypothesis polset $HYP = \{h_1, \cdots, h_6\}$ as given below:

$$\begin{aligned} h_1 =&\ x_1^2 - x_1^2 * x_7^2 - 1, \\ h_2 =&\ x_2^2 - x_2^2 * x_8^2 - x_2^2 * x_9^2 - 1, \\ h_3 =&\ x_3^2 - x_3^2 * x_{10}^2 - x_3^2 * x_{11}^2 - 1, \\ h_4 =&\ x_4 - x_1 * x_2 + x_1 * x_2 * x_7 * x_8, \\ h_5 =&\ x_5 - x_1 * x_3 + x_1 * x_3 * x_7 * x_{10}, \\ h_6 =&\ x_6 - x_2 * x_3 + x_2 * x_3 * x_8 * x_{10} + x_2 * x_3 * x_9 * x_{11}. \end{aligned} \tag{1.16}$$

The first pol C_1 of the char-set of HYP is found to be given by

$$\begin{aligned} C_1 =&\ (x_1^2 + x_2^2 + x_3^2 + x_4^2 + x_5^2 + x_6^2) - (x_1^2 * x_6^2 + x_2^2 * x_5^2 + x_3^2 * x_4^2) \\ &- 2 * (x_1 * x_3 * x_5 + x_1 * x_2 * x_4 + x_2 * x_3 * x_6 + x_4 * x_5 * x_6) \\ &+ 2 * (x_1 * x_4 * x_3 * x_6 + x_1 * x_2 * x_5 * x_6 + x_2 * x_4 * x_3 * x_5) \\ &- 1. \end{aligned} \tag{1.17}$$

The equation $C_1 = 0$ is then the relation to be sought for.

Example 1.6. Determine the area $\triangle$ of a triangle in a hyperbolic plane in terms of its 3 sides.

For the sides of the triangle $A_0A_1A_2$ and the area $\triangle$ let us set

$$\begin{aligned} x_1 = cosh\ A_0A_1, \quad x_2 &= cosh\ A_0A_2,\ x_3 = cosh\ A_1A_2, \\ x_4 &= cos\ \triangle. \end{aligned} \tag{1.18}$$

We have then

$$\pi - \triangle = \alpha_0 + \alpha_1 + \alpha_2, \tag{1.19}$$

where $\alpha_0, \alpha_1, \alpha_2$ are the 3 internal angles of the triangle. Let us take a coordinate system such that

$$A_0 = (0,0),\ A_1 = (x_5, 0),\ A_2 = (x_6, x_7).$$

Set

$$\begin{aligned} x_{10} &= cos\ \alpha_0, \quad x_{11} = sin\ \alpha_0, \\ x_{12} &= cos\ \alpha_1, \quad x_{13} = sin\ \alpha_1, \\ x_{14} &= cos\ \alpha_2, \quad x_{15} = sin\ \alpha_2. \end{aligned} \tag{1.20}$$

Then we have a hypothesis polset $HYP = \{h_1, \cdots, h_{12}\}$ with h_i given below:

$$\begin{aligned} &h_1 = x_1^2 - x_1^2 * x_5^2 - 1, \\ &h_2 = x_2^2 - x_2^2 * x_6^2 - x_2^2 * x_7^2 - 1, \\ &h_3 = x_3 - x_1 * x_2 + x_1 * x_2 * x_5 * x_6, \\ &h_4 = x_8^2 - x_6^2 - x_7^2, \\ &h_5 = x_9^2 - x_7^2 - x_1^2 * (x_5 - x_6)^2, \\ &h_6 = x_8 * x_{10} - x_6, \\ &h_7 = x_9 * x_{12} - x_1 * (x_5 - x_6), \\ &h_8 = x_8 * x_{11} - x_7, \\ &h_9 = x_9 * x_{13} - x_7, \\ &h_{10} = x_8 * x_9 * x_{14} - x_1 * x_7^2 + x_1 * x_6 * (x_5 - x_6), \\ &h_{11} = x_2 * x_8 * x_{15} - x_1 * x_5 * x_{13}, \\ &h_{12} = x_{10} * x_{13} * x_{15} + x_{11} * x_{12} * x_{15} + x_{11} * x_{13} * x_{14} - x_{10} * x_{12} * x_{14} - x_4. \end{aligned} \tag{1.21}$$

For the first char-pol C_1 the equation $C_1 = 0$ can be written in the form

$$\begin{aligned} &(x_1 + 1) * (x_2 + 1) * (x_3 + 1) * (x_4 - 1) \\ =&x_1^2 + x_2^2 + x_3^2 - 2 * x_1 * x_2 * x_3 - 1. \end{aligned} \tag{1.22}$$

Replacing x_i, $i = 1, \cdots, 4$, by the respective hyperbolic cosines of the sides and cosine of the area, we see that the last equation is equivalent to a form as given in Ex. 27 of [Gr], p.361, which is now found in an automatic manner here.

Remark 1.3. In recent years the determination of the volume of a tetrahedron in the hyperbolic space in terms of its 6 edges has been seriously discussed in the literature. However, as was pointed by Dehn, (cf. e.g. [Kl], p.203) the volume of such a tetrahedon, unlike the case of a triangle in the hyperbolic plane, cannot be expressed in terms of the dihedral angles of the edges of the tetrahedron. Moreover, as was already proved before, the expression of the hyperbolic tetrahedron should involve highly transcendental functions, viz. the so-called Lobachevsky functions. So whether one can determine the volume in terms of the 6 edges by means of the method developed here as in Ex. 1.6 is in doubt. This remains a problem to be studied later.

Automatic Determination of Unknown Algebraic Relations.

Our general principle and method are not only applicable to geometry problems but also to algebraic ones. The following is one of the simplest examples for illustration.

Example 1.7. Newton's Formulae. Determine the k-th power sum of roots of an algebraic equation of degree n in terms of the coefficients of the equation.

As in Remark 1.1, we can in principle determine automatically such an expression for each pair of *fixed* integers k, n but by no means for *general* pair. To fix the ideas, let us take $k = 4, n = 4$. Following the general underlying principle let us write the algebraic equation E_4 in a single unknown z of degree $n = 4$ in the form

$$E_4 : \ z^4 + x_1 * z^3 + x_2 * z^2 + x_3 * z + x_4 = 0. \tag{1.23}$$

Let x_5, x_6, x_7, x_8 be now the k-th power-sum of the roots for $k = 1, 2, 3, 4$. The problem is to find an inductive relation in reducing x_8 to x_7, x_6, x_5 and then to $x_1, \cdots, x_4$.

For this purpose let us take the four roots of the equation E_4 to be $x_9, x_{10}, x_{11}, x_{12}$. Then we have a hypothesis set $HYP = \{h_1, \cdots, h_7\}$ with h_i given below:

$$\begin{aligned}
h_1 =& \ x_9 + x_{10} + x_{11} + x_{12} + x_1, \\
h_2 =& \ x_9 * x_{10} + x_9 * x_{11} + x_9 * x_{12} \\
& + x_{10} * x_{11} + x_{10} * x_{12} + x_{11} * x_{12} - x_2, \\
h_3 =& \ x_9 * x_{10} * x_{11} + x_9 * x_{10} * x_{12} + x_{10} * x_{11} * x_{12} + x_3, \\
h_4 =& \ x_9 * x_{10} * x_{11} * x_{12} - x_3, \\
h_5 =& \ x_6 - (x_9^2 + x_{10}^2 + x_{11}^2 + x_{12}^2), \\
h_6 =& \ x_7 - (x_9^3 + x_{10}^3 + x_{11}^3 + x_{12}^3), \\
h_7 =& \ x_8 - (x_9^4 + x_{10}^4 + x_{11}^4 + x_{12}^4).
\end{aligned} \tag{1.24}$$

Form the char-set CS of HYP we see that the first three char-pols are:

$$\begin{aligned}
C_1 =& \ x_6 + 2 * x_2 - x_1^2, \\
C_2 =& \ x_7 + 3 * x_3 - 3 * x_2 * x_1 + x_1^3, \\
C_3 =& \ x_8 + 4 * x_4 - 4 * x_3 * x_1 - 2 * x_2^2 + 4 * x_2 * x_1^2 - x_1^4.
\end{aligned} \tag{1.25}$$

The equations $C_i = 0, \ i = 1, 2, 3$, are then the Newton Formulae for the case of $k = 2, 3, 4, \ n = 4$ to be sought for.

An interesting problem has been studied in some detail by X.S.Gao in his doctoral thesis [Gao1], (cf. also the paper [WWD]). See the example below.

Example 1.8. Discriminant Systems.

Suppose given an algebraic equation E_n in a single unknown z of degree n and a partition $\Pi = (\pi_1, \cdots, \pi_k)$ of n. To determine the conditions for the coefficients of the equation E_n such that the roots will be partitioned in equal ones as the given partition Π.

Let us take for example fixed $n = 4$ and fixed partition $\Pi = (2, 2)$. Write the equation E_4 as in (1.23). Let the 4 roots be x_5, x_5, x_6, x_6. Then we have a hypothesis set $HYP = \{h_1, h_2, h_3, h_4\}$ with h_i given by

$$\begin{aligned} h_1 &= 2 * (x_5 + x_6) + x_1, \\ h_2 &= (x_5^2 + 4 * x_5 * x_6 + x_6^2) - x_2, \\ h_3 &= 2 * x_5 * x_6 * (x_5 + x_6) + x_3, \\ h_4 &= x_5^2 * x_6^2 - x_4. \end{aligned} \tag{1.26}$$

The first two char-pols C_1, C_2 of the char-set of HYP will give the conditions to be found:

$$\begin{aligned} C_1 &\equiv 8 * x_3 - 4 * x_2 * x_1 + x_1^3 = 0, \\ C_2 &\equiv 54 * x_4 - 16 * x_2^2 + 8 * x_2 * x_1^2 - x_1^4 = 0. \end{aligned} \tag{1.27}$$

Note that we have not excluded the possibility $x_5 = x_6$ so that the conditions include the partitions (2,2) as well as (4).

Remark 1.4. Recently L.Yang had extended the above discriminant systems to the case for which *reality* of roots were taken into acount. Cf. e.g. [YL].

Automatic Discovering of Geometry Theorems.

Our general method furnishes us also a method of automatic discovering of geometry theorems eventually new. The following example is a simple illustration.

Example 1.9. Let us consider again the Feuerbach Theorem of Ex. 2.2 in Chap.7. Suppose we are ignorant of any possible relation which may exist between the nine-point circle and the in- or es-cribed circles of a triangle. In order to discover any such possible relations let us set the radii of these two circles to be x_2 and x_1 respectively and the distance between their centers to be x_3. Take coordinates so that

$$I = (0, x_1),\ A = (x_4, 0),\ B = (x_5, 0),\ C = (x_6, x_7),\ N = (x_8, x_9).$$

Then the hypothesis polset becomes $HYP' = \{h'_1, \cdots, h'_6\}$ with h'_i given below:

$$\begin{aligned}
h'_1 &= (x_1^2 + x_4 * x_5) * x_6 \ - \ x_1^2 * (x_4 + x_5),\\
h'_2 &= (x_1^2 + x_4 * x_5) * x_7 \ - \ 2 * x_1 * x_4 * x_5,\\
h'_3 &= 4 * x_8 \ - \ (x_4 + x_5 + 2 * x_6),\\
h'_4 &= 4 * x_7 * x_9 \ - \ x_7^2 \ - \ (x_5 - x_6) * (4 * x_8 - 2 * x_4 - x_5 - x_6),\\
h'_5 &= 4 * x_2^2 \ - \ (2 * x_8 - x_4 - x_5)^2 \ - \ 4 * x_9^2,\\
h'_6 &= x_3^2 \ - \ x_8^2 \ - \ (x_9 - x_1)^2.
\end{aligned} \tag{1.28}$$

It turns out that *wsolve* for $x_1 * x_2 * x_7 * x_9 \neq 0$ will give us 4 char-sets $CS_i,\ i = 1, 2, 3, 4$, of HYP' of which the first pols are C_{i1} given below:

$$\begin{aligned}
C_{11} &= -x_3 + x_2 + x_1, \ C_{21} = x_3 - x_2 + x_1,\\
C_{31} &= -x_3 - x_2 + x_1, \ C_{41} = x_3 + x_2 + x_1.
\end{aligned} \tag{1.29}$$

The relation $C_{i1} = 0$ says that the nine-point circle and the in- or es-cribed circle is tangent to each other in some contact manner. From the general theory we have

$$Zero(HYP'/x_1 * x_2 * x_7 * x_9) \subset \bigcup_i Zero(CS_i) \subset \bigcup_i Zero(C_{i1}).$$

We get thus Feuerbach Theorem which is now discovered in an automatic way.

Automatic Deduction of Geometric Loci.

The determination of the geometric loci of a moving point under some geometric conditions is just the determination of the relation between the coordinates of that point under these conditions. It follows that our general method of automatic determination of unknown relations may be applied in the present case. Let us consider the problem below.

Definition 1.1. A *4-bar linkage* := A mechanism consisting of 3 rods AU, UV, VB joined at the pivots U, V with the ends A, B fixed while the rods AU, BV can be turned around the fixed ends A, B respectively.

Coupler plane := The plane containing the linkage.

Coupler rod := The rod UV.

Coupler point := Some fixed point Z marked on the moving plane of the coupler rod UV.

Coupler curve := The locus of the coupler point Z when the rods AU, BV are turned around A, B.

Example 1.10. Coupler Curve. To determine the geometric loci of the coupler point Z, viz. the coupler curve.

In order to simplify the computations let us denote the points in the linkage plane by complex numbers. Thus, for each point P we shall denote by P_1, P_2 the corresponding complex number and its complex conjugate. Let the inclination of the rod UV oriented in the direction from U to V w.r.t. the real axis be θ. Denote the oriented angles $<(UV, UZ), <(UV, VZ)$ by α, β and the side-lengths of UZ, VZ by s_b, s_a respectively. Set $(\sqrt{-1} = \mathbf{i})$

$$\begin{aligned} e^{\mathbf{i}*\theta} &= \tau_1, \quad s_b * e^{\mathbf{i}*\alpha} = \mu_1, \quad s_a * e^{\mathbf{i}*\beta} = \nu_1, \\ e^{-\mathbf{i}*\theta} &= \tau_2, \quad s_b * e^{-\mathbf{i}*\alpha} = \mu_2, \quad s_a * e^{-\mathbf{i}*\beta} = \nu_2. \end{aligned} \tag{1.30}$$

If we drop a perpendicular ZW from Z to UV with foot W, then we have

$$\begin{aligned} &Im(\mu_1) = Im(\mu_2) = \mathbf{i} * h, \\ &Re(\mu_1) = s, \ Re(\nu_1) = s - r, \end{aligned} \tag{1.31}$$

in which Re, Im means the real and imaginary parts respectively and the oriented lengths of UW, VW on the oriented line UV are $s, s-r$ respectively. See the diagram Fig. 1.2 above.

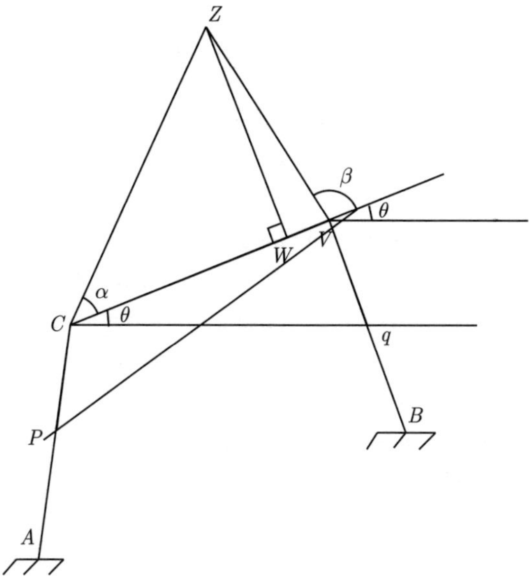

Fig. 1.2.

Let the absolute lengths of the rods AU, BV, UV be p, q and r respectively. Then we have to determine the loci of the point Z or the equation of the conjugate complex numbers Z_1, Z_2 in terms of the known values $A_1, A_2, B_1, B_2, p, q, r, h, s$, which may be turned into 9 parameters in real values. We have now the hypothesis polset $HS = \{h_1, \cdots, h_6\}$ with

$$
\begin{aligned}
h_1 &= (U_1 - A_1) * (U_2 - A_2) - p^2, \\
h_2 &= (V_1 - B_1) * (V_2 - B_2) - q^2, \\
h_3 &= U_1 + \tau_1 * \mu_1 - Z_1, \\
h_4 &= U_2 + \tau_1 * \mu_1 - Z_2, \\
h_5 &= V_1 + \tau_1 * \nu_1 - Z_1, \\
h_6 &= V_2 + \tau_1 * \nu_1 - Z_2.
\end{aligned}
\tag{1.32}
$$

Let us arrange the variables in the ordering

$$
\begin{aligned}
&p \prec q \prec r \prec A_1 \prec A_2 \prec B_1 \prec B_2 \prec \mu_1 \prec \mu_2 \prec \nu_1 \prec \nu_2 \\
&\prec Z_1 \prec Z_2 \prec U_1 \prec U_2 \prec V_1 \prec V_2 \prec \tau_1 \prec \tau_2.
\end{aligned}
$$

Then the first char-pol C_1 of the char-set will give the loci-equation in the form

$$
C_1 \equiv Z_1^3 * Z_2^3 * (\mu_1 - \nu_1) * (\mu_2 - \nu_2) + \cdots = 0. \tag{1.33}
$$

The equation $C_1 = 0$ expresses Z_1, Z_2 in terms of the known parameters with number of terms equal to 362. If we express the equation in real coordinates

$$
Z = (x, y),\ A = (a_x, a_y),\ B = (b_x, b_y),
$$

and the real parameters p, q, r, h, s, then C_1 will have number of terms equal to 2016 and is of degree 6 in each of the running coordinates x, y with terms of highest degree equal to $(x^6 + y^6) * r^2$.

Remark 1.5. The above geometric locus, as a point set in the plane, is in fact only a part of $Proj\ Zero(HS) \subset Zero(C_1)$, with $Proj$ the natural projection of the space in $Z_1, Z_2, etc.$ onto $\mathbf{R}^2(x, y)$, A complete determination of the locus by means of the method based on the Projection Theorem as described in Sect. 5.4 of Chap. 5 would be undesirably too complicated to be acceptable. The present method, though quite incomplete and even inexact, is a rather preferable short cut to the solving of the problem. This is in fact the usual way adopted universally in the practice and in the literature.

8.2 Applications to Problems involving Inequalities, Optimization Problems, and Non-Linear Programming.

Ler $\mathbf{R} = real\ field,\ X = (x_1, \cdots, x_n)$, etc. as before.

Problems involving Inequalities.

Perhaps the CAD method of Collins (cf. for literature [Col1], [Ar1]) is the most general complete method which has been applied to problems involving inequalities.

On the other hand in Chap. 5, Sect. 5 we have exhibit some method englobed in the two Theorems 5.5.4 and 5.5.5 which, though not so general and complete as that of Collins, seems to be quite efficient for certain kinds of problems. Besides the proving of geometry theorems involving inequalitie already considered in Chap. 7, Sect. 3, the range of applications covers also the proving of inequalities, algebraic or geometric, the non-linear programming, the optimization problems, etc. We are interested in all such problems which may be reduced to one of the following form:

Problem Ineq. Let $\mathbf{R}^n(X)$ be the *real* Euclidean space of dimension n in coordinates $X = (x_1, \cdots, x_n)$ and D a domain in E^n. Let $f, h_i, (i \in I = \{1, \cdots, r\}, r < n)$, and g be all real polynomials in $\mathbf{R}^n[X]$ over the domain D. To determine for what real value c we shall have always

$$f \geqslant c \text{ or } f > c \text{ or } f \leqslant c \text{ or } f < c \tag{2.1}$$

under the conditions

$$HS = 0, \text{ where } HS = \{h_i \mid i \in I\} \text{ and} \tag{2.2}$$

$$g \neq 0. \tag{2.3}$$

For convenience of readers we reproduce below the basic formulae in Theorems 5.5.4 and 5.5.5 for the determination of optimal (greatest or least) values, supposed exist, of the Problem Ineq (D open or closed domain, O open domain):

$$Optm\ \{f, D, HS\} = Optm\ Ker\{f, D, HS\}. \tag{$*$}$$

$$\begin{aligned} &Optm\ \{f, g, O, HS\} = Optm_{X \in O}\ \{f(X) \mid X \in LJVal\}, \text{ where} \\ &LJVal = LProj\ \tilde{O}Zero_{\mathbf{R}}(LS/g) + OZero_{\mathbf{R}}(JS/g). \end{aligned} \tag{$**$}$$

In $(**)$ LS and JS are respectively the Lagrangian polset and the Jacobian polset of the relavant problem.

Definiteness of Pols.

Among the simplest applications we may cite the study of definiteness or semi-definiteness of polynomials. In this respect we refer besides the relevant papers of the author also a paper of D.M.Wang, [WDM1].

Example 2.1. (Motztkin Polynomial) The Motztkin polynomial M given by

$$M = 1 + x_1^2 * x_2^4 + x_1^4 * x_2^2 - 3 * x_1^2 * x_2^2 \tag{2.4}$$

is semi-definite positive in the whole plane $\mathbf{R}^2(x_1, x_2)$.

Proof. First let us remark that from simple degree considerations the least value of M in the plane necessarily exists. Introduce now a new variable x_0 and set

$$h_0 = M - x_0. \tag{2.5}$$

The problem is then reduced to show that the least value of x_0 under the constraint condition $h_0 = 0$ is $\geqslant 0$. There are two methods to follow for the determination of the least value.

For the first method based on Theorem 5.5.5, let us form with Lagrange multiplier λ the Lagrange pol

$$L = x_0 + \lambda * (M - x_0). \tag{2.6}$$

The Lagrangian polset is

$$\begin{aligned} LS = & \left\{ \frac{\partial L}{\partial x_k}, h_0 \mid k = 0, 1, 2 \right\} \\ = & \{1 - \lambda,\ 2 * x_1 * x_2^2 * \lambda * (x_2^2 + 2 * x_1^2 - 3), \\ & 2 * x_1^2 * x_2 * \lambda * (2 * x_2^2 + x_1^2 - 3),\ M - x_0\}. \end{aligned} \tag{2.7}$$

It is readily found that

$$Zero_{\mathbf{R}}(LS) = \bigcup_{k=1,2,3} Zero_{\mathbf{R}}(CS_k), \text{ with} \tag{2.8}$$

$$\begin{aligned} CS_1 &= \{x_0 - 1, x_1, \lambda - 1\}, \\ CS_2 &= \{x_0 - 1, x_2, \lambda - 1\}, \\ CS_3 &= \{x_0, x_1^2 - 1, x_2^2 - 1, \lambda - 1\}. \end{aligned} \tag{2.9}$$

Denote by $Proj_0$ the projection of $\mathbf{R}^3(x_0, x_1, x_2)$ to $\mathbf{R}(x_0)$. Then we have

$$Proj_0\ Zero_{\mathbf{R}}(LS) = \{0, 1\}. \tag{2.10}$$

On the other hand for the Jacobian polset

$$JS = \left\{ \frac{\partial h_0}{\partial x_k},\ h_0 \mid k = 0, 1, 2 \right\} \tag{2.11}$$

it is clear that

$$Zero_{\mathbf{R}}(JS) = \varnothing. \tag{2.12}$$

By Theorem 5.5.5 it follows that

$$Least\{x_0, \mathbf{R}^3, h_0\} = Least\{0, 1\} = 0 \tag{2.13}$$

or $M \geqslant 0$ as to be proved. Moreover, the least value $x_0 = 0$ comes from CS_3 so that it is attained for $x_1 = \pm 1, x_2 = \pm 1$. Remark that for the greatest value of M we get no information from the set $\{0, 1\}$ since the greatest value of M does not exist so that the above method cannot be applied.

Alternatively, we may also determine the least value by the Finite Kernel Theorem or Theorem 5.5.4 as follows.

It is readily found that

$$Zero_{\mathbf{R}}(h_0) = \bigcup_{k=1,\cdots,5} Zero_{\mathbf{R}}(CS'_k/*), \tag{2.14}$$

in which * means the relevant initial-separant product and

$$\begin{aligned} CS'_1 &= \{h_0\}, \\ CS'_2 &= \{x_1^2 * (x_1^2 - 3)^2 + 4 * (1 - x_0), 2 * x_2^2 + x_1^2 - 3\}, \\ CS'_3 &= \{x_0 - 1, x_1\}, \\ CS'_4 &= \{x_0 - 1, x_2\}, \\ CS'_5 &= \{x_0, x_1^2 - 1, x_2^2 - 1\}. \end{aligned} \tag{2.15}$$

In the right-hand side of (2.14) the first two zero-sets furnish no contributions to the finite kernel set while the other three furnish the values 1, 1, 0 respectively. It follows that

$$K = Ker\{x_0, \mathbf{R}^3, h_0\} = \{0, 1\}. \tag{2.16}$$

Hence we have again

$$Least(M) = Least\{x_0, \mathbf{R}^3, h_0\} = Least\{K\} = 0, \tag{2.17}$$

as to be proved. Moreover, the least value $x_0 = 0$ comes from CS'_5 so that it is attained for $x_1 = \pm 1, x_2 = \pm 1$ as before. □

Proving of Algebraic and Trigonometric Inequalities.

A lot of algebraic, trigonometric, and geometric inequalities which occur so frequently in mathematics may be proved by means of our method. Examples of geometric inequalities have been given in Chap. 7, Sect. 4. Below are some such simplest examples for algebraic and trigonometric inequalities which serve for illustration. More examples may be found in [WDK1] and also some relevant papers of the author.

Example 2.2. For $a, b, c > 0$ we have

$$(b + c) * (c + a) * (a + b) \geqslant 8 * a * b * c. \tag{2.18}$$

Moreover, the equality is attained for any triple of values $a = b = c > 0$.

Proof. Let us rewrite a, b, c by x_1, x_2, x_3 and introduce by Seidenberg tactic y_1, y_2, y_3 such that $h_i = 0, i = 1, 2, 3$, where

$$h_i = y_i^2 * x_i - 1,\ i = 1, 2, 3. \tag{2.19}$$

Let us define x_0 by $h_0 = 0$ where

$$h_0 = y_1^2 * y_2^2 * y_3^2 * (x_2 + x_3) * (x_3 + x_1) * (x_1 + x_2) - x_0. \tag{2.20}$$

Then we come to Problem Ineq with

$$\begin{aligned} &D:\ 0 \leqslant x_i < +\infty,\ 0 \leqslant y_i\ < +\infty,\\ &f = x_0,\ g = 1,\ HS = \{h_0, h_1, h_2, h_3\}. \end{aligned}$$

It is clear that least value of x_0 necessarily exists so that our method may be applied. The inequality to be proved is then equivalent to say that the least value of x_0 given by $h_0 = 0$ is 8. With Lagrange multipliers $\lambda_i,\ i = 0, 1, 2, 3$, the Lagrangian pol is given by

$$L = x_0 + \lambda_0 * h_0 + \lambda_1 * h_1 + \lambda_1 * h_1 + \lambda_3 * h_3. \tag{2.21}$$

The Lagrangian polset is given by

$$LS = \left\{\frac{\partial L}{\partial x_0}, \frac{\partial L}{\partial x_i}, \frac{\partial L}{\partial y_i}, h_0, h_i \mid i = 1, 2, 3\right\}.$$

With the ordering

$$x_0 \prec x_1 \prec x_2 \prec x_3 \prec y_1 \prec y_2 \prec y_3 \prec \lambda_1 \prec \lambda_2 \prec \lambda_3,$$

it is readily found that the char-set of LS is consisting of 9 pols with the first 3 char-pols given below:

$$\begin{aligned} C_1 &= x_0 * (x_0 - 8),\\ C_2 &= x_2 - x_1,\\ C_3 &= x_3 - x_1. \end{aligned} \tag{2.22}$$

The solution $x_0 = 0$ of $C_1 = 0$ is impossible because of $h_i = 0,\ i = 1, 2, 3$. It follows that the least value of x_0 is 8 which is achieved by any value $x_1 = x_2 = x_3 (> 0)$ as to be proved. □

Our method based on the Finite Kernal Theorem of Sect. 5.5 may be applied also to the proof of trigonometric inequalities. The following example is one of the simplest which has been proved in this way. Many other trigonometric inequalities proved in this way may be found in the doctoral thesis of D.K.Wang, cf. [WDK1].

Example 2.3. Sine-Sum Theorem. For a triangle ABC we have

$$sin\ A + sin\ B + sin\ C \leqslant \frac{3}{2} * \sqrt{3}. \tag{2.23}$$

Proof. Let us set

$$\begin{aligned} &sin\ A = x_1,\ \ sin\ B = y_1, sin\ C = z_1, \\ &cos\ A = x_2,\ \ cos\ B = y_2, cos\ C = z_2. \end{aligned} \tag{2.24}$$

Then we have $HS = 0$, where $HS = \{h_1, \cdots, h_5, h_0\}$ with h_i given below:

$$\begin{aligned} h_1 &= x_1^2 + x_2^2 - 1, \\ h_2 &= y_1^2 + y_2^2 - 1, \\ h_3 &= z_1^2 + z_2^2 - 1, \\ h_4 &= x_1 * y_2 + x_2 * y_1 - z_1, \\ h_5 &= x_2 * y_2 - x_1 * y_1 + z_2, \\ h_0 &= x_0 - (x_1 + y_1 + z_1). \end{aligned} \tag{2.25}$$

Let us take the closed domain

$$D : 0 \leqslant x_1, y_1, z_1 \leqslant +1$$

and ordering of variables

$$x_0 \prec x_1 \prec y_1 \prec z_1 \prec x_2 \prec y_2 \prec z_2.$$

We have to consider the determination of the greatest value of x_0 in D under the restraint conditions $HS = 0$. It is found that $Zero_{\mathbf{R}}(HS)$ is the union of 29 zero-sets $Zero_{\mathbf{R}}(CS_k/ISP_k)$, $k = 1, \cdots, 29$, of which the explicit form may be found in the doctoral thesis of D.K.Wang, [WDK1]. It turns out that most of the solutions are either outside of the domain D (e.g. $x_2 < 0$) or furnish no contribution to the finite kernel set. The kernel set due to the Finite Kernel Theorem is found to be

$$K = \left\{0, 2, \sqrt{2} \pm 1, \frac{3}{2} * \sqrt{3}\right\}. \tag{2.26}$$

It follows that the greatest value of x_0, i.e. the greatest value of $sin\ A + sin\ B + sin\ C$ is $\frac{3}{2} * \sqrt{3}$ due to the char-set

$$\begin{aligned} CS = \{&4 * x_0^2 - 27, 3 * x_1 - x_0,\ 3 * y_1 - x_0,\ 3 * z_1 - x_0, \\ &2 * x_2 - 1,\ \ 2 * y_2 - 1,\ \ 2 * z_2 - 1\}. \end{aligned} \tag{2.27}$$

The values of x_1, y_1, z_1 are thus all equal to $\frac{1}{2} * \sqrt{3}$ and those of x_2, y_2, z_2 are all equal to $\frac{1}{2}$. It follows that (2.23) is true and the equality occurs when and only when the triangle is an equilateral one. We remark that the least value of x_0 in D under the conditions $HS = 0$ is 0 for which $x_1 = y_1 = z_1 = 0$ or the triangle collapse to a line. □

A Quartic Problem and an Ellipse Problem.

The two problems have been studied extensively by various authors. Cf. e.g. [Ar1], [A-M], [A-S], [C-H], [Kah], [Laz], [Mig], etc. In applying our method to the first problem we require the following lemma.

Lemma 2.1. Let $\triangle$ be the discriminant of the equation

$$c_0 * y^3 + c_1 * y^2 + c_2 * y + c_3 = 0, \ c_0 > 0. \tag{2.28}$$

Then for y satisfying the equation to be always positive it is necessary and sufficient that either (a) or (b) below be true:

(a) $\triangle < 0$ *and* $c_3 < 0$.

(b) $\triangle \geqslant 0$ *and* $c_1 < 0, \ c_2 > 0, \ c_3 < 0$.

For y satisfying the equation (2.28) to be always non-negative it is necessary and sufficient that *(a)*, or *(b)* above, or *(c)* below be true:

(c) $c_3 = 0, \ c_1 \leqslant 0, \ c_2 \geqslant 0$, and $c_1^2 - 4 * c_0 * c_2 \geqslant 0$.

Proof. See e.g. [Fr], p.188. □

Example 2.4. The Quartic Problem. Consider a real quartic

$$y = x^4 + p * x^2 + q * x + r. \tag{2.29}$$

To find the conditions on p, q, r such that y is always $\geqslant 0$ or always > 0.

Our method leads to the consideration of a Problem Ineq for which

$$D = \mathbf{R}^2(x, y), \ f = y, \ g = 1, \text{ and}$$

$$h_1 = -y + x^4 + p * x^2 + q * x + r.$$

It is clear that the least value of y exists and Theorems 5.5.4 or 5.5.5 may be applied. The corresponding set $LJVal$ is readily found to be the set of real zeros of the cubic equation (2.28) with

$$\begin{aligned}
c_0 &= 256, \\
c_1 &= 128 * (p^2 - 6 * r), \\
c_2 &= 16 * (p^4 - 16 * p^2 * r + 48 * r^2 + 9 * p * q^2), \\
c_3 &= -16 * p^4 * r + 4 * p^3 * q^2 + 128 * p^2 * r^2 - 144 * p * q^2 * r + 27 * q^4 - 256 * r^3.
\end{aligned} \tag{2.30}$$

The discriminant $\triangle$ of (2.28) is given by

$$\triangle = -q^2 * (8 * p^3 + 27 * q^2)^3. \tag{2.31}$$

The original quartic problem is thus equivalent to the determination of conditions about c_i so that (2.28) will have no negative roots or only positive roots. Explicit conditions are then determined by Lemma 2.1.

Example 2.5. The Ellipse Problem. Consider the unit circle $C : x^2+y^2 = 1$, and an ellipse $E(x, y) = 0$ with

$$E = a_1 * x^2 + 2 * a_2 * x * y + a_3 * y^2 + 2 * a_4 * x + 2 * a_5 * y + a_6. \tag{2.32}$$

To ensure that $E = 0$ is really an ellipse, we shall impose the imequality-conditions below:

$$a_2^2 - a_1 * a_3 < 0, \ a_1 > 0, \ a_3 > 0.$$

The problem is to find the conditions in a_i such that the ellipse will lie wholly inside C (with touching allowed or not allowed).

This problem had also been extensively studied along with the quartic problem restricted mainly to the case $a_2 = a_5 = 0$, i.e. the case for which the ellipse has one of its axis coincident with the X-axis. Without any such restrictions our method leads to the consideration of a Problem Ineq for which

$$D = \mathbf{R}^3(x, y, r), \ f = r, \ g = 1, \ \text{and}$$

$$h_1 = E(x, y), \ h_2 = x^2 + y^2 - r.$$

Owing to the condition that $E = 0$ is an ellipse, it is clear that greatest value of f exists and again Theorem 5.5.4 or 5.5.5 may be applied. We may dispose off the uninteresting case for which E is a circle, or the case with $a_2^2+(a_1-a_3)^2 = 0$, as well as some degeneracy cases which we do not give explicitly here. We find then readily in the generic case that the greatest value of r possible is the greatest real root of a quartic equation $Q(r) = 0$ in r. The quartic Q has all its coefficients pols in a_i and has a number of terms $= 123$. The original problem is thus equivalent to one of finding conditions in a_i such that $Q(r) = 0$ has no real roots > 1 or $\geqslant 1$ according to whether touching is allowed or not. This latter problem may then be settled in reducing it to Ex. 2.4. See the diagram Fig.2.1 below.

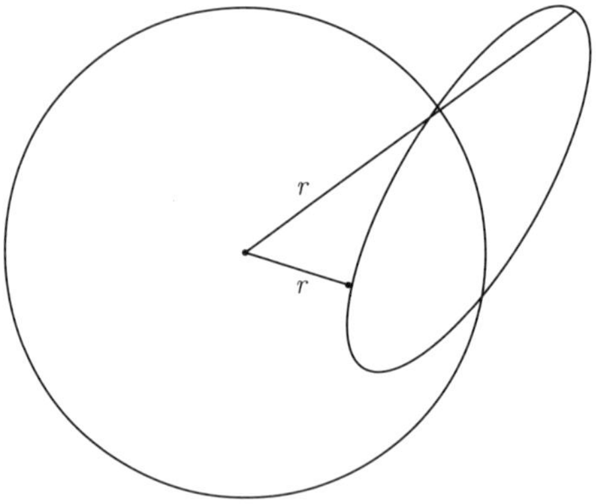

Fig. 2.1.

Problems of Non-Linear Programming.

The non-linear programming problems for which the objective function and the constraint conditions are all *polynomial* ones can clearly be treated by our method. The following example is one of the simplest which serves for illustration. More examples may be found in the paper [WTJ1].

Example 2.6. A Non-linear Programming Problem. To minimize

$$f = 1000 - x_1^2 - 2 * x_2^2 - x_3^2 - x_1 * x_2 - x_1 * x_3 \tag{2.33}$$

under constraints

$$\begin{aligned} h_1 &= x_1^2 + x_2^2 + x_3^2 - 25 = 0, \\ h_2 &= 8 * x_1 + 14 * x_2 + 7 * x_3 - 56 = 0, \ \text{and} \end{aligned} \tag{2.34}$$

$$x_i \geqslant 0, \ i = 1, 2, 3. \tag{2.35}$$

First of all let us remark that the least value of f necessarily exists owing to the first condition of (2.34). Let us split now the domain defined by (2.35) into 8 open domains in taking each x_i either > 0 or $= 0$. It is readily found that the only E-zero in open domain D_1 with all $x_i > 0$ is

$$Z_1 = (3.512, \ 0.217, \ 3.552). \tag{2.36}$$

For the other open domains there is only one other E-zero from the open domain D_2 defined by $x_1 = 0, \ x_2 > 0, \ x_3 > 0$, viz.

$$Z_2 = \left(0, \ \frac{32 - 2\sqrt{61}}{10}, \ \frac{8 + 2\sqrt{61}}{5}\right). \tag{2.37}$$

The corresponding E-values of Z_1 and Z_2 are respectively

$$f(Z_1) \approx 961, \text{ and } f(Z_2) \approx 972. \tag{2.38}$$

We see thus the least value of f under the stated conditions is approximately 961 taking place at the point Z_1 of (2.36). For details we refer to the paper [WTJ1] cited above.

A Collision Problem.

We consider now a problem below which has been studied by Collins and Hong in some special case, cf. e.g. [C-H].

Example 2.7. Collision Problem. Suppose that two cars of given form are moving respectively along the X-axis and the Y-axis in positive directions with known velocities $v_1 > 0$ and $v_2 > 0$. To decide whether the cars will collide or not, and to determine in the colliding case the time and place of first collision.

Let the starting positions of the two cars be given respectively by

$$f_1(x, y) = 0, \text{ and } f_2(x, y) = 0.$$

Then this is seen to be a Problem Ineq for which

$$D = \{t > 0\} \subset \mathbf{R}^3(t, x, y), \ f = t, \ g = 1,$$

$$h_1 = f_1(x - v_1 * t, y), \ h_2 = f_2(x, y - v_2 * t).$$

In particular, let us consider the case for which the two cars are both of elliptical form given by $(c_1 < 0, c_2 < 0)$

$$\begin{aligned} f_1(x, y) &= b_1^2 * (x - c_1)^2 + a_1^2 * y^2 - a_1^2 * b_1^2, \\ f_2(x, y) &= b_2^2 * x^2 + a_2^2 * (y - c_2)^2 - a_2^2 * b_2^2. \end{aligned}$$

Then in the case of *generic* values of a_i, b_i, c_i, and v_i our method will give rise to an irreducible polynomial equation of degree 8 in t with number of terms = 696. Leaving aside some uninteresing cases, the two cars will collide if and only if this equation has some positive root. The time and place of first collision can be easily determined if numerical values of a_i, b_i, c_i, and v_i (meeting the genericity condition) are substituted. Consider the further special case for which the two elliptical forms are similar, viz.

$$a_1^2 * b_2^2 = a_2^2 * b_1^2,$$

which occurs in particular when both cars are circular in form. Then we will arrive at a set of linear and quadratic equations in t. The conditions for possible collision as well as time and place of first collision are all expressible by explicit equations in a_i, b_i, c_i, and v_i. For details see the paper [WTJ2]. See also Fig. 2.2 below.

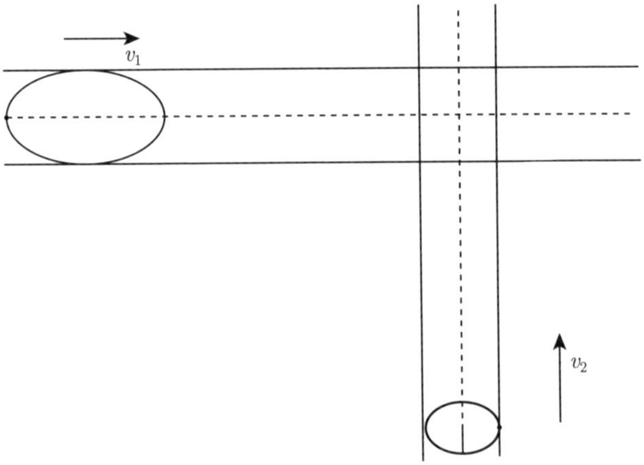

Fig. 2.2.

8.3 Applications to 4-Bar Linkage Design.

Problems of 4-Bar Linkage Design.

The description of a 4-bar linkage has been given in Sect. 8.1 in Def. 1.1 and Ex. 1.10. A 4-bar linkage in a given plane depends on 9 parameters: 4 parameters for fixing the positions of the fixed ends $A = (a_1, a_2),, B = (b_1, b_2)$; 3 parameters p, q, r for the lengths $|AU|, |BV|, |UV|$; and 2 parameters for fixation of the triangle UVZ which we have taken to be the height h of the triangle UVZ on the side UV, and the orthogonal projection s of the side UZ on UV with sign conveniently taken into account, for which we refer to Sect. 8.1. Thus, a first problem for the 4-bar linkage design is the following one:

Problem L1. Knowing 9 positions of the coupler point Z in the linkage plane, to determine the 4-bar linkage in that plane.

Denote the coupler point Z by (x, y), then in Ex. 1.10 we have determined the coupler equation $H(x, y) = 0$ of Z in terms of the 9 parameters. Let the 9 known positions of Z have coordinates $Z_i = (c_i, d_i)$. Then

$$H_i \equiv H(c_i, d_i) = 0, \ i = 1, \cdots, 9$$

will give us 9 equations in the 9 unknowns $a_1, b_1, a_2, b_2, p, q, r, h, s$. To solve such a complicated polynomial equation-system is out of the range of our present char-set method of polynomial equations-solving, though in principle we can always do.

Even if we use complex variables instead of real ones as in Sect. 8.1 for which the equations to determine the unknown parameters are much simpler, it seems that the solving will still be out of the range of our *symbolic* method. In concrete cases with numerical datas our hybrid method as described in Chap. 6, Sect. 2 may be applied but trials are yet to be made. However, in spite of these there are still problems which are important in technologies and can be solved completely by our *symbolic* methods. The fillowing is such a one.

4-Bar Linkage Design in Knowing 5 *Coupler-Bar Planar Positions.*

Instead of Problem L1 we shall study the following equally important 4-bar linkage design problem:

Problem L2. Knowing 5 planar positions of the moving coupler planes over the fixed coupler plane, to determine the 4-bar linkage.

In year 1995 Liu and Guan gave a simple and elegant method of solving the above Problem L2 in reducing it to the solving of some polynomial equation-system which was then done in applying our general method, cf. their paper [L-G]. It turns out that in the classics there were already solutions of the Problem L2 based on geometry theory of Burmester points which were also reduced to some problem of polynomial equations-solving. In comparing the two methods the present author had made a simplification of the above second method for which much less knowledge of geometry is required. We shall present all the three methods in what follows.

Recall of Burmester Geometric Theory of 4-*Bar Linkage.*

Notation 3.1.

$E :=$ fixed plane of the 4-bar linkage $AUVB$.

$E_i,\ i = 0, 1, \cdots, 4 :=$ 5 moving coupler planes over the fixed plane E.

$U_iV_i,\ i = 0, 1, \cdots, 4 :=$ 5 moving coupler rods in the moving coupler planes E_i.

$P_{ij} = P_{ji},\ i, j = 0, 1, \cdots, 4,\ i \neq j :=$ Poles of rotation from plane E_i to E_j.

$\theta_{ij} = -\theta_{ji}\ mod\ 2 * \pi,\ i, j = 0, 1, \cdots, 4,\ i \neq j :=$ Angle of rotation about P_{ij} from plane E_i to E_j.

Definition 3.1. Two points P_i, P_j in the moving planes E_i, E_j respectively are *related* $:=$ P_i moves to P_j when the plane E_i moves to the plane E_j.

Definition 3.2. *Center point* C_{ijkl} w.r.t. four mutually distinct moving planes E_i,E_j,E_k,E_l $:=$ The center of a circle for which there are 4 related points in the moving planes E_i, E_j, E_k, E_l respectively lying on that circle.

Definition 3.3. *Center curve* Γ_{ijkl} $:=$ The locus of all center points C_{ijkl} w.r.t. the moving planes E_i, E_j, E_k, E_l.

Definition 3.4. *Pole point* $P_{ij,kl}$ w.r.t. four mutually distinct moving planes

$E_i, E_j, E_k, E_l :=$ Point in the fixed plane E which subtends same angles to $P_{ik}P_{il}$ and $P_{jk}P_{jl}$ such that

$$\begin{aligned} &< (PP_{ik}, PP_{il}) =< (PP_{jk}, PP_{jl}), \text{ or} \\ &< (PP_{ik}, PP_{jk}) =< (PP_{il}, PP_{jl}). \end{aligned} \tag{3.1}$$

Remark 3.1. The angles in this section are all oriented angles as defined in Chap. 7, Sect. 2. Moreover, all equations in oriented angles, like (3.1) above and (3.2) below, are understood to be congruences *mod* π, unless otherwise explicitly stated or expressed.

Remark 3.2. For known positions of the moving planes $E_i, i = 0, 1, \cdots, 4$, the poles P_{ij}, and the angles θ_{ij} *mod* $2 * \pi, i, j = 0, 1, \cdots, 4, i \neq j$, are all known too.

Definition 3.5. *Pole curve* $\Pi_{ij,kl}$ w.r.t. four mutually different moving planes $E_i, E_j, E_k, E_l :=$ The locus of all pole points $P_{ij,kl}$ w.r.t. these moving planes.

Proposition 3.1. For $i \neq j$, the points P_{ij}, A are on the perpendicular bisector of U_iU_j so that

$$< (P_{ij}U_i, P_{ij}A) = \frac{\theta_{ij}}{2}.$$

Proposition 3.2. For distinct moving planes E_i, E_j, E_k:

$$< (P_{ij}P_{ik}, P_{ij}P_{jk}) = \frac{\theta_{ij}}{2}. \tag{3.2}$$

Proposition 3.3. (Burmester) The center curve Γ_{ijkl} and the pole curve $\Pi_{ij,kl}$ are cubic curves which coincide to each other.

Proposition 3.4. (Burmester) The center curve Γ_{ijkl} or the pole curve $\Pi_{ij,kl}$ passes through the 6 poles $P_{ij}, P_{ik}, P_{il}, P_{jk}, P_{jl}, P_{kl}$ as well as the two circular points at infinity.

For the proofs of these Propositions we refer to classical texts on geometric theory of mechanisms or linkages, e.g. [Bey]. See also [LXY].

Solving of Problem L2 Based on Burmester Geometry Theory.

It is clear that the fixed end A is a center point since the joint points $U_i, i = 0, 1, \cdots, 4$ are related points in the moving planes E_i which lie on one and the same circle with center A. It follows that A to be determined is one of the common intersections of the two cubic curves $\Pi_{01,23}, \Pi_{01,24}$. Since two cubic curves intersect in general in 9 points of which 3 are the poles P_{01}, P_{02}, P_{12} and 2 are the circular points at infinity which have no interest to our problem, we may expect that there will be in general 4 solutions for the position A, real or complex. This forms the theoretical basis for the solving of Problem L2. See the diagram Fig. 3.1 above.

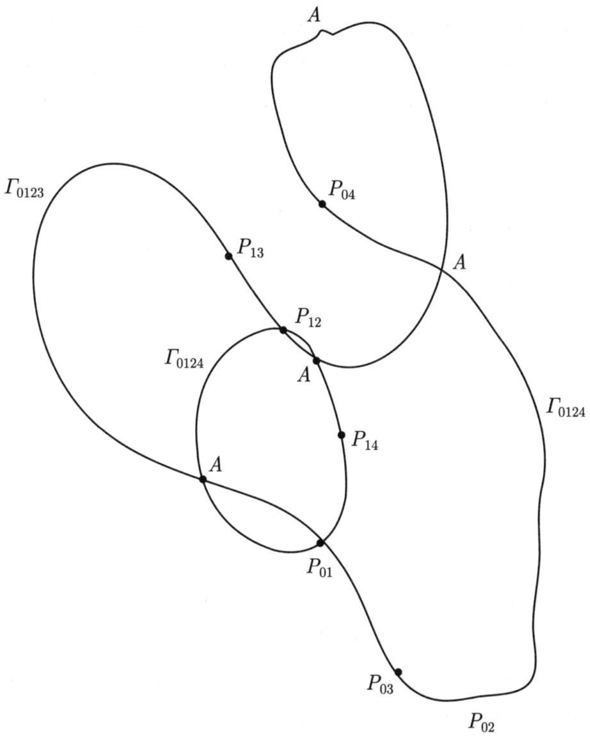

Fig. 3.1.

Let us thus take coordinates such that

$$P_{ij} = (p_{ij}, q_{ij}),\ i, j = 0, 1, \cdots, 4,\ i \neq j,$$

$$A = (x, y).$$

As in Remark 3.2, p_{ij}, q_{ij} are all known values. We get then from (3.1) for $(ij, kl) = (01, 23), (01, 24), (01, 34)$:

$$< (AP_{02}, AP_{12}) =< (AP_{03}, AP_{13}) =< (AP_{04}, AP_{14}). \tag{3.3}$$

Denote the common angle by ω and set $tan\ \omega = w$, then we get 3 equations

$$\begin{aligned} &w((x^2 + y^2) - x * (p_{0j} + p_{1j}) - y * (q_{0j} + q_{1j}) + (p_{0j} * p_{1j} + q_{0j} * q_{1j}) \\ &= x * (q_{0j} - q_{1j}) + y * (p_{1j} - p_{0j}) + (p_{0j} * q_{1j} - p_{1j} * q_{0j}),\ j = 2, 3, 4. \end{aligned} \qquad (3.4)_j$$

The equations $(3.4)_j$, $j = 2, 3, 4$, are 3 in number in the 3 unknowns w, x, y. Arrange the unknowns in the order

$$w \prec x \prec y$$

and proceed in the usual way by our general method we get a char-set for which the first pol is in w of degree 4 and the two other pols are linear in the leading variables x, y respectively. This gives the solutions for the end A of our Problem L2. Owing to the particular form of the equations $(3.4)_j$ we may also solve these equations in the following more direct way.

Let us subtract $(3.4)_2$ from $(3.4)_3, (3.4)_4$ we get two equations of the form

$$(\alpha_{i1} * w + \alpha_{i2}) * x + (\alpha_{i3} * w + \alpha_{i4}) * y(\alpha_{i5} * w + \alpha_{1i}) = 0, \ i = 1, 2, \qquad (3.5)_i$$

in which all α_{ik} are definite pols in p_{ij}, q_{ij}. Solve $(3.5)_1, (3.5)_2$ for x, y we get equations of the form

$$\begin{aligned} x &= \frac{\beta_{11} * w^2 + \beta_{12} * w + \beta_{13}}{\beta_{14} * w^2 + \beta_{15} * w + \beta_{16}}, \\ y &= \frac{\gamma_{11} * w^2 + \gamma_{12} * w + \gamma_{13}}{\gamma_{14} * w^2 + \gamma_{15} * w + \gamma_{16}}, \end{aligned} \qquad (3.6)$$

in which all the β's and the γ's are definite pols in the α's and are thus definite pols in the p's and q's. Substitute now the last expressions of x, y in (3.6) into any one of (3.4), we will get the same equation in w alone of degree 4 whose coefficients are all definite pols in the β's, the γ's and the p's and q's and are thus all definite pols in the known parameters p's and q's.

From the last equation in w and the equations (3.6) in x, y we see that there are in general 4 solutions, real or complex, for the fixed end $A = (x, y)$, say $A_i = (x_i, y_i), i = 1, \cdots, 4$. Now the fixed end B is in same situation as the end A, so the 4 points A_i may equally well be taken for the end B. Hence in general there will be 6 possible pairs for (A, B), real or complex. For any such pair, so far it is *real*, the joint points U_0, V_0 may then be determined by means of Proposition 3.1 and the 4-bar linkage will be completely determined. We see thus the actual possible solutions of 4-bar linkages with 5 given positions of the moving coupler-planes is in general atmost 6 in number. Of course in degenerate cases there may be no such solutions or an infinite number of solutions. In any way a detailed analysis may be given in each concrete case.

<u>*A Simplified Solution of Problem L2.*</u>

The preceding classical solution requires a lot of knowledge about geometry facts involved in the linkage movements. We present below a simplified version of the solution which requires only some rudimentary knowledge of such geometrical facts.

As in Remark 3.1, the poles P_{0j} and the angles θ_{0j} *mod* $2 * \pi$ for the rotation of plane E_0 to planes $E_j, j = 1, \cdots, 4$ are all known. By Prop. 3.1 P_{0j} and the end A

are both on the perpendicular bisector of U_0U_j. It follows that, in leaving aside the uninteresting and hardly possible degenerate cases $P_{0j} = U_0$ or $P_{0j} = A$, we have

$$< (P_{0j}U_0, P_{0j}A) = \frac{1}{2} < (P_{0j}U_0, P_{0j}U_j) = \frac{1}{2} * \theta_{0j},\ j = 1, \cdots, 4. \tag{3.7}$$

Let us set $tan\left(\frac{1}{2} * \theta_{0j}\right) = t_j$ and take coordinates such that

$$A = (x, y),\ U_0 = (x_0, y_0),\ P_{0j} = (c_j, d_j),\ j = 1, \cdots, 4.$$

The values of c_j, d_j, t_j are known while those of x, y, x_0, y_0 are to be determined. See the diagram Fig. 3.2 below.

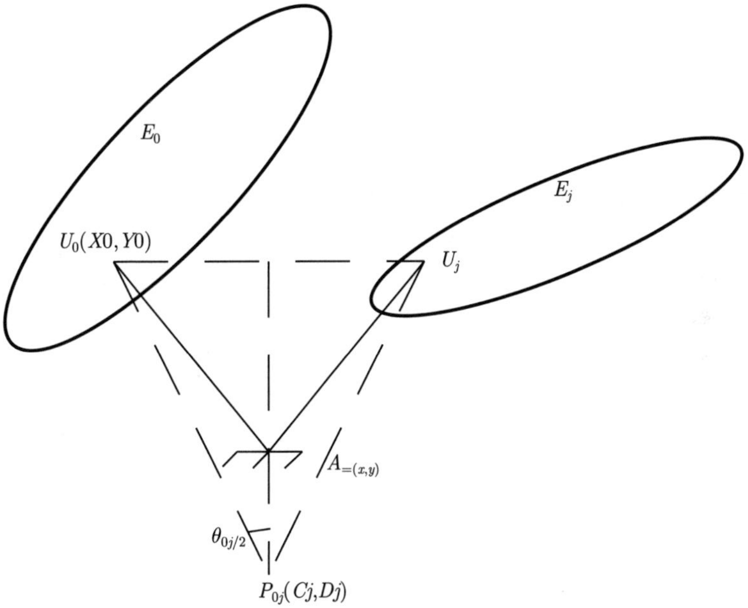

Fig. 3.2.

From (3.7) we get then equations $P_j = 0,\ j = 1, \cdots, 4$, where

$$\begin{aligned} P_j = &(y_0 * x - x_0 * y) - (y_0 * y + x_0 * x) * t_j \\ &+y_0 * (t_j * d_j - c_j) + x_0 * (t_j * c_j + d_j) \\ &+y * (t_j * d_j + c_j) + x * (t_j * c_j - d_j) - t_j * (c_j^2 + d_j^2). \end{aligned} \tag{3.8}$$

The four pols P_j are linear in the four unknowns $y_0 * x - x_0 * y,\ y_0 * y + x_0 * x,\ y_0$ and x_0. Let us denote the determinant of the coefficient matrix of these four unknowns by D. Then in solving the equations $P_j = 0$ we get

$$\begin{aligned} &D * (y_0 * x - x_0 * y) = r_{11} * x + r_{12} * y + r_{10}, \\ &D * (y_0 * y + x_0 * x) = r_{21} * x + r_{22} * y + r_{20}, \\ &D * x_0 = r_{31} * x + r_{32} * y + r_{30}, \\ &D * y_0 = r_{41} * x + r_{42} * y + r_{40}, \end{aligned} \tag{3.9}$$

in which D and the r's are all definite pols in the known parameters t_j, c_j, d_j. Substitute the values of $D * x_0, D * y_0$ of the last two equations of (3.9) in the first two equations, we get two equations of the form

$$s_{i1} * y^2 + (s_{i2} * x + s_{i3}) * y + (s_{i4} * x^2 + s_{i5} * x + s_{i6}) = 0, \ i = 1, 2, \tag{3.10}$$

in which all the s's are definite pols in the r's and hence also definite pols in t_j, c_j, d_j. Let us take the ordering of variables to be

$$x \prec y \prec x_0 \prec y_0.$$

A char-set of the polset $PS = \{P_1, \cdots, P_4\}$ is then seen to be consisting of 4 pols $C_1, \cdots, C_4$ of which C_1 is a pol in x alone of degree 4, while C_2, C_3, C_4 are all linear in the leading variables y, x_0, y_0 for which C_3, C_4 may be taken to be the pols deduced in the last two equations of (3.9). The coefficients in the pols $C_1, \cdots, C_4$ are all definite pols in the known parameters t_j, c_j, d_j. It follows that there are in general atmost 4 real solutions of the pair (A, U_0). As before we may take a pair of two such real solutions to form a 4-bar linkage AU_0V_0B which are thus in general atmost 6 in number.

The Liu-Guan Solution of Problem L2.

The Liu-Guan method for solving of Problem L2 requires still less geometry knowledge about the linkage movements than the preceding one, see their joint paper [L-G]. For their method let us take an arbitrary point P_0 in the moving plane E_0 and also its related points P_j in the moving planes $E_j, j = 1, \cdots, 4$. Let us complete the parallelograms $U_0U_jP_jP_j'$ as in Fig. 3.3 below.

The points P_0, P_j are known so that the vectors $\overrightarrow{P_0P_j}$ are known too. The points U_0, U_j, A are all unknowns to be determined. However, as U_0, U_j are clearly related points as the points P_0, P_j, the angles $< (U_0P_0, U_jP_j) = \alpha_j$ are actually known. Let the unknown angles $< (AU_0, AU_j) = \phi_j$. Take now coordinates such that

$$\overrightarrow{AU_0} = (x_1, x_2), \ \overrightarrow{U_0P_0} = (x_3, x_4), \ \overrightarrow{P_0P_j} = (c_j, d_j), \ j = 1, \cdots, 4.$$

Then it is sufficient to determine x_1, x_2, x_3, x_4 in terms of ϕ_j, c_j, d_j for the determination of A, U_0 which may then be completed to a 4-bar linkage as in the two previous solutions.

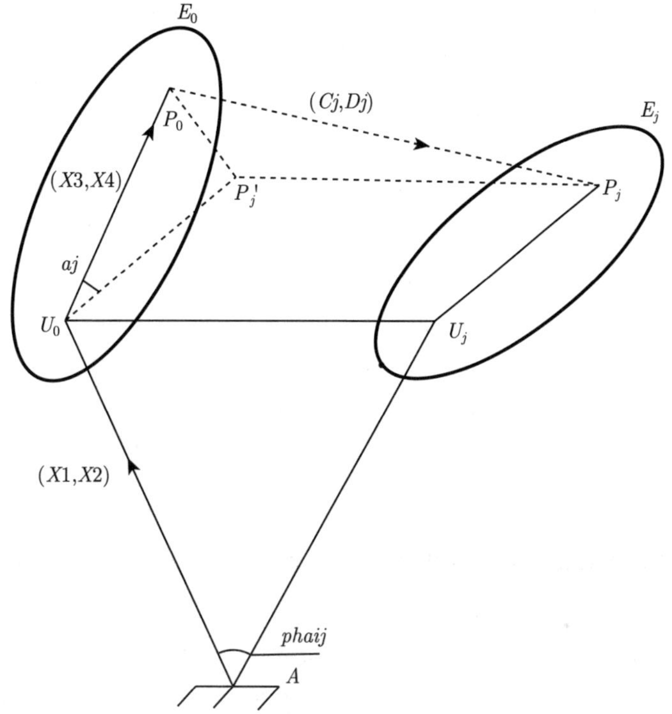

Fig. 3.3.

To determine $x_1, \cdots, x_4$ in terms of ϕ_j, c_j, d_j we see from Fig. 3.3 that

$$\begin{aligned}
\overrightarrow{P_0P_j} &= \overrightarrow{P_0P_j'} + \overrightarrow{P_j'P_j} = \overrightarrow{P_0P_j'} + \overrightarrow{U_0U_j},\\
\overrightarrow{P_0P_j'} &= \overrightarrow{U_0P_j'} - \overrightarrow{U_0P_0} = \overrightarrow{U_0P_0}.(R_{\alpha_j} - 1), \qquad (j = 1, \cdots, 4). \qquad (3.11)_j\\
\overrightarrow{U_0U_j} &= \overrightarrow{AU_j} - \overrightarrow{AU_0} = \overrightarrow{AU_0}.(R_{\phi_j} - 1),
\end{aligned}$$

In (3.11) R_{α_j}, R_{ϕ_j} denote rotations of angles α_j, ϕ_j about U_0, A respectively. Set $cos\ \phi_j = y_j$, $sin\ \phi_j = z_j$, $cos\ \alpha_j = a_j$, $sin\ \alpha_j = b_j$. Then we have in all 12 equations below $(j = 1, \cdots, 4)$:

$$\begin{aligned}
x_1 * (y_j - 1) - x_2 * z_j + x_3 * (a_j - 1) - x_4 * b_j - c_j = 0,&\\
x_1 * z_j + x_2 * (y_j - 1) + x_3 * b_j + x_4 * (a_j - 1) - d_j = 0,& \qquad (3.12)_j\\
y_j^2 + z_j^2 - 1 = 0.&
\end{aligned}$$

In (3.12) the x's, y's, z's are all unknowns while the a's, b's, c's d's are known parameters verifying the conditions $a_j^2 + b_j^2 = 1$.

Let us arrange the unknown variables in the order

$$x_1 \prec x_2 \prec x_3 \prec x_4 \prec y_1 \prec z_1 \prec \cdots \prec y_4 \prec z_4.$$

Then in eliminating the variables y_j, z_j in an evident way we get a polset $PS = \{P_1, \cdots, P_4\}$ in x_j, $j = 1, \cdots, 4$, alone with

$$\begin{aligned} P_j =& f_{j0} * (x_3^2 + x_4^2) \\ &+x_4 * (f_{j1} * x_1 + f_{j2} * x_2 + f_{j3}) \\ &+x_3 * (f_{j4} * x_1 + f_{j5} * x_2 + f_{j6}) \\ &+(f_{j7} * x_1 + f_{j8} * x_2 + f_{j9}), \ j = 1, \cdots, 4, \end{aligned} \tag{3.13$_j$}$$

in which all the f's are definite pols in the known parameters a, b, c, d's. By our general method of polynomial equations-solving we get then finally a char-set $CS = \{C_1, \cdots, C_4\}$ with C_1 a pol in x_1 alone of degree 4 and C_2, C_3, C_4 linear in the respective leading variables x_2, x_3, x_4. As before there are in general atmost 4 real solutions for the pair A, U_0 and combining two of them there are in general atmost 6 solutions for the 4-bar linkage AU_0V_0B.

8.4 Applications to Surface-Fitting Problem in CAGD.

Let $\mathbf{K}, X = (x_1, \ldots, x_n)$ be as before.

Surface-Fitting Problem in Computer Aided Geometry Design.

In the present section we shall give a general method of solving the following surface-fitting problem in computer-aided geometry design (CAGD):

Problem SF. Given in real 3-space R^3 three sets of real irreducible algebraic curves C_i, C_j, C_k. with $i \in I$, $j \in J$, $k \in K$ respectively, I, J, K being all finite sets of indices. Given also two sets of real irreducible algebraic surfaces $S_j, S_k, (j \in J, k \in K)$ containing C_j, C_k respectively. To determine a real irreducible algebraic surface S of given degree m verifying the following conditions:

(1) S contains all the curves C_i, C_j, C_k, for $i \in I, j \in J, k \in K$.

(2) S is in smooth contact with each of S_j, S_k along the curves C_j, C_k respectively, for $j \in J$, $k \in K$. More precisely, for each point on C_j or C_k which is regular for C_j, S_j or for C_k, S_k, the surface S to be found is regular at the point too, and moreover, S and S_j or S and S_k will have same tangent planes at that point.

(3) S possesses same curvature as S_k along the curves C_k, for each $k \in K$. More precisely, for each point on C_k which is regular for C_k and S_k, the surface S and S_k will have not only the same tangent plane but also the same (Gaussian) curvature at that point.

We may also replace (3) by more stringent conditions which we shall not enter.

Remark 4.1. The precise meaning of real curves, real surfaces, and regular points, will be clarified in paragraphs below.

Remark 4.2. The problem for the requirements (1), (2) and some further smooth requirements have already been solved by Bajaj et al by the method of interpolation based on the theorem of Bezout and its extensions, cf. [Baj] and [B-I]. We shall solve the above problem in all its generality based on some entirely different principle and method to be described in what follows.

Connection between Algebraic Geometry and Differential Geometry.

Remark 4.3. In practice only *real* pieces of curves and surfaces C_i, S_j, etc. are actually given and only piece of *real* surface S are required. For this sake the basic field $\mathbf{K}$ will be the *real* field $\mathbf{R}$. For real varieties we shall rely on Whitney's theory as exhibited in Sect. 5.1 of Chap.5. Thus, by a *real* variety of *real* dimension d in euclidean space $\mathbf{R}^n(X)$ will be meant henceforth one with a real-complete basis $PS \in \mathbf{R}[X]$ so that its complexification $Zero_c(PS)$ is of complex dimension d. The real variety $V = Zero_{\mathbf{R}}(PS)$ will have then a Whitney decomposition of which real manifold-components are all of real dimension less than or equal to d and those of real dimension d actually occur.

Definition 4.1. For a real variety $V = Zero_{\mathbf{R}}(PS) \subset \mathbf{R}^n(X)$ of real dimension d with PS a real-complete basis any real manifold-component of highest real dimension d in the Whitney decomposition of V := A *dominant* component of the real variety V.

Definition 4.2. For a real variety $V = Zero_{\mathbf{R}}(PS) \subset \mathbf{R}^n(X)$ of real dimension d with PS a real-complete basis, V is *irreducible* := $Zero(PS)$ is irreducible or $Zero(PS)$ is *absolutely irreducible* in the usual sense.

Remark 4.4. In Problem SF above and in what follows by *real curves* or *real surfaces* we shall mean such real varieties as in Def. 4.1 of real dimension 1 or 2 so that they actually contain dominant components or real pieces of dimension 1 or 2 respectively. We assume also that by real irreducible is mean irreducible in the sense of Def. 4.2. Henceforth these assumptions will be supposed to hold but not explicitly stated.

Remark 4.5. From Whitney's decomposition we see that for a real algebraic variety the real points will form a set of real differential manifolds for which the concept of regular points, tangent space, normal space, curvature, etc. are well-defined as in ordinary differential geometry.

Some Basic Theorems about Real Varieties.

Theorem 4.1. A regular point in a dominant component of an irreducible real variety $V = Zero_{\mathbf{R}}(PS) \subset \mathbf{R}^n(X)$ with a real-complete basis PS is a simple point of the corresponding irreducible variety $Zero(PS)$.

Proof. Let the real dimension of V be d and X^0 be a regular point in a dominant component of V. It is clear that the equations of the tangent space at X^0 in the usual sense will define the tangent space at X^0 of the variety $Zero(PS)$ in the sense of Sect. 5.1. Hence X^0 is a simple point of $Zero(PS)$ as to be proved. □

Theorem 4.2. For an irreducible real variety $V = Zero_{\mathbf{R}}(PS) \subset \mathbf{R}^n(X)$ with a real-complete basis $PS \subset \mathbf{R}(X)$, a pol $P \in \mathbf{R}[X]$ will vanish on whole of the variety $Zero(PS)$ if and only if P vanishes at all the regular points of any one dominant component of the Whitney decomposition of V.

Proof. Let the real dimension of V be d. If P does not vanish on the whole of the variety V, then $Zero(P)$ and $Zero(PS)$ will intersect in a variety of dimension less than or equal to $d-1$ for which the real part will be of real dimension less than or equal $d-1$ too. Hence in any dominant component there will be points which are not zeros of P. As the converse part is evident, the theorem is proved. □

Theorem 4.3. Let $V = Zero_{\mathbf{R}}(PS)$ be as in Theorems 4.1 and 4.2 and $Zero(PS)$ is irreducible. Then, a *real* polynomial will be 0 at all *real regular* points of any dominant component of V if and only if the polynomial is 0 at a generic point of that variety, or, if and only if the remainder of the polynomial w.r.t. a defining irreducible asc-sec of the generic point is 0.

Proof. This follows from the previous Theorem and Theorem in Sect. 3.6. □

Remark 4.6. Theorem 4.3 forms the basis of the Principle and Method for the solving of the Problem SF. See Theorems 4.4 ~ 4.6.

Some Basic Theorems for Solving of Problem SF.

Let $\mathbf{K} = \mathbf{R}$ be the real field and let $n = 3$ so that we are considering real curves and real surfaces in the ordinary real space $\mathbf{R}^3$. Let us replace x_1, x_2, x_3 by the usual x, y, z with the ordering

$$x \prec y \prec z.$$

Then a real irreducible algebraic surface in $\mathbf{R}^3$ in the sense of Defs. 4.1, 4.2 will be defined by an equation $P = 0$ with P irreducible in $\mathbf{C}$ while P is a real polynomial and $P = 0$ has real traces ofreal dimension 2. The leading variable of P will be either z or y or x. If the leading variable of P is x, then P is necessarily linear in x and the surface is a plane $x =$ const. On the other hand a real irreducible algebraic curve in $\mathbf{R}^3$ will be defined by an irreducible asc-set $IRR = \{P_1, P_2\}$ with $P_1, P_2 \in \mathbf{R}[x, y, z]$.

The leading variables of P_1, P_2 will be either x, y or x, z or y, z. In case the leading variable of P_1 is x, then P_1 is necessarily linear in x so that the real part of the curve lies wholly in some plane$x =$ const. Remark that the real irreducible curve C determined by the irreducible asc-set IRR, viz. the real part of $Var[IRR]$, is in general different from real part of the curve $Zero(IRR)$.

In what follows let C be such a real irreducible algebraic curve, viz. the real part of $Var[IRR]$ defined by the irreducible asc-set $IRR = \{P_1, P_2\}$ with $P_1, P_2 \in \mathbf{R}[x, y, z]$ and S, S' be real irreducible algebraic surfaces defined by $P = 0$ and $P' = 0$ respectively with $P, P' \in \mathbf{R}[x, y, z]$.

As a particular case of Theorem 4.3 we have the following

Theorem 4.4. The surface S will contain whole of C if and only if

$$Remdr(P/IRR) = 0. \tag{4.1}$$

Theorem 4.5. Suppose that C is contained wholly in S and also in S' but is not contained wholly in the singular locus of S or S'. Set $P_x = \dfrac{\partial P}{\partial x}, P'_x = \dfrac{\partial P'}{\partial x}$, *etc.* and form the pols

$$\left.\begin{aligned} D_1 &= P_x * P'_y - P_y * P'_x, \\ D_2 &= P_x * P'_z - P_z * P'_x, \\ D_3 &= P_y * P'_z - P_z * P'_y. \end{aligned}\right\} \tag{4.2}$$

Then S and S' will touch smoothly along C at all points of C which are regular for all of C, S, S' if and only if

$$Remdr(D_i/IRR) = 0, \quad i = 1, 2, 3. \tag{4.3}$$

Proof. From ordinary differential geometry we know that S and S' will touch smoothly along C at all points of C which are regular for all of C, S, S' if and only if $D_i = 0, \quad i = 1, 2, 3$ at all these points. By Theorem 4.3 the latter is however true if and only if (4.3) is true. The theorem is thus proved. □

For the proof of Lemma 4.1 below we refer to elementary treatises on differential geometry, e.g. [Ei].

Lemma 4.1. For the real irreducible algebraic surface S given by $P = 0$ the curvature κ at a regular point is given by

$$\kappa = \frac{V}{H^2}, \tag{4.4}$$

where

$$H = \sqrt{P_x^2 + P_y^2 + P_z^2}. \tag{4.5}$$

$$\begin{aligned} V = & Cyc(P_{xx} * P_{yy} * P_z^2 - 2 * Cyc(P_x * P_y * P_{xy} * P_{zz})) \\ & +2 * Cyc(P_x * P_y * P_{xz} * P_{yz}) - Cyc(P_x^2 * P_{yz}^2). \end{aligned} \tag{4.6}$$

In (4.6) Cyc means the cyclic sum gotten from the corresponding expression over the variables x, y, z and all the partial derivatives $P_x, P_{xx} = \dfrac{\partial^2 P}{\partial x^2}$, *etc.* which take values at that regular point.

Theorem 4.6. Let C, S, S' be as before which satisfy the conditions in Theorem 4.5. Then S, S' will have same curvatures at points regular to C, S, S' if and only if

$$Remdr(H^2 * V' - H'^2 * V/IRR) = 0. \tag{4.7}$$

In (4.7) H' and V' are pols for P' similar to H and V for P.

Proof. At a point regular for C, S, S' we have $H \neq 0, H' \neq 0$. Hence by Lemma 4.1 the curvatures κ, κ' to the surfaces are well-defined by (4.4) and will be equal to each other if and only if

$$H^2 * V' = H'^2 * V \tag{4.8}$$

is true at all points of C regular for C, S, S'. The latter is however equivalent to (4.7) by Theorem 4.3. The theorem is proved. □

Method for Solving of Problem SF.

Based on Theorem 4.4~4.6 the solving of the Problem SF is reduced to the verification of the vanishing of the remainders which will be further reduced to the solving of polynomial equations as described below.

Method. For the irreducible algebraic surface S of degree m defined by $P = 0$ let us write P in the form

$$P = \sum_m u_{abc} * x^a * y^b * z^c, \tag{4.9}$$

in which $\sum_m$ is to be extended over triples (a, b, c) verifying

$$0 \leqslant a, \qquad 0 \leqslant b, \qquad 0 \leqslant c, \qquad a + b + c \leqslant m. \tag{4.10}$$

We shall denote by U_m the set of all u_{abc} verifying (4.10).

Consider e.g. the problem that S contains as a whole the irreducible algebraic curve C defined by the irreducible asc-set IRR. By Theorem 4.4 this will be equivalent to the condition below:

$$R = Remdr(P/IRR) = 0. \tag{4.11}$$

Let the polset formed of the coefficients of various powers of x, y, z in R be $US_0 \subset \mathbf{R}[U_m]$. Then the condition (4.11) is identical to the conditions

$$US_0 = 0. \tag{4.12}$$

For the requirement (1) in Problem SF let us form, analogous to US_0 of (4.12), the polsets $US_{ai}, US_{aj}, US_{ak}$ for each of the curves C_i, C_j, C_k. Similarly for the requirement (2) in Problem SF Theorem 4.5 will give us polsets US_{bj}, US_{bk} and for the requirement (3) Theorem 4.6 gives us polsets US_{ck} whose vanishing are the corresponding conditions to be verified. Remark that all pols in $US_{ai}, US_{aj}, US_{ak}, US_{bj}, US_{bk}$ are linear in the u', while those in US_{ck} are in general non-linear in the $u's$. Combining now all the above polsets into a single one $US \subset \mathbf{R}[U_m]$, then all possible solutions S verifying the requirements of Problem SF are given by the real zeros of $Zero_{\mathbf{R}}(US)$.

Some Examples.

A general cubic surface S in $\mathbf{R}^3$ is of the form

$$\begin{aligned} f(x,y,z) = & u_{300} * z^3 + z^2(u_{210} * y + u_{201} * x + u_{200}) + z * (u_{120} * y^2 \\ & + u_{111} * y * x + u_{102} * x^2 + u_{110} * y + u_{101} * x + u_{100}) + u_{030} * y^3 \\ & + u_{021} * y^2 * x + u_{012} * y * x^2 + u_{003} * x^3 + u_{020} * y^2 \\ & + u_{011} * y * x + u_{002} * x^2 + u_{010} * y + u_{001} * x + u_{000} = 0, \end{aligned} \tag{4.13}$$

in which the u' are all in the real field $\mathbf{R}$. We give below some examples illustrating our method of determining such cubic surfaces S meeting some requirements as described in the precise statements (1), (2), (3) of the Problem SF.

Example 4.1. In $\mathbf{R}^3$ consider two circular cylinders CYL_1, CYL_2 with x-axis and y-axis as their axis and two circular sections C_1, C_2 by planes orthogonal to these axis respectively. Then these circular sections will have their defining asc-sets $AS_1 = \{C_{11}, C_{12}\}$ $AS_2 = \{C_{21}, C_{22}\}$ given by

$$C_{11} = x - d_1, \qquad C_{12} = z^2 + y^2 - r_1^2. \tag{4.14}$$

$$C_{21} = x - d_2, \qquad C_{22} = z^2 + x^2 - r_2^2. \tag{4.15}$$

The two circular cylinders will be given by equations $C_{12} = 0, C_{22} = 0$ respectively. Naturally we shall assume that d_1, d_2, r_1, r_2 are all non-zero and positive. We now ask for the determination of cubic surfaces (4.13) which will contain the circles C_1, C_2 and touch the cylinders CYL_1, CYL_2 smoothly along these circles. Our method gives now a set of 28 equations in U_3 for the solutions. It is readily found by the package *wsolve* of D. K. Wang that such solutions will exist only if

$$r_1^2 + d_1^2 = r_2^2 + d_2^2. \tag{4.16}$$

In that case the only possible cubic surface is then given by

$$\begin{aligned} &z^2 * (y * d^2 + x * d_1 - d_2^2 - d_1^2) + (y^3 * d_2 + x^3 * d_1) + y * x * (y * d_1 \\ &+x * d_2 - 2 * d_2 * d_1) - (y^2 + x^2) * (d_2^2 + d_1^2) + y * d_2 * (d_1^2 - r_1^2) \\ &+x * d_1 * (d_2^2 - r_2^2) + (r_1^2 + d_1^2) * (r_2^2 + d_2^2) - 2 * d_1^2 * d_2^2 = 0. \end{aligned} \tag{4.17}$$

See the diagram 4

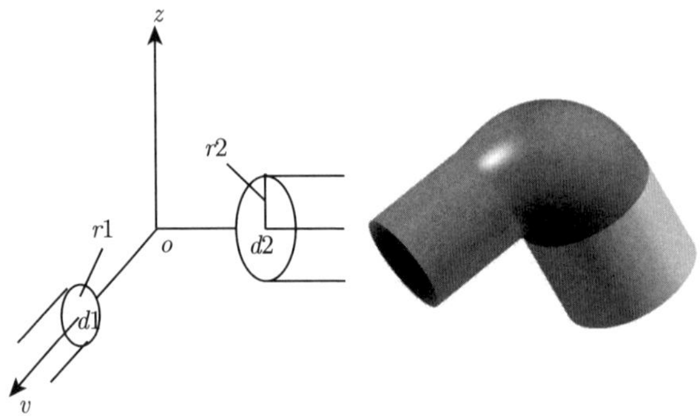

Fig. 4.1.

Example 4.2. Consider besides circles C_1, C_2 as in Ex. 4.1 a third circle C_3 with center on z-axis and orthogonal to that axis. Let the defining asc-set of C_3 be $\{C_{31}, C_{32}\}$ with

$$C_{31} = y^2 + x^2 - r_3^2, \qquad C_{32} = z - d. \tag{4.18}$$

As before we assume that all d_i, r_i are positive. The package *wsolve* will give a complete set of possible solutions of this problem. Suppose first that these d_i, r_i are otherwise entirely arbitrary, to be considered as independent parameters. Then our method shows that the only cubic surface containing all the 3 circles is the trivial one which degenerates into 3 planes containing the 3 circles respectively. If the d_i, r_i satisfy the relation (4.16) but otherwise arbitrary, then besides the trivial surface of 3 planes, there is the only one family of cubic surfaces which degenerate into the plane $z = d_3$ containing the circle C_3 and quadrics through the circles C_1, C_2 given by the equation below, u_0 being a parameter:

$$\begin{gathered} (d_1 * d_2 + u_0) * (x^2 + y^2 + z^2) + (r_1^2 + d_1^2) * (y * x - d_1 * y - d_2 * x) \\ -u_0 * (r_1^2 + d_1^2) = 0. \end{gathered} \tag{4.19}$$

Suppose now there exists the relation

$$r_1^2 + d_1^2 = r_2^2 + d_2^2 = r_3^2 + d_3^2 = k^2 \tag{4.20}$$

between d_i, r_i but otherwise arbitrary. Then clearly the sphere of center the origin and radius $|k|$ will contain all the 3 circles C_i. Together with an arbitrary plane, there will be a family of ∞^3 degenerate cubic surfaces containing all the 3 circles. However, our method shows that there are in fact 4 families of $\infty, \infty^2, \infty^3, \infty^4$ cubic surfaces containing the 3 circles. The cubic surfaces consisting of the sphere and an arbitrary plane form in fact only a subfamily of the family of ∞^4 surfaces. The last family depends on the parameter ratios $u_{100} : u_{002} : u_{010} : u_{001} : u_{000}$. Setting

$$\left.\begin{aligned} u_{111} = u_{110} = u_{101} = 0, \\ u_{201} = u_{003} = u_{021} = a, \\ u_{210} = u_{030} = u_{012} = b, \\ u_{300} = u_{120} = u_{102} = c, \\ u_{200} = u_{020} = u_{002} = -d. \end{aligned}\right\} \tag{4.21}$$

then we get the subfamily of ∞^3 degenerate cubic surfaces consisting of the sphere and the plane $a * x + b * y + c * z = d$. For a cubic surface in the ∞^4-family but not in this degenerate ∞^3-family the diagram Fig 4.2 may serve to give a rough idea.

Our method shows that the above are the only possible cubic surfaces which contains the 3 circles. As each non-degenerate family of cubic surfaces given above depends on several parameters, we may determine, if we like, such ones among them which will meet further requirements by our method. However, if $CYK_i, i = 1, 2, 3$ are the cylinders with x, y, z-axis as axis and bounded by $C_i, i = 1, 2, 3$ respectively, then in view of the form of (4.16), there cannot exist any cubic surface containing all the 3 circles and touching smoothly the 3 cylinders along these circles.

Fig. 4.2.

In the above two examples the curves like circles and surfaces like circular cylinders are easily parameterized. With such parametrizations the problems in these examples can be dealt with by other known methods, cf. e.g. [Baj]. However, curves and surfaces are generally non-parametrizable except in the very rare case of rational ones. This is already so for general cubic curves and cubic surfaces. The following is such an example which cannot be treated by means of parametrizations.

Example 4.3. Let $f_1(x,y,z), f_2(x,y,z)$ be as $f(x,y,z)$ in (4.13) with coefficients $u's$ replaced by $a's$ and $b's$ respectively. Let S_1, S_2 be the irreducible cubic surfaces defined by $f_1 = 0$ and $f_2 = 0$ respectively and C_1, C_2 be the irreducible cubic curves with respective defining asc-sets $IRR_1 = \{y, g_1\}$, $IRR_2 = \{y, g_2\}$, where $g_1 = f_1(x,0,z)$, $g_2 = f_2(x,y,0)$. Our problem is to determine cubic surfaces S as in (4.13) which will contain both C_1 and C_2 and meet eventually further requirements. Now C_1, C_2 will intersect the x-axis $y = z = 0$ in points with x-coordinates given by the respective equations

$$\left.\begin{aligned} a_{003} * x^3 + a_{002} * x^2 + a_{001} * x + a_{000} = 0, \\ b_{003} * x^3 + b_{002} * x^2 + b_{001} * x + b_{000} = 0. \end{aligned}\right\} \tag{4.22}$$

For a cubic surface S containing C_1, C_2 to exist it is necessary that these two triples of intersection points should be the same. So we assume at the outset that

$$a_{003} = b_{003}, \qquad a_{002} = b_{002}, \qquad a_{001} = b_{001}, \qquad a_{000} = b_{000}. \tag{4.23}$$

For the sake of simplifying the computations we shall assume that none of the coincident triple of intersection points will be the origin or the point at infinity on the x-axis so that

$$a_{000} = b_{000} \neq 0, \qquad a_{003} = b_{003} \neq 0. \tag{4.24}$$

We assume further that neither the z-axis will be asymptotic to the curve C_1 nor the y-axis will be asymptotic to the curve C_2 so that

$$a_{300} \neq 0, \qquad b_{030} \neq 0. \tag{4.25}$$

Under the conditions (4.23)~(4.25) our method, in applying the package *wsolve*, shows that the only cubic surfaces S of (4.13) containing the two cubic curves will form a family of ∞^4 cubic surfaces defined by the equations below:

$$\left.\begin{array}{ll} u_{000}*a_{300}-u_{300}*a_{000}=0 & u_{001}*a_{300}-u_{300}*a_{001}=0 \\ u_{010}*a_{300}-u_{300}*b_{010}=0 & u_{002}*a_{300}-u_{300}*a_{002}=0 \\ u_{011}*a_{300}-u_{300}*b_{011}=0 & u_{020}*a_{300}-u_{300}*b_{020}=0 \\ u_{003}*a_{300}-u_{300}*a_{003}=0 & u_{012}*a_{300}-u_{300}*b_{012}=0 \\ u_{021}*a_{300}-u_{300}*b_{021}=0 & u_{030}*a_{300}-u_{300}*b_{030}=0 \\ u_{100}*a_{300}-u_{300}*a_{100}=0 & u_{101}*a_{300}-u_{300}*a_{101}=0 \\ u_{102}*a_{300}-u_{300}*a_{102}=0 & u_{200}*a_{300}-u_{300}*a_{200}=0 \\ u_{201}*a_{300}-u_{300}*a_{201}=0. & \end{array}\right\} \quad (4.26)$$

The above family depends on the parameter ratios

$$u_{110}:u_{111}:u_{120}:u_{210}:u_{300} \quad (4.27)$$

and is thus ∞^4 in number. We may determine subfamilies meeting further requirements by our method if required.

8.5 Some Miscellaneous Complements and Extensions.

Solving of Quartic Equation.

Let us consider the solving of a quartic equation

$$f(z) \equiv z^4 + a*z^2 + b*z + c = 0.$$

The classical solving method of L.Ferrari(+1522, +1565) requires some artifice in reducing the solving to that of a cubic equation. The following method, which is a modification of one originating from the papers [Lyon] and [Ku], seems to be more natural.

Let us suppose, as in the method of Hu-Wang on factorization in Sect. 4.5, that $f(z)$ may be factored as

$$f(z) = (z^2 + x_1*z + x_2)*(z^2 - x_1*z + x_3)$$

with x_1, x_2, x_3 as coefficients yet to be determined. Comparing various z-powers in both side, we get a polset $PS = \{P_1, P_2, P_3\}$ for which we have to solve $PS = 0$ for x_1, x_2, x_3:

$$\begin{aligned} P_1 &= x_3 + x_2 - {x_1}^2 - a, \\ P_2 &= x_1*(x_3 - x_2), \\ P_3 &= x_2*x_3 - c. \end{aligned}$$

We get readily

$$Zero(PS) = Zero(CS_1/x_1) + Zero(CS_2 + \{b\})$$

in which $CS_i = \{C_{i1}, C_{i2}, C_{i3}\}, i = 1, 2$, with

$$\begin{aligned}
C_{11} &= x_1^6 + 2 * a * x_1^4 + (a^2 - 4 * c) * x_1^2 - b^2, \\
C_{12} &= 2 * x_1 * x_2 - x_1(x_1^2 + a) + b, \\
C_{13} &= x_3 + x_2 - x_1^2 - a, \\
C_{21} &= x_1, \\
C_{22} &= x_2^2 - a * x_2 + c, \\
C_{23} &= x_3 + x_2 - a.
\end{aligned}$$

Leaving away the uninteresting case of CS_2 for which $b = 0$, $C_{11} = 0$ will give us a cubic equation in x_1^2 playing analogous role as the cubic equation in Fernrai's solution. That the cubic equation is in x_1^2 is quite natural since the two factors of $f(z)$ may be interchanged. Remark also that this simple example shows already how powerful it is the naive Principle and Method of Undetermined Coefficients which we learned early in high schools. A more sophisticated problem on partial differential equations treated by means of this naive Principle and Method will be given below.

The Pyramid Problem.

The following problem, proposed by J. Z. Zhang, was solved by D. K. Wang, see e.g. [WDK2].

Let P-$ABCD$ be a regular square-pyramid with vertex P and square base $ABCD$. Let π be a plane which cuts the edges AB, BC, CP, PD, PA respectively in points E, F, G, H, I.

Pyramid Problem. To decide whether π can be so chosen such that $EFGHI$ is a regular pentagon.

To solve the problem let us take the coordinate system with the center O of the square base as the origin and OC, OD as the positive x-and y-axis. It is immaterial to take $|OC|$ as the unit length so that we have

$$A = (-1, 0, 0), \quad B = (0, -1, 0), \quad C = (1, 0, 0), \quad D = (0, 1, 0).$$

For $EFGHI$ to be a regular pentagon we should have $GI \| EF \| AC$. Hence following D. K. Wang we may take coordinates such that

$$\begin{aligned}
&E = (-x_1, x_2, 0), \quad F = (x_1, x_2, 0), \quad G = (x_3, 0, x_5). \\
&I = (-x_3, 0, x_5), \quad H = (0, x_6, x_7), \quad P = (0, 0, x_9).
\end{aligned}$$

That $EFGHI$ is a regular pentagon will lead us to a polset $PS = \{P_1, \cdots, P_8\}$, for which the real zeros correspond to pyramids and planes π with regular-pentagon cuts. The pols P_i are given by:

$$
\begin{aligned}
P_1 &= x_2 - x_1 + 1, \\
P_2 &= x_9 * (x_3 - 1) + x_5, \\
P_3 &= x_9 * (x_6 - 1) + x_7, \\
P_4 &= x_7 * x_2 + x_5 * (x_6 - x_2), \\
P_5 &= x_3^2 + x_6^2 + (x_7 - x_5)^2 - 4 * x_1^2, \\
P_6 &= (x_3 - x_1)^2 + x_2^2 + x_5^2 - 4 * x_1^2, \\
P_7 &= 4 * x_3^2 - x_1^2 - (x_6 - x_2)^2 - x_7^2, \\
P_8 &= x_3^2 - x_1 * (x_3 + x_1).
\end{aligned}
$$

The real zero-set $Zero_R(PS)$ is readily determined by the general char-set method. A detailed analysis of $Zero_R(PS)$ thus determined leads to the following theorem:

Theorem 5.1. Pyramid Theorem. For a regular square pyramid $P - ABCD$ which does not collapse to $P = O$ to admit regular- pentagon cuts $EFGHI$ it is necessary and sufficient that the height PO of the pyramid should be equal to half of the diagonal of the square base. In that case E, F, G, H, I will be completely determined for which ($\varepsilon = \pm 1$)

$$
\begin{cases}
x_1 = x_5 = x_6 = \dfrac{3 + \varepsilon * \sqrt{5}}{2}, \\
x_2 = -x_3 = -x_7 = \dfrac{1 + \varepsilon * \sqrt{5}}{2}.
\end{cases}
$$

In the formulae $\varepsilon = -1$ gives a regular-pentagon cut while $\varepsilon = +1$ will give a further regular-star cut(See Figs. 5.1, 5.2).

A Problem of Zassenhaus on Geometrical Construction.

It is trivial to construct the bisectors of a given triangle by means of ruler and compass. The reverse problem is however not a trivial one:

Zassenhaus Problem. To decide whether a triangle with known triples of bisectors (internal or external) may be geometrically constructed by means of ruler and compass alone.

The problem was completely solved by X. S. Gao and D. K. Wang with answer in the negative. See e.g. [WDK2]. Their method runs as follows.

Let the sides of the triangle to be constructed be $a, b, c (> 0)$ with, to fix the ideas, corresponding known *internal* bisectors $a_i, b_i, c_i (> 0)$. Then we have to determine

a, b, c from a_i, b_i, c_i in solving $PS = 0$ with $PS = \{P_1, P_2, P_3\}$ given below:

$$P_1 = (b+c)^2 * a_i^2 - b * c * (b+c-a) * (a+b+c),$$
$$P_2 = (c+a)^2 * b_i^2 - c * a * (c+a-b) * (a+b+c),$$
$$P_3 = (a+b)^2 * c_i^2 - a * b * (a+b-c) * (a+b+c).$$

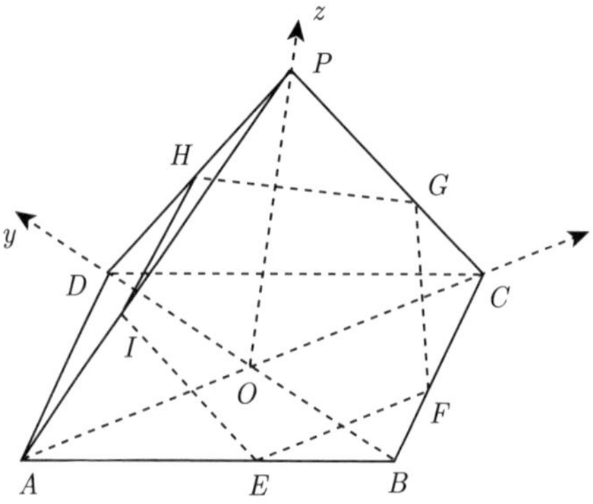

Fig. 5.1.

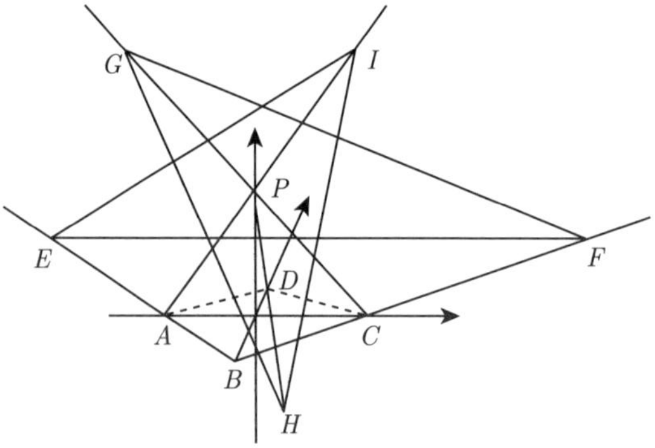

Fig. 5.2.

Consider the generic case for which a_i, b_i, c_i are indeterminates and take the ordering of variables to be $a_i \prec b_i \prec c_i \prec a \prec b \prec c$. It turns out that the first char-pol of PS is an irreducible pol of degree 10 in a^2 and of term-number 331. As the degree of the pol in a is not a power of 2, it is known from general theory that the triangle cannot be constructed from a_i, b_i, c_i by ruler and compass alone, as to be proved. The cases for which one or more of the three bisectors are external ones are similarly treated with same conclusions.

A Prblem of Erdös.

From oral communications of R. S. Boyer to S. C. Chou and X. S. Gao, we know the following problem proposed by Erdös:

Erdös 10-Piont Problem. In the plane is given 10 poins $A_{ij}, 1 \leqslant i < j \leqslant 5$. To determine 5 points $B_i, i = 1, \cdots, 5$, such that each A_{ij} is on the line joining B_i and B_j.

It is easy to turn the problem to one of polynomial equations-solving. Based on the char-set method in applying some further techniques of Clifford algebra, Shi He and Li Hongbo have essentially solved the problem in the generic case with result stated below (see their joint paper [L-S])

Theorem 5.2. For generic A_{ij}'s there are a most 6 possible quintuples of points B_i such that each A_{ij} lies on the line B_iB_j.

Poncelet Porism in Enumerative Geometry.

Let us consider Problem EG of enumerative geometry in Sect. 6.3 again.

By introducing suitable coordinates, such a problem will be turned into finding the number of solutions of certain systems of equations. However, the result of such a problem can often be expressed in the form of a theorem. It can thus be considered either as a problem of equations-solving or one of theorem-proving. As an illustration, let us consider the following

Poncelet Porism. Let C and C' be two given conics in the complex projective plane. Then, if there is one triangle inscribed in one of the two conics, say C, and circumscribing the other, C', there will be an infinity of such triangles, inscribed in C and circumscribing C'.

We may also state a problem related to the above theorem in the form of equations-solving:

Problem P. Find the number N of triangles inscribed in one conic C and circumscribing another conic C'.

According to the theorem of Poncelet we have:

(P1) The number N is either 0 or infinity.

One of the method of modern algebraic geometry for dealing with such problems is the method of algebraic correspondence. For the above problem P, let us consider an algebraic correspondence as follows. Take an arbitrary point p_0 on C. Through p_0 draw a tangent to C' that meets C in a further point p_1. Though p_1 draw a further tangent to C' meeting C in a further point p_2. Through p_2 draw a further tangent to C' meeting C in a further point p_3. Then $p_0 \to p_3$ establishes an algebraic correspondence of C onto itself. This is a (2, 2)-correspondence and the general theory shows that there are either 4 or an infinity of fixed points. Each such fixed point $p_3 = p_0$ will give rise to a triangle $p_0p_1p_2$ inscribed in C and circumscribing C' (see Fig. 5.3). It follows that we have:

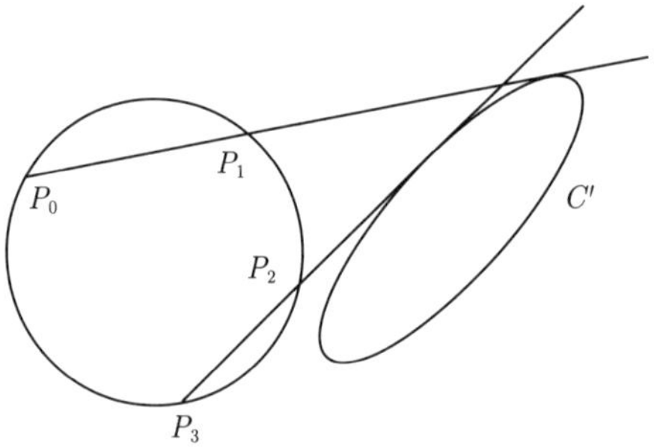

Fig. 5.3.

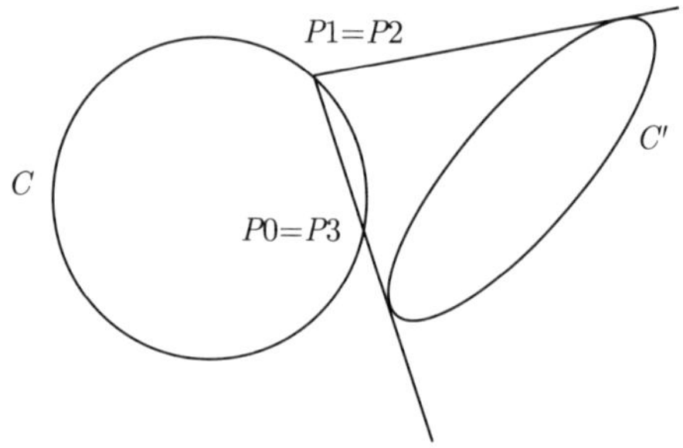

Fig. 5.4.

(P2) The number N is either 4 or infinity.

The apparent discrepancy in the answers to problem P, namely, (P1) and(P2), is due to the ambiguities in the meaning attached to the geometrical concepts involved in the problem. Thus, let us consider two conics C, C' in the general position and let L_0 be a common tangent touching C at a point p_1. Through p_1 let the second tangent to C' be L_2, which meets C in another point p_0. Starting from p_0 we will get the series of points $p_0, p_1, p_2(= p_1), p_3(= p_0)$ and the above point p_0 is a fixed point of the algebraic correspondence $p_0 \to p3$. This fixed point p_0 will give rise to a degenerate triangle $p_0p_1p_2$ counted as one inscribed in C and circumscribing C'. As there are four common tangents of the conics C and C', there are 4 such degenerate triangles, which acounts for the number 4 in answer (P2). On the other hand such degenerate triangles are not to be considered as a solution of problem P for the Poncelet Porism and this acounts for the number 0 in answer (P1). See Fig. 5.4.

To rectify the matter, we have first to attach precise meanings to the various geometrical concepts involved. Thus:

By a *triangle* we shall mean a triple of points p_0, p_1, p_2 called *vertices*, and a triple of lines, L_0, L_1, L_2 called sides, such that p_0 lies on L_1 and L_2, p_1 lies on L_0 and L_2, and p_2 lies on L_0 and L_1. Equivalently, we require that L_0 passes through p_1 and p_2, that L_1 passes through p_0 and p_2, and that L_2 passes through p_0 and p_1. Note that the points p_0, p_1, p_2 need not be distinct and that, similarly, the lines L_0, L_1, L_2 need not be distinct. So, we alow degenerate tingles. The pairs of vertices and sides $(p_0, L_0), (p_1, L_1)$, and (p_2, L_2) are then said to be *opposite* ones.

The above triangle will be said to be *inscribed* in the conic C, if the pair p_1, p_2, whether coincident or not, constitutes the pair of points, at which the line L_0 meets the conic C, and if similarly the pair p_0, p_2 is the pair of points at which the L_1

meets the conic C, and the pair p_0, p_1 is the pair at which the line L_2 meets C.

The above triangle will be said to be *circumscribing* the conic C', if the pair L_1, L_2, whether coincident or not, constitutes the pair of tangent lines to C' that passes through p_0 and the analogous conditions hold for the pair L_0, L_2, and point p_1, and the pair L_0, L_1 and point p_2.

In the example described above, let us denote by L_1 the tangent line to C' different from L_0 that passes through $p_2(= p_1)$. Then L_1 coincides with L_2, and the triple of points (p_0, p_1, p_2) and triple of lines (L_0, L_1, L_2) form a triangle in the above sense. This triangle is inscribed in the conic C but does not circumscribe the conic C' in the above sense since the tangent lines to C' through p_0 consist not of L_1 and L_2 but L_1(or L_2) and some other tangent to C'. This triangle is therefore not to be considered as a solution of problem P. It shows also that the method of algebraic correspondence, though an extremely powerful one, may give rise sometimes to extraneous solutions to a geometrical problem.

We now give the solution to Problem P using the method introduced in previous chapters.

First let us consider the conic C as a point conic, and the conic C' as a line conic, since only tangents to C' will be involved. Without loss of geometrical interest, let us assume neither that C degenerates into a pair of lines nor that C' degenerates into a pair of points. We assume also that C and C' are not dual to each other. Under these assumptions, it is always possible to find a fundamental triangle consisting of a triple of distinct vertices a_0, a_1, a_2, and a triple of distinct sides S_0, S_1, S_2, with (a_i, S_i) being opposite pairs, such that S_0, S_2 form the pair of tangents to C' through a_1, S_0, S_1 form the pair of tangents to C' through a_2, and a_1 and a_2 form the pair of points at which S_0 meets C.

Let the homogeneous point coordinates w.r.t. the above fundamental triangle and with arbitrary scale be $(z_1 : z_2 : z_3)$, and let its dual line coordinates be $(u_1 : u_2 : u_3)'$. Then the point conic C and the line conic C' will be expressed in the form of equations in point and line coordinates as shown below:

$$
\begin{aligned}
C &: \quad x_{10} * z_1^2 + x_{11} * z_2 * z_3 + x_{12} * z_1 * z_3 + x_{13} * z_1 * z_2 = 0, \\
C' &: \quad x_{21} * u_2 * u_3 + x_2 2 * u_1 * u_3 + x_{23} * u_1 * u_2 = 0.
\end{aligned}
$$

Let us call a triangle with triple of points (p_0, p_1, p_2), triple of lines (L_0, L_1, L_2), and opposite pairs $(p_0, L_0), (p_1, L_1), (p_2, L_2)$, a *Ponelet triangle* w.r.t. the pair of conics C, C' if L_0 meets C in the pair p_1, p_2, through p_0 passes the pair of lines L_1, L_2 of C', and so forth. we note that for the fundamental triangle to be a Poncelet triangle it is necessary and sufficient that a_0 be on C or that $x_{10} = 0$.

Our problem is then to determine the number N of such Poncelet triangles w.r.t. C, C'.

Suppose there is a Poncelet triangle w.r.t. C, C' with vertices (p_1, p_2, p_3) and sides (L_1, L_2, L_3), as above. Then in introducing point coordinates for p_i, and line coordinates for L_i w.r.t. the fundamental triangle, we are led to a polset PS consisting of 24 pols P_i in 18 unknown ratios and 7 known parameters $x_{10}, x_{1i}, x_{2i}, i = 1, 2, 3$. That the number 18 of actual unknowns is less than the number 24 of actual equations $P_i = 0$ means simply that the 24 equations are not independent. This is of no consequence to our general method.

The problem is then reduced to the determination of $Zero_C(PS)$ whose zeros, if exist, will correspond to possibly existent Poncelet triangles. The actual determination of PS and $Zero_C(PS)$ is a little involved. We shall leave the details to the paper [WWT29] and restrict ourselves to state the final result in the form of the following theorem:

Theorem 5.3. (Poncelet). The number of Poncelet triangles w.r.t. a pair of a point conic and a line conic not dual to each other and non-degenerate is either 0 or infinity, unless the conics are in some special relative positions.

As an example of such special relative positions, we mention the following: Take any non-degenerate point conic C and any two different points p_0, p_1 on it. Take $p_2 = p_1$ and denote the lines p_0p_1, p_0p_2 by L_2 and L_1 which are in fact identical. Let L_0 be the tangent to C at p_1. Let C' be any non-degenerate line conic containing L_0 and $L_1 = L_2$, and touching $L_1 = L_2$ in p_0. Then C, C' form a pair of conics in special relative positions. In that case we see that the triples (p_0, p_1, p_2), and (L_0, L_1, L_2) form a Poncelet triangle w.r.t. C, C'. The number of Poncelet triangles is thus neither 0 nor infinity, as long as C' is chosen generically. See Fig. 5.5 above. Note that the above is essentially a reproduction of some relevant paragraphs in [WWT29].

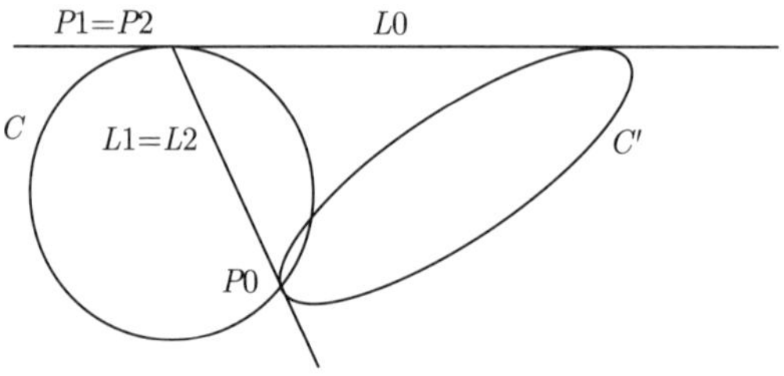

Fig. 5.5.

Theorem-Proving in Finite Geometries.

Throughout the present book we have restricted ourselves to the case of the basic field K being of characteristic 0. There are however interests in finite characteristic both in theory and applications, the latter for coding theory, cryptography, block design, etc. Lin Dongdai and Liu Zhuojun have considered the case of finite characteristic and applied their methods to geometry theorem-proving in finite geometries. See their joint paper [L-L].

Let us consider therefore the finite projective plane $PG(2, \mathbf{F}_q)$. where $\mathbf{F}_q$ is the finite field of $q = p^m$ elements, p being a prime number. The Well-Ordering Principle and methods of Geometry Theorem- Proving may be established in a way essentially the same as in the characteristic 0 case, only in that the computations should be carried out over the *integers* rather than the rationals, and care should be paid to the possible occurrence of integer factors divisible by the finite characteristic p.

It is well known that in the planar Fano geometry for $p = 2$ there are mutually different points A, B, C, D, E, F, G for which each of the 7 triples of points $ADC, AGE, AGC, AFB, CGF, CEB, DEF$ is on the same line while no four of the 7 points are on the same line. Clearly this cannot occur in the case of characteristic 0. Lin and Liu found also some interesting results which are peculiar to the Case of *finite* characteristic. Moreover, their results show a clear distinction between the case of p even and that of p odd. See their theorems below:

Theorem 5.4. The diagonal points of a complete quadrangle in $PG(2, \mathbf{F}_q)$ are always collinear for $p = 2$.

Theorem 5.5. No three distinct points on a conic can be collinear.

Theorem 5.6. If p is odd, then no three distinct tangents of a conic can be concurrent.

Theorem 5.7. If p is even, then all the tangents of a conic are concurrent in a single point.

Genus of Algebraic Curves.

There are several definitions of Chern classes for a projective (irreducible) algebr5aic variety with arbitrary singularities, due to R. MacPherson, J. Mather, J. P. Brasselet, Mad. Schwartz, et al. Among these definitions the one due to the present author (cited below as Wu) has the peculiarity of being computational, see Sect. 5.3 in Chap. 5. Liu Xian-fang (cf. [LXF1, 2]) has made a comparison of these definitions and proved the following

Theorem 5.8. The Chern classes given by Wu coincide with those given by Mather, but are different from those given by MacPherson.

On the other hand. while Chern numbers have been defined in a computational manner by Wu for arbitrary irreducible varieties, they are not even defined by any others. By means of Wu's computational definition Liu has given a neat explicit formula for the genus of a complex projective irreducible curve, whether planar or not, viz.

Theorem 5.9. Let C be a complex algebraic irreducible curve with singularities p_i of multiplicities m_i. Let $\pi : \tilde{C} \to C$ be a smooth resolution of C. then the genus $g(C)$ of C is given by

$$g(C) = 1 - \frac{1}{2} * [C_{W1}(C) + \Sigma_i(|\pi^{-1}(p_i)| - m_i)].$$

In the formula $C_{W1}(C)$ is the first Chern number of C in the sense of Wu which may be computed as $C_{W1}(C) = 2 * Ord(C) - Rank(C)$, and $|\pi^{-1}(p_i)|$ is just the number of the local analytic components of C at p_i.

Remark that for a singular curve formulae for genus are known only in the planar case and in some form not easily computed in concrete cases. Remark that Wu's definition has also been studied and compared with other definitions by J. Y. Zhou, see [Zhou].

Solitary-Wave Solutions of Non-Linear Evolution Equations.

Non-linear evolution PDE (partial differential equations) prevailed in mathematical physics and the seeking of exact solutions of some definite type is of utmost importance in applications to high technology. There is no general method of seeking such exact solutions which require usually intricate techniques to discover. Z. B. Li and his collaborators have devised however a general method applicable to a wide variety of non-linear evolution PDE. The method permits to get *all* possible solutions of some special type in reducing to polynomial equations-solving which will furnish a lot of solitary wave solutions. The method runs briefly as follows.

Li noted that for most non-linear evolution PDE their solitary-wave solutions can be written in terms of the hyperbolic tangent functions. Consider thus a PDE of unknown wave function $u(x, t)$ in position and time variables x and t. A travelling-wave solution will be of the form $u(x, t) = v(z)$ where $z = k * x - \omega * t$ with k the wave-number and $c = \frac{\omega}{k}$ the wave velocity. Introduce further the variable $y = \tanh(z)$ so that $v(z) = w(y)$ for some function w. The problem is then reduced to the extraction of solitary wave solutions from all possible wave solutions of certain type in terms of w which can be dealt with by our general method. Let us consider therefore the travelling-wave solutions w of the form below:

$$w = \Sigma_i a_i * y_i$$

in which the summation runs from $i = 0$ to $i = m$ and the a_i's are coefficient-functions in k. ω yet to be determined. Substitute now $u(x,t) = w(y)$ in the given PDE and set, in accordance to the naive Method of Undetermined Coefficients, the coefficients of terms of various y-powers to 0. We get then the value of m as well as a set of polynomial equations in the coefficients a_i, which will give the totality of all possible solutions of the given PDE in the above form of w.

As a simple illustration let us consider the classical KdV-equation

$$u_t + u * u_x + p * u_{xxx} = 0.$$

To find solutions w of the above type let us set $z = k * x - \omega * t, y = \tanh(z)$, and $w = \sum_i a_i * y_i$ as above. We get readily $m = 2$ and a polset $PS = \{P_1, \cdots, P_6\}$ with

$$\begin{aligned}
P_1 &= (2 * k^3 * p + \omega - k * a_0) * a_1, \\
P_2 &= k * a_1^2 + 16 * k^3 * p * a_2 + 2 * \omega * a_2 - 2 * k * a_0 * a_2, \\
P_3 &= (8 * k^3 * p + \omega - k * a_0 + 3 * k * a_2) * a_1, \\
P_4 &= -k * a_1^2 + 40 * k^3 * p * a_2 + 2 * \omega * a_2 - 2 * k * a_0 * a_2 + 2 * k * a_2^2, \\
P_5 &= -3 * k * a_1 * (2 * k^2 * p + a_2), \\
P_6 &= -2 * k * a_2 * (12 * k^3 * p + a_2).
\end{aligned}$$

For $a_2 \neq 0$, the char-set of PS is $CS = \{C_1, C_2, C_3\}$ with

$$\begin{aligned}
C_1 &= k * a_0 - 8 * k^3 - \omega, \\
C_2 &= a_1, \\
C_3 &= a_2 + 12 * k^2 * p.
\end{aligned}$$

The char-set CS will give us an infinity of possible solutions in terms of parameters k, ω(and also p) of the form

$$u(x,t) = \frac{8 * k^3 * p + w}{k} - 12 * k^2 * p * \tanh^2(k * x - \omega * t).$$

In particular, in setting $\omega = 4 * k^3 * p$, we will get the will-known classical soliton-solution

$$u(x,t) = 12 * k^2 * p * \operatorname{sech}^2(k * (x - 4 * k^2 * p * t)).$$

It follows that the known soliton solution is only a very special one among an infinity of possible solutions which have been found almost in a trivial way. For more details and more examples as well as some extensions see e.g. the papers [LZB], [ZSQ] and [Li-Ch].

Chemical Equilibrium Problem.

Many non-linear programming problems leading to non- linear polynomial equations-solving arise from chemical engineering. Among such chemical engineering problems the chemical equilibrium problem has been considered as the most difficult one deserving special attention. Let us consider thus a mixture of n chemical species containing x_i mols of i-th species in equilibrium. Set $x_0 = \Sigma_i x_i$. Then the free energy of such a mixture at temperature T and pressure P is given by

$$F(X) = \Sigma_i x_i * \left(c_i + \ln \frac{x_i}{x_0}\right).$$

In the sum $c_i = \dfrac{F_i}{R*T} + \ln P$ with $\dfrac{F_i}{R*T}$ the Gibbs free energy of the i-th species are all known constants. For the mixture to be in equilibrium the values $x_i > 0$ should render $F(X)$ to reach the least value under the mass balance equations

$$\Sigma_i a_{ij} * x_i = b_j, \qquad j = 1, \cdots, m.$$

In these equations m is the number of different type of atoms, a_{ij} is the number of atoms of element j in a molecule of species i, and b_j is the total number of atomic weights of element j originally present in the mixture. Set now

$$\begin{aligned} h_0 &= x_0 - \Sigma_i x_i. \\ h_j &= \Sigma_i a_{ij} * x_i - b_j, \qquad j = 1, \cdots, m. \end{aligned}$$

We may assume without loss of generality that the coefficient matrix of h_j in x_i has exactly the rank m since otherwise we may remove the redundant ones at the outset. This being understood let us form the Lagrangian pol with Lagrangian multipliers λ's:

$$L = F(X) + \lambda_0 * h_0 + \Sigma_j \lambda_j * h_j.$$

Set

$$\begin{aligned} e_j &= \exp(-c_i), \qquad i = 1, \cdots, n. \\ y_j &= \exp(-\lambda_j), \qquad j = 1, \cdots, m. \end{aligned}$$

Then $\dfrac{\partial L}{\partial x_i} = 0, i = 0, 1, \cdots, n$, become $\lambda_0 = 1$, and

$$P_i \equiv e_i * x_0 * \Pi_j (y_j^{a_{ij}}) - x_i = 0, \qquad i = 1, \cdots, n.$$

For Lagrangian polset we have $Zero_R(LAG) = Zero_R(PS)$, where $PS = \{P_i, h_0, h_j\}$ is a polset in x_0, x_i, y_j. As the rank assumption implies $Zero_R(JAC) = \phi$ for the

Jacobian polset JAC the method in Sect. 5.5 reduces the problem to the determination of $DZero_R(PS)$ where D is the oopen domain defined by $x_0 > 0, x_i > 0, y_j > 0$ with $e_i > 0$ considered as known parameters.

As a concrete example we have considered the equilibrium problem about the combustion of stoichiometric mixture of $\frac{1}{2}$ mol of hydrazine N_2H_4 and $\frac{1}{2}$ mol O_2 at known values of T and P. The mixture in equilibrium is of ten species so that $n = 10,\ m = 3$:

$$H, H_2, H_2O, N, N_2, NH, NO, O, O_2, OH.$$

The problem has been widely studied in the literature by means of various numerical methods. To apply our methods developed in Sect. 5.5, let us first introduce an ordering of variables

$$y_1 \prec y_2 \prec y_3 \prec x_1 \prec \cdots \prec x_{10} \prec x_0.$$

Our method then leads to a polset with a char-set CS of index-sets

$$[1118\ \ y_1\ \ 12], [210\ \ y_2\ \ 1], [8\ \ y_3\ \ 1], [6\ \ x_1\ \ 1], \cdots$$

The initials and factors removed during the procedure are in general non-zero so that $DZero_R(CS)$ furnishes all the possible solutions from which the solution with least free energy $F(X)$ may be determined. This is in contrast to the usual numerical methods which in general furnish for optimization problems only one of the possible solutions with locally extremal values but not necessarily the optimal value. For more details cf. e.g. [WWT15].

Extensions to Differential Case.

The char-set theory and method have been extended to the differential case with similar notions of d-pols, d-polset, d-bas-set, d-char-set, etc. Cf. in particular the paper [wWT33]. Remark that our theory relies heavily on the two classical books of Ritt [Ritt1, 2] as well as the earlier works of Riquier and Janet. The theory in the case of a single independent variable is actually parallel to the case of ordinary polynomials and has been applied to theorem proving of curve-theory in differential geometry. In particular, we have studied in details the theorems about a Bertrand pair of curves in both ordinary space and affine space, see e.g.[WWT31.34]. It has been even applied to the automated deduction of Newton's laws from Kepler's laws as a consequence of the general method of automated deduction of unknown relations in the polynomial case described in Sect. 8.1. It has also been applied to the solving of Kepler's equation in celestial mechanics. On the other hand the theorem-proving for surfaces and higher dimensional differential manifolds are yet only fragmentally touched. In this respect we refer to, besides[WWT34], some relevant papers of S. C. Chou, X. S. Gao, Z. M. Li, and notably one due to H. B. Li, see his papers [LHB1, 2].

Kepler Equation in Celestial Mechanics.

The following is essentially a reproduction of some relevant paragraphs in [WWT34]. Cf. also the paper [W-W].

In celestial mechanics, Kepler's equation plays an important role in astronomical computations. As an example of illustrating our general method, we shall show how to solve Kepler's equation in a manner entirely different from the usual one.

For this purpose, let us consider a planet P moving around the Sun S in an elliptical orbit with eccentricity e and period T. Let t be the time elapsed in starting from the perihelion $\prod$. The angle to be determined, E, is the angle spanned by the major axis toward the perihelion, and the radius vector from the centre of the elliptical orbit to that point Q on the circumscribing circle which projects to the same point R on the major axis, as does the planet P. See the Fig. 5.6.

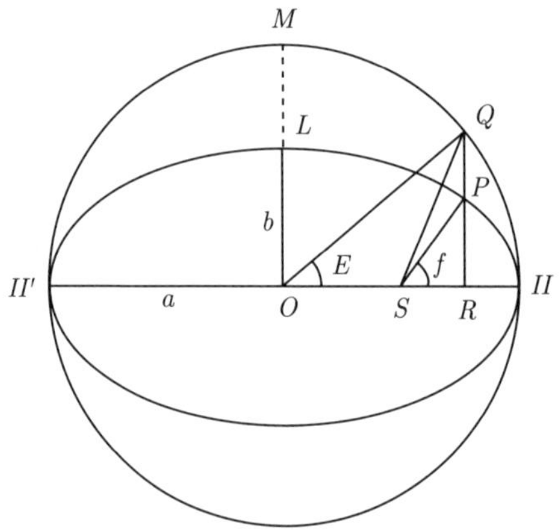

Fig. 5.6.

Kepler's equation, which permits us to determine E, is then of the form

$$E - e * \sin E = M$$

with $M = \dfrac{2 * \pi * t}{T}$. This equation is usually solved by numerical or graphical methods. A solution is also furnished by the *Lagrange series* of E in e with coefficients in terms of derives of powers of $\sin M$ w.r.t M, namely

$$E = M + \Sigma_n \frac{e^n}{n!} * \frac{d^{n-1}(\sin^n M)}{dM^{n-1}}$$

in which the summation runs from n=1 to ∞.

The derivation of he Lagrange series is somewhat intricate and the series converges only for $e(< 1)$ sufficiently small, namely $e < 0.6627\cdots$ We shall now give a simple device turning the equation into a form which permits us, by our general method, to determine E, as well as other simple functions of E like $\sin E$, etc., in the form of a convergent series of M.

The underlying principle of the method is quite simple. Let us consider $sinB$ as a separate indeterminate function by setting

$$F = \sin E.$$

The sine function, as also with most transcendental functions occurring in mathematics, will satisfy a differential equation. This differential equation, together with the original Kepler equation, will furnish us with a d-polset DPS consisting of two d-pols in E and F on the independent variable M. Applying our general method, we then get equations in E or F which may be solved in the form of a series in M.

The computation is quite simple and runs as follows. Considering E and F as functions of M with primes denoting derivatives w.r.t. M, we get from $F = \sin E$

$$\begin{aligned} F' &= \cos E * E', \\ F'' &= -\sin E * E'^2 + \cos E * E''. \end{aligned}$$

Removing $\sin E$, $\cos E$in the Kepler equation and the equation for F" by means of the equations for F and F', we get the equations $DP_1 = 0,\ DP_2 = 0$ with $DP_1,\ DP_2$ given below:

$$\begin{aligned} DP_1 &= E - e * F - M, \\ DP_2 &= E' * F'' - F' * E'' + F * E'^3. \end{aligned}$$

Applying our general method to the d-polset $DPS = \{DP_1, DP_2\}$, we then get equations in E or F as required.

For example, in order to solve for E, let us take the ordering $E \prec F$. To form the d-char-set of DPS w.r.t. this ordering, we just form the d-remainder of DP_2 w.r.t. DP_1, thus getting a d-pol

$$DC_1 = E'' + (E - M) * E'^3$$

which, together with $DC_2 = DP_1$ will form a d-char-set of the d-polset DPS w.r.t. the above ordering. The equation $DC_1 = 0$, or

$$E'' + (E - M) * E'^3 = 0$$

is then the equation of E in terms of M, as required. The initial values of E and E' at $M = 0$ are obviously

$$E(0) = 0, \qquad E'(0) = \frac{1}{1-e}.$$

We, therefore, get a series of the form

$$E = \frac{1}{1-e} * M - \frac{e}{(1-e)^4} * \frac{M^3}{3!} + \cdots$$

which is convergent in the neighborhood of M=0 by the general theory of differential equations. It can, in fact, be proved that the series is convergent for all values of M.

If we are interested in the determination of $F = \sin E$, we may take the ordering $F \prec E$ instead of $E \prec F$, Proceeding as in the previous case in forming the d-remainder of DP_2 w.r.t. DP_1 but this time w.r.t. the new ordering, we will get a d-pol of the form

$$DC_1' = F'' + F * (e * F' + 1)^3$$

which, together with $DC_2' = DP_1$ will form a d-char-set of the d-polset DPS w.r.t. the present ordering. The equation $DC_1 = 0$, or

$$F'' + F * (e * F' + 1)^3 = 0$$

then gives $F = \sin E$ in terms of M. With the obvious initial values

$$F(0) = 0, \qquad F'(0) = \frac{1}{1-e}.$$

we then get the series of $\sin E$ as required.

$$\sin E = \frac{1}{1-e} * M - \frac{1}{(1-e)^4} * \frac{M^3}{3!} + \cdots$$

The series can again be proved to be convergent for all values of M. The above method is quite general and may be used to determine e.g. $\cos E$, the Sun-planet radius vector, the anomaly, etc. in terms of M. It can also be applied to the case of parabolic or hyperbolic orbits.

Automatic Derivation of Newton's Laws from Kepler's Laws.

As in the case of Kepler equation, the following is essentially a reproduction of some relevant paragraphs in [WWT34]. Cf. also the paper [WWT32].

It was an important historical event when Newton derived his gravitational laws from Kepler's laws. During a visit to Argonne National Laboratory in 1986, the author was told by Prof.J.R. Gabriel about the significance of deducing Newton's gravitational laws from Kepler's Laws in a mechanical way. Prof.Gabriel had already been quite successful in applying his own automated reasoning method to achieve this, cf. [GAB]. As a further example of illustrating our general method, we shall show below how our method can be applied to deal with such problems.

For this purpose, let us first formulate Kepler's laws (K) and Newton's laws (N) in the follwing manner:

(K_1) The planets move in elliptic orbits around the Sun as the focus.

(K_2) The vector from the Sun to the planet sweeps equal areas in equal times.

(K_3) The squares of periods of planet motions are proportional to the cube of the major axis of the elliptic orbits.

(N_1) The acceleration of a planet is inversely proportional to the square of the distance from the Sun to the planet.

(N_2) The acceleration vectors of planets are directed toward the Sun.

(N_3) The proportionality factor of the inverse square law (N_1) is independent of the different planets.

In order to deduce mechanically, or even discover automatically Newton's laws from Kepler's laws, let us first take coordinates and transform the various laws into equation forms as follows.

Let us take e.g., the rectangular coordinates (x, y) with the Sun at the origin and the major axis of the elliptic orbit as the X-axis. Let r be the radius vector from the Sun to the planet. Then the orbit will have an equation of the form

$$r = p + e * x. \tag{5.28}$$

The Kepler's law(K_1) will then correspond to Equation(5.28), and also to (5.29)–(5.31) taken together:

$$r^2 = x^2 + y^2. \tag{5.29}$$

$$p = \text{const}, \qquad p' = 0. \tag{5.30}$$

$$e = \text{const}, \qquad e' = 0. \tag{5.31}$$

in which the prime means derivation w.r.t. time t. Similarly, Kepler's law (K_2) will correspond to Equations (5.32), (5.33)

$$x * y' - y * x' = h. \tag{5.32}$$

$$h' = 0. \tag{5.33}$$

For (K_3), let T be the period for a planet to turn once around the Sun on its elliptical orbit and $2 * a, 2 * b$ be the major and minor axis of this orbit. Then Kepler's law (K_3) means that $\frac{T^2}{a^3}$ is a universal constant independent of the planets, although possibly dependent on the Sun. Now, according to the meaning of h, we have

$$h * T = 2 * \pi * a * b.$$

Since $\frac{b^2}{a} = p$, we have

$$\frac{T^2}{a^3} = 4 * \pi^2 * \frac{p}{h^2}.$$

It follows that

$$c = \frac{p}{h^2} \tag{5.34}$$

is also a universal constant independent of the planets with $c' = 0$. On the other hand Newton's laws (N_1), (N_2), with A as the acceleration, will correspond to the following set of equations:

$$A^2 = x''^2 + y''^2, \tag{5.35}$$

$r^2 * A =$ const., or since $r \neq 0$,

$$2 * r' * A + r' * A' = 0, \tag{5.36}$$

$$x * y'' = y * x''. \tag{5.37}$$

For Newton's law (N_3), the constant $r^2 * A$, as asserted by (N_1), is a universal one and thus there may exist relations connecting this constant with c which we have to try to find out.

As Newton's law (N_2) corresponding to (5.37) follows directly from Kepler's law (K_2) corresponding to (5.32)(5.33), we shall consider only the laws (N_1) and (N_3).

For this purpose, let us now introduce the indeterminates in replacing the various functions by x's as given below:

$$(r, A, c, p, e, x, y, h) = (x_{11}, x_{12}, x_{15}, x_{21}, x_{22}, x_{31}, x_{32}, x_{51}).$$

The various functions are so arranged that r, A, and c come first in order to discover the possible relationships between them. With this change of notations, Equations

(5.28) ∼ (5.37) will turn out to be the equations $P_i = 0$ with P_i given by (5.38) ∼ (5.47), as shown below (prime means derivative w.r.t. t):

$$P_1 = x_{11} - x_{31} * x_{22} - x_{21}, \tag{5.38}$$

$$P_2 = x_{31}^2 + x_{32}^2 - x_{11}^2, \tag{5.39}$$

$$P_3 = x_{21}', \tag{5.40}$$

$$P_4 = x_{22}', \tag{5.41}$$

$$P_5 = x_{31} * x_{32}' - x_{32} * x_{31}' - x_{51}, \tag{5.42}$$

$$P_6 = x_{51}', \tag{5.43}$$

$$P_7 = x_{15} + x_{51}^2 - x_{21}, \tag{5.44}$$

$$P_8 = (x_{31}'')^2 + (x_{32}'')^2 - x_{12}^2, \tag{5.45}$$

$$P_9 = 2 * x_{11}' * x_{12} + x_{11} * x_{12}', \tag{5.46}$$

$$P_10 = x_{31} * x_{32}'' - x_{32} * x_{31}''. \tag{5.47}$$

Now take the d-polset DPS to consist of the d-pols (5.38)∼(5.45) of the above set corresponding to Kepler's laws. We remark that the planets move in true non-degenerate elliptic and non-circular orbits, so that we have

$$\left.\begin{array}{l} x_{21} = p \neq 0, \quad x_{22} = e \neq 0 \\ x_{11} = r \neq 0, \quad x_{11}' = r' \neq 0 \end{array}\right\} \tag{5.48}$$

and also

$$x_{15} = c \neq 0, \quad x_{12} = A \neq 0. \tag{5.49}$$

In applying our algorithm for finding d-char-set DCS of DPS, we can remove any such factors during the procedure. The DCS is found to be the 7-th d-bas-set consisting of the 8 d-pols C_i with the following index sets:

$$[7\ \ r^{(iv)}\ \ 1], [4\ \ A\ \ 2], [3\ \ c\ \ 1], [3\ \ p\ \ 1], [7\ \ e\ \ 2], [3\ \ x\ \ 1], [3\ \ y\ \ 2], [3\ \ h\ \ 1]$$

of the 8 d-pols C_i, the first one is in $C_i x_{11} = r$ alone and the second in $x_{12} = A$. The first thus gives us a differential equation of derivative order 4 observed by the radius vector r. This equation and the second one between A and r are both too complicated and so are of little interest. However, during the process, the 4th d-bas-set, DBS, appears to consist of 6 d-pols B_{4i} of which the first one, B_{41}, is given by

$$B_{41} = 2 * x_{12} * x_{11}' + x_{12}' * x_{11} = 2 * r' * A + r * A'.$$

By our general Principle, $B_{41}=0$ should be a consequence of the original d-polset, i.e. a consequence of Kepler's laws. The equation $B_{41}=0$ is, however, nothing else but Newton's inverse square law $r^2 * A=$ const. We have thus discovered, in an automatic manner, Newton's law (N_1) from Kepler's laws by means of our general principle. Furthermore, the 3rd d-bas- set during the process consists of 6d-pols B_{3i}, the first of which is given by

$$B_{31}=x_{15}^2 * x_{12}^2 * x_{11}^4-1=c^2 * r^4 * A^2-1.$$

The equation $B_{31}=0$ is a consequence of Kepler's laws, too, which implies

$$c=\pm\frac{1}{r^2 * A}. \tag{5.50}$$

It follows that the proportionality constant$r^2 * A$, as already asserted by Newton's law (N_1), is equal to the reciprocal of the universal constant c up to a sign and is, therefore, a universal constant. The Newton's law (N_3) is, thus, also discoveredin an automatic manner.

The ambiguity of the sign in (5.50) comes from our introduction of acceleration A by means of (5.35) which gives its magnitude but not its direction. If we take the positive value of $\sqrt{A^2}$ as A with the corresponding acceleration vector pointing to the Sun, then the +sign in (5.50) is to be taken so that

$$r^2 * A=+\frac{1}{c}=+\frac{h^2}{p}$$

is the universal constant independent of the planets, although eventually dependent on the Sun.

Let us now assume that the attractive force between two masses M and m is proportional to both M and m. Let us also assume that the mass m under the action of a force F will undergo a motion with an acceleration A given by the general law $F=m * A$, also due to Newton. Then we immediately deduce from Newton's laws $(N_1)\sim(N_3)$ the following *general gravitational law*of Newton.

The gravitational force between the Sun of mass M and a planet of mass m is given by

$$F=G * \frac{M * m}{r^2}$$

in which r is the distance between the Sun and the planet, while G is a universal constant independent of the planet. The universal constant G is, in fact, connected with the above c by the relation $G=+\dfrac{1}{M * c}$ which should be independent of the Sun if the roles of Sun and the planet are to be considered as symmetric.

Remark. In the above formulation, we have supposed that Newton's gravitational laws are not known *a-priori* and it turns out that they will be discovered in an automatic manner by our general method. The computations are carried out on a SUN3/140 with a running time= 15'58" and $maxt = 342$, where $maxt$ means the maximum number of terms of d-pols occurring during the procedure. In the paper [WWT32] some different sets of coordinates and equations were tried. Comparing the various trials it is shown that the mere mechanical proving of Newton's laws which are supposed to be already known would be somewhat simpler than automatically discovering laws which are supposed to be unknown.

Some Further Complements and Extensions.

It is impossible to exhibit all the contributions what the members of MMRC as well as their collaborators have done in these years concerning mathematics mechanization and its related fields. Below is a very incomplete list of titles of their studies. We shall not give the references and leave these to their relevant papers mostly in the MM-Res. Preprints.

List of research titles and contributors:

Proving by examples, traditional and readable proofs, Clifford-algebra technique.(J. W. Hong, S. C. Chou, X. S. Gao, J. Z. Zhang, L. Yang, M. D. Cheng, H. B. Li, *et al.*).

Parametrization Problem, Curve-Isomorphism Problem. (X. S. Gao, H. Du, *et al.*).

General algebraic systems, real discriminant systems, invariant and covariant theory. (L. Li, L. Yang, X. R. Hou, F. M. Zou, *et al.*).

Eigen-value method of polynomial equations solving, linear system with quantifiers. (W. D. Wu, Y. Z. Huang, G. C. Feng, L. Li, K. Huang, S. G. Zhang, *et al.*).

Planar differential systems, limit cycles, stability. (D. M. Wang, Z. Y. Lu, S. L. Ma, S. C. Ning, S. M. Zhu, Z. J. Liu, *et al.*).

Differential systems, partial differential equations, chaos. (Z. M. Li, Z. B. Li, M. L. Wang, S. J. Wu, H. Q. Zhang, J. Q. Fang, *et al.*).

Yang-Baxeter equations in statistical mechanics, quantum groups. (H. Shi, K. Wu, S. K. Wang, *et al.*).

Coding theory, cryptography. (Z. J. Liu, H. Du, *et al.*).

Stewart platform problem, spring system, electric-engineering system, spline system. (W. D. Wu, Y. Z. Huang, C. G. Liang, C. Chen, R. H. Wang, Y. S. Zhou, *et al.*).

Control theory, pole assignment problem. (H. Shi, S. Y. Li, *et al.*).

Computer vision, data compression, pattern recognition, wavelet theory.(M. D. Cheng, Q. Y. Shi, C. X. Xu, W. W. Fang, *et al.*).

Remainder method of first order logic, many-valued logic, generalized resolution, proving verification, probable inference, infinite logic. (J. Z. Wu, Z. J. Liu, X. H. Liu, Q. Y. Huang, G. D. Hu, *et al.*).

Bibliography

[A-H] Appel, K., & Haken, W., Every map is four colorable, Illinois J. Math., 21 (1977) 421-490, 491-567.

[Ak] Akritas, A.G., Elements of Computer Algebra with Applications, John Wiley & Sons, 1989.

[Ar] Arnon, D.S., A bibliography of quantifier elimination for real closed fields, J. Symb. Comp., 5 (1988) 267-274.

[A-M] Arnon, D.S., & Mignotte, M., On mechanical quantifier elimination for elementary algebra and geometry, J.Symb.Comp., 5 (1988) 237-260.

[Baj] Bajaj, C.L., Surface fitting using implicit algebraic surface patches, in Curve and Surface Modelling (ed. H.Hagen), SIAM Publ., (1991) 1-31.

[B-I] Bajaj, C.L., & Ihm, I., Algebraic surface design with interpolation, ACM Trans. on Graphics, 19 (1992) 61-91.

[B-W] Becker, T. & Weispfennig, V., Groebner basis, Springer, (1993).

[Bey] Beyer,R.A., The kinematic synthesis of mechanisms, McGraw-Hill, (1963).

[B-M] Blaschke,W., & Mueller,H.R., Ebene Kinematik, Muenchen (1956).

[B-L] Bledsoe, W.W., & Loveland, D.W., (Eds.), Automated theorem proving, after 25 Years, Amer.Math.Soc., (1984).

[Bo] Bocher, M., Introduction to Higher Algebra, Macmillan, 1907.

[B-C-R] Bochnak, J., Coste, M., & Roy, M-F., Géomètrie algébrique réelle, Springer, (1986).

[B-S] Brasselet, J.P., & Schwartz, M.H., Sur les classes de Chern d'un ensemble analytique complexe, Astéristique, $N^0$82-83, (1981).

[B-T] Brown, W.S., & Traub, J.F., On Euclid's algorithm and the theory of subresultants, J.ACM, 18 (1971) 505-514.

[Bu1] Buchberger, B., Ein algorithmitisches Kriterium fuer die Loesbarkeit eines algebraishes Gleichungssystems, Aeq.Math, 4 (1970) 374-383.

[BU2] Buchberger, B., Some properties of Groebner bases for polynomial ideals, ACM SIGSAM Bull., 10 (1976) 19-24.

[BU3] Buchberger, B., An algorithmic method in polynomial ideal theory, in N.K.Bose (Ed.), Recent Trends in Multidimensional Systems Theory, Reidel (1985) 184-232.

[B-L-C] Buchberger, B., Loos, R., & Collins, G.E., Computer Algebra – Symbolic and Algebraic Computation, Springer Verlag (1982).

[Chern1] Chern, S.S., On the characteristic classes of complex sphere bundles and algebraic varieties, J. Amer. Math. Soc., 75 (1953) 565-597.

[Chern2] Chern, S.S., Characteristic classes of Hermitian manifolds, Annals of Math., 47 (1946) 88-121.

[Chou1] Chou, S.C., Proving elementary geometry theorems using Wu's algorithm, in [B-L], 243-286.

[Chou2] Chou, S.C., Proving and discovering geometry theorems using Wu's method, Ph.D Thesis, UTA, (1985).

[Chou3] Chou, S.C., Mechanical geometry theorem-proving, Reidel, (1988).

[C-G] Chou,S.C., & Gao,X.S., On the mechanical proof of geometry theorems involving inequalities, Preprints TR-89-31, UTA, (1989).

[C-G-M] Chou,S.C., Gao,X.S., & Mcphee, N., A combination of Ritt-Wu's method and Collins' method, Preprint TR-89-28, UTA, (1989).

[C-G-Z] Chou,S.C., Gao,X.S., & Zhang,J.Z., Mechine proofs in geometry, World Scientific, Singapore, (1994).

[C-S-Y] Chou, S.C., Schelter, W.F., & Yang, J.G., An algorithm for constructing Groebner bases from characteristic sets and its applications to geometry, Algorithmica, 5 (1990) 147-154.

[Chow1] Chow, W.L., Algebraic varieties with rational dissections, Proc. Nat. Acad. Sci., 42 (1956) 116-119.

[Chow2] Chow, W.L., On the equivalence classes of cycles in an algebraic variety, Annals of Math., 64 (1956) 450-479.

[Ch-VdW] Chow,W.L., & Van der Waerden, Ueber Zugeordnete Formen und algebraische Systeme von Mannigfaltigkeiten, Math. Ann., 113 (1937) 692-704.

[Col1] Collins, G.E., Subresultants and reduced polynomial sequences, J.ACM, 14 (1967) 128-142.

[Col2] Collins, G.E., Quantifier elimination for real closed fields: a guide to the literature, in [B-L-C], p.79-81.

[Col3] Collins, G.E., Quantifier elimination for real closed fields by cylindrical algebraic decomposition, in 2nd GI Conf. Automata Theory & Formal Languages, Springer, (1975) 134-183.

[C-H] Collins, G.E., & Hong, H., Partial cylindrical algebraic decomposition for quantifier elimination, J. Symb. Comp., 12 (1991) 299-328.

[Cou] Courant, R., Differential and Integral Calculus, vol.2, London & Glascow, (1936).

[C-L-S] Cox, D., Little, J., & O'Shea, D., Ideals, varieties, and algorithms, Springer, (1992).

[C-G] Czapor, S.R., & Geddes, K.O., On implementing Buchberger's algorithm for Groebner basis, Proc. SYMSAC'86, (1986) 233-237.

[Dehn1] Dehn, M., Ueber raumliche Polyeder, Nachr.Akad.Wiss.Goettingen, Math.-Phys.Kl., (1900) 345-354.

[Dehn2] Dehn, M., Ueber der Raum Inhalt, Math.Ann., 55 (1902) 465-478.

[Eh] Ehresmann,C., Sur la topologie de certaines espaces homogènes, Ann. of Math., 35 (1934) 396-443.

[Eisb] Eisenbud, D., Commutative algebra with a view toward algebraic geometry, Springer, (1995).

[Eish] Eisenhart,L.P., A treatise on the differential geometry of curves and surfaces, (1909).

[F-W-Z] Feng, G.C., Wu, W.D., & Zhang, S.G., The eigenvalue problem equivalent to multivariate polynomial system, MM-Res.Preprints, MMRC, 8 (1992) 134-142.

[Fr] Fricke,R., Lehrbuch der Algebra, Bd 1 (1924).

[Gab] Gabriel, J.R., SARA - A small automated reasoning assistant, Preprint, Argonne Nat. Lab., (1986).

[Gam] Gamkrelidze,P.B., Chern cycles of complex algebraic manifolds, Izv. Acad. Scis., CCCP, math. ser., 20 (1956) 685-706 (in Russian).

[Gao1] Gao Xiao-shan, The discriminant systems of unary equations and their computation, MM-Res.Preprints, MMRC, No.1 (1987) 13-32.

[Gao2] Gao Xiao-shan, Constructive methods for polynomial sets and their applications, (in Chinese) PhD Thesis, Inst. of Systems Science, Acad. Sinica, (1988).

[G-L-L-L] Gao, X.S., Li, Y., Lin, D.D., & Liu, X.S., A prover based on Wu's method, Proc.Intern.Workshop on Math.Mech. in Beijing, Intern.Acad.Publishers, Beijing, (1992) 201-205.

[Gau1] Gauss,C.F., Bestimmung der groessten Ellipse welche die vier Seiten eines gegebenen Vierecks beruehrt, Zach. Monatliche Correspondenz fuer Erd- und Himmelskunde, Bd 22, (1810) 112-121, or Werke, Bd.4 (1882) 385-392.

[Gau2] Gauss,C.F., Das vollstaendige Fuenfeck und seine Dreiecke, Astro. Nachrichten, Nr.42, (1923), or Werke, Bd.4 (1882) 406-407.

[Gau3] Gauss,C.F., Supplementum theorias combinations observationum erroribus minimis obnoxiae, in Werke, Bd 4 (1880) 8-93.

[Gre] Greenberg, M.J., Euclidean and non-euclidean geometries, development and history, 2nd Ed., Freeman, (1980).

[Gro] Groebner, W., Moderne algebraische Geometrie, Springer, (1949).

[G-L] Guan, A.W., & Liu, H.L., The machine proof of five positions plane righy body guidance mechanism synthesis, Proc.ASCM'95 in Beijing, Scientists Incorporated, Tokyo, (1995) 145-152.

[H-D] Hartenberg,R.S., & Denavit,J., Kinematic synthesis of linkages, McGraw-Hill, (1964).

[Hen] Hensel,K., Eine neue Theorie der algebraichen Zahlen, Math.Z., 2 (1918) 433-452.

[Her] Hermann,G., Die Frage der endlich vielen Schritte in der Theorie der Polynomideale, Math.Ann., 95 (1925) 736-788.

[Hil1] Hilbert, D., Grundlagen der Geometrie, Goettingen, (1899).

[H-P1] Hodge, W.V.D., & Pedoe, D., Methods of algebraic geometry, vol.1, Cambridge, (1947).

[H-P2] Hodge, W.V.D., & Pedoe,D., Methods of algebraic geometry, vol.2, Cambridge, (1952).

[H-W] Hu, S., & Wang, D.M., Fast factorization of polynomials over rational number field or its extension fields, Kuxue Tongbao, 31 (1986) 150-156.

[H-WWD] Huang, Y.Z., & Wu, W.D., A modified version of an algorithm for solving multivariate polynomial systems, MM-Res.Preprints, MMRC, 5 (1990) 23-29.

[Jan] Janet,M., Lecons sur les systèmes d'équations aux dérivées partièlles, Gauthier Villars, (1920).

[Kal] Kalkbrenner, M., A generalized Euclidean algorithm for computing triangular representations of algebraic varieties, J. Symb. Comp., 15 (1993) 143-167.

[Kap] Kapur, D., Using Groebner bases to reason about geometry problems, J. Symb. Comp., 2 (1986) 399-408.

[K-M] Kapur, D. & Mundy, J.L., Wu's method and its applications to perspective viewing, Art.Int., 37 (1988) 15-36.

[Kl] Klein, F., Vorlesungen ueber nicht-Euklidische Geometrie, (1927).

[Knu] Knuth, D.E., The art of computer programming, v.2., Addison-Wesley, (1969).

[K-F-F] Kobayashi, H., Fujise, T., & Furakawa, A., Solving systems of algebraic equations by a general elimination method, J.Symb.Comp., 5 (1988) 303-320.

[Kol] Kolchin, E.R., Differential algebra and algebraic groups, Academic Press, (1973).

[Kr] Kronecker, L., Grundzuege einer arithmetischen Theorie der algebraischen Groessen, J.f.Math., 92 (1882) 1-122.

[Ku] Ku, Y.H., Note on a method of evaluating the complex roots of a quartic equation, J.Math. and Physics, Cambridge, Mass., 5 (1926) 126-128.

[K-S] B.Kutzler & S.Stifter, On the application of Buchberger's algorithm to automated geometry theorem proving, J.Symb.Comp., 2 (1986) 389-398.

[Laz1] Lazard, D., Solving zero-dimensional algebraic systems, J.Symb.Comp., 13 (1992) 117-131.

[Laz2] Lazard, D., A new method for solving algebraic systems of positive dimension, to be published in Discr.Appl.Math.

[Lef] Lefschetz, S., Topology, Amer.Math.Soc.Coll.Publs., (1930).

[LHB1] Li, H.B., New explorations on mechanical theorem proving of geometries, PhD Thesis, Beijing Univ., Beijing, (1994).

[LHB2] Li, H.B., Automated reasoning with differential forms, Proc.ASCM'95 in Beijing, Scientists Incorporated, Tokyo, (1995) 29-32.

[L-C] Li, H.B. & Cheng, M.T., Proving theorems in elementary geometries with Clifford algebra method, to be published in Chinese Math. Progress.

[L-S] Li, H.B. & Shi, H., On Erdös′ ten point problem, MM-Res.Preprints, MMRC, 15 (1997) 53-63. Also in Acta Math. Sinica, New Ser., 13 (1997) 221-230.

[LTY] Li, T.Y., Solving polynomial systems by homotopy continuation methods, Proc. Nankai on Computer Math. in 1991 Tianjin, World Sientific, Singapore, (1993) 18-35.

[LXY] Li, X.Y., Synthetic introduction to four-bar mechanism, (in Chinese), Mech. Industry Press, Beijing, (1965).

[LZB] Li, Z.B., Wu method and solitons, Proc.ASCM′1995, Scientists Incorporated, Tokyo, (1995) 157-164.

[Li-Ch] Li, Z.B., & Chen, S.K., Finding solitary wave solutions of non-linear wave equations using symbolic computation, Proc. ASCM'96 in Kobe, Scientific Incorporated, Tokyo, (1996) 63-72.

[LZM1] Li, Z.M., A new proof of Collins's theorem, MM-Res.Preprints, MMRC, No.1 (1987) 33-37.

[LZM2] Li, Z.M., On the trianguration for any finite polynomial set (I), MM-Res. Preprints, MMRC, No.2 (1987) 48-54

[LZM3] Li, Z.M., Determinant polynomial sequences, MM-Res.Preprints, MMRC, No.3 (1989) 51-55.

[Liang] Liang, C.G., Input-output equation of 7R mechanism and inverse solution of general 6R manipulator, Proc.1992 Intern.Workshop on Math.Mech., (Eds. W.T.Wu & M.D.Cheng), Intern.Acad.Publishers, Beijing, (1992) 189-200.

[L-L] Lin, D.D., & Liu, Z.J., Some results on theorem proving in finite geometry, Proc.1992 Intern.Workshop on Math.Mech. in Beijing, Intern.Acad.Publishers, Beijing, (1992) 222-235.

[LXF1] Liu, X.F., Singularities of algebraic varieties, PhD Thesis, Eastern China Normal Univ., Shanghai, (1995).

[LXF2] Liu, X.F., The genera of algebraic curves with singularities, Proc.ASCM'95, Beijing, (1995) 23-28.

[Liu] Liu, Z.J., Mechanical approach for deciding the construction of system equations' solutions and its related problems, (in Chinese) PhD Thesis, Inst. of Systems Science, Beijing, (1988).

[Lyon] Lyon, W.V., Note on a method of evaluating the complex roots of a quartic equation, J.Math. and Physics, Cambridge, Math., 3 (1924).

[Mac] Macaulay, F.S., Some properties of enumeration in the theory of modular systems, Proc.London Math.Soc., 26 (1927) 531-555.

[MP] MacPherson, R.D., Chern classes for singular algebraic varieties, Ann. of Math., 100 (1974) 423-432.

[Mig] Mignotte, M., Computer versus paper and pencil, Proc. CALSIF 4 (1986) 63-69. Also in Proc.Nankai on Comp.Math. in 1991 Tianjin, World Scientific, Singapore, (1993) 112-117.

[Miy] Miyaoka, Y., On the Chern numbers of surfaces of general type, Invent. Math., 42 (1977) 225-237.

[Mou] Moulton, F.R., The straight line solutions for the problem of n-bodies, Ann. of Math., 12 (1910) 1-17.

[O1] Ostrowski, A.M., Solution of equations and systems of equations, Acad.Press, (1960).

[O2] Ostrowski, A.M., Recherches sur la méthode de Graeffe et les zeros des polynomes et des series de Laurent, Acta Math., 72 (1940) 99-155.

[Paul] Paul, R.P., Robot manipulators: Mathematics, programming, and control, MIT Press, (1981).

[Pet] Petersen, J., Die Steinersche Loesung der Malfattischen Aufgabe, J.f.Math., 89 (1880) 127-135.

[Riq] Riquier, C.H., La méthode des fonctions majorantes et les systèmes d'équations aux dérivées partièlles, Gauthier Villars, (1928).

[Ritt1] Ritt, J.F., Differential equations from the algebraic standpoint, Amer. Math. Soc., (1932).

[Ritt2] Ritt, J.F., Differential algebra, Amer.Math.Soc., (1950).

[S-S-K-S] Sasaki,T., Suzuki, M., Kolar, M., & M.Sasaki, Approximate factorization of multivariate polynomials and absolute irreducibility testing, Japan J.Ind. Appl. Math., 8 (1991) 359-375.

[S-R] Semple, J.G., & Roth, L., Introduction to algebraic geometry, Oxford, (1949).

[Shi1] Shi, H., On Chern characters of algebraic hypersurfaces with arbitrary singularities, Acta Math.Sinica, New Ser., 4 (1988) 289-300.

[Shi2] Shi, H., Chern classes of algebraic varieties with singularity, Singularity theory, World Scientific, Singapore, (1995) 705-730.

[Sm1,2] Smale, S., Topology and mechanics, I, II, Invent. Math., 10 (1970) 305-331; 11 (1970) 45-64.

[St] Steiner, J., Geometrische Betrachtungen, J.f.Math., 1 (1826) 178-180.

[Syd] Sydler,J.P., Conditions nécèssaires et suffisantes pour l'équivalence des polyédres de l'espace euclidien à trois dimensions, Comm.Math.Helv., 40 (1965) 436-480.

[Syn] Synge, J.L., On the motion of three vortices, Canad.J.Math., 1 (1949) 257-270.

[Tai] Tai, S.S., A class of symmetric functions and Chern numbers of algebraic varieties, Lect.Notes in Math., No.1369, Springer, (1989) 261-274.

[Tar] Tarski, A., A decision method for elementary algebra and geometry, 2nd Edn., Univ.Calif.Press, Berkeley, (1951).

[Th1] Thomas, J.M., Differential systems, Amer.Math.Soc., (1937).

[Th2] Thomas, J.M., Riquier's existence theorems, Annals of Math., 30 (1929) 286-310; 35 (1934) 306-311.

[Todd] Todd, J.A., The geometrical invariants of algebraic loci, Proc. London Math. Soc. (2), 43 (1937) 127-138.

[Tr] Traeger, B.M., Algebraic factoring and rational function integration, Proc. 1976 ACM Symp. Symb. and Alg. Comp., (1976) 219-226.

[VdV] Van de Ven, On the Chern numbers of surfaces of general type, Invent. Math., 36 (1976) 285-293.

[VdW1] Van der Waerden, Moderne Algebra, Teil 1, (1930), Teil 2, (1931).

[VdW2] Van der Waerden, Einfuerung in die algebraischen Geometrie, Dover Publ., (1945).

[Wald] Waldvogel, J., Note concerning a conjecture by A.Wintner, Celestial Mechanics, 5 (1972) 37-40.

[Wall] Wall, B., On the computation of syzygies, ACM SIGSAM Bull. 23 (1989) 5-14.

[Wan] Wan, Z.X., Geometry of classical groups over finite fields, Studentliteratur, Lund, (1993).

[WDK1] Wang, D.K., A mechanization proving of a group of space geometry problems, MM-Res.Preprints, MMRC, No.5 (1990) 66-81.

[WDK2] Wang, D.K., Polynomial equations solving and mechanical geometric theorem proving, PhD.Thesis, MMRC, (1993).

[WDM1] Wang, D.M., Mechanical approach for polynomial set and its related fields (in Chinese), Ph.D thesis, Inst.Sys.Sci., (1987).

[WDM2] Wang, D.M., On Wu's method for proving constructive geometric theorems, MM-Res.Preprints, MMRC, No.3 (1989) 89-101.

[WDM3] Wang, D.M., A new theorem discovered by computer prover, Journal of Geometry, 36 (1989) 173-182.

[WDM4] Wang, D.M., A method for factoring multivariate polynomials of successive algebraic-extension fields, Preprint, RISC-LINZ Johannes Kepler University, Australia, (1995).

[WDM5] Wang, D.M., A decision method for definite polynomial, MM-Res. Preprints, MMRC, No.2 (1987) 68-74.

[W-G] Wang, D.M., & Gao, X.S., Geometry theorems proved mechanically using Wu's method — Part on Euclidean geometry, in MM-Res.Preprints, MMRC, No.2, (1987) 75-106.

[WH1] Wang, H., Toward mechanical mathematics, IBM Journal, 4 (1960) 2-22.

[WH2] Wang, H., A survey of mathematical logic, Science Press, Beijing, (1962).

[WPS1] Wang, P.S., Factoring multivariate polynomials over algebraic number fields, Math. Comp., 30 (1976) 324-336.

[WPS2] Wang, P.S., An improved multivariate polynomial factoring algorithm, Math. Comp., 32 (1978) 1215-1231.

[W-R] Wang, P.S., & Rothschild, L.P., Factoring multivariate polynomials over the integers, Math.Comp., 29 (1975) 935-950.

[Weil] Weil, A., Foundations of algebraic geometry, Amer.Math.Soc. Colloq.Publ., 29 (1962).

[Whit] Whitney, H., Elementary structure of real algebraic varieties, Annals of Math., 66 (1957) 545-556.

[Wint] Wintner, A., Analytic foundations of celestial mechanics, (1941).

[W-L] Wu, J.Z., & Liu, Z.J., On first order theorem proving, Proc. ASCM'96 in Kobe, Scientists Incorporated, Tokyo, (1996) 163-186.

[WTJ1] Wu, T.J., Some test problems on applications of Wu's method in nonlinear programming problems, MM-Res. Preprints, MMRC, No.6 (1991) 144-155. Also in Ch.Q.J.Math., 9 (1994) 8-17.

[WTJ2] Wu, T.J., On a collision problem, MM-Res. Preprints, MMRC, No.7 (1992) 96-104. Also in Acta Math.Scientia, 15 (Supp.) (1995) 32-38.

[WWD] Wu, W.D., A note to the discriminant systems for the unary equations, MM-Res.Preprints, MMRC, No.3 (1989) 33-38.

[WWT1] Wu, W.T., On the decision problem and the mechanization of theorem-proving in elementary geometry, Scientia Sinica 21 (1978) 159-172, re-published in [B-L], (1984) 235-242.

[WWT2] Wu, W.T., Toward mechanization of geometry — Some comments on Hilbert's "Grundlagen der Geometrie", Acta Math.Scientia, 2 (1982) 125-138

[WWT3] Wu, W.T., Some remarks on mechanical theorem-proving in elementary geometry, Acta Math.Scientia 3 (1983) 357-360.

[WWT4] Wu, W.T., Some recent advance in mechanical theorem-proving of geometries, in [B-L], (1984) 235-242.

[WWT5] Wu, W.T., Basic Principles of Mechanical Theorem Proving in Geometries (Part on Elementary Geometries), (in Chinese), Science Press, (1984). English translation by D.M.Wang et al, Springer, (1994).

[WWT6] Wu, W.T., Basic principles of mechanical theorem-proving in elementary geometries, J.Sys.Sci. & Math.Scis., 4 (1984) 207-235. Re-published in J. of Automated Reasoning, 2 (1986) 221-252.

[WWT7] Wu, W.T., On zeros of algebraic equations — an application of Ritt principle, Kexue Tongbao, 31 (1986) 1-5.

[WWT8] Wu, W.T., A mechanization method of geometry and its applications, I. Distances, areas, and volumes, J.Sys.Sci. & Math.Scis.,6 (1986) 204-216

[WWT9] Wu, W.T., A mechanization method of geometry and its applications, I. Distances, areas, and volumes in euclidean and non-euclidean geometries, Kuxue Tongbao 32 (1986) 436-440

[WWT10] Wu, W.T., A zero structure theorem for polynomial-equations-solving and its applications, MM-Res.Preprints, MMRC, No.1 (1987) 2-12.

[WWT11] Wu, W.T., A mechanization method of geometry and its applications, 3. Mechanical proving of polynomial inequalities and equations-solving, MM-Res. Preprints, MMRC, No.2, (1987) 1-17.

[WWT12] Wu, W.T., A mechanization method of geometry and its applications, 4. Some theorems in planar kinematics, J.Sys.Scis & Math.Sci., 2 (1989) 97-109.

[WWT13] Wu, W.T., Some remarks on characteristic-set formation, MM-Res. Preprints, MMRC, No.3 (1989) 27-29.

[WWT14] Wu, W.T., On the generic zero and Chow basis of an irreducible ascending set, MM-Res. Preprints, MMRC, No.4 (1989) 1-21.

[WWT15] Wu, W.T., On the chemical equilibrium problem and equations-solving, MM-Res. Preprints, MMRC, 4 (1989) 22-39. Also in Acta Math.Sci., 10 (1990) 361-374.

[WWT16] Wu, W.T., On a projection theorem of quasi-varieties in elimination theory, MM-Res.Preprints, MMRC, No.4 (1989) 40-48. Also in Chin.Ann. of Math., 11B (1990) 220-226.

[WWT17] Wu, W.T., A mechanization method of geometry and its applications, 6. Solving inverse kinematic equations of Puma-type robots, MM-Res.Preprints, MMRC, 4 (1989) 49-53.

[WWT18] Wu, W.T., On the construction of Groebner basis of a polynomial ideal based on Riquier-Janet theory, MM-Res.Preprints, MMRC, No.5 (1990) 5-22. Also in Sys.Sci.& Math.Scis., 4 (1991) 193-207.

[WWT19] Wu, W.T., On a linear equations method of non-linear polynomial equations-solving, MM-Res.Preprints, MMRC, No.6 (1991) 23-36. Also in Sys.Sci. & Math.Scis., 6 (1993) 1-12.

[WWT20] Wu, W.T., Equations-solving and theorem-proving: Zero-set formulation and ideal formulation, Proc.Asian Math.Conf.1990 in Hongkong, (1992) 1-10.

[WWT21] Wu, W.T., A mechanization method of equations-solving and theorem- proving, Adv.in Comp. Res., 6 (1992) 103-138.

[WWT22] Wu, W.T., On problems involving inequalities, MM-Res.Preprints, MMRC, No.7 (1992) 1-13.

[WWT23] Wu, W.T., On a finiteness theorem about optimization problems, MM-Res. Preprints, MMRC, No.8 (1992) 1-18.

[WWT24] Wu, W.T., A report on mechanical geometry theorem proving, Progress in Natural Science, 2 (1992) 1-17.

[WWT25] Wu, W.T., On a hybrid method of polynomial equations solving, MM-Res. Preprints, MMRC, No.9 (1993) 1-10.

[WWT26] Wu, W.T., On surface-fitting problem in CAGD, MM-Res. Preprints, MMRC, No.10 (1993) 1-10.

[WWT27] Wu, W.T., On a finiteness theorem about problems involving inequalities, Sys.Sci. & Math.Scis., 7 (1994) 193-200.

[WWT28] Wu, W.T., Some remarks on factorization and GCD of multivariate polynomials, MM-Res.Preprints, MMRC, No.11 (1994) 1-14.

[WWT29] Wu, W.T., Polynomial equations-solving and its applications, Proc. ISAAC'94, or Algorithm and Computations, (Eds. D.Z.Du & X.S.Zhang), Springer (1994) 1-9.

[WWT30] Wu, W.T., Central configurations in planet motions and vortex motions, MM-Res. Preprints, MMRC, No.13 (1995) 1-14.

[WWT31] Wu, W.T., A mechanization method of geometry and its applications, 2. Curve pairs of Bertrand type, Kexue Tongbao, 32 (1987) 585-588.

[WWT32] Wu, W.T., Mechanical derivation of Newton's gravitational laws from Kepler's laws, MM-Res. Preprints, MMRC, No.1 (1987) 53-61.

[WWT33] Wu, W.T., On the foundation of algebraic differential geometry, MM-Res. Preprints, MMRC, No.3 (1989) 1-26.

[WWT34] Wu, W.T., Mechanical theorem-proving of differential geometries and some of its applications in mechanics, J.Aut.Reasoning, 7 (1991) 171-191.

[WWT35] Wu, W.T., On Chern characteristic systems of an algebraic variety, (in Chinese) Shuxue Jinzhan, 8 (1965) 395-401.

[WWT36] Wu, W.T., On algebraic varieties with dual rational dissections, (in Chinese) Shuxue Jinzhan, 8 (1965) 402-409.

[WWT37] Wu, W.T., On Chern numbers of algebraic varieties with arbitrary singularities, Acta Math. Sinica, New Series, 3 (1987) 227-236.

[W-L] Wu, W.T. & Lu, X.L., Triangles with equal bisectors, (in Chinese), People's Edu.Press, Beijing, (1986).

[W-W] Wu, W.T. & Wu, T.J., A mechanization method of geometry and its applications, 5. Solving transcendental equations by algebraic methods, MM-Res.Preprints, MMRC, No.3, (1989) 30-32.

[Yau] Yau, S.T., Calabi's conjecture and some new results in algebraic geometry, Proc. Nat. Acad. Sci., 74 (1977) 1798-1799.

[Y-H-Z] Yang, L., Hou, X.R., & Zeng, Z.B., An alternative algorithm for determining the number of real roots of a polynomial, Intern.Workshop on Logic and software Engineering, Beijing, (1995).

[Y-Z-H] Yang, L., Zhang, J.Z., & Hou, X.R., Nonlinear algebraic equation system and automated theorem-proving, (in Chinese), Shanghai Sci. and Tech. Edu. Publ.

House, Shanghai, (1996).

[ZSQ] Zhang, S.Q., Wu elimination and its application in partial differential equations. Solitary wave solutions for non-linear evolution equations using symbolic computation, PhD Thesis, Lanzhou Univ., Lanzhou, (1995).

[Zhi] Zhi, L.H., Polynomial factorization over algebraic fields and applications, PhD. Thesis, MMRC, (1996).

[Zhou] Zhou, J.Y., Classes de Chern pour les variétès singulières. Classes de Chern en théorie bivariant, PhD Thesis, Marseille, (1995).

[Zou] Zou, F.M., Applications of the characteristic set method in celestial motions and invariant theory, PhD Thesis, MMRC, (1997).

Index

E

F

J

K

L

M

N

O

P

Q

R

S

W

X

Y

Z